ON THE COVER: @Photo by Ceyhun/Adobe Stock- A tank in motion on an urban street, passing through a crowd of people and cars.

AF878442

DIVIDER PAGES:

Part 1/Chapter 1: Photo by reewungjunerr/Adobe Stock
Chapter 2: Photo by zef art/Adobe Stock
Chapter 3: Photo by reewungjunerr/Adobe Stock
Chapter 4: Photo by KKF/Adobe Stock
Part 2/Chapter 5: Photo by Meysam Azarneshin/Adobe Stock
Chapter 6: Photo by May Chanikran/Adobe Stock
Chapter 7: Photo by narong/Adobe Stock
Appendices: Photo by zef art/Adobe Stock

BACK COVER: This artistic rendition of the Chinese Communist Party attempting to erase the Tiananmen Square protests from history exemplifies actions supporting resiliency. Source. D. Thompson, U.S. State Department. Illustration.

JSOU PRESS
Melanie Casey, Editor in Chief
Beth DeGeorge, Editor
Alina Alvarez, Editor
Book layout and design by Jackie Lucal

This work was cleared for public release; distribution is unlimited.

July 2025

ISBN 978-1-941715-71-0

JSOU PRESS
JOINT SPECIAL OPERATIONS UNIVERSITY

RESILIENCE AND RESISTANCE:

Interdisciplinary Lessons in Competition, Deterrence, and Irregular Warfare

Edited and with an introduction by Robert S. Burrell
Foreword by Charles T. Cleveland

JOINT SPECIAL OPERATIONS UNIVERSITY
MacDill Air Force Base, Florida | July 2025

JSOU Report 25-16

CONTENTS

Dedication...x

Acknowledgements...xii

Foreword...xiv

Charles T. Cleveland

Introduction...xviii

Robert S. Burrell

PART I: FOUNDATIONS IN RESILIENCE AND RESISTANCE 1

CHAPTER 1: MILITARY SCIENCE 1

I. The Resilience and Resistance Model: An Overview

Robert S. Burrell and John Collison...3

II. Measuring Resiliency and Resistance

Robert S. Burrell and John Collison...35

III. Redefining the Bang: A Full Spectrum of Conflict Design

Robert S. Burrell...53

CHAPTER 2: POLITICAL SCIENCE AND INTERNATIONAL RELATIONS 65

I. Research Methods for Area Studies

A. Jackson...67

II. Appreciating the World: A Framework for Doing Socio-Political Analysis

David P. Oakley...83

III. Building Resistance in Georgia

Brian Petit...107

IV. Measuring and Quantifying State Fragility: Why Governments Lose Internal Conflicts and What that Means for Counterinsurgency

Chris Mason...119

V. Understanding Institutional Resilience

John H. Mongan...137

CHAPTER 3: SOCIOLOGY **155**

I. Social Movements: What Causes Resistance to Succeed or Fail?

A. Jackson......157

II. The Arab Spring as a Case Study

David DiOrio, PhD......181

CHAPTER 4: HISTORY **203**

I. Support to Resistance: Case Study of Cuba, 1959–1961

Robert S. Burrell......205

II. Support to Resilience: Case Study of Laos, 1954–1975

Robert S. Burrell......223

III. Support to Resistance: Case Study of Tibet, 1951–1975

Robert S. Burrell......239

IV. Support to Resilience: Case Study of Uribe's Colombia (2002–2006)

Thomas A. Marks......261

PART II: APPLICATION OF RESILIENCE AND RESISTANCE **289**

CHAPTER 5: MILITARY SCIENCE **289**

I. Assessing Resistance for the Purpose of Informing International Policy

Robert S. Burrell and John Collison......291

II. Classical and Modern Deterrence Theories

Aaron Baty......309

III. Integrated Deterrence as a National Strategy

Aaron Baty......337

IV. Allied Security Strategies: NATO Special Operations as a Case Study

Gabriele Pierini......355

CHAPTER 6: INTERNATIONAL RELATIONS **369**

I. China's Evolving Way of War: From Deception to Deterrence

Christopher Marsh....371

II. Russia's "Special" Way of War: Special Operations Forces, *Spetsnaz*, and Irregular Warfare

Christopher Marsh....387

III. More than a Terror State: The Democratic People's Republic of Korea is a Misunderstood and Neglected Global Malign Actor that Requires an Innovative New Strategy

David Maxwell....411

IV. Hegemon in the Middle East: Iran's Islamic Revolutionary Guards and the Quds Force

David R. DiOrio....431

V. Defeated? Violent Extremist Organizations and Non-State Actors

Namrata Goswami....453

CHAPTER 7: ENGINEERING **479**

I. Cyber, Artificial Intelligence, and Space: Impact on the Future of Special Operations

Namrata Goswami....481

PART III: APPENDICES **511**

Appendix A. A 2024 Guide to Assessing Resiliency and Resistance in Iran

Robert S. Burrell and David R. DiOrio....513

Appendix B. Utilizing Resilience and Resistance as a Textbook

Robert S. Burrell....541

Appendix C. Acronyms....547

DEDICATION

Paul J. Tompkins Jr.

This book is dedicated to Paul J. Tompkins Jr. For two decades after his retirement from activity duty in Special Forces, Tompkins served as the Chief of U.S. Army Special Operations Command G3X, Special Programs Division. During his more than 20 years in that office, he was the unconventional warfare authority for both the U.S. Army and U.S. Special Operations Command. He led most of the Department of Defense's unconventional warfare and support to resistance capability development efforts. He also led the effort to update the historical Special Operations Research Office (SORO) studies on insurgencies, resistance movements, and unconventional warfare. Tompkins promoted and obtained command support and funding for a joint partnership with Johns Hopkins University, resulting in more than 30 published products during a 10-year effort. This collaboration with Johns Hopkins University/Applied Physics Laboratory enabled the update of the SORO studies into the Assessing Revolutionary and Insurgent Strategies (ARIS) project, which is available online at **https://www.soc.mil/ARIS/ARIS.html.** ARIS remains a foundational source for this book.

ACKNOWLEDGEMENTS

Embarking on a project of this magnitude requires recognizing the many contributors who sacrificed their time and talents to create this body of new knowledge. In no particular order, I have to start with the authors, including Dr. Namrata Goswami—an interdisciplinary scholar; Dr. David DiOrio–former deputy director at the Joint Forces Staff College of National Defense University; Mr. Brian Petit–retired Special Forces colonel and JSOU faculty; Mr. David Maxwell–retired Special Forces colonel, vice president of the Center for Asia Pacific Strategy, and editor of *Small Wars Journal*; Mr. Aaron Baty–retired Special Forces officer and JSOU faculty; Dr. David P. Oakley–academic director, Global and National Security Institute; Mr. John Collison–former Civil Affairs officer and contracted staff at U.S. Special Operations Command; Mr. Gabriele Pierini–staff at NATO Special Operations Headquarters; Dr. Chris Mason–faculty at the U.S. Army War College; Dr. Christopher Marsh–faculty at National Defense University and editor of *Inter Populum: Journal of Irregular Warfare and Special Operations*; Mr. John Mongan–a foreign service officer with the U.S. State Department; and Dr. Thomas A. Marks–senior faculty at National Defense University. Finally, many thanks to Charles T. Cleveland, Lieutenant General, retired, for supporting this book and compiling the forward.

In addition to the authors, several other honorable mentions are required. I would like to thank the early founders of the Support to Resilience and Resistance (SRR) Program of Study at JSOU, including former university president Dr. Isaiah Wilson III, as well as other faculty supporting his vision, including Dr. Scott Smitson and Dr. David Dudas. JSOU librarians including Laurie Talerico, Gisela Gonzalez, and Stacy Harn professionally responded to dozens of inquiries for interlibrary loans of books and articles. Initial reviewers of the draft included Christian Ramthun and Frank Reidy, both faculty at JSOU. Last, but not least, many thanks to the JSOU Press, including Cameron Cobb and Melanie Casey. Again, I owe a dept of gratitude to Paul J. Tompkins Jr. and U.S. Army Special Operations Command for sponsoring the Assessing Revolutionary and Insurgent Strategies (ARIS) project. ARIS provides a wealth of scholarship on resistance written by scholars from Johns Hopkins University/Applied Physics Laboratory. This book attempts to operationalize many of those lessons.

FOREWORD

Charles T. Cleveland

Over the past 75 years, the most significant national security failing of the United States has been its unwillingness to generate the needed capacity to win irregular wars. Our dependence on a defense apparatus built for traditional or conventional conflicts has demonstrated that we cannot achieve U.S. objectives in today's global competition, particularly when adversaries choose strategies of protracted war, a hallmark of Mao Zedong and his subsequent acolytes. In short, cost imposition over time erodes political U.S. will, which eventually reaches an indeterminable point whereupon American leadership often calculates that regaining political advantage means leaving the field of combat. In the case of Vietnam, it took just over 10 years; in Afghanistan it was 20. Both wars ended in eerily similar defeats with our enemies triumphant after untold billions are expended and thousands of young American lives shattered. We owe the American people better.

This Joint Special Operations University book, Resilience and Resistance, comprises an excellent primer concerning a core mission of the USSOCOM enterprise – low intensity conflict. Rob Burrell and his talented set of colleagues explain the military science and historical case studies related to asymmetric approaches to counter irregular threats and use them as an instrument to U.S policy, which will educate both the novice soldier and the professional warfighter.

Truth be told, there has been a significant amount of scholarship written about irregular war. What has yet to occur is the maturing of that body of work into a canon from which to educate professionals in the business. This volume and other initiatives at JSOU and elsewhere show promise in initiating discussion over best practices, including a more scientific approach than current doctrine allows. The requirement to truly educate, and not indoctrinate, compels such debate. Noteworthy examples in this book include Chris Mason's excellent evidence on "the five factors" that determine success or failure in a counterinsurgency. These provide a useful yardstick in measuring policy responses to foreign interventions. Additionally, Brian Petit outlines operations designed to blunt Russian influence in the Republic of Georgia, demonstrating the power of long-term, tailored, and nuanced engagements. The content contained in Petit's essay should provoke conversation and debate in the force—in the team rooms where education of SOF professionals really takes place. Other readings on the Arab Spring, Cuba, Laos, Tibet, Colombia, North Korea, and Russia are well written, informative, and highly relevant to today's international security environment. Aaron Baty's essays on deterrence theories and integrated deterrence as a U.S. strategy also prove insightful. All these arguments will be further honed and refined by the ongoing irregular campaigns in Ukraine, Gaza, Yemen, Taiwan, and the emerging crisis in Haiti.

Perhaps most important is Burrell's use of the resilience and resistance methodology in a case study on Iran. It is just an example, but it brings to mind the reality that the fight in which we find ourselves is one that will end favorably for the U.S. only if the globe's clutch of dangerous authoritarian regimes and their ideologies are defeated.

JSOU is uniquely postured to research, curate, and assemble what should become part of a liberal democratic canon on irregular warfare. Ideally, the U.S. government would use such a body of work to educate a dedicated cadre of civilian and military professionals in irregular warfare. Such a cohort would fill important positions within headquarters and government offices whose mission it is to conceive of an accepted concept for our conduct of irregular war, as well as defense against such strategies of our adversaries. The concept then frames the competent development of what the Army calls the DOTML-PF (doctrine, organization, training, material, leadership and education, personnel, and facilities). A new U.S. national security enterprise dedicated to irregular warfare is now an existential requirement. We must have a world-class capacity to conduct irregular war in pursuit of our objectives. The DoD needs compulsory education on irregular warfare, not only to defend against irregular war as waged by great powers, regional tyrants, and their surrogates and proxies but also increasingly important is to employ a world-class irregular warfare capability on behalf to further our objectives. Through such professional military education, this new breed of irregular warfare commanders and practitioners can provide informed military advice that allows general officers a competent understanding of low-intensity conflict and helps policymakers avoid national humiliation and the waste of blood and treasure.

In 2021, Congress passed the Consortium to Study Irregular Warfare Act. This legislation comes in recognition of our military leaders' poor track record with population-centric struggles, our failed conventionally minded strategies, and all-too predictable campaigns. However, Congress has made many such attempts to refocus the DoD on special warfare. It took Congress to act after the missteps of Desert One in 1980 and the invasion of Grenada in 1983. Arguably, the Goldwater-Nichols Act of 1986 and Nunn-Cohen Amendment of 1987 did not go far enough. Enamored with direct action and counterterrorism, the DoD has paid scant attention to irregular forms of deterrence, competition, and war until recently.

So, it is with increasing frequency that Congress has recommended that the DoD pay more attention to irregular warfare. In 2021, it recommended the creation of an Irregular Warfare Center, resulting in an interim organization at the Defense Security Cooperation University. A final location was to be decided by a Department of the Air Force-basing study, but two years on there is still no permanent basing solution. In the interim, U.S. Army has directed U.S. Army Special Operations Command for proponency in irregular warfare, who in turn has directed the John F. Kennedy Special Warfare Center and School to stand up an Irregular Warfare Academy.

As the DoD, the Services, and most of its special operations enterprise turn to winning a war with the pacing threat of China, irregular warfare once again risks being seen as a distraction. A properly supported Irregular Warfare Center, with JSOU as one of its primary educational pillars, should change the boom or bust cycle to a more sustainable one. The goal must be to harvest the very best thinkers from the many fields outside the DoD that are the key elements of today's irregular war. The national security leaders, technocrats, and military units designed for conventional war and nuclear deterrence need irregular warfare counterparts.

This volume is a necessary good start on such an education. That it has been compiled and published by U.S. Special Operations Command's Joint Special Operations University is especially important as it is in this command that the military's irregular warfare maneuver forces reside and the need for a compulsory irregular warfare education is most urgent.

Lastly, the book rightly recognizes Paul Tompkins, a great Green Beret, an honorary warrant officer of the Regiment, and a friend. He, more than anyone in contemporary times kept the embers hot on unconventional warfare, resilience and resistance, and the role of special operations activities in the fight below the threshold of traditional war. He recognized the need to educate leaders on this unique form of warfare. If U.S. Special Operations Command can be at the forefront of creating the capability the country and the West needs, it will be to no small degree because Paul never gave up.

About the Author

Charles T. Cleveland (lieutenant general, retired) is a native of Arizona, son of a career Army NCO, and a 1978 graduate of the U.S. Military Academy, West Point. As the commander of 10th Special Forces Group (Airborne), he led the largest and most successful U.S. unconventional warfare campaign during Task Force Viking in Northern Iraq in 2003. He also served as the commanding general of the U.S. Army Special Operations Command from 2012 to 2015. Following his retirement, he was the Senior Mentor to the Chief of Staff of the Army's Strategic Studies Group from 2015-2017. He is currently a Senior Mentor at the U.S. Army War College and is an Adjunct at Rand. In 2018 he coauthored "Military Strategy in the 21st Century" with Drs Ben Jensen and Susan Bryant and Lt. Col. Arnel David, and in 2020 he coauthored "The American Way of Irregular War" with Rand's Dr. Daniel Egel. He serves on the Advisory Boards of the nonprofit organizations Spirit of America Foundation and the Global SOF Foundation, and is one of the honorary chairs of the OSS Society.

INTRODUCTION

Robert S. Burrell

In accordance with the *Consortium to Study Irregular Warfare Act of 2021*, House Resolution 5130 (H.R. 5130), Joint Special Operations University (JSOU) dedicated its talents and resources to revising and creating modern curriculum for Special Operations Forces (SOF) as well as the interagency and conventional forces supporting irregular activities. This book specifically facilitates research and analysis "into operational capabilities, as appropriate, by supporting applied research and developing tools to counter irregular threats" (H.R. 5130). As well as an "understanding of irregular threats," it provides "the integration of data-based responses to such threats" through a unique 12-step process (H.R. 5130). Meeting congressional intent on irregular warfare education remains at the heart of this effort. Efforts began in August 2021, with research and curriculum developed in JSOU Support to Resilience and Resistance (SRR). SRR was built to complement the U.S. Special Operations Command's (USSOCOM) SOF *2040 Concept* to prepare SOF for future irregular warfare, competition, and deterrence. Support to resilience and resistance concepts, education, and training support the USSOCOM Commander's, General Bryan Fenton, priorities of *Win* and *Transform*.

For my part, I was assigned as the SRR Academic Program Director in April 2022 and immediately set about building an interdisciplinary team tasked with creating a new curriculum. The team was cemented with leading world academic experts in numerous disciplines, as well as recognized professors of practice—particularly those with recent experience in Europe, Africa, the Middle East, and the Pacific. These talented scholars and practitioners, both American and foreign, included those from U.S. Army Special Operations Command, the U.S. Army War College, the U.S. Department of State's Bureau of Conflict and Stabilization Operations, the U.S. Agency for International Development (USAID), University of South Florida's Global and National Security Institute, National Defense University, Allied Special Operations Forces Command (SOFCOM), U.S. Cyber Command, and the Center for Asia Pacific Strategy. The diversity of these academics and other government experts ensured a comprehensive and whole of government (and partners) approach.

From the start, it quickly became obvious that current scholarship fell short in addressing innovative education in irregular warfare—particularly its intersection with competition and deterrence. Consequently, the faculty agreed to conduct this research and translate academic scholarship and concepts into laymen terms—resulting in this book. These efforts generated new knowledge that practitioners can utilize for persistent engagement (or campaign planning) as well as contingency plans. The edited volume allows students and practitioners to understand the dynamics of intrastate conflict in which irregular warfare takes place but also adds new research that highlights the strategies of adversaries to influence the outcome of those same conflicts—including the activities of the Chinese Communist Party, Russia, North Korea, Iran, and violent extremist organizations.

This *Resilience and Resistance* volume is broken down into two parts. The first section provides fundamentals on resilience and resistance to accentuate the human domain in which security professionals (intelligence agents, foreign service officers, and military personnel) operate. It introduces an interdisciplinary conceptual framework for understanding aspects of resilience and resistance within a state because intrastate conflict is the primary setting for irregular warfare. The initial chapters illustrate novel applied methodologies to assess the resilience of current authority as well as resistance to the same. Part I also presents a full spectrum of conflict design to help practitioners better visualize aspects of international conflict in terms of the interaction between deterrence, conventional war, irregular warfare, and competition. From this perspective, gray zone conflict and hybrid warfare become readily visible and transparent—indeed, they are the central forms of conflict, literally at the epicenter between these four forms.

In addition to military science, Part I explains irregular challenges around the globe in terms of political science and international relations. Several experts provide examples by which aspects of resilience and resistance interplay in contemporary case studies, to include discussions about Israel, Georgia, Syria, and Ukraine. Next, Part I utilizes sociology (and social movement theory) to examine the growing influence of non-state organizations in influencing social and governmental change. New research from Dr. A. Jackson explores social movements in Iran as case studies to argue why these types of grassroots organizations succeed or fail. The discussion of social movements is followed by an expansive chapter on the Arab Spring as another social movement case study. Finally, Part I concludes with historical examples of U.S. support to reinforcing state resiliency or sponsoring resistance groups. These 20th and 21st-century cases include two in the Pacific region and two in Latin America.

Part II of this volume builds on the foundational lessons in Part I to further operationalize methods of analysis related to resilience and resistance. It begins by outlining the remaining steps required to fully understand a state in terms of intrastate conflict. This data-driven and human-centric approach analyzes resistance movements for the purposes of (a) potentially supporting them, or (b) better understanding them to help partner governments neutralize them. Part II then explores the military science of deterrence from aggressor states by explaining what deterrence is and how it works conceptually. Part II also explores the strengths and weaknesses of a coalition approach to achieving collective security, with NATO as an example. Part II also provides the latest research from several experts on the threats posed by Communist China, Russia, North Korea, Iran, and violent extremist organizations. Subsequently, as a look into the future, recognized scholar Dr. Namrata Goswami explores how cyber, artificial intelligence, and space will intersect with the deployment of special operations teams. Finally, the case study in Appendix A offers insight into Iran, utilizing the full scope of resilience and resistance methodologies offered in Part I and Part II. Figure A outlines the content of this book in terms of topic with percentages of content by number of chapters.

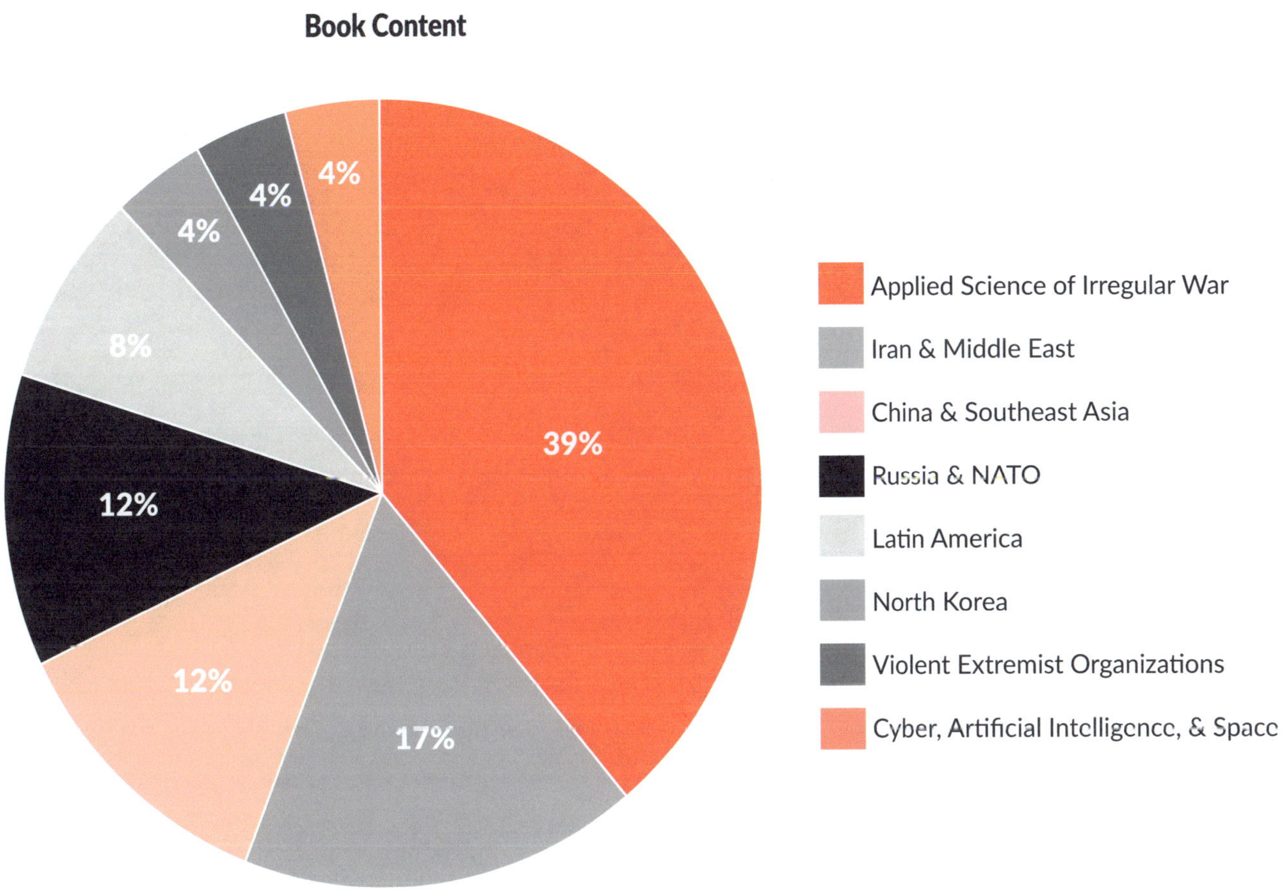

Figure A. percentage of volume content by topic. Source: Author

This book intentionally indicates a first edition. The authors fully expect to revise the research and implement changes to improve applicability for practitioners. If you have any comments or recommendations, we greatly value them. These can be directed to Dr. Robert S. Burrell, University of South Florida, Global and National Security Institute, 4202 E. Fowler Avenue, Tampa, FL, 33620.

PART I:

FOUNDATIONS IN RESILIENCE AND RESISTANCE

CHAPTER 1: MILITARY SCIENCE

I. The Resilience and Resistance Model: An Overview

By Robert S. Burrell and John Collison

ABSTRACT: This section details an applied methodology to operationalize an irregular approach to conflict, in particular external support to intrastate resilience or resistance. It introduces two foundational learning concepts: (a) the *Resilience and Resistance Model* and (b) the *Resistance Continuum.* The Resilience and Resistance Model consists of four components: (1) resistance to current governance, (2) resilience to change, (3) support and legitimacy from the population, and (4) international support from external sponsors. Using this model, scholars and practitioners can better understand the dynamics of change occurring within the context of intrastate conflict. Of the four components, the resistance node requires the most scrutiny, as it represents the challenge to the status quo and thus the driver of instability. Using the Resistance Continuum, analysts can categorize the general nature of resistance movements across a spectrum from non-violent protest to violent rebellion. Parts of this chapter were previously published in *Small Wars & Insurgencies* journal, *Expeditions with Marine Corps University Press*, and the *Irregular Warfare Initiative.*[1]

INTRODUCTION

In the 20th and 21st centuries, irregular conflicts (including those in Indochina and the Middle East) have caused the U.S. a great deal of angst.[2] Accordingly, the U.S. Congress in 2021 demanded that the Department of Defense develop education to prepare for these types of future irregular struggles.[3] To address the academic gap in preparing for irregular forms of conflict, this chapter proposes new conceptual methods for analyzing resiliency and resistance. Subsequent chapters detail how to utilize these concepts to gather data and propose strategies for supporting resilience or resistance within intrastate conflict. This enhanced methodology provides a more informed approach to assist decision and policymakers to consider irregular approaches to address competition, deterrence, and war.[4]

| The Resilience and Resistance Model consists of four components: (1) resistance to current governance, (2) resilience to change, (3) support and legitimacy from the population, and (4) international support from external sponsors. |

Throughout the proposed methodology, interdisciplinary methods remain vital, as military science cannot alone prepare practitioners for future conflict.

One needs no further example of the need for an interdisciplinary approach to studying intrastate conflict than to examine the Ukraine/Russia struggle. Since at least 2004, Russia has aggressively employed hybrid warfare against Ukraine with various means, some violent and some non-violent. With respect to international relations, Russia's rigging of the 2004 Ukrainian election for Viktor Yanukovych could be the starting point for analyzing political warfare. In terms of sociology, perhaps the Orange Revolution in 2004 or the Maidan Revolution in 2014 are climatic events for creating social change.[5] In political science, perhaps the law "On the Foundations of National Resistance" adopted by the Verkhovna Rada in July 2021 holds prominence.[6] From a military science approach, a scholar or practitioner might commence analysis with the insertion of Russian proxy forces during the 2014 annexation of Crimea.[7] Additionally, the discipline of history, of course, offers excellent insight into how and why the tensions within Ukraine developed over hundreds of years.[8] In summation, without a deliberate analysis based on multiple disciplines, a comprehensive understanding of intrastate conflict will remain elusive.

| The study of external support to intrastate resilience or resistance requires this prescribed interdisciplinary approach—one that includes, but is not limited to history, political science, religion, ethnology, law, sociology, international relations, and military science. |

The study of external support to intrastate resilience or resistance requires this prescribed interdisciplinary approach—one that includes, but is not limited to history, political science, religion, ethnology, law, sociology, international relations, and military science. By expanding analysis to include multiple disciplines, the methodology of intrastate conflict materializes along a continuum, one that includes nonviolent dissent on one end and violent confrontation between resistance elements and the recognized authority on the other. Within this domestic affair, the role of external sponsors can also prove quite important to the outcome of conflict. However, to effectively consider support to resilience or resistance elements, sponsors should understand intrastate conflict through the Resistance Continuum.

Human society contains elements of resistance to current governance or foreign occupation. Resistance can span a spectrum of activities from nonviolent and legal forms to illegal or violent means. In contrast, the regime or occupying power typically attempts to brace the resolve of the population against internal political, economic, or social change—even revolution. Concurrently, sponsors can make a deliberate choice to: (a) help stabilize another nation's governance through support to resilience, (b) aid aggrieved populations against other states through support to resistance, or (c) not directly interfere in the sovereignty of another state. The *yin* and *yang* of resilience and resistance, and their interaction with the population, each other, and external supporters is continuous and dynamic in every society.

This chapter explains how resistance movements and the recognized authority conceptually interact with each other, with their shared population, and among international benefactors. While external sponsors can choose to support resilience of governance or resistance to it in other societies, identifying the nature of resistance movements remains critical to ascertaining the most appropriate and effective support for them or the most suitable means to counter them. Using the Resistance Continuum, analysts can categorize the general nature of resistance movements across a spectrum, from non-violent protest through violent civil war.

Figure 1.1. The yin and yang of resilience and resistance. Source: Microsoft

THE RESILIENCE AND RESISTANCE MODEL

An important framework to consider is the Resilience and Resistance Model, adapted from Gordon McCormick's model for insurgency.[9] We have modified his model so that it now includes the full spectrum of the Resistance Continuum from peaceful demonstration to belligerency. In this model, the resilience node represents the recognized governance and authority, and the resistance node represents opposition to existing governance or occupation. The existing governing authority and resistance movements can directly confront each other while simultaneously struggling to garner both domestic and international support. Concurrently, they each attempt to counter the efforts of the other. The following describes the Resilience and Resistance Model and offers definitions for the four nodes. See Figure 1.2.

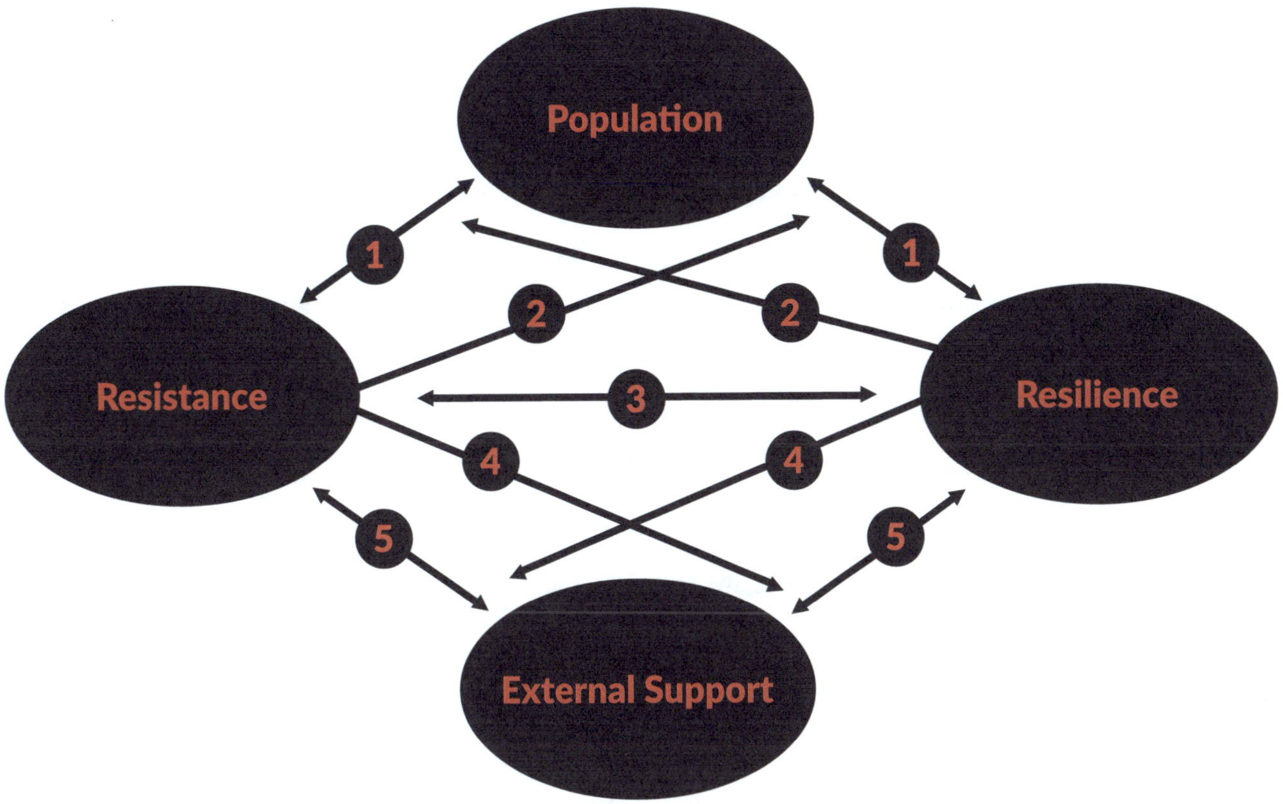

Figure 1.2: Diagram of the Resilience and Resistance Model. Source: Author

In this model, there are four primary nodes: (a) the population node, (b) the resilience node, (c) the resistance node, and (d) the external support node. As represented by the resistance and resilience nodes, the governing body and organized resistance to it perform five basic actions in opposition to each other:

(1) attempt to gain support from the population,
(2) disrupt the other's efforts to garner support from the population,
(3) perform violent and/or nonviolent actions directly against one another,
(4) attempt to interrupt their opponent's attempts to garner international support, and
(5) attempt to garner international support.

Both the population and the external support nodes have agency and can initiate actions to influence the resilience and/or the resistance nodes as well (hence the dual arrows on lines 1 and 5). The power of the Resilience and Resistance Model is that it applies in nearly every intrastate conflict, no matter the scale or level of violence.

THE POPULATION NODE

The discipline of sociology studies human societies and processes that preserve or change them. One textbook states "sociologists study all aspects and levels of society. A society is a group of people whose members interact, reside in a definable area, and share a culture. A culture includes the group's shared practices, values, beliefs, norms and artifacts."[10] Many important sociological factors make up the character of a population. This includes institutions (like religion), communities, ethnicities, languages, geography, and social classes. Although populations share distinct practices, culture, and often geographical boundaries, there are many variations within each society that make populations less than homogeneous.

Populations have vulnerabilities that resistance movements often utilize to create divisions between opposing groups. Simultaneously, the state can use government programs to better bond various sorts of peoples together. For instance, Belgium has three official languages: French, Dutch, and German. Language acts as a division and strong factor in local resistance, and French communities in Belgium strongly enforce the use of their language in education, media, and business. The Belgium government addresses this vulnerability by adopting all three as official national languages.[11] As illustrated, any population can be defined in terms of sociological and competing factors. Considering the preceding, we propose the following definition of the population node:

> ***The population node within the resilience and resistance model typically describes the peoples living within defined national geographic boundaries and is characterized by various aspects of institutions, communities, ethnicities, languages, and social class; this node can support and/or influence both the resilience and resistance nodes.***

Population spirit. As a caveat, every population has factors that help influence its ability to resist foreign aggression or endure internal conflict, which could be described as spirit, mental health, or even pain tolerance. There are many variables, including the manner and outlying factors influencing the stressors on a society, which are never the same and make evaluation difficult. However, one mental health study proposes ***seven characteristics of highly resilient peoples who can adapt or rebound from adversity and stress.***[12] ***These factors include: (1) innovation, (2) decisive action, (3) tenacity, (4) interpersonal connectedness, (5) honesty, (6) self-control, and (7) optimism.*** Assessing these characteristics at a societal level may provide important insight into how long a population can endure adversity, particularly at the more violent end of the Resistance Continuum.

THE RESILIENCE NODE

The resilience node represents governing authority—either the internationally recognized national government or the government of an occupying power. Both instances are confined by geographic boundaries, which are those recognized by international conventions. The common understanding of a nation derives from the Treaty of Westphalia in 1648.[13] Membership in the official list of countries is exclusive, and, consequently, countries do not normally recognize the authority of non-state actors. The following definition of resilience is offered:

> ***Resilience is a measure of the sustained ability of governance to utilize resources and processes to respond to, withstand, and recover from natural or man-made shocks to its environmental norms and sustain systems of order.***[14]

Hyper-resilience. A government may attempt to prepare its population to endure "distress, impairment, or dysfunction" in advance of hostilities.[15] In boxing terminology, hyper-resilience prepares a population and government to "take a punch." See Figure 1.3. A number of countries have institutionalized such preparedness. For instance, Singapore has adopted a "Total Defense" strategy that blends social resilience with collectivism and national identity.[16] Other nations with similar strategies include Finland, Israel, and Switzerland. Sweden, for instance, publishes educational documents for its citizens to prepare for armed conflict.[17] Hyper-resilience can also increase deterrence by presenting a hard target.

Figure 1.3: Boxing analogy of hyper-resilience in interstate conflict. Source: Author

THE RESISTANCE NODE

In the 1960s, the term *resistance* in sociology referred to forms of dissent or civil disobedience to current authority.[18] In the humanities disciplines, armed rebellion typically used different terminology. In stark contrast, during this same period, military science called armed groups fighting foreign occupation a "resistance" while domestic armed groups were categorized by "insurgency." The "French Resistance" in World War II provides one example of this type of labeling. A good example of the bifurcation in military science is evident in this title *Undergrounds in Insurgent, Revolutionary, and Resistance Warfare* published as recently in 2013.[19] The title of the book, clarifying similar types of undergrounds while insisting three separate distinctions (insurgency, revolution, and resistance) in terms of movements, demonstrates the baffled approach. In short, until the 21st century, there existed three related terms to deconflict: (1) resistance in sociology referring to dissent or civil disobedience, (2) resistance in military science referring to armed groups opposing foreign occupation, and (3) domestic armed groups opposing the regime referred to as insurgency.

Perhaps the first publication to use resistance as a broader term to encompass all forms was *Legal Implications of the Status of Persons in Resistance* by Erin N. Hahn and W. Sam Lauber, published in 2012.[20] Additionally, as the lead doctrinaire in the drafting of Joint Publication 3-05.1 *Unconventional Warfare*, Robert Burrell attempted to codify "resistance" as a term embodying the entire spectrum of dissent; he published this conceptual framework in 2013.[21] The groundbreaking scholarship from Hahn and Lauber is discussed later in some detail, as it created the framework for resistance categories used in the Resistance Continuum as presented in this volume. This broader and comprehensive version of the term "resistance" has begun to mean encompassing the entire spectrum of groups or organizations opposing current governance or foreign occupation, particularly in military science. The phenomenon can be seen in publications recently sponsored by U.S. Army Special Operations Command under the guidance of Paul J. Tompkins Jr., a thought leader in this field. This is predominantly true regarding knowledge authored by scholars from Johns Hopkins University Applied Physics Laboratory. These publications include *Understanding States of Resistance* circa 2017,[22] *Threshold of Violence* circa 2018,[23] *Science of Resistance* in 2019,[24] *Resistance Manual* in 2019,[25] and *Resistance and the Cyber Domain* in 2019.[26]

In the broader sense of the term, resistance comprises one or many organizations that generate dissent toward current governance or occupation and include the employment of various means. The reasons for opposition to governance or occupation vary greatly, but may include socioeconomic inequality, political grievances, nationalism, or religious opposition. These groups may be highly organized or even intentionally leaderless. Some remain small and disciplined while others are large and disorganized. These groups may endorse a variety of means from nonviolent legal protest to irregular or traditional warfare. Examples of groups in the resistance node can include a spectrum, such as opposition political parties, social movements, and armed groups. Considering the preceding, the resistance node depicted in the Resilience and Resistance Model represents the following definition:

> ***Resistance is a measure of an actor's, group's, or population's will and ability to withstand external pressure and influences and/or recover from the effects of those pressures or influences.***[27]

Nationalized resistance. A state can also plan for its own transition from the resilience node to the resistance node. Normally, such a transition occurs following defeat by a foreign power and subsequent military occupation. This concept is called national resistance but is probably more aptly named state-sponsored resistance. In these situations, the government can prepare for the organization and execution of resistance activities in advance—creating plans for underground networks, shadow governance, government-in-exile, clandestine communication systems, arms caches, and much more. A state-sponsored resistance strategy can asymmetrically resist, or "block" occupational forces. Addressing this strategy, in 2020 both JSOU and the

Swedish Defence University published the *Resistance Operating Concept*. This study "explores actions that a sovereign state can take to broaden its national defense strategy and prepare to defend itself against a partial or full loss of national sovereignty."[28] A number of nations have instituted national resistance strategies, including Ukraine, Estonia, Latvia, and Lithuania. In 2021, Ukraine passed a resistance law that outlines irregular force organization, militia training, and civil disobedience.[29] A component to national resistance could include designing urban landscapes deliberately to support resistance activities. Scholar Sandor Fabian recommends constructing "large-scale pre-conflict infrastructural preparations in urban areas to turn their towns and villages into 'man-made jungles' to enhance the effectiveness" of targeting the invaders.[30] A recognized national resistance strategy can also contribute to deterrence. For this purpose, the definition of nationalized resistance from the *Resistance Operating Concept* works well.

> ***A nation's organized, whole-of-society effort, encompassing the full range of activities from nonviolent to violent, led by a legally established government (potentially exiled/ displaced or shadow) to reestablish independence and autonomy within its sovereign territory that has been wholly or partially occupied by a foreign power.***[31]

Figure 1.4 demonstrates the relationships between resilience, resistance, and external support in a state-sponsored resistance strategy. While state-sponsored resistance strategies are rare, when a state employs this strategy, combined with strong external sponsors, it can offer an asymmetric advantage to counter the ambitions of aggressive neighbors. In such cases of asymmetric advantage, an external sponsor offers support to resilience primarily prior to conflict but prepares for and executes support to nationalized resistance following hostilities. While support to resilience or support to resistance is the norm for external support within intrastate conflict, supporting a state-sponsored resistance strategy in a partner may constitute a premier choice, the *crème de la crème*, during strategic competition between great powers, and hence the possibility is worthy of examination, despite being a less likely alternative.

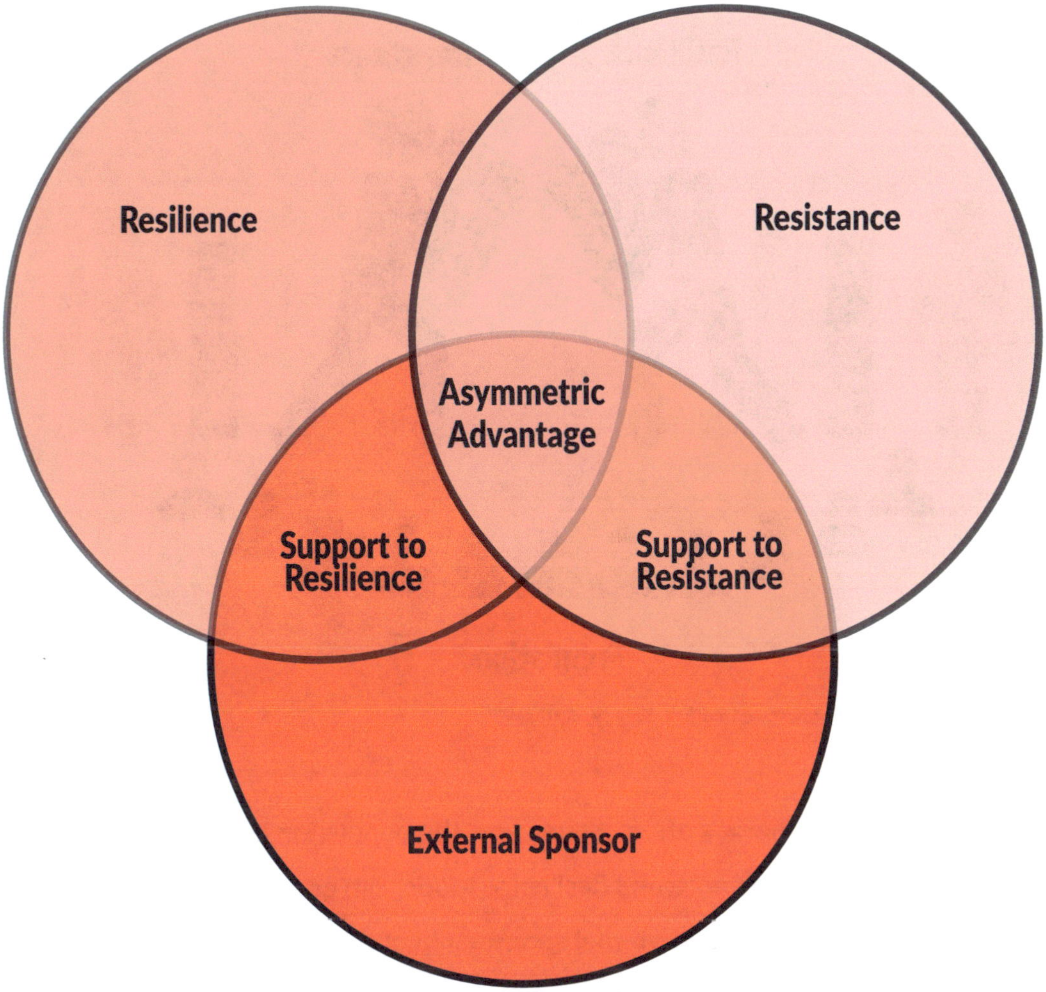

Figure 1.4. Conceptual framework of resilience, resistance, and the external sponsor. Source: Author

EXTERNAL SUPPORT TO RESILIENCE OR RESISTANCE

Currently popularized constructs due to Russian aggression against Ukraine, as well as China's ambitions to integrate Taiwan—"nationalized resistance" and "hyper-resilience"—address unconventional approaches to interstate conflict. However, these circumstances pale in comparison to the escalating instances of strategic competitors intervening within *intrastate* conflict. From 1945 to 1991, the actions of the U.S. and the Soviet Union during the Cold War prolifically demonstrated the phenomenon of external opponents supporting participants of intrastate conflict. Historians often refer to these instances as proxy wars.[32] In today's era of U.S. competition with China, Russia, Democratic People's Republic of Korea (DPRK), Iran, and violent extremist organizations, the study of external support to resilience or resistance within intrastate conflict retains its primacy.[33] Figure 1.5 illustrates the relationships between resilience and resistance competing in a boxing ring while the population supports either opponent or remains neutral; simultaneously, external supporters work the corners to support the respective antagonists.

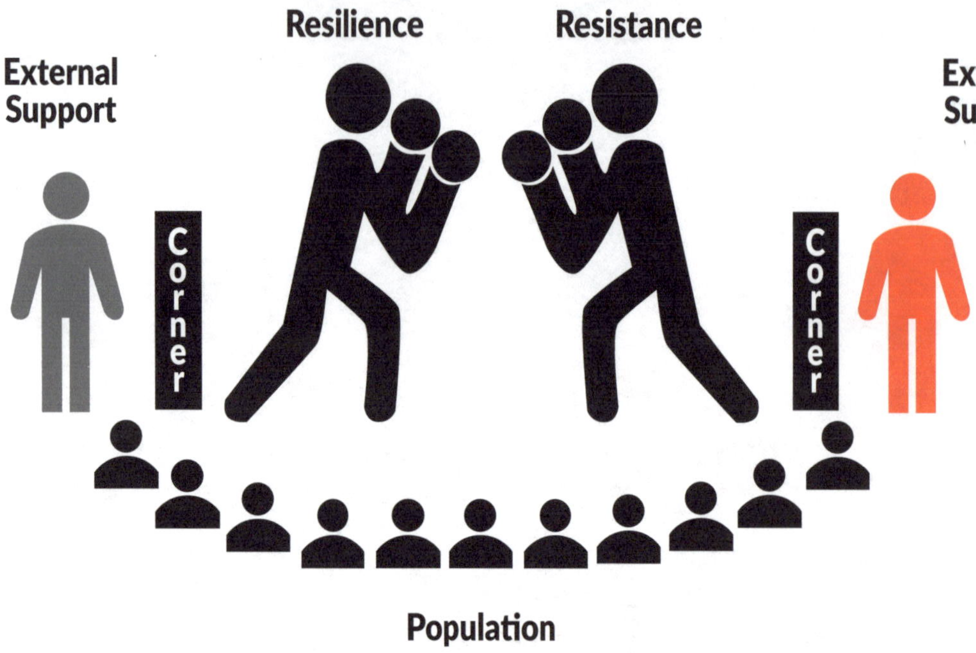

Figure 1.5: Boxing analogy of intrastate conflict. Source: Author

Irregular warfare practitioners often convey how the population is the center of gravity, but all four nodes of the Resistance and Resilience Model can prove critical in the success of resilience or resistance objectives. External sponsorship of resistance or resilience elements can be accomplished through multiple ways and with multiple means. In terms of military aspects, support to resilience normally includes activities like foreign internal defense, counterinsurgency, and stabilization. The means might include arms, equipment, and training. In contrast, the military way to support resistance is typically referred to as *unconventional warfare or special warfare* (depending on the national doctrinal variances). The military ways of supporting resistance include covert and overt military assistance to enhance the subversion of the opposing state. Nonetheless, military ways and means are but one methodology to support a partner.

Just as the ways and means of support are diverse, so too are the sponsors in intrastate conflict. A sponsor can consist of a nation-state or even a collection of states. One country recognizing another nation's sovereignty can prove powerful support to its legitimacy. As of January 2024, NATO represents 31 countries that support each other's resilience. However, nations are only one type of sponsor. Corporations can also take sides in intrastate struggles, as can religious organizations. A diaspora can prove quite powerful in external support. For instance, the Irish Republican Army was greatly funded by the Irish diaspora in the United States.[34] Accordingly, there are many external sponsors to consider in the Resilience and

Resistance Model. Note that three of the nodes (population, resistance, and resilience) typically reside within or in the proximity of a nation–state. The external support comes from outside the state and may include the diaspora, another neighboring state, illicit organizations, religious institutions, international corporations, regional associations, and many others. The following definition describes the external support node:

> ***External support* in intrastate conflict can consist of governments, regional organizations, nongovernmental organizations, diasporas, corporations, religious groups, global influencers, transnational criminal organizations, and non-state actors who offer support to either resilience or resistance entities.**

An external partner has three options in regard to adopting policies in relation to another nation's internal conflict: (a) support a regime's resilience in order to free and protect its society from threats like subversion, lawlessness, and insurgency; (b) support indigenous resistance against an adversary's governance to coerce, disrupt, or overthrow the regime or occupying power; or (c) choose to prepare the environment for a future policy supporting resiliency or resistance. In the first two cases, the external sponsor can utilize the combination of several methods to support a partner or surrogate. For simplicity, we will narrow down the types of support to either *diplomatic, information, military,* or *economic* (DIME).[35] In each situation, the "DIME cocktail" provided by an external sponsor will likely have varying sized ingredients based on desired outcomes. Legitimacy can be a deciding factor in achieving success for either the resistance movement or the current regime; consequently, diplomatic support from a recognized sovereign state can prove quite valuable. As a population-centric struggle, the battle for the narrative and information activities can also prove important. Economic support can assist in stabilizing a partner and enhancing their legitimacy, including facets of food, medical care, and/or employment opportunities. Figure 1.6 illustrates one concept of various options in support.

Military support, or lethal aid, can enhance human security and possibly eliminate rivals through violence, creating stability (at least, in the short term). Over time, the more destructive the conflict, the more unstable the region potentially becomes.[36] In many cases, violence can exacerbate human security needs. Consequently, deciding if and when to offer military

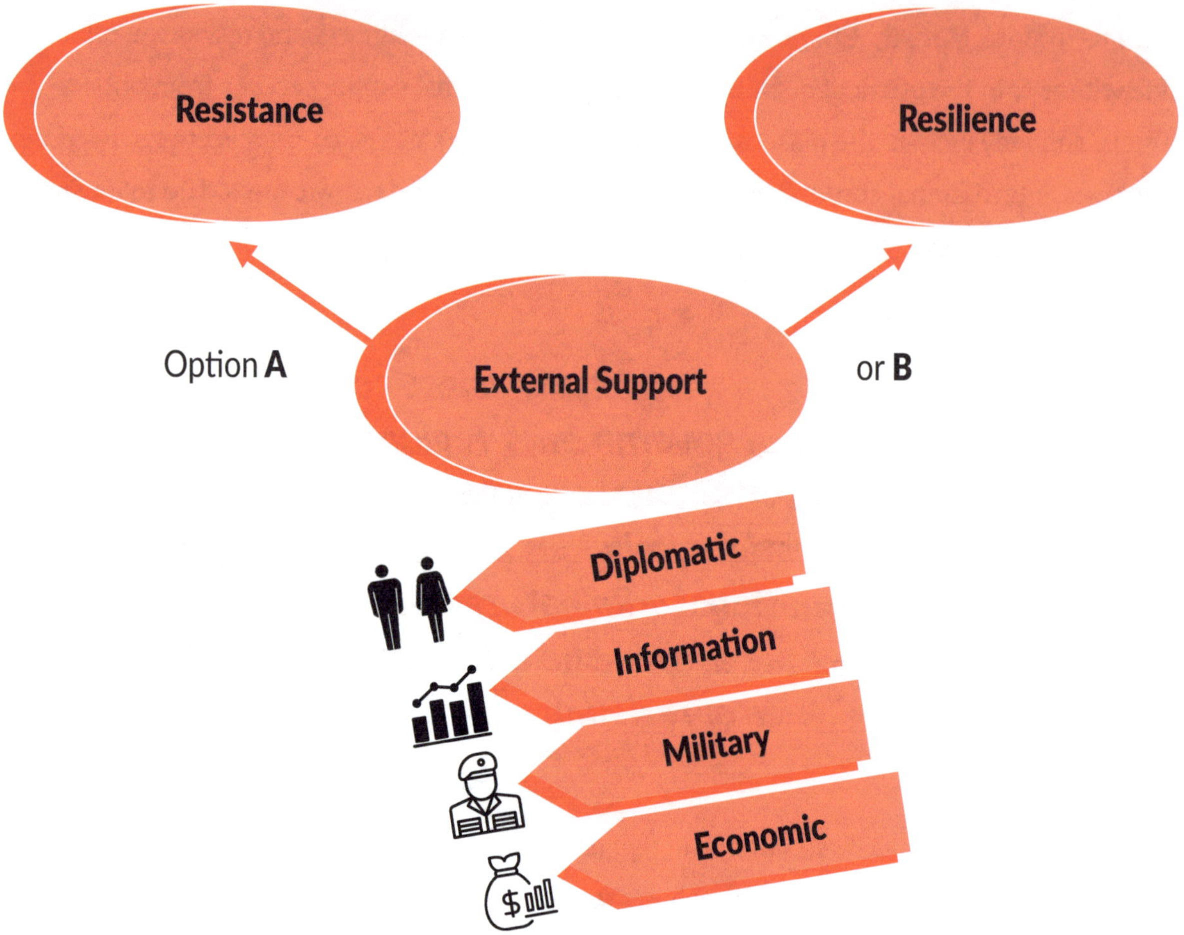

Figure 1.6. Examples of external support to resilience or resistance. Source: Author

assistance involves great care and analysis, including a full evaluation into what types of military aid are most appropriate. There are several factors to consider when evaluating resistance movements in a particular region, including religion, demographics, ethnicities, and social hierarchies. However, understanding the distinct categories of resistance can help determine what types of military support, if any, are appropriate for external support to any particular resilience partner or resistance movement.

BIFURCATED SPONSORSHIP

As an additional option, an external sponsor can attempt to sponsor both nodes of resilience and resistance simultaneously. This is not common, but it does have some pragmatic applications. A nation or corporation might carry out the sale of military equipment to both sides of a conflict simply for economic benefit. A more nuanced diplomatic strategy might include acknowledging the legitimacy of a government while also demonstrating support to organizations seeking reforms. For example, in 2016 President Barack Obama met with the

General Secretary of Vietnam Nguyen Phu Trong; he concurrently entertained the idea of visiting with pro-democracy groups in the country. Such bifurcation of sponsoring a resilience and resistance movement simultaneously can result in a falling out with one side or the other. In the case of Obama's visit to Vietnam, the General Secretary opposed his meeting with resistance so vehemently that it was cancelled.[37] In short, bifurcated sponsorship is possible—but tricky and unusual.

THE RESISTANCE CONTINUUM

Within the Resilience and Resistance Model, particular care should be taken in accessing the resistance node. Human populations inherently develop opposition to endogenous governance or foreign occupation. Afghanistan is perhaps the archetypal example across its history. Afghan societies have successfully opposed centralized indigenous governance as well as that imposed by invaders via occupation and have done so for centuries. Despite inherent resistance, each regime attempts to steel the resolve of the population from reform. Although resistance is a commonality around the globe, resistance movements themselves are quite distinct. The purpose of a movement might rely on factors of social injustice, ethnic tensions, or ideologic and religious differences. Consequently, resistance movements develop unique approaches to motivate regime change. Essentially, no resistance movement is the same. Irrespective of this diversity, five general typologies of resistance are illustrated in the following figure.

Resistance generally occurs along a continuum (see Figure 1.7). The continuum indicates a scale of protest and conflict, but resistance movements often incorporate more than one of these methods over time. Resistance often begins with nonviolent protest, conducted legally or at least within established international norms. In fact, the United Nations (UN) recognizes lawful assembly and protest as a universal human right.[38] One example of nonviolent legal resistance is Martin Luther King Jr.'s movement to protest Jim Crow laws and achieve voting equality for African Americans in the 1960s. King carefully staged protests conducted within established norms, but still achieved national attention via television—particularly due to the violence imposed on the peaceful protesters, which King anticipated, as demonstrated in Birmingham in 1963.[39] Another form of protest is nonviolent but remains inherently illegal. Those who supported the American antebellum underground railroad for transporting slaves, like Harriet Tubman, fall into this category because the underground railroad directly opposed congressional law pertaining to the rights of slave owners.[40]

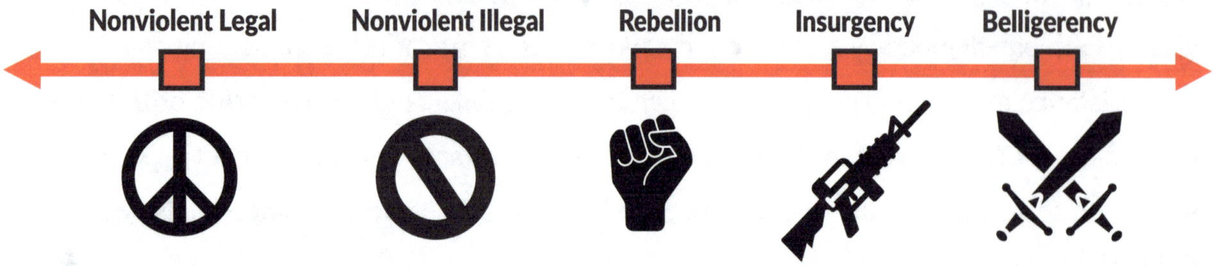

Figure 1.7: Diagram of the Resistance Continuum. Source: Author

When protest turns violent, it becomes rebellion. Small outbreaks of violence such as Nat Turner's revolt against slave owners in 1831 illustrates rebellion.[41] This category of rebellion is generally defined by the scale of violence, meaning that in most cases the local government can attempt to counter the violence with ready means available, like law enforcement. In contrast, insurgency is also violent, but the civil government can no longer address the resistance through rule of law. This type of violent conflict can result in insurgency. For example, in the period called Bleeding Kansas (1855–1861), pro-slave and free-state communities used coercion and violence against one another with the objective of controlling voting outcomes and securing their disparate visions of statehood.[42] Both parties also had external support outside the territory in terms of weapons and funding; these included the New England abolitionist societies as well as the slavocracy in Missouri. Bleeding Kansas could not be addressed by local civil authorities. In the final form of resistance, the American Civil War is a classic example of belligerency. In belligerency, the resistance demonstrates such autonomy that it resembles its own nation–state. Typically, belligerency results in a bloody civil war with the resistance and the regime fighting through conventional military tactics for the ultimate stakes of controlling the future state.

These terms of (1) nonviolent legal, (2) nonviolent illegal, (3) rebellion, (4) insurgency, and (5) belligerency were introduced by Erin N. Hahn and W. Sam Lauber, both lawyers working at Johns Hopkins University Applied Physics Laboratory.[43] Hahn and Lauber attempted to create legal categories for participants in, and external actors supporting, resistance activities. From the perspective of external support to intrastate conflict, these legal categories consist of an important advancement for discussion. However, these five categories do not neatly fit within either the social movement taxonomy, law, nor those of military science. They do, however, offer a construct from which all three taxonomies can perhaps combine into a cohesive understanding and subsequent useful application for military strategy and foreign policy. To further understand and operationalize the Resistance Continuum, this section defines and illustrates examples of all five categories.

NONVIOLENT LEGAL

Nonviolent legal and nonviolent illegal forms of protest are often lumped together and treated the same. As one example, Gene Sharp in 1973 created 198 ways to perform nonviolent action without any distinction to legality.[44] However, mixing these two methodologies, one legal and other illegal, ultimately makes the entire resistance organization an illicit one and subject to arrest and prosecution. Legal forms of protest have unique advantages in moral and ethical supremacy, and careful consideration should be followed before negating this benefit.

The quintessential example of nonviolent legal protest comes from Mohandas Karamchand Gandhi and his methods utilized in South Africa and India. A religious guru, Gandhi rose in the political ranks and eventually became a national symbol of resistance to foreign influence, particularly the British. One of his most frequent means of protest included hunger strikes, but perhaps the most iconic was his Salt March. Brilliantly, Gandhi meant to protest the British monopoly on salt. At the time, England placed taxes on salt, which were consequently paid by all Indians of every social class. To make this monopoly lucrative, the British outlawed Indians making salt. While other Indian nationalists found Gandhi's protest idea ridiculous, opposing the exploitation of salt quickly grew into a symbolic demonstration with shared national interest across all of India's diverse populations.[45]

Gandhi planned his protest with one main objective—to garner domestic and international support for Indian independence. He prefaced his march with a cultivated narrative, working with reporters to spread his message about Indian independence from Britain and the importance of salt as a symbol of British exploitation. He also leveraged his religious status to issue provocative statements during prayer meetings. Soon, the impending Salt March was receiving international attention. Then, from 12 March to 6 April 1930, Gandhi executed his march. Participation grew along the route, with more than 2,500 taking part. The march covered 240 miles and traveled through four major cities and dozens of villages. At each location, Gandhi stopped to make speeches to increase attention and support for independence. The Salt March route culminated on the coast. When Gandhi reached these natural salt flats (which the British had destroyed prior to his arrival), he reached through the mud and held a lump of natural salt in the air, symbolically breaking British law. In response, hundreds of thousands of Indians nationwide began protests, resulting in more than 60,000 arrests—including Gandhi himself.[46]

Many resistance movements have learned from and adhered to the nonviolent and legal approach used by Gandhi. Martin Luther King Jr. saw Ghandi as a mentor and visionary, inspiring his trip to India in what can best be described as a pilgrimage.[47] King's methods have been described in many words, including "militant nonviolent direct action."[48] Both King's and Ghandi's approaches can also be described as *civil disobedience*. While both leaders incorporated illegal methods at times, they generally attempted to use approaches that their intended audiences would consider justified. As Henry David Thoreau stated, "under a government which imprisons any unjustly, the true place for a just man is also a prison." As such, Ghandhi's and King's methods generally met what their audiences considered as legal (or at least justified) nonviolent methods, which likely would also have generally met the parameters identified as a universal human right by recent United Nations consensus.[50] Consequently, the actions of both Ghandi and King offer examples of good case studies, largely, for *nonviolent legal protest*.

Several other terms exist that are similar to nonviolent legal protest. *Nonviolent action* is one, *nonviolent collective action* is another, and, as introduced previously, *nonviolent direct action*. According to the King Center, nonviolent action consists of "marches, boycotts, picketing, sit-ins and prayer vigils."[51] Fortunately, protest has a useful legal definition as "a public remonstrance or demonstration (usually expressing dissent) centered around an act or policy of a governing body, such as sending troops overseas, use of the death penalty, or any other regulation or law."[52] After consideration, this section proposes a definition for nonviolent legal resistance as:

> ***Nonviolent legal forms of resistance consist of public actions taken to demonstrate dissent with the acts or policies of the governing body; it may include nonviolent action and forms of protest, which the United Nations considers a fundamental human right.***

NONVIOLENT ILLEGAL

The premier example of nonviolent but illegal resistance comes from Nelson Mandela and his campaign against racial segregation. The white South African government instituted a system of exclusion of blacks (and other non-whites) from representative government and equal opportunities—a system called apartheid. Through the African National Congress, Mandela demonstrated several methods to resist authority, primarily through nonviolent legal methods characterized by Gandhi and King. However, in 1961, the *Umkhonto we Sizwe* paramilitary group, headed by Mandela, began planning acts of sabotage in South Africa. These specifically targeted infrastructure to avoid loss of life.[53]

Perhaps what became Mandela's most inspirational form of dissent surfaced as a result of his trial and incarceration in 1962. A South African court found him guilty of conspiring to overthrow the government, for which he received a life sentence. Leveraging the global media, the mantra of "Free Mandela" became universal in the popular culture of Great Britain and the United States. An international movement grew, which forced many corporations and governments to impose sanctions on South Africa. Several prominent American artists took up the banner, including Whitney Houston, who organized a Free Mandela concert in the 1988 FreedomFest. After three decades of pressure, South Africa was compelled to release Mandela from prison in 1990. Subsequently, in 1994, South Africans elected Mandela as the new president, and he dismantled apartheid.[54]

This category of *nonviolent illegal* probably best serves as also embracing social movement terms like *nonviolent insurrection*, as well as warfare terminology like *subversion* and *sabotage*. Nonviolent insurrection or *unarmed insurrection* refers to a general uprising against a regime or occupying power, but largely without utilizing violent means. This is probably best described by Stephen Zunes as activities like "strikes, boycotts, mass demonstrations, the popular contestation of public space, tax refusal, destruction of symbols of government authority (such as official identification cards), refusal to obey official orders (such as curfew restrictions), and the creation of alternative institutions for political legitimacy and social organization."[55] A recent example of nonviolent insurrection might describe the *Orange Revolution* in 2003 and 2004. The protests were massive, including a large segment of the Ukrainian population. While these demonstrations remained nonviolent, the protestors brought the city of Kyiv to a halt, effectively, and unlawfully, shutting down the government.[56] Another example of nonviolent illegal includes the *Umbrella Movement* seeking change to Chinese policies in Hong Kong in 2014, an interesting case study that also stands apart in terms of innovation by leveraging technology and social media in modern resistance.[57]

Two important activities often associated with military science include subversion and sabotage. It is important to differentiate between legal forms of protest and illegal forms like subversion.[58] Unlike protest, subversion crosses a line of legality by "advocating overthrow of the government," which U.S. Congressional law clearly (if not lengthily) describes.[59]

Congressional law defines these activities as planning, organizing, or advocating with the intent to use violent means. A fundamental distinction is that while subversion itself remains a nonviolent activity, it is illegal.

Sabotage is another important method used to resist authority, but it does not necessarily include lethal means. The definition of sabotage is twofold. In military science, it specifically refers to damage done solely to war utilities and war materials.[60] For example, resistance to Nazi occupation in France often targeted infrastructure with military purposes, like railroads. Many resistance movements to Nazi rule also organized workers to damage factories that produced war goods.

Sabotage can also occur as a form of protest, entirely unrelated to military objects, and it can occur within intrastate conflict to undermine or defund authority. For instance, unionized workers may sabotage a factory "to inflict a production or profit loss for the targeted organization."[61] A discussed previously, *Umkhonto we Sizwe* in South Africa advocated the sabotage of mines and factories to damage the economy and delegitimize the government.[62] Groups called *ecoterrorists*, like Earth Liberation Front (ELF) and the Animal Liberation Front (ALF), reportedly use tree-spiking and arson of facilities.[63] Within the context of intrastate conflict, sabotage includes damage or destruction to civilian or government infrastructure or material to support policy aims.

After considering the preceding, the following statement proposes a definition for nonviolent illegal resistance.

> ***Nonviolent illegal forms of resistance consist of nonlethal acts, which undermine the legitimacy of the governing authority or occupying power through nonviolent insurrection, sabotage, or subversion.***

REBELLION

On the Resistance Continuum, rebellion clearly indicates a marked change in methodology, which includes lethality. However, any movement embracing the use of violence might also include nonviolent forms of protest discussed previously. The distinction in rebellion is that the resistance either deliberately uses lethality, or lethality evolves from what was originally intended as nonviolent protest. Violence can result as a choice to evade capture, to protect oneself or other members from the reprisal of authority, or it may be a deliberately planned for and implemented to achieve desired ends.

The key constraint in rebellion remains the extent of the violence and how it is addressed by authority. Hahn and Lauber explain rebellion as violence in which a "state's law enforcement mechanisms are able to suppress," or, in other words, military force is not mobilized to suppress it.[64] This is an excellent threshold: lethality remains limited, either by choice, in the case of a large demonstration preferring other nonviolent methods, or simply by the limited size of the participants—even if lethality remains the primary tool. Two case studies illustrate these two situations.

In rebellious organizations of limited size, the movement can form an armed component that is dedicated to the use of violence as the primary means of resistance. Using a small armed force against a large state apparatus rarely achieves dramatic success if done directly. For instance, in 1831, Virginia state authorities found and executed Nat Turner and his followers in just six weeks. Often, enduring organizations attempt asymmetric methods familiar to military science, like raids, ambushes, and assassination. The state, in turn, may label a violent movement a terrorist organization, after which the state can justifiably mobilize its military to destroy it. Terrorism thus must be included as an insurgency (discussed later in this volume).

Large-scale social revolts without a major armed component have become more prevalent in the 21st Century.[65] Beginning in late 2010, the Arab Spring encompassed revolutions in more than a dozen countries. In each case, the resistance fell into various categories, to include nonviolent illegal, insurgency, and belligerency. However, the resistance in Egypt most likely fits the case of rebellion because deaths were less than 1,000 and the military did not directly intervene. In fact, when security forces could not contain the crowds, the Egyptian military refused orders to put down the protests and interjected forces only to save the resistance from harm.[66] Thus, the Arab Spring in Egypt fits the category of a rebellion, where violence did occur but on a limited scale despite very large numbers of protestors. Another similar example of a large resistance movement using limited violence remains the *Euromaidan Revolution* in Ukraine. During four months of conflict, the numbers killed remained minor in comparison to the number of participating protestors and security forces, and no military response occurred to put down the rebellion.[67]

Rebellion includes methods that intentionally, or unintentionally, cause fatalities due to resistance activities. For instance, the unintended result of a protest could be its development into a destructive riot. Taking the preceding discussion into account, the concise definition offered by Hahn and Lauber appears well constructed:

> ***Rebellion* is a category of resistance governed by domestic criminal law and falls short of an armed conflict but features violence.**

INSURGENCY

The distinction between rebellion and insurgency remains the use of a nation's military: Once a military begins operations against a resistance movement, the threshold has been crossed from rebellion to insurgency. The classification of insurgency has a long history in military science while terrorism does not. However, sovereign states also identify certain resistance organizations as terrorists, which also allows for the use of military force.

In terms of legal status, unfortunately, insurgent groups are not clearly differentiated from terrorist organizations in terms of the law. Both can be argued as utilizing illegal lethal means against the state—either the nation's military or its citizens. Ethically, there are distinct differences in these two methodologies, as one method can avoid killing non-combatants and the other can completely dehumanize the use of violence.[68]

The bifurcated approach, and resulting confusion between terrorism and insurgency, has an extended historiography; no consensus exists on the topic within any discipline. Lawyer Ranbir Singh argues that "there is a very thin line of distinction between 'terrorism,' 'insurgency' and 'belligerency'; and in almost all cases these are terms denoting the various stages of the same process."[69] Singh goes on to use examples of the *Jammu Kashmir Liberation Front* and the *Hizbul Mujahideen* in Afghanistan to substantiate his case.

However, these examples of terrorism and insurgency embodying the same organizations are only a few. In studying the *Kurdistan Workers' Party* (PKK), Mustafa Coşar Ünal demonstrates that the PKK consisted of both a terrorist and an insurgent organization during its evolution.[70] This blur between terrorism and insurgency is also illustrated in the case of the *Front de Liberation Nationale* (FLN) against the French in Algeria between 1954–1962.[71] Additionally, the *Provisional Irish Republican Army*, from 1969–2005, also encompassed a good example of a movement comprising both an insurgent group and a terrorist organization.[72] Perhaps Ünal best describes the difference between the two as: (a) one comprising a method, and (b) and the other embodying an organization. He states, "terrorism is considered to be a method of pursuing a political goal, while insurgency is a political movement aimed at realizing a specific political goal, which is generally to overthrow a regime."[73]

Undoubtedly, this volume will not resolve the legal or moral distinctions between terrorism and insurgency, yet they both undoubtedly fall within the Resistance Continuum in the insurgency category. A unified definition describing this grouping is offered below.

> ***Insurgency* and *terrorism* both use illegal forms of lethal violence against a state and/or its citizens to attain political goals or regime change, a threat to the government's sovereignty considered grievous enough to require addressing with military force.**

BELLIGERENCY

In belligerency, a resistance organization emerges to make conventional war against a state. Sometimes, belligerency occurs when a successful insurgency evolves to maintain state-like functions in a region and sustains human security responsibilities over a segment of the population. Unfortunately, belligerency as a resistance category does not have international agreement, primarily because sovereign nation-states do not want to recognize lethal forms of resistance as anything but illegal.

Nevertheless, belligerency, or civil war, has a long history of resistance developing into a sovereign power. The formulation of the U.S. is case in point, as the American Revolution primarily gained its legitimacy after recognition from France. In the current international global order, only other recognized nation-states can facilitate the belligerent's need to achieve their desired sovereign status; this makes it extremely hard for revolutions to gain belligerent status. Nevertheless, foreign powers have conducted military interventions into other states' civil wars despite this act constituting "*de jure* interference in another state's internal affairs."[74] In 1957, the United Kingdom argued the following conditions must be met for a belligerency to be officially recognized:[75]

a. There must exist an armed conflict of a general (as distinguished from a purely local) character.

b. The contesting party which is not the legitimate government must occupy and administer a substantial portion of the national territory.

c. The above-named contesting party must conduct the hostilities in accordance with the rules of war and through organized armed forces acting under a responsible authority

d. There must exist circumstances which make it necessary for outside states to define their attitude by means of recognition of belligerency."

A classic example of belligerency is the American Civil war, where the Confederacy met most of the preceding conditions. However, try as it might, the Confederacy could not receive its desired recognition by Great Britain.[76] This situation remains in a modern context, violent resistance stemming from an insurgency must receive official recognition by a nation-state to be considered a belligerent. In such cases, the more states that provide recognition the better, or even recognition from an international or regional body.

A great and recent example of belligerency includes the Syrian Civil War, the factional violence of which drew international attention in 2011. Regional instability combined with the oppressive governance of the Syrian Arab Republic fostered doubts in its legitimacy. Public demonstrations soon grew from peaceful protest to violent insurrection. A complex belligerency based on entangled religious, tribal, alliance, and international politics emerged, with multiple factions attempting to wrestle control of the nation from each other. Another challenger for legitimacy, the Islamic State, began to operate in Syria and created its own shadow government in 2014. Within this contested space, many groups vied for control over Syria without a dominant resistance emerging. The growth and power of the Islamic State of Iraq and Syria (ISIS) created anxiety in the international community, creating opportunities for internal organizations to garner external support; the U.S. formed a coalition of anti-ISIS and anti-regime elements; Russia, Iran, and Hezbollah supported the Syrian Arab Republic; and many non-state groups supported ISIS.[77]

As the Syrian conflict escalated, three major competitors emerged: (a) the Syrian Arab Republic, (b) a coalition of organizations opposing the Republic, and (c) ISIS. Each vied for public support and external sponsorship. The National Coalition for Syrian Revolutionary and Opposition Forces achieved status as a recognized belligerent by multiple sponsors. In contrast, ISIS did not receive international recognition and lost its bid to seize Syria. In summation, Syria rose to the level of belligerency due to the scale of the violence and the opportunity for outside nations and non-state actors to offer support to respective belligerents.

Considering the preceding information, this section proposes the following description:

> ***Belligerency*, as a status of resistance, occurs in a civil war when a state-like-entity, typically recognized internationally as a legitimate authority, engages in direct warfare against a sovereign state to overthrow the existing government or to gain independence from it.**

LEVERAGING THE CONSTRUCTS

Both the Resilience and Resistance Model and the Resistance Continuum adhere to a multi-disciplinary approach to better appreciate intrastate conflict. Often exclusively used after violence begins, military science can overmilitarize analysis of combatants without examining the underlying factors that create tensions or cohesion. Conversely, during discernable periods of nonviolent conflict, other important disciplines (like international affairs, political science, and sociology) provide excellent context to understand the human dynamics of change, but these disciplines remain equally important after violence occurs. Without the use of a broad interdisciplinary approach, external assessments of another country's intrastate affair impart a myopic interpretation of the complexity existing in in the human environment. Instead, a comprehensive interdisciplinary analysis provides a more holistic assessment from which to support foreign policy decisions.

Practitioners can operationalize the Resilience and Resistance Model and the Resistance Continuum to identify the most appropriate partner, and also to identify the most complimentary support packages in intrastate conflicts. External sponsors to resilience or resistance can use the interdisciplinary and comprehensive approach via the Resistance Continuum to identify the most prevalent nature of resistance movements. While resistance movements change and evolve, understanding their fundamental character at any point in time provides insight into how they might best be supported, influenced, or defeated. Each nation-state or region contains numerous resistance movements, and evaluating each along the resistance continuum can help a sponsor find the most appropriate partner—or the most dangerous organization to neutralize.

This chapter introduced two foundational interdisciplinary constructs regarding foreign support to indigenous resilience or resistance. The Resilience and Resistance Model demonstrates the interactive way resilience and resistance interrelates with each other, the population, and with external sponsors. This model applies in all cases, extending from nonviolent legal protest through belligerency. Before entertaining sponsorship of a partner, advocates should first consider the Resistance Continuum to generally classify different forms of protest and violence. These five categories along the continuum (nonviolent legal, nonviolent illegal, rebellion, insurgency, and belligerency) generally differentiate from one another in terms of their size and scale of violence, from legal protest through belligerency. The terms in the Resistance Continuum potentially have internationally recognized legal and ethical implications, which are important to consider when an external sponsor contemplates support to another government's resilience or its indigenous resistance. Utilizing both these models, scholars can study important case studies or current intrastate conflicts.

Each resistance movement is best countered or supported by unique methods. For instance, providing lethal force to a resilience or resistance partner is not always the best method of fomenting desired change. In contrast, the type of external support offered can also deliberately change the methods used by a resistance and the nature of a conflict. For instance, Iran primarily offers military support to Ansar Allah. This ensures the conflict will remain violent and aligns with Iran's regional goals of fostering instability in and around Saudi Arabia. Thus, external partners can deliberately or incidentally change the nature of the intrastate conflict through the means delivered.

CONCLUSION

As outlined in this book, one can apply the presented research methodologies to complete a comprehensive analysis of a nation-state in terms of resiliency and resistance for the purposes of advocating external support options for international policy. As discussed in the following chapter, we advise starting with data-centric analysis of state resiliency and resistance to determine potential in reinforcing either. Subsequently, and discussed in Part Two, analysts can leverage the research efforts of several prominent institutions to identify various resistance movements within the state and categorize them along the Resistance Continuum. After this, analysts can leverage an approach advocated by Johns Hopkins University/Applied Physics Laboratory and U.S. Army Special Operations Command to provide a deeper understanding of the resistance organizations themselves. In the final step, analysts can determine if external support to resilience or resistance comprises a viable strategy. The recommended data-driven and human-centric approach, prescribed in a 12-step process, was designed to address occasions of irregular warfare, competition, and deterrence, and is long overdue.

Methods of Analysis

1. **Analyze state resilience and resistance data.**
2. **Identify and categorize resistance movements.**
3. **Determine if external support is viable.**

About the Authors

Dr. Robert S. Burrell is a resilience and resistance interdisciplinary scholar using data-driven and human-centric methodologies to analyze intrastate conflict ranging from nonviolent protest through belligerency. He is a Senior Research Fellow at the Global and National Security Institute of the University of South Florida. From 2020–2024, he taught irregular warfare at Joint Special Operations University and was the editor in chief of special operations doctrine from 2011–2014.

John Collison is a contractor with Huntington Ingalls Industries and works in J59-Concept Development and Integration (CD&I) at U.S. Special Operations Command.

NOTES

1. Robert S. Burrell and John Collison, "A Guide for Measuring Resiliency and Resistance," *Small Wars & Insurgencies* 35, no.1 (December 14, 2023): 147–172, https://doi.org/10.1080/09592318.2023.2292942; Robert S. Burrell and John Collison, "Assessing Resistance for the Purpose of Informing International Policy," *Expeditions with Marine Corps University Press*, January 9, 2024, https://doi.org/10.36304/ExpwMCUP.2024.01. Also see Robert S. Burrell, "A Guide for Measuring Resiliency," *Irregular Warfare Initiative*, January 16, 2024.

2. David H. Ucko and Thomas A. Marks, "Redefining Irregular Warfare: Legitimacy, Coercion and Power," *Modern War Institute* (October 18, 2022), https://mwi.usma.edu/redefining-irregular-warfare-legitimacy-coercion-and-power/

3. Consortium to Study Irregular Warfare Act of 2021, H.R. 5130, 117th Cong. (2021).

4. For a description of irregular competition, see Robert Burrell, "How to Integrate Competition and Irregular Warfare," *Modern War Institute* (August 5, 2021), https://mwi.usma.edu/how-to-integrate-competition-and-irregular-warfare/

5. Taras Kuzio, "The Orange and Euromaidan Revolutions: Theoretical and Comparative Perspectives," *Kyiv-Mohyla Law and Politics Journal 2* (2016): 91–115, https://doi.org/10.18523/kmlpj88183.2016-2.91-115

6. Nolan Peterson, "In Ukraine, National Resistance Is Now the Law," *Coffee or Die* (January 7, 2022). https://coffeeordie.com/ukraine-national-resistance-law/

7. U.S. Army Special Operations Command, *"Little Green Men": A Primer on Modern Russian Unconventional Warfare, Ukraine* 2013–2014 (U.S. Army Special Operations Command, 2015).

8. See the series of volumes by Mykhaïlo Hrushevs'kyĭ published by Canadian Institute of Ukrainian Studies.

9. Gordon McCormick, *The Shining Path and Peruvian Terrorism*, (RAND Corporation, 1987), https://www.rand.org/pubs/papers/P7297.html

10. William Little, *Introduction to Sociology—1st Canadian Edition* (BCcampus, 2014), 3.

11. See Ludo Veny and Brecht Warnez, "Techniques for Protecting Minority Languages under Belgian Federalism." *International Journal on Minority and Group Rights* 23, no. 2 (2016): 211–236, https://doi.org/10.1163/15718115-02302002

12. George S. Everly Jr., Dennis K McCormack, and Douglas A Strouse, "Seven Characteristics of Highly Resilient People: Insights from Navy SEALs to the 'Greatest Generation,'" *International Journal of Emergency Mental Health* 14, no. 2 (2012): 137–143, https://ssrn.com/abstract=2118935

13. Derek Croxton, "The Peace of Westphalia of 1648 and the Origins of Sovereignty," *The International History Review* 21, no. 3 (September 1999): 569-592, https://doi.org/10.1080/07075332.1999.9640869

14. For the purpose of this essay, *resilience* specifically describes governmental resilience within the context of intrastate and interstate conflict. For a definition of community resilience, see Chandra et al., "Understanding Community Resilience in the Context of National Health Security: A Literature Review," *RAND Working Paper*, (February 2010). For a definition of economic resilience, see the U.S. Economic Development Administration's definition as "economic resilience aims to better prepare regions to anticipate, withstand, and bounce back from any type of shock, disruption, or stress it may experience," found at https://www.eda.gov/grant-resources/comprehensive-economic-development-strategy/content/economic-resilience

15. Michael O. Kaminsky et al., "An Evidence-Informed Model of Human Resistance, Resilience, and Recovery: The Johns Hopkins' Outcome-Driven Paradigm for Disaster Mental Health Services," *Brief Treatment and Crisis Intervention* 7, no. 1 (2007): 3, https://doi.org/10.1093/brief-treatment/mhl015

16. Ron Matthews and Fitriani Bintang Timur, "Singapore's Total Defence' Strategy," *Defence and Peace Economics* (March 9, 2023): 1–21, https://doi.org/10.1080/10242694.2023.2187924

17. Swedish Civil Contingencies Agency, *If Crisis or War Comes (Om krisen eller kriget kommer)* (Karlstad: MSB, May 2018).

18. James W. Vander Zanden, "The Non-Violent Resistance Movement Against Segregation," *American Journal of Sociology* 68, no. 5 (1963): 544–550.

19. Paul J. Tompkins Jr., and Robert R. Leonhard, *Undergrounds in Insurgent, Revolutionary, and Resistance Warfare*, (U.S. Army Special Operations Command, 2012).

20. Erin N. Hahn and W. Sam Lauber, *Legal Implications of the Status of Persons in Resistance.* (United States Army Special Operations Command, 2013).

21. Robert S. Burrell, "Joint Doctrine for Unconventional Warfare," *Air Land Sea Bulletin*, January 2013.

22. W. Sam Lauber et al., *Understanding States of Resistance*, (U.S. Army Special Operations Command, 2017).

23. Guillermo Pinczuk, *Threshold of Violence*, (U.S. Army Special Operations Command, 2018).

24. Summer D. Agen, *The Science of Resistance*, (U.S. Army Special Operations Command, 2019).

25. Meg Keiley-Listermann, W. Sam Lauber, and Christine Martin, *Resistance Manual*, (U.S. Army Special Operations Command, 2019).

26. Kristen Ryan, *Resistance in the Cyber Domain*, (U.S. Army Special Operations Command, 2019).

27. See the U.S. Army definition for a resistance organization, Army Technical Publication 3-05.1 *Unconventional Warfare at the Combined Joint Special Operations Task Force Level*, (USAJFKSWCS, April 2016).

28. Otto Fiala, *Resistance Operating Concept* (Joint Special Operations University Press, 2020), xv.

29. Peterson, Ukraine National Resistance, 2022.

30. Sandor Fabian, "Professional Resistance Forces as a Defence Framework for Small Countries," *Bezpieczeństwo. Teoria i Praktykay*, 3 (2022): 193, https://doi.org/10.48269/2451-0718-btip-2022-3-013

31. Fiala, *Resistance Operating Concept*, 238

32. For a good study on this, see Michael A. Innes, *Making Sense of Proxy Wars: States, Surrogates and the Use of Force*, (Potomac Books, 2012).

33. For a list of U.S. opponents and competitors, see White House, *National Security Strategy*, October 12, 2022.

34. Danielle A. Zach, "'It Was Networking, All Networking': The Irish Republican Movement's Survival in Cold War America," *Journal of Ethnic & Migration Studies* 47 no. 9 (2021): 2218–2236.

35. The acronym DIME has been used for many years. See Donald M. Bishop, "DIME, not DiME: Time to Align the Instruments of U.S. Informational Power," *Strategy Bridge*, 20 June 2018. https://thestrategybridge.org/the-bridge/2018/6/20/dime-not-dime-time-to-align-the-instruments-of-us-informational-power

36. Kendra Dupuy, Scott Gates, and Håvard Mokleiv Nygård, "State Fragility and Armed Conflict," *State Fragility and Armed Conflict, Conflict Trends* 7, 2016.

37. Reuters Staff, "Obama Says Several Vietnam Activists Prevented From Meeting Him," *Reuters*, 24 May 2016. https://www.reuters.com/article/us-vietnam-obama-dissidents/obama-says-several-vietnam-activists-prevented-from-meeting-him-idUSKCN0YF0ES

38. OHCHR, "UN Human Rights Committee Provides New Guidance on the Right of Peaceful Assembly," *United Nations*, 15 December 2020, https://www.ohchr.org/en/stories/2020/12/un-human-rights-committee-provides-new-guidance-right-peaceful-assembly

39. "Civil Rights Movement," *John F. Kennedy Library*, accessed 18 April 2023. https://www.jfklibrary.org/learn/about-jfk/jfk-in-history/civil-rights-movement

40. Janell Hobson, "Harriet Tubman: A Legacy of Resistance," *Meridians: Feminism, Race, Transnationalism* 12, no. 2 (1 July 2014): 1–8.

41. Breen Patrick H, *Shall Be Deluged in Blood: A New History of the Nat Turner Rebellion*, (New York: Oxford University Press, 2015).

42. Brie Swenson Arnold, "'To Inflame the Mind of the North': Slavery Politics and the Sexualized Violence of Bleeding Kansas," *Kansas History* 38, no. 1 (2015): 22–39.

43. Erin N. Hahn and W. Sam Lauber, *Legal Implications of the Status of Persons in Resistance*. (North Carolina: United States Army Special Operations Command, 2013).

44. Robert L. Helvey, *On Strategic Nonviolent Conflict: Thinking About the Fundamentals* (The Albert Einstein Institution, 2004).

45. Kwabena P. Slaughter, "The Nonviolent Imagination: Trickster Gandhi Revisited," *Philosophy & Social Action*, 29, no. 3 (2003): 41–50.

46. Thomas Weber, "Gandhian Nonviolence and the Salt March." *Social Alternatives* 21 no. 2 (May 2002): 46–51.

47. Martin Luther King, Jr. "My Trip to the Land of Gandhi," *Ebony Magazine*, July 1959, 84–92.

48. James A. Colaiaco, "Martin Luther King, Jr. and the Paradox of Nonviolent Direct Action," *Phylon (1960-)* 47, no. 1 (1986): 16, https://doi.org/10.2307/274691

49. Henry David Thoreau, *Civil Disobedience*, 1849. Available online from Columbia University at https://blogs.law.columbia.edu/uprising1313/files/2017/10/Civil-Disobedience-by-Henry-David-Thoreau.pdf, accessed on 24 July 2024. Another copy under separate title includes *On the Duty of Civil Disobedience*, (London: The Simple Life Press, 1903).

50. United Nations, "General Comment No. 37 on Article 21 (Right of peaceful assembly)," 17 September 2020.

51. "Nonviolent Action," *King Center*, accessed on 18 April 2023. https://thekingcenter.org/about-tkc/glossary-of-nonviolence/

52. Cornell Law School, "Protest," accessed 18 April 2023. https://www.law.cornell.edu/wex/protest

53. Simon Stevens, "Violence, Political Strategy and the Turn to Guerrilla Warfare by the Congress Movement in South Africa," *Journal of Southern African Studies* 47, no 6. (2021): 1011–1028. Also see Thula Simpson, "Mandela's Army: Urban Revolt in South Africa, 1960–1964." *Journal of Southern African Studies* 45, no 6 (2019): 1093–1110.

54. J. Bierman and C. Erasmus, "Tearing down the System," *Maclean's* 103, no. 7 (1990): 38. Also see, Margot Dougherty, Jonathan Cooper, and Janine Di Giovanni. "South African FreedomFest: Music's Most Powerful Dynamos Rev up in London to Honor South Africa's Imprisoned Nelson Mandela," *People* 29, no. 25, (27 June 1988): 30.

55. Stephen Zunes, "Unarmed Insurrections Against Authoritarian Governments in the Third World: A New Kind of Revolution." *Third World Quarterly* 15, no. 3 (1994): 403.

56. Mark R. Beissinger, "The Semblance of Democratic Revolution: Coalitions in Ukraine's Orange Revolution," *The American Political Science Review* 107, no. 3 (August 2013): 574–592.

57. Colin Agur and Nicholas Frisch, "Digital Disobedience and the Limits of Persuasion: Social Media Activism in Hong Kong's 2014 Umbrella Movement," *Social Media + Society* 5, no. 1 (2019), https://doi.org/10.1177/2056305119827002

58. R. J. Spjut, "Defining Subversion," *British Journal of Law and Society* 6, no. 2 (1979): 254–261, https://doi.org/10.2307/1409771

59. 18 U.S. Code Chapter 115 §2385, *Treason, Sedition, and Subversive Activities*, "Advocating Overthrow of the Government," 1994.

60. 18 U.S. Code Chapter 105 §2151, *Definitions*, "Sabotage," 2011.

61. Robert A. Giacalone and Paul Rosenfeld, "Reasons for Employee Sabotage in the Workplace," *Journal of Business and Psychology* 1, no. 4 (1987): 367.

62. Simon Stevens, "Violence, Political Strategy and the Turn to Guerrilla Warfare by the Congress Movement in South Africa," *Journal of Southern African Studies* 47, no. 6 (2021): 1011–1028.

63. United States Senate Committee on Environment and Public Works, "Eco-Terrorism Specifically Examining the Earth Liberation Front and the Animal Liberation Front," Washington D.C., 18 May 2005.

64. Hahn and Lauber, 49.

65. See Zunes. Also see, Sahan Savas Karatasli, "The Twenty-First Century Revolutions and Internationalism: A World Historical Perspective," *Globalizations* 16, no. 7 (22 August 2019), https://doi.org/10.1080/14747731.2019.1651525

66. Zoltan Barany, "Comparing the Arab Revolts: The Role of the Military," *Journal of Democracy* 22, no. 4 (2011): 24–35.

67. Andrii Gladun, "Impact of Repression on Mobilization: The Case of the Euromaidan Protests in Ukraine," *Social Justice* 46, no. 2/3 (156) (2019): 29–50, https://www.jstor.org/stable/26952572. Also see *Winter on Fire: Ukraine's Fight for Freedom*, directed by Evgeny Afineevsky (2015, Netflix).

68. Gjorjji Veljovski, "Terrorism or Insurgency? The Most Common Strategic Fallacy," *Contemporary Macedonian Defence* 20, no. 39 (2020): 107–119.

69. Ranbir Singh, "Insurgency and International Law & its Legal Consequences," *Student Advocate* Vol.8, 1996: 134-141, http://dx.doi.org/10.2139/ssrn.3182341. Quote from page 134.

70. Mustafa Cosar Ünal, *Counterterrorism in Turkey: Policy Choices and Policy Effects toward the Kurdistan Workers' Party* (Routledge, 2010).

71. Martha Crenshaw Hutchinson, *Revolutionary Terrorism: The FLN in Algeria, 1954-1962* (Stanford, Calif.: Hoover Institution Press, 1978).

72. Gregory M. Maney, Michael A. McCarthy, and Grace B. Yukich, "Explaining Political Violence Against Civilians in Northern Ireland: A Contention-Oriented Approach," *Mobilization: An International Journal*, 17, no. 1 (February 2012): 27–48. https://doi.org/10.17813/maiq.17.1.k074707202065h35.

73. Mustafa Coşar Ünal, "Terrorism Versus Insurgency: A Conceptual Analysis," *Crime, Law, and Social Change* 66 (2016): 27. https://doi.org/10.1007/s10611-015-9601-7

74. Joe Clare and Vesna Danilovic, "The Geopolitics of Major Power Interventions in Civil Wars," *Political Research Quarterly* 75, no.1 (2022): 20–34.

75. Rob McLaughlin, "Whither Recognition of Belligerency?" *Articles of War*, 17 September 2020. https://lieber.westpoint.edu/whither-recognition-of-belligerency/

76. Henry Clews, "Great Britain and the Confederacy," *The North American Review* 149, no. 393 (1 August 1889): 215–222.

77. *The Consequences of Inaction: US and NATO Syria Policy*, (Middle East Institute and Etana Syria, November 2020). Also see Clare.

| 33 |

II. Measuring Resiliency and Resistance

By Robert S. Burrell and John Collison

ABSTRACT: This section addresses the critical need to apply a more advanced human-centric model to assess current levels of governmental and societal resilience and resistance to subversion and coercion and internal and/or external aggression. It proposes a method of analysis and assessment of the will, potential and sustainability for national and/or sub-national resilience, as well as the potential for resistance to malign indigenous governance or external aggression. Finally, it examines, from a prospective external supporter's perspective, the likely success of support to a partner's resiliency or the potential to enable resistance to inspire change in adversary behaviors. Parts of this chapter were previously published in *Small Wars and Insurgencies* journal.[1]

INTRODUCTION

The question of state resiliency remains essential to the three biggest U.S. foreign policy questions of today. Why did Afghanistan fall so rapidly to the Taliban in 2021? Why, in contrast, has Ukraine stood firm against Russian aggression since 2014? And what will the Republic of China (Taiwan) do in the face of the Chinese Communist Party's coercion and perhaps military force tomorrow? In response to inaccurate assessments of operating environments in Iraq, Afghanistan, and the Levant from 2003 to 2021, the Defense Intelligence Agency compiled a 40-page trade craft note in 2023 "which reexamines how the agency measures a country's will to fight."[2] Such a confession indicates that, in the past, the Department of Defense has struggled to properly assess the operational environments in which resiliency and resistance act as vital factors in national outcomes.[3]

In military doctrine, the operational environment consists of "a composite of the conditions, circumstances, and influences that affect the employment of capabilities and bear on the decisions of the commander."[4] Framing the operational environment remains an essential requirement, but the same doctrine principally describes the assessment of physical characteristics of air, land, maritime, space, and cyberspace—aspects essential to planning a conventional campaign but inadequate for describing human-centric struggles. Out of need, and over time, military planners created a series of mnemonics to frame and analyze the environment and inform campaign design. Operationally focused examples include DIGEST, SWEATMSG, and MIDLIFE.[5] More institutionally assimilated examples include ASCOPE, METT-TC, and PMESII-PT.[6] These mnemonics continue a trend of identifying the quantifiable and material aspects of operating environments. Each focuses inadequately on the human terrain.[7] In other words, the military's design process lacks sufficient anthropocentric considerations essential for measuring governmental resilience or societal resistance—critical criteria for operational and strategic campaign development and decision-making.

We suggest a more advanced human-centric model specifically for assessing resilience and resistance potential within nation-states. Consequently, this section recommends a guide for measuring resilience and resistance in terms of the will, potential, and sustainability for resilience to subversion, coercion and aggression, and the potential for resistance to indigenous governance or foreign occupation. Finally, it examines, from a potential external actor's perspective, the likely success of support to a partner's resilience or support to resistance to compel change in adversary behaviors. The **five categories** recommended for framing the operational environment in support of irregular conflict include:

(a) measuring **Resiliency** of a prospective partner's government and society,
(b) assessing the potential of a **State-sponsored** resistance strategy to address external threats,
(c) assessing the potential for external support to **Reinforcing resiliency** in a partner,
(d) measuring current **Resistance potential** to malign governance or occupation, and
(e) assessing the potential for external **Support to resistance**.
The acronym is **RSRRS**.

RESILIENCY MEASUREMENT

A state or population's resiliency can be measured by its propensity to resist subversion, coercion, and aggression from internal or external threats. Some key factors to consider when measuring resiliency include: (a) the physical aspects of the nation or region in question, (b) the fragility of current governance, (c) cultural resilience to change, (d) the will of the population to fight, and (e) the perceived effectiveness of current governance.

A student, scholar, or practitioner might reference many sources when creating this assessment. Humans live in a physical environment impacted by geography, weather, and infrastructure (both ASCOPE and PMESII-PT address this, but it remains important for RSRRS as well). Keep in mind that these physical aspects affect the population in many ways and can change over time. Characteristics to consider are the size and makeup of the population in terms of ethnicity, linguistics, religion, type of governance, and socioeconomic classes. One place to get a general idea of this type of data is from the Central Intelligence Agency's *World Factbook*. This database covers relevant metrics regarding geography, population makeup, natural environment, type of government, economy, energy, communications, transportation, and military capabilities.[8]

Another key criterion involves analyzing the fragility of current governmental systems. The *Fund for Peace* has established just such a database, which considers multiple metrics binned into four categories: (a) cohesion, (b) economic, (c) political, and (d) social.[9] Finally, analyzing resiliency requires a comprehensive study of governance. Some of the factors include the regime's accountability to the people, the stability of the regime, how effective the regime's activities and efforts are, the types of regulations in place and how they are enforced, the rule of law, and government corruption. Fortunately, the World Bank maintains such a database for nations and investors to consider, and it works equally well for resiliency metrics.[10]

Two additional sources are worthy of consideration. One reference with insights into cultural resilience includes *Swiss Re Institute's* resilience index.[11] This index ranks 31 nations by their ability to absorb a range of challenges, including natural disasters. It measures nations against nine benchmarks. However, while the index provides a good example of investigation, it contains only a fraction of states worthy of analysis. Another source, from Johns Hopkins University, provides a methodology to estimate national morale during violent conflict.[12] This scholarship provides methods for evaluating both a state's military's confidence and its ability to maintain combat operations. Another study from RAND defines national will to fight as "the determination of a national government to conduct sustained military and other operations for some objective even when the expectation of success decreases or the need for significant political, economic, and military sacrifices increases."[13] Both military morale and national will to fight prove useful metrics for analysis. Table 1.1 demonstrates this previously described methodologies for measuring resilience.

TABLE 1.1. RESILIENCY MEASUREMENT

ASSESSMENT	DATA FIELDS	SOURCES
Resiliency Measurement Describe the country and population in terms of physical aspects, state fragility, cultural resilience, and governance. Remember that the human terrain is the focus of study but that physical aspects may shape will and perceptions.	Physical	https://www.cia.gov/the-world-factbook/
		Population size, ethnic groups, languages, religions, government type, GDP per capita. All this context is important to consider but not necessarily a measurable statistic.
	State Fragility	https://fragilestatesindex.org/
		Ranking of the country's fragility in comparison to others.
	Cultural Resilience	https://www.swissre.com/dam/jcr:3f36e9da-fe0f-401d-8648-9a12770ffc0f/2022-june-sigma-resilience-index-en.pdf
		This index ranks 31 countries in order of their ability to absorb a range of "shocks", including natural disasters and financial collapse. It measures nations against nine benchmarks.
	National Morale	https://muse.jhu.edu/pub/419/article/851424/pdf
		This article offers models to analyze a nation's will to fight. Table one offers methods to access a military's will to fight, while Table 2 examines national will factors.
	Governance	https://info.worldbank.org/governance/wgi/
		Ranking of the country's governance in comparison to others, using accountability, stability, effectiveness, regulations, rule of law, and corruption.

By way of caveat, another element to categorize a nation, or more specifically a population's resilience, might include how they rate in terms of human security. The United Nations agreed on seven categories of human security; these include (a) economic, (b) food, (c) health, (d) environmental, (e) personal, (f) community, and (g) political. In 2012, United Nations Resolution 66/290 recognized human security as a fundamental human right. Populations lacking elements of human security engender grievances and dissatisfaction with the status quo, potentially motivating popular action to changes in governance. A few organizations are in the process of indexing human security, but none have completed a consistent database yet. David Hastings, an influential scholar on the topic, examined aspects of human security by evaluating more than 200 countries using the same metrics. He argues in his 2013 study, "The Human Security Index: Pursing Enriched characterization of Development," that human security "is arguably the reason for the state," and as such it may be a serious metric to consider regarding the measurement of resilience and resistance all in and of itself.[15] Importantly, Hastings demonstrated that numerical values could be utilized to index this topic, a process which proves informative for the efforts of this section. Indeed, we have leveraged many of his metrics and integrated them into our proposed methodology.[16]

STATE-SPONSORED RESISTANCE POTENTIAL

Only a few nations have instituted a state-sponsored (or national) resistance strategy that plans for leveraging asymmetric methods of resistance in occupied or partially occupied states. The main reason for a state-sponsored resistance remains to deter aggression and if required defend one's nation asymmetrically when a conventional defense fails. Some nations employing such strategies, motivated by historical experience and geographic realities, include Switzerland, Finland, Sweden, Poland, Estonia, Latvia, Lithuania, and Ukraine. One reason for the scarcity of this strategy is that organizing a resistance organization poses inherent risks to the current regime from potential revolutionary activities; another is that such a strategy might be perceived as admitting the government incapable to defend its population from attack, bringing its legitimacy into question. Because of this, one of the first things to consider when formalizing a nationalized resistance approach is the state's relationship with its population—specifically, the state's security relationship and legitimacy with the population. The next consideration includes an evaluation of a state's international relations with its neighbors. Smaller nations bordering expansionist ones (like China or Russia) prove good candidates for this strategy.

Both international diplomacy and popular support remain vital to a state-sponsored resistance. The relationship between any respective nation and its neighbors proves essential. Not only should an existential threat to its sovereignty exist, but it likely needs the assistance of a neighboring state for establishing a sanctuary (and hosting a government in exile, if required). There exists no repository to find such information. Cursory knowledge might come from *Wikipedia* or the *CIA Handbook*. A database for popular support of national institutions includes one managed by the *Organisation for Economic Co-operation and Development* (OECD).[17] While this site has a limited sample size of 41 developed nations outside those listed might not be likely candidates for a state-sponsored resistance anyway due to factors of instability and reduced legitimacy. OECD employs measurable statistics in developing its trust in governance estimate, which includes government revenue, debt, reserves, and spending.

Using data collected from multiple sites, the researcher should also analyze the subgroups within a particular state, as homogeneous societies rarely exist. One should determine if any of the subgroups share cultural relationships with a likely aggressor state. If so, that could indicate an inherent threat to a nationalized resistance strategy. Conversely, analysis should consider if any subgroups have a sense of national character or social cohesion that would prove additive to state-sponsored resistance potential. Are there any groups with historic rivalry to an aggressor (indicating a potential foe) and/or historic ties to neighboring populations (indicating a potential ally)? Table 1.2 shows one methodology for assessing state-sponsored resistance potential. Some of these metrics are quite subjective.

Homogeneous societies rarely exist—subgroup dynamics can define the success or failure of a resistance.

TABLE 1.2. STATE-SPONSORED RESISTANCE POTENTIAL		
ASSESSMENT	**DATA FIELDS**	**SOURCES**
State-Sponsored Resistance Potential Assess the existence or potential of establishing the infrastructure for a state-sponsored resistance organization prior to conflict. Organizing a resistance organization has many inherent risks to the current regime; this strategy requires a great deal of trust to invest in and most nations choose not to.	International Relations	Multiple sites
		Analyze each of the neighboring countries and what the state of relations is. Look for potential geographic regions for possible sanctuary inside the national boundaries or with a neighboring country.
	Sub-Group Cultures	Multiple sites
		Analyze the subgroups within the state to determine if any share cultural relationships with a likely aggressor state. If so, that provides an inherent threat to a nationalized resistance strategy. Conversely, look for subsets with a sense of national character or social cohesion which would prove additive to state sponsored resistance potential. Are there any groups with historic rivalry to an aggressor?
	Trust in Governance	https://data.oecd.org/gga/trust-in-government.htm
		This site measures trust in governance. Use the overall percentage indicated in the OECD database. If not in database, leave empty.

REINFORCING RESILIENCY POTENTIAL

External sponsors can support the resilience of another nation-state through many types of means. There are several reasons this occurs. Primarily, external sponsors to resilience prefer the status quo or stability of an existing regime. Another reason stems from strategic competition between rivals. When a competitor attempts to undermine and subvert the legitimacy of country, these actions almost compel its rival to support resilience in the same state as a counterbalance.

Examining the strength of a partner to combat both internal and external threats can prove essential when planning asymmetric or irregular security strategies. There are a few categories to consider when evaluating potential levels of support to resilience, including foreign relations, foreign aid, and security assistance. One U.S. Government site that spells out current diplomatic relations with foreign governments is managed by the U.S. Department of State (in the case of another state or partner, use a similar foreign relations site or data, as applicable). [18] The State Department maintains up to five years of data about official relations between the U.S. and other nations. Another useful U.S. Government site is foreignassistance.gov.[19] This site maintains comprehensive information over decades about both nation-states and other international organizations that provide foreign aid to a particular country. Examples include funding by USAID, World Food Program, United Nations, and many others. Finally, the Center for International Policy maintains another useful database that monitors arms sales, training, and other lethal means provided by the U.S. to other countries—with data going back almost two decades to 2007.[20] Table 1.3 illustrates several valuable sources to assess current support to resilience.

| When a competitor subverts a state's legitimacy, its rival is almost compelled to respond in kind—with support for resilience. |

TABLE 1.3. REINFORCING RESILIENCY POTENTIAL		
ASSESSMENT	**DATA FIELDS**	**SOURCES**
Reinforcing Resiliency Potential Identify key elements of resilience for a partner to support/enable/leave alone. Given the current partner relationships with this country, which activities strengthen this nation's resilience to internal or external threats? What types of support does the U.S. and other partners offer to resiliency? Are there any perceived gaps in support to resilience?	U.S. Relations	https://www.state.gov/u-s-bilateral-relations-fact-sheets/
		This site spells out the current USG diplomatic relations with foreign governments and used to answer the question of if support to resilience is politically feasible.
	Nonviolent Aid	https://www.foreignassistance.gov/cd
		This site shows the current aid to each country by the U.S. and other partners. Subjectively analyze the types of aid provided to the respective country. Will more aid increase resiliency or does the amount and type of aid indicate fundamental weaknesses in national systems which remain unabated?
	Security Assistance	https://securityassistance.org/
		Database for security force assistance, arms sales, and foreign military training for countries worldwide. Subjectively analyze the security aid provided to the respective country.

RESISTANCE MEASUREMENT

From an intrastate perspective, every nation has some level of resistance to current governance, often not comprising a direct or existential threat, but, instead, couched in terms of political opposition. The absence of formal, legal, process-based venues for venting grievances can increase nonviolent and violent opposition to current authority. Some nation-states, though, possess more dangerous, violent, and/or criminal elements, which may threaten the current government or authority. Resistance movements frequently include many functions and activities below the threshold of discernibility. As such, measuring the capabilities of an often opaque, and sometimes illicit organization resembles more of an art than a science—akin to adducing the unseen mass of an iceberg below the surface of the water. That stated, active resistance to current governance challenges the status quo and, consequently, draws a lot of attention by organizations seeking to preserve national sovereignty and the current international order.

In terms of measuring resistance to current state governance, several databases provide valuable information related to a regime's capacity to protect its society from "subversion, lawlessness, insurgency, terrorism, and other threats to its security."[21] The *Global Economy* has developed a source to analyze: (a) rule of law, (b) political stability, (c) corruption, (d) political rights, (e) civil liberties, and (f) the level of shadow governance.[22] Additionally, the Armed Conflict Location & Event Data Project offers detailed reports that catalogue political violence and protests around the world. While not always comprehensive, most countries with significant popular unrest or even small levels of violent protest generate reports by watch organizations to consider. One such source includes a database designed by the Carnegie Endowment for International Peace.[23] This site maintains links to news and information about global peace and security, including great power competition. The "protest tracker" provides details about specific acts of civil disobedience, where they occurred, what triggered them, and the motivation behind the unrest.

When rebellion develops, the University of Maryland maintains a global terrorism database.[24] This site maintains information over decades and is searchable by country. In addition, the Armed Conflict Location & Event Data Project collects information on the dates, actors, and types of global political violence and protests, but subsequently provides the analysis to describe and explore such conflicts.[25] The Institute for Economics and Peace maintains one more useful database in terms of indexing internal peace versus domestic violence.[26] Its "Global Peace Index" ranks each nation alongside one another, and it also offers reports on specific threats. Freedom House offers another dataset for most nations that assesses political rights and civil liberties worldwide.[27] Finally, the CIA Facebook offers another beneficial indicator of potential unrest by analyzing the amount of food security in a nation.[28] Poor access to food has been linked to political instability.[29] All this data, and more, can provide a solid estimate of current and potential resistance to governance or occupation. Table 1.4 summarizes these data collection methods.

TABLE 1.4. RESISTANCE MEASUREMENT		
ASSESSMENT	**DATA FIELDS**	**SOURCES**
Resistance Measurement Assess resistance levels or resistant potential to governance or occupation. Such an exercise comprises more of an art than a science, but there are many datasets which researchers can potentially utilized to provide an analysis.	Stability	www.theglobaleconomy.com
		Use the country rankings in the "governance and business environment" section; utilize an inverse approach to measure potential resistance levels; use data to analyze demonstrated capacity to evade rule of law, inherent political instability, uncontrolled corruption, perceived unaddressed political rights, unprotected civil liberties and current shadow governance.
	Armed Conflict	https://acleddata.com/
		Use the "analysis search" to find reports on the country or region, reports which provide contexts to battles, riots, explosions and violence against civilians. Subjective data to consider and not a percentage.
	Protest	https://carnegieendowment.org/publications/interactive/protest-tracker
		Shows major protests over recent years and the triggers. Subjective data to consider and not a percentage.
	Terrorism	https://www.start.umd.edu/gtd/
		Graphs the incidents of terrorism over decades. Subjective data to consider and not a percentage.
	Unrest	https://www.visionofhumanity.org/maps/#/
		Rank of state in terms of internal peace versus inherent violence. Again, use the inverse to measure instability.
	Freedom	https://freedomhouse.org/report/freedom-world
		Rank of the population's freedom compared to global norm. Use the map to click on the country of interest.
	Food Insecurity	https://www.cia.gov/the-world-factbook/field/food-insecurity/
		Assess the population's access to food.

SUPPORT TO RESISTANCE POTENTIAL

The student, scholar, or practitioner can also measure the potential for supporting resistance within a rival's nation–state as an act of foreign policy and as a part of military strategy. Obviously, subversive activities to another nation's sovereignty potentially destabilizes the international order, a mandate which the U.S. and many others endeavor to uphold. That said, the U.S. has exercised unconventional warfare to coerce, disrupt, or overthrow adversaries, most often recently (in Iraq and Afghanistan) as a means toward greater, long-term, stability. As such, unconventional warfare remains a possible option for U.S. policymakers.

One of the first aspects worthy of consideration centers on whether supporting a resistance movement poses a politically viable option. One might start with examining current U.S. relations with a particular foreign government of interest. As stated previously in the support to resilience potential analysis, the State Department maintains up to five years of data on the official relations between the U.S. and other nations.[30]

Resistance spans a spectrum from nonviolent protest to belligerency. In terms of violent opposition, the U.S. Army War College has developed perhaps the most useful dataset to forecast the possible success of insurrection. Its "Study of Internal Conflict" has analyzed dozens of insurgency case studies and developed four primary criteria which bear critical impact on the success of a violent resistance organization:

(a) if 15 percent or more of the population does not primarily identify itself as a citizen of the nation–state, then an insurgency maintains success 96 percent of the time;

(b) if 15 percent or more of the population does not acknowledge the legitimacy of the regime or occupying power, then insurgency maintains success 94 percent of the time;

(c) if 15 percent or more of the population has meaningful contact with resistance movements, then an insurgency maintains success 96 percent of the time; and

(d) if a sanctuary exists within the nation or a bordering nation from which armed components can seek refuge, then insurgency maintains success consistently.[31]

In this Army War College study, each of these variables are considered self-standing (i.e., they do not stack).

As the existence of a sanctuary poses such an important element to successful insurgency or belligerency, identifying the potential of a sanctuary remains essential. No one website or database will provide such information. By using multiple sources, one should attempt to gather information about neighboring countries or regions that may be sympathetic to a resistance, either by virtue of shared ethnicity, similar culture, or potential political gain. If a sanctuary does not appear feasible, the long-term success of violent resistance remains doubtful but not necessarily unachievable. Table 1.5 lays out methodologies for assessing support to resistance potential.

TABLE 1.5. SUPPORT TO RESISTANCE POTENTIAL		
ASSESSMENT	**DATA FIELDS**	**SOURCES**
Support to Resistance Potential The potential external support to a resistance movement within another state or occupied area maybe categorized as subversion. Are there factors in USG relations which allow this strategy to remain a feasible option? Could a resistance movement survive the regime's countermeasures? Which resistance capabilities should a sponsor support/enable/ leave alone?	U.S. Relations	https://www.state.gov/u-s-bilateral-relations-fact-sheets/
		This site spells out the current USG diplomatic relations with foreign governments. In the case of foreign partner, utilize foreign relations sites or data as applicable.
	Probability of Violent Resistance Success	https://ssi.armywarcollege.edu/soic-study-methodology/
		Ask the following questions: (1) does 15percent or more of the population not identify as a citizen of the current state? (2) does 15percent or more of the population not acknowledge the legitimacy of the regime? (3) does 15percent or more of the population have meaningful communication with a resistance movement? (4) is there a ready sanctuary for an armed component in a neighboring state?
	International Relations	Multiple sites
		Use multiple points of reference to analyze each of the neighboring countries to the case study and what their relations are. Look for potential geographic regions for possible sanctuary inside the national boundaries or with a neighboring country.

CONCLUSION

This chapter offers a guide for analyzing resiliency and resistance potential within sovereign nation–states. For an illustration of this method in a case study, see the example compiled on Iran in Appendix A. This methodology has clear relevance to the three biggest U.S. foreign policy questions of today. Why did Afghanistan fall so rapidly to the Taliban in 2021? Why, in contrast, has Ukraine stood firm against Russian aggression since 2014? And what will the Republic of China (Taiwan) do in the face of the Chinese Communist Party's coercion and perhaps military force tomorrow? The proposed assessment framework utilizes many databases. Still, the conclusions from that data remain subjective to the analyst or practitioner who uses such metrics for analysis—which must also consider proper context. Nevertheless, because the State Department, USAID, the U.S. Treasury Department, and non-governmental organizations generate, utilize, and manage much of the data used in RSRRS, this type of analysis inherently comprises a more holistic and whole-of-government assessment of the operational environment. Further, the possibility exists to quantify many aspects of the following approach, and that effort remains under further research, analysis, and revision. We readily assume that the RSRRS methodology will evolve as practitioners and scholars test it. As a starting point, however, in cases of irregular warfare, competition, and deterrence, analyzing resilience and resistance potential within sovereign states remains a crucial step in framing the operational environment as the initial stage of assessment in supporting policy decisions and subsequent military strategies.

| In cases of irregular warfare, competition, and deterrence, analyzing resilience and resistance potential within sovereign states remains a crucial step in framing the operational environment as the initial stage of assessment in supporting policy decisions and subsequent military strategies. |

About the Authors

Dr. Robert S. Burrell is a resilience and resistance interdisciplinary scholar using data-driven and human-centric methodologies to analyze intrastate conflict ranging from nonviolent protest through belligerency. He is a Senior Research Fellow at the Global and National Security Institute of the University of South Florida. From 2020–2024, he taught irregular warfare at Joint Special Operations University and was the editor in chief of special operations doctrine from 2011–2014.

John Collison is a contractor with Huntington Ingalls Industries and works in J59-Concept Development and Integration (CD&I) at U.S. Special Operations Command.

NOTES

1. Robert S. Burrell and John Collison, "A Guide for Measuring Resiliency and Resistance," *Small Wars & Insurgencies* 35, no. 1 (2024): 147-172, https://doi.org/10.1080/09592318.2023.2292942

2. How the Pentagon Assesses Ukraine's Progress: A Rare Interview with America's Defense Intelligence Agency," *The Economist* 9 no. 15 (September 2023): 25.

3. How the Pentagon Assesses Ukraine's Progress, 25-26.

4. Army Technical Publication 5-0.1: *Army Design Methodology*, (Headquarters: Department of the Army, July 2015), 3-1.

5. DIGEST (drugs, infrastructure, governance, economics, structures, and tribes) was developed by Provincial Reconstruction Teams in Afghanistan in 2004 and 2005 to analyze the environment; SWEATMSG (sewer, water, electricity, agriculture, trash, medical, schools, and governance) was used by the 1st Cavalry Division in Iraq in the 2005 timeframe; and MIDLIFE (military, information, diplomatic, law enforcement, intelligence, financial, and economic) comes from an article by Eric P. Wendt, "Strategic Counterinsurgency Modeling," *Special Warfare* 18, no. 2 (September 2005): 2–13.

6. ASCOPE (areas, structures, capabilities, organizations, people, and events) is primarily used by the U.S. Army. METT-TC (mission, enemy, terrain, terrain and weather, troops, time available, and civil considerations) is another used by the Army and Marine Corps. PMESII-PT (political, military, economic, social, information, infrastructure, physical environment, and time) represents another popular Army mnemonic.

7. The term human terrain "is analogous to geographic terrain, but in the terms of people in sociogeographic space... Human dynamics over space can be conceptualized as a multilayered surface of hills and valleys corresponding with sociocultural attributes. Human terrain is often focused on cultural/ethnic traits but should also include characteristics of all systemic components affecting human behavior (e.g., demographics, economics, politics, human-environment interactions). Humans operate in open social and physical systems, and thus are affected by, and affect, surrounding elements of society and their physical environment." Quote from Median, "From Anthropology to Human Geography: Human Terrain and the Evolution of Operational Sociocultural Understanding," 140.

8. See the Central Intelligence Agency Factbook at https://www.cia.gov/the-world-factbook.

9. See the "Fragile States Index" by The Fund for Peace at https://fragilestatesindex.org.

10. See the governance statistics provided by the World Bank at https://info.worldbank.org/governance/wgi.

11. For one version, see "Resilience Index 2022: Risks to Resilience on the Rise Again After a Year of Respite," *ResSwiss Re Institute*, 2022. Found at https://www.swissre.com/dam/jcr:3f36e9da-fe0f-401d-8648-9a12770ffc0f/2022-june-sigma-resilience-index-en.pdf.

12. Connable, "Structuring Cultural Analyses: Applying the Holistic Will-to-Fight Models," 153-167.

13. See McNerney et. al., "National Will to Fight: Why Some States Keep Fighting and Others Don't."

14. Table 1 includes: (a) the Central Intelligence Agency factbook found at https://www.cia.gov/the-world-factbook; (b) the fragile state index found at https://fragilestatesindex.org; (c) the resilience index by Swiss Re Institute, (d) military moral calculations by Connable; (e) national will to fight calculations by McNerney et. al.; and (f) the World Bank statistics on evaluating governance.

15. David A Hastings, "The Human Security Index: Pursuing Enriched Characterization of Development," *Development* 56 (2013): 67, https://doi.org/10.1057/dev.2013.7

16. As an additional resource, in 2022 the United Nations released as special report entitled *New Threats to Human Security in the Anthropocene: Demanding Greater Solidarity*. Instead of a country-to-country comparison, this U.N. study shows global trends on human security and their impact. A link to the United Nations report is found at https://hs.hdr.undp.org/pdf/srhs2022.pdf. Another aspect of resiliency that one might consider is legitimacy of current governance. Scholars for decades have attempted to measure a regime's legitimacy. Christian von Haldenwang published one such methodology titled "Measuring Legitimacy–New Trends, Old Shortcomings?" Haldenwang advises five primary methods for surmising legitimacy, including: (a) polling/surveys of the population, (b) data analysis on both violence and political participation, (c) investigation of local media discourse, (d) the opinions of subject matter experts, and (e) reading qualitative secondary works based on data.

17. See the Organisation for Economic Co-operation and Development trust in governance data at https://data.oecd.org/gga/trust-in-government.htm.

18. State Department official relations information can be found at https://www.state.gov/u-s-bilateral-relations-fact-sheets.

19. This site on foreign assistance is established by both the Department of State as well as USAID and found at https://www.foreignassistance.gov/cd.

20. The Security Assistance Monitor is maintained by the Center for International Policy and found at https://securityassistance.org.

21. Quote from the Department of Defense definition of foreign internal defense, found in Joint Publication 3-22, *Foreign Internal Defense.*

22. The website TheGlobalEconomy.com was initially established as an extension of The Global Economy class at Georgia State University and set up by professor Neven Valev. It currently has three professors who maintain it at https://www.theglobaleconomy.com/

23. The Carnegie Endowment for International Peace manages a protest tracker at https://carnegieendowment.org/publications/interactive/protest-tracker.

24. The University of Maryland manages terrorist statistics at https://www.start.umd.edu/gtd.

25. The Armed Conflict Location & Event Data Project employees over 200 team members to maintain data on political violence, found at https://acleddata.com.

26. The Institute for Economics and Peace produces the Global Peace Index at https://www.visionofhumanity.org/maps/#.

27. Freedom House manages this dataset at https://freedomhouse.org/report/freedom-world.

28. See the Central Intelligence Agency Factbook.

29. Chase Sova and Eilish Zembilci, "Dangerously Hungry: The Link between Food Insecurity and Conflict," *Center for Strategic and International Studies*, 21 Apr 2023. Found at https://www.csis.org/analysis/dangerously-hungry-link-between-food-insecurity-and-conflict.

30. https://www.state.gov/u-s-bilateral-relations-fact-sheets/

31. For more information, see the U.S. Army War College website at https://ssi.armywarcollege.edu/soic-study-methodology/. Also see Mason, "COIN Doctrine is Wrong," 19-34.

III. Redefining the Bang: A Full Spectrum of Conflict Design

By Robert S. Burrell

ABSTRACT: This chapter offers a conceptual framework in which to illustrate the relationship between conventional warfare, irregular warfare, competition, and deterrence. Through this lens, one can also see the connections between gray-zone activities and hybrid warfare. In so doing, it breaks the linear fashion constructed by U.S. Joint doctrine to offer a highly needed two-dimensional framework. Not only can one visualize how conventional warfare, irregular warfare, competition, and deterrence interact with one another, but hybrid warfare and gray zone conflict materialize at the epicenter. Portions of this essay were previously published by *Modern Warfare Institute* and by the *Irregular Warfare Initiative.*[1]

INTRODUCTION

China's *gray zone* conflict...Russian *hybrid warfare*. These academic terms have emerged to describe belligerent activities that standard U.S. military operations struggle to address. Although these two adversarial tactics remain central to today's security environment, they are absent from the current Joint doctrinal framework. Even the new Joint Doctrine Note on competing fails to address hybrid warfare at all and relegates the mention of gray zone to a single instance.[2] In short, the Department of Defense has demonstrated an inability to address the asymmetric approaches taken by its adversaries, when, in fact, these methods of conflict should remain front and center in U.S. national military strategy.

> ***| In short, the Department of Defense has demonstrated an inability to address the asymmetric approaches taken by its adversaries, when, in fact, these methods of conflict should remain front and center in U.S. national military strategy. |***

In other words, China and Russia's strategies do not fit into conventional paradigms of conflict. Instead, since the inception of Joint doctrine, the U.S. has generally envisioned conflict in a linear fashion where peace and full-scale war occupy opposite sides of a continuum, with varying degrees of each in between. Doctrine's evolution has made little change in its standard conflict continuum over time, even though 21st Century conflict warrants a more complex conceptual methodology.

The competitive strategies adopted by two rivals have the Department of Defense especially concerned. The first is what the Chinese Communist Party introduced as *unrestricted warfare.*[3]

In basic terms, this approach involves a broad array of nonmilitary types of antagonism, including the use of illicit networks, industrial espionage, cyberattacks, propaganda, bribery of public officials, and intimidation—all of which intentionally avoid direct military confrontation with the United States. An example of this approach includes the deployment of a huge Chinese maritime militia composed of hundreds of commercial fishing vessels to drop anchor on disputed islands in the South China Sea. These provocative actions are meant to intimidate the region in order to enforce China's territorial claims without deploying the People's Liberation Army Navy.[4]

Russia's *new generation warfare* is the second approach to competition that has alarmed the United States. This strategy combines nuclear, conventional, and nonmilitary instruments to produce strategic deterrence, while at the same time offering tools to accomplish strategic goals without international interference. During Russia's forcible annexation of Crimea from Ukraine in 2014, it used subversion, proxy forces, direct military intervention, and coercion to achieve its objectives. With a blend of intimidation, soft power, and irregular activities, Russia accomplished its strategic objectives through actions short of traditional war, leaving the U.S. and its European allies incapable of a swift and strong response.[5]

These novel approaches to warfare by the Chinese Communist Party and Russia (unrestricted and new generation) are only two categories in the ever-growing list of terms to describe modern interstate conflict, which includes gray-zone conflict, hybrid warfare, legal warfare, drug (or crime) warfare, cyberwarfare, economic warfare, trade warfare, religious warfare, proxy warfare, and many others. None of these terms fit nicely into the current Joint taxonomy of concepts regarding war. The primary constraint remains that Joint doctrine generally addresses military capabilities in terms of violent action (or support for violent action), while many of the evolving approaches to competition use nonviolent means. In the U.S., the Department of Defense has played an increasingly influential role in security cooperation, but it is still playing catch up when it comes to the inclusion of nonviolent competition into doctrine.

In his 2020 article, Eric Robinson brilliantly outlines "the missing, irregular half of great power competition," suggesting that conventional warfare and irregular warfare definitions may not need significant revision. Instead, Robinson argues that the Department of Defense should incorporate the missing half of related nonviolent conflict by adding both irregular competition and conventional deterrence to the spectrum of nonviolent conflict (versus changing the definition of irregular warfare to include nonviolent activities).[6] Examples of conventional deterrence include establishing alliances, forming coalitions, fostering partnerships, creating trade agreements, and supporting international law. These nonviolent tools, managed primarily by other government agencies, deter overt aggression against the U.S. from opponents.

Equally important to this new sphere of conventional deterrence, irregular competition addresses interstate conflict short of war (commonly referred to as *left of bang* activities). Irregular competition includes the use of statecraft, military engagements, and economic incentives to either partner with or intimidate middle powers and developing countries. From a Department of Defense perspective, it describes military activities deliberately used to gain influence, from freedom-of-navigation operations in the South China Sea to multilateral training activities in the Baltics. Expanding warfare to include conventional deterrence and irregular competition addresses the totality of competition terminology.

The 2022 U.S. *National Security Strategy* and *National Defense Strategy* made progress in detailing a more comprehensive approach to conflict. But much more can be done to incorporate both deterrence and competition, as well as irregular warfare, into operational design.[7] As Robinson detailed in an *Irregular Warfare Initiative* essay in 2020, there are four foundational *ways* of responding to belligerents: conventional warfare, deterrence, irregular warfare, and competition. If planners adopted this framework, they would have a *full spectrum of conflict design* that would help them to integrate *ends, ways*, and *means* into a coherent campaign plan. And they would also be better able to understand the activities of the U.S.' most important competitors.

UNDERSTANDING CONFLICT—FROM ONE DIMENSION TO TWO

The U.S. Joint force has attempted to define different forms of conflict for decades, but it has failed to move beyond a linear understanding. A competition continuum was introduced in the 2019 Joint Doctrine Note that begins with cooperation on the left, evolves to adversarial competition below armed conflict in the middle, and concludes with war on the right. This one-dimensional concept shown in Figure 1.8, has been adopted by other doctrinal publications, such as Joint Publication 1 (JP-1), the Joint Doctrine Note on *Warfighting* and Joint Publication 3-0 *Joint Campaigns and Operations*. While this is the most recent version, for the past two decades doctrine has only slightly redefined categories between war and peace through minor revisions to descriptions, but these changes have added little significance.[9]

Figure 1.8: Competition Continuum in Joint doctrine. Source: Author

Connecting ways and means to strategic ends remains the essential task of strategy. Thus, adding a second means-based dimension to the existing continuum creates a full spectrum of conflict design, which dramatically improves our understanding of conflict. This second dimension—which serves as the y-axis in Figure 1.9—is similar to Herman Kahn's work on escalation.[10] While escalation of means varies with circumstances, means can generally be employed directly (and overtly), indirectly (including covertly and clandestinely), or somewhere in between.[11] The full spectrum of conflict creates a holistic framework that clarifies the relationship between means (the resources used in conflict) and ways (how those means are employed). Figure 1.9 illustrates the two-dimensional framework where the x-axis delineates the continuum between war and peace while the y-axis delineates the military means utilized. Together, these axes produce four ways to conduct conflict: (a) deterrence (b) conventional warfare, (c) irregular warfare, and (d) competition.

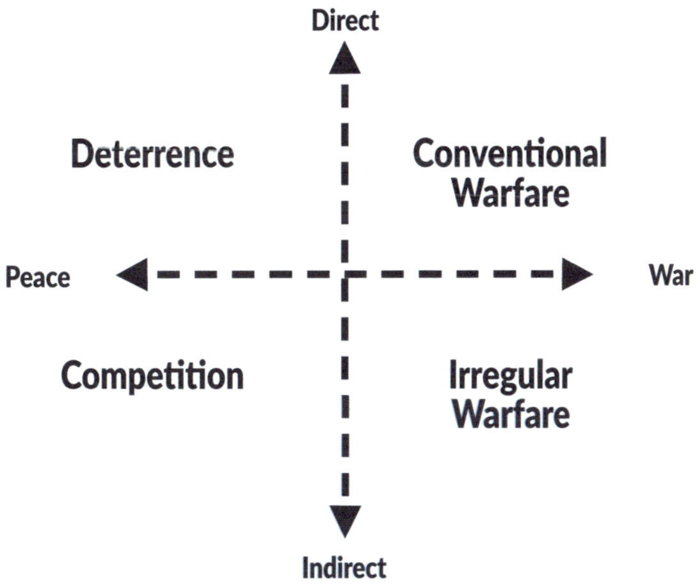

Figure 1.9: Full spectrum of conflict design. Source: Author

a. Deterrence. In this quadrant, a government prevents adversaries' aggressive actions by threatening the use of traditional means. In addition, a peacetime coalition of treaty allies or the buildup of weapons of mass destruction can also deter aggression. The deterrence quadrant is characterized by nonviolent and low-intensity conflict between competitors—reinforced and backstopped by hard power.

b. Conventional warfare. This quadrant is characterized by the authorized use of conventional military force to defeat an adversary's army. Conventional warfare also includes the implementation of a diplomatic treaty for the purposes of combined warfare. Conventional forces primarily dominate in this quadrant.

c. Irregular warfare. This quadrant identifies violent activities waged against an opponent through nontraditional and indirect methods, such as foreign internal defense and unconventional warfare (either support to resilience or resistance). Special operations typically dominate in irregular warfare, and the means utilized are typically more population-centric than those used in the direct approach.

d. Competition. This quadrant is characterized by rivalry between nation-states or non-state actors where the means applied fall short of violence. This includes nonproliferation activities to deny competitors arms or equipment and sanctions imposed on persons, groups, and nations. It also includes persistent engagement by special operations forces to prepare an environment for violent irregular warfare if/when required. Finally, humanitarian assistance and stability activities occur in this space as they alleviate human security demands in unstable regions where rivals operate (like al-Qaeda and ISIS).

THE RELATIONSHIP AMONG QUADRANTS

These four ways directly relate to one another; they are not independent, and a nation can employ all of them simultaneously. For instance, deterrence fundamentally connects to conventional warfare because they both use the same means, from nuclear arsenals to aircraft carriers. In other words, weapons of mass destruction or conventional arms can be used for war or deterrence equally. One could also use conventional warfare and deterrence simultaneously, such as employing military forces while threatening the use of a nuclear response.

Likewise, competition and irregular warfare are directly related. Each can take deliberate actions in opposition to a state (or non-state) competitor with indirect application of military means. The main distinction between the two is that irregular warfare simply uses more violence than competition. For instance, a foreign military force might be utilized to ensure fair elections in a fragile state. This nonviolent activity occurs in the space of competition. However, that same military force can be used in a different way—to counter irregular threats with violence.

The five recognized components of irregular war (foreign internal defense, unconventional warfare, stabilization activities, counterterrorism, and counterinsurgency) span the breadth of both competition and irregular warfare—a subject I published on in 2021.[12] In fact, activities like stabilization primarily exist in nonviolent competition in most situations, rather than irregular war. The full spectrum of conflict design better allows the visualization of indirect use of military means in both competition and irregular warfare. Indirect uses of military power include (a) support to another regime's *resilience* to protect its society from threats like subversion, lawlessness, and insurgency; and (b) supporting indigenous *resistance* against an adversary's governance to coerce, disrupt, or overthrow the regime. Neither of these two methods (support to resilience or resistance) are constrained to violent means and concurrently exist in both the irregular war and competition quadrants. It is precisely for this reason that Joint Special Operations University has transitioned to the terms "support to resilience" and "support to resistance" as opposed to "irregular warfare."

Military planners can more holistically address irregular threats by including both competition and irregular warfare as separate but related *ways* of dealing with violent and nonviolent threats. With this framework, planners and operators will be prepared for nonviolent activities to evolve into irregular warfare, or for violent activities to revert to competition. Operation RESTORE HOPE in Somalia is a good example of military force that was initially used for peaceful humanitarian assistance that quickly evolved to employ violent means to establish rule of law. My 2021 article further illustrates such shifts.[13]

In short, the full spectrum of conflict framework should not be used to compartmentalize any nation's activities into a single quadrant or type of military means, as conflict rarely relies exclusively on one type of means. Instead, a comprehensive security strategy would leverage all four quadrants – deterrence, conventional warfare, irregular warfare, and competition. Figure 1.9 uses dotted lines to indicate that conflict might (and likely will) include activities in more than one space. *As further explained below, when activities span multiple ways, this results in hybrid warfare and/or gray zone competition – which explains their commonality!*

CLARIFYING GRAY ZONE AND HYBRID WARFARE

By employing multiple ways illustrated in the full spectrum of conflict framework, nations may specifically operate near the axes where activities are blurred to intentionally complicate an adversary's response. Fortunately, the *Full Spectrum of Conflict* framework allows scholars and practitioners to visually identify gray zone conflict and hybrid warfare as shown in Figure 1.10.

a. Gray zone conflict occurs along the y-axis because the gray zone blurs the line between war and peace. For example, the Chinese Communist Party tends to operate aggressively in the gray zone, as this gives them freedom of action without consequences. In the South China Sea, they often employ a maritime militia, instead of their navy, to enforce recognition of their territorial claims and to disguise their intentions, purposely construing actions as not clearly warlike or peaceful.

b. Hybrid warfare occurs along the x-axis by purposefully implementing both direct and indirect means. In 2014, Russia invaded Crimea by sponsoring proxies and inserting Russian operatives without uniforms. At the same time, Russia used its air force and navy to enforce air and maritime exclusion zones. Further, Russia used economic pressure and nuclear threats to ensure noninterference from NATO. Russia also established and maintained its narrative through propaganda with domestic and international audiences to justify its aggression.

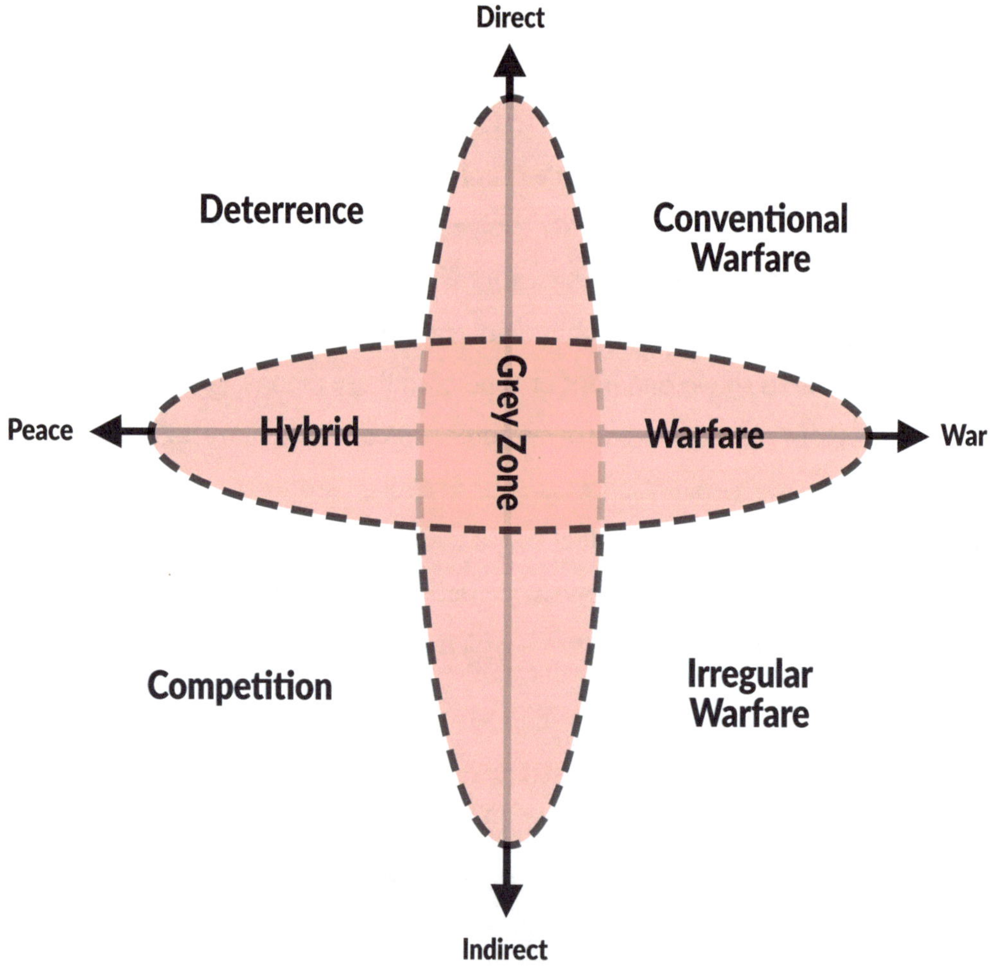

Figure 1.10. Identifying gray zone and hybrid warfare. Source: Author.

APPLYING THE FULL SPECTRUM OF CONFLICT FRAMEWORK

One *application* for the full spectrum of conflict design is to map out major events in a struggle, to better appreciate the *ways* and *means* utilized, and to identify the inherent nature of the conflict or the strategies of the participants, particularly over time. Figure 1.11 illustrates major Russian actions taken against Ukraine from 2004 through 2022.

In general, the graph demonstrates that Russian aggression has been escalatory in terms of military *means* over time. Russia began its campaign with influence, cyber, and economic coercion (2004–2013), but then transitioned to irregular warfare supported by conventional warfare, or hybrid warfare, (2013–2021), and finally transitioned to conventional warfare (2022) while simultaneously making threats of a nuclear response for the purposes of deterrence (2022). Russia's activities occupy all four quadrants, generally employing a hybrid warfare approach (particularly in 2014) and leveraging multiple *ways* and *means* in achieving national objectives. However, Russia achieved dramatically successful outcomes only when leveraging multiple quadrants simultaneously during the annexation of Crimea in 2014—with the use of irregular warfare, conventional warfare, information warfare, and cyber-attacks. This indicates the potential advantage of hybrid warfare (and probably gray zone warfare) beyond the isolated use of conventional warfare and deterrence alone—even though the latter uses the most direct force. Aiming "for the bull's eye," so to speak, between the quadrants of deterrence, conventional warfare, irregular warfare, and competition, may result in the most effective application of means utilizing multiple ways to achieve desired ends.

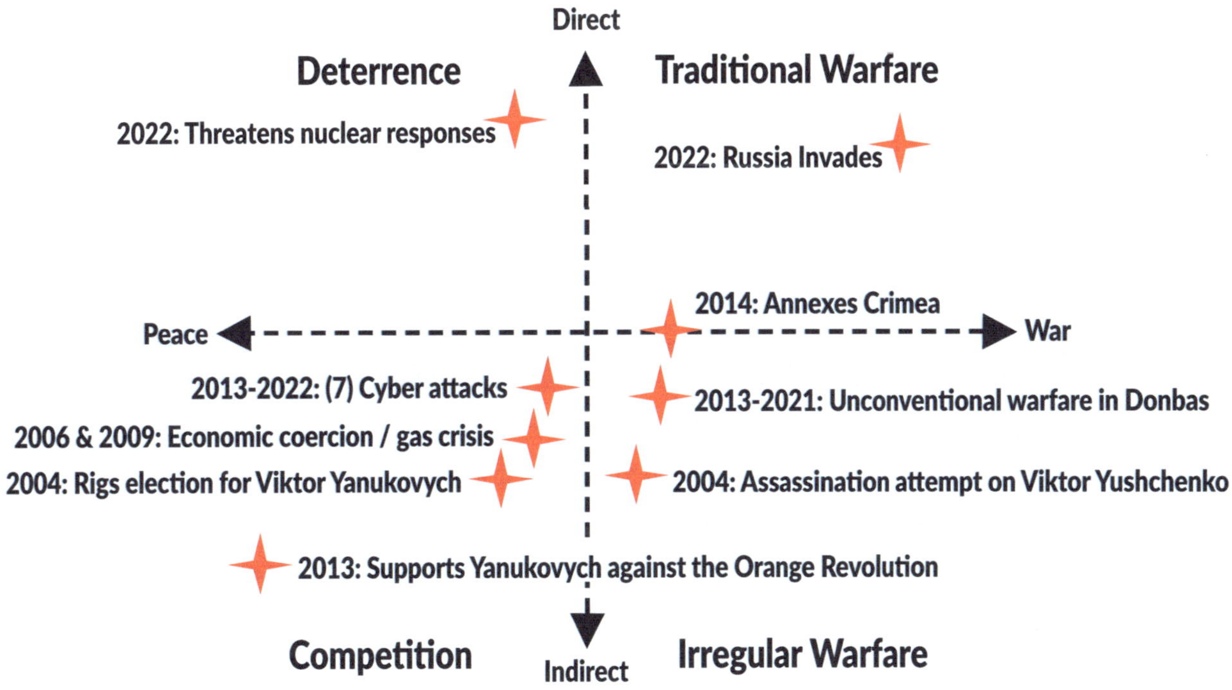

Figure 1.11. Full spectrum of conflict design, Russian Aggression, 2004–2022. Source: Author.

THE WAY AHEAD

Today, both the State Department and the Department of Defense often identify belligerent aggression that does not fit neatly along a linear conflict continuum, resulting in a bewildered and delayed response. Hybrid warfare and gray zone aggression will likely continue to be the foremost methods for employing ways and means in 21st Century conflict. As such, a full spectrum of conflict design framework better allows for collaboration within the U.S. Government in matters of national security. As Philip Kapusta argued in 2015, understanding the gray zone can "enable early application of U.S. instruments of power…by shaping the arc of change closer to its origins"—as long as they are supported by a whole-of-government approach.[14]

Employing means in multiple ways creates the advantages of hybrid warfare and gray zone competition, but current Joint doctrine and its linear approach to conflict limits the U.S. military's ability to detect or employ these methodologies. Joint doctrine can broaden conflict analysis by adopting a framework that integrates the distinct ways of (a) deterrence, (b) conventional warfare, (c) irregular warfare, and (d) competition. The full spectrum of conflict design allows for increased understanding of the linkages and relationships between direct and indirect means across the continuum of peace and war. This framework provides a fuller picture of multiple efforts oriented toward the same desired outcome. The Ukraine/Russian conflict is a prime example of how the use of the full spectrum of conflict design can be applied to better illustrate and articulate aspects of conflict in one graph.

Military planners can adopt the full spectrum of conflict design to synchronize activities throughout engagement and contingency plans, as well as major operations and campaigns. This simple framework allows practitioners to analyze and integrate multiple *ways* and *means* into one planning frame, as well as to synchronize resources and actions to achieve strategic *ends.* Finally, the *full spectrum of conflict design* can help scholars, planners, and theorists graphically appreciate, and potentially anticipate, the strategies of opponents.

About the Author

Dr. Robert S. Burrell is a resilience and resistance interdisciplinary scholar using data-driven and human-centric methodologies to analyze intrastate conflict ranging from nonviolent protest through belligerency. He is a Senior Research Fellow at the Global and National Security Institute of the University of South Florida. From 2020–2024, he taught irregular warfare at Joint Special Operations University and was the editor in chief of special operations doctrine from 2011–2014.

NOTES

1. See Robert S. Burrell "How to Integrate Competition and Irregular Warfare," *Modern Warfare Institute*, 8 Aug 2021, found at https://mwi.usma.edu/how-to-integrate-competition-and-irregular-warfare/, as well as Robert S. Burrell, "A Full spectrum of conflict design: How Doctrine Should Embrace Irregular Warfare," Irregular Warfare Initiative, 14 Mar 2023; found at https://irregularwarfare.org/articles/a-full-spectrum-of-conflict-design-how-doctrine-should-embrace-irregular-warfare/.

2. See JDN 1-22 *Joint Force in Strategic Competition*, (Washington D.C.: Department of Defense, 2 Feb 2023), I-1.

3. Qiao Liang and Wang Xiangsui, *Unrestricted Warfare*, (People's Liberation Army Literature and Arts Publishing House, 1999).

4. Masaaki Yatsuzuka, "How China's Maritime Militia Takes Advantage of the Gray Zone," *The Strategist*, 19 Jan 2023.

5. Phillip A. Karber, "Russia's New Generation Warfare," *National Geospatial-Intelligence Agency*, 4 Jun 2015. Found at https://www.nga.mil/news/Russias_New_Generation_Warfare.html; accessed on 5 Jul 2023.

6. Eric Robinson, "The Missing Half of Great Power Competition," *Modern Warfare Institute*, 8 Sep 2020. Found at https://mwi.usma.edu/the-missing-irregular-half-of-great-power-competition/

7. Joe Biden, *National Security Strategy*, (Washington D.C.: White House, 12 Oct 2022), and Lloyd Austin, *National Defense Strategy*, (Washington D.C.: Department of Defense, 27 Oct 2022).

8. See JDN 1-19 *Competition Continuum*, (Washington D.C.: Department of Defense, 3 Jun 2019), I-1.

9. Illustration by author. For a similar depiction, see JP 3-0 *Joint Campaigns and Operations*, (Washington D.C.: Department of Defense, 18 Jun 2022), I-5.

10. For a synopsis on Herman Khan's work and its application to 21st Century strategic design, see Michael Fitzsimmons, "The False Allure of Escalation Dominance," *War on the Rocks*, 16 November 2017, found at https://warontherocks.com/2017/11/false-allure-escalation-dominance

11. To see how Kahn's work can be modified to prevalent circumstances, see Jonah Lo, Ng Kang Jie and Hannah Lo, Reconstructing the Ladder: Toward a More Considered Model of Escalation," *Strategy Bridge*, 1 Sep 2022. Found at https://thestrategybridge.org/the-bridge/2022/9/1/reconstructing-the-ladder-toward-a-more-considered-model-of-escalation

12. Burrell, "How to Integrate Competition and Irregular Warfare."

13. Burrell, "How to Integrate Competition and Irregular Warfare."

14. Philip Kapusta, "The Gray Zone," *White Paper: U.S. Special Operations Command*, 9 Sep 2015. Found at https://info.publicintelligence.net/USSOCOM-GrayZones.pdf

CHAPTER 2: POLITICAL SCIENCE AND INTERNATIONAL RELATIONS

I. Research Methods for Area Studies

By A. Jackson

ABSTRACT: There is a growing recognition across the Department of Defense of a need to build collective expertise in the social sciences to meet the security challenges of the 21st century. This requirement includes renewed national security and defense interest in foreign societies and cultures; the role of social networks in irregular war; the nature of tribal systems; the nature of power and authority in weak states; and the role of ethnic and religious conflict in fueling insurgency, among other topics. A better understanding of all these topics is crucial for policy, planning and operations focused on improving governance, developing sustainable economic systems, and communicating effectively because these topics involve social change. The social sciences offer the Department of Defense a well-honed set of tools it can use to describe societies and to develop methods to affect social behavior and social systems.

WHAT IS SOCIAL SCIENCE?

Social science[1] is the scientific study of the social life of human groups, which includes the study of social structure[2], culture[3], politics[4], and economics[5], among other topics. Beginning with the philosophical inquiries of Socrates, Thucydides, and Plato, the human and social sciences have developed clearly defined methods to describe or explain human and social behavior. Social science methods are designed to improve the accuracy of descriptions and explanations of social life. The rigorous application of social science methods is the best way to improve descriptions and explanations of social phenomena such as insurgent and terrorist groups, social movements, state failure, civil war, cultural systems, and political systems.

The academic field of social science is composed of a set of inter-related academic disciplines[6]: sociology, anthropology, political science, economics, and psychology. There are also a number of emergent interdisciplinary social science fields that take a cross section of the core methodologies of the above disciplines and apply them to specific subject areas, including communication studies, religious studies, rural sociology, and geography. Finally, many area studies programs, such as Middle Eastern Studies, African Studies, or Asian Studies, deploy a cross section of these social science methods. The social science disciplines share a general commitment to accurately describing or explaining social life by collecting facts we do know with certainty (e.g., data or observations) to learn about things that we do not yet know with certainty (e.g., theories or hypotheses).[7] Social science seeks to describe social life or to identify the causes of social phenomena. There are two major processes in social science: data collection and analysis.

One of the most important differences between social science methods and physical sciences (e.g., physics) methods is that the subject of study is a society and the researcher and anyone they study are social beings. Error and bias are more complex in the social sciences because the objects being studied and the people studying them are social beings who are affected by the process of research. The most common types of error—or the difference between reality and a description of that reality—are related to the social relationship between the researcher and the subject. *Reactive error* results from the effects of the observer's presence on the phenomenon under study.[8] For instance, if the ethnic identity of the researcher causes an interview subject to distrust the interviewer, the information they give that researcher may not be entirely accurate. Perceptual and interpretive error result from the researcher having their own personal characteristics and perspective. For instance, an American researcher might observe two Arab men holding hands and infer that they are homosexuals. In this case, the analyst's culture may cause them to misinterpret the relationship between these two men, who are very likely friends. Finally, there is a likelihood of sampling errors, in which the observer does not collect a reasonably representative sample of a social phenomenon to be able to draw valid conclusions about the topic.[9] For instance, if, when studying a society, a researcher only interviews and observes university students. This group is younger, better educated, and predominantly from the middle and upper class. Thus, their beliefs and behaviors are therefore not representative of those of the entire society.

As the description of these errors indicates, there is no perfect method for measuring social reality,[10] but there are methods for identifying and clearly stating or reducing the major types of error. These methods for reducing error are the heart of the disciplines of the social sciences and there is no substitute for the years of painstaking self-criticism and self-assessment that these disciplines teach. It is for this reason that it is preferable that the Department of Defense (DoD) researchers and analysts incorporate social science expertise into projects in which they describe social groups and behaviors or attempt to identify the causes of social phenomena.

WHY SHOULD THE DEPARTMENT OF DEFENSE CARE ABOUT SOCIAL SCIENCE?

Social science methods provide a tool for the DoD to improve descriptions of foreign societies and identify the causes of the behaviors of social groups. Understanding the factors that drive social behavior will better enable the DoD to plan operations intended to essentially change social behavior—like counterinsurgency operations or strategic communications. Social science methods will be helpful for those trying to gain deeper understanding of the culture, social structure, politics, economics, and behaviors of a society or social group with which DoD is interacting; to identify the causes of a given social action (e.g., smuggling, participation in armed militia group); or to compare across cases to develop a theory that can be applied in multiple operations about how a social phenomenon functions in order to produce a plan to effect it. This information can be used to support planning and assessment for counterinsurgency, counterterrorism, building partner capacity, and influencing the behavior of social groups.

BEST PRACTICES: PLANNING AND EVALUATING SOCIAL SCIENCE RESEARCH AND ANALYSIS

To use this tool kit well, DoD personnel will need a basic understanding of social science standard operating procedures (SOP) so they can assess the value of research and analysis as well as the capabilities of the researchers with whom they interact. The following are a set of widely shared methods and practices used in the social sciences.

1. DETERMINE WHAT IS ALREADY KNOWN ABOUT THE QUESTION BEING ASKED.

A literature review is SOP for finding what is already known and identifying gaps in knowledge. Once researchers/analysts identify their research question(s), they typically conduct a literature review. A literature review is an exhaustive search for academic books and articles on the subject of the research.

For DoD personnel conducting research on social phenomena, the literature review allows them to determine: What is already known about the research question? Where are the current gaps/what is left to be learned? And who are the relevant experts that DoD can consult on this topic?

The person conducting the literature review should identify what is already known about the topic through library research using library databases of books and social science journals (such as WorldCat or Social Science Full Text). It is particularly useful to identify the Library of Congress Subject Headings for the topic being researched because scholarly databases are designed around these headings.[11] A thorough review of books and articles written by social scientists on the topic will allow the researcher to identify what is already known about the topic and what remains to be studied. It is also useful for identifying people in different fields who are currently working on the topic (as a starting point for reaching out to the academic community).

2. FIND THE RIGHT RESEARCH EXPERT FOR THE JOB.

Not all genres of social science are best suited to answer a specific question. While a literature review can help researchers narrow down the kinds of disciplines that might pertain to a specific question, they wouldn't ask a mechanical engineer to tell them the best form of dentistry. Similarly, the title "social scientist" isn't interchangeable to answer all questions related to "people." (e.g., if you are interested in a description of the key beliefs of a particular social group, an anthropologist with the appropriate language skills and research experience might be helpful. However, they will be less helpful for determining factors that drive social group assistance for insurgent groups across multiple countries; a sociologist or political scientist might be more helpful in this case). The following is a cursory description of the social science disciplines:

When looking for information that is very specific to a particular place, time, and group of people, look for an *area studies specialist.* An area specialist is a social scientist with language skills in a specific area who has spent a lot of time in that specific area doing research. Note: this is NOT the same thing as being from the region. Don't assume that someone from the region is an expert on that region. A gas station attendant, dentist, and lawyer in America are American citizens, but that does not make them experts on American political culture. The same is true for other societies. Individuals trained in fields such as area studies, history, and cultural anthropology are almost by definition area specialists of this kind. Many political scientists[12] also have considerable area specialization depending on what they study.

When looking for broader cross-referenced cases to address functional issues (e.g., radicalization, counterinsurgency best practices, state failure), choose someone with a good grasp of comparative case methodology, that is, someone who has studied a single phenomenon in multiple settings and in a variety of contexts—someone who has studied ethnic conflict in Rwanda, Kosovo, and Iraq and makes generalizations about the phenomena across cases, for example. Comparative study is a common technique in sociology, political science, economics, social psychology, and some kinds of anthropology.

As is evident in these descriptions, there is ample room for overlap, so someone may be an Afghan regional specialist who has a broad understanding of communication systems in developing contexts, or an expert on ethnic conflict who has spent multiple years in the Middle East and speaks Arabic and Farsi. For this reason, it is critical to review the publications of each social scientist under consideration to ascertain whether they have the appropriate skills and experience. Table 2.1. provides a quick guide to the disciplines in the social sciences.

TABLE 2.1. A QUICK GUIDE TO THE SOCIAL SCIENCE DISCIPLINES
• Anthropology (cultural, social and physical anthropology): studies the "human animal"
• Social reality is described or explained through field research in the specific society (ethnography)
• Results of field research are normally expressed in narrative form
• Often single-case studies that describe in great detail the context of that culture and society
• Sociology: studies social groups
• Social reality is described through collection of data, which is then compared to develop theories
• Theories are often used to build models of a world where the theories operate
• Economics: studies production and consumption of goods and services
• Individual interest can be aggregated in order to describe the behavior and causes of behavior of both individuals and groups
• Emphasis on models over empirical evidence
• Political Science: studies power
• Historically applied all methods, but currently focused on:
• Rational choice theory
• Standard causal analysis* versus case studies
• Emphasis on modeling to generate hypotheses and predict behavior
• To develop covering laws for political behavior
• Psychology (including cognitive science): studies individual behavior
• There are universal laws of individual human behavior
• History: studies the past
• Narrative explanation of specific periods, events or people

Source: Author.

To appropriately apply social science methods, any social scientist needs to have received academic training in research and analysis methods. In the social sciences, that research training occurs after the master's program during the doctorate program. For this reason, it is important that DoD researchers seek out people with a doctorate in one of the social science disciplines listed in Table 2.1 to assist them when conducting social science research and analysis. There is a tendency to assume that terminal degrees, like a Doctorate in Philosophy (PhD) in History, for example, are overly specialized. While the subject matter of a specific PhD may be very specialized, the difference between doctoral, masters, and bachelor's degrees is training in research and analysis methods. If intending to conduct a research and analysis project on a social topic, it is best to consult or hire someone with a PhD because they can implement the methods that put the "science" in social science to improve the accuracy of the project's outcomes.

One of the principal mechanisms that ensures high quality social science is the concept of *peer review*. Once a literature review is conducted and a social scientist is selected to assist in developing a research and analysis plan, the next step is to convene a panel of experts from multiple fields (some of whom might come from the literature review) who can review the plan to ensure it contains all the elements that might be of use to the DoD. When selecting people to conduct the peer review, it is important to ensure that they are disinterested, outside parties who cannot benefit from either a positive or negative review. Instead, they should be provided with incentives that cause them to provide an objective review.

Finally, many people working in the DoD assume that social scientists do not wish to work with the DoD. While this may be true for some social scientists, for others it is not—and DoD personnel will never know unless they ask. Social scientists are often concerned that by working with or in the government, they may sacrifice their academic freedom to criticize what the government is doing. They are also concerned about working with the military because they fear that the use of violence in a combat theater may injure the subjects of their studies.

The key to working well with academic social scientists is to understand their interests and the constraints under which they work. Social scientists are committed to enhancing knowledge through research and analysis that is then published and shared throughout their community, thereby generating more discussion of and research on the topic. What a social scientist publishes is one of the criteria for getting jobs, promotions, or further funding. For this reason, putting measures in place that allow them to maintain independence in analysis and publish their findings in unclassified books or journals enhances the likelihood that they will work with you.

Additionally, social scientists are concerned about maintaining trust-based relationships with the subjects of their research. For this reason, they are concerned about ethical behavior toward them and about military applications of information about the subjects of research. Further, academics are concerned with their ability to maintain academic and intellectual independence. Understanding these interests, and bearing them in mind when interacting with academics, will help create a relationship that will be more likely to last and produce results. Finally, putting measures in place like peer review of research methods ensures that social scientists can prove they are meeting their discipline's ethical standards.

3. DETERMINE WHICH COMBINATION OF RESEARCH AND ANALYSIS METHODS TO USE

The next step in any effort to describe or explain the causes of a social phenomenon is the decision about which methods of research and analysis are appropriate for the task. A researcher needs to select methods that are a good fit for the question being studied. In that regard, there are four major combinations of methods of data collection and analysis that have been successful historically in the social sciences:

Ethnographic field research and analysis. In this method, researchers immerse themselves in a social setting to understand its culture and social structure by participating in the social life of the community.

| Ethnographic field research and analysis. In this method, researchers immerse themselves in a social setting to understand its culture and social structure by participating in the social life of the community. |

This participant observation[13] is normally augmented by informal interviews with people in the society under study. Ethnography produces a mass of detailed field notes that the ethnographer then analyzes it to discern patterns that allow the ethnographer to describe the group and identify the most important aspects of their social structure, beliefs, and behavior and generate hypotheses about the causes of social phenomena. The goal of many ethnographies is to translate the symbol system of the culture of another social group for their readers.

Small n comparison. While ethnographic research normally focuses on one social group, the small n comparison method of data collection is a detailed study of a small number of social groups. The areas are normally selected because of some social characteristic or lack thereof (for example, high level of insurgent recruiting versus low level of insurgent recruiting) and then compared so that the causes of selection characteristics can be identified. Methods of collecting information in each case can be ethnographic field research, interviews, or analysis of documents. Often, a combination of these research methods is the most effective for providing a detailed description of the case. Small n comparison normally involves the collection of a consistent set of categories of information about each case so cases can be compared. While this standardization strips away some of the specificity of each case by starting with a hypothesis, the detailed study still allows the researcher to explore topics outside of the analytical framework. The purpose of small n comparison is normally to test a hypothesis about what causes certain social phenomena across groups.

Large N surveys and standard causal analysis. In this data collection method, researchers devise a standardized questionnaire that they use to ask respondents the same questions, with mostly pre-formulated responses. The researchers interview a large number of respondents who have an equal probability of selection (i.e., random or probability sample). Standard causal analysis is a statistical analysis method that takes the measures of various responses to the survey and employs statistical models to draw inferences about the relationship among those variables.[14] Surveys are built to measure specific constructs, normally to test a hypothesis about the causes of a particular social phenomenon. They are sometimes designed to describe how much of a population shares a set of characteristics.

Formal modeling. In this data collection method, analysts create models of how they believe the social world functions. These models are an expression of theories the analyst has about how a social phenomenon operates. A model can take a theory about how a social group will behave and create a space in which analysts can experiment by changing variables within the social setting model they have created to determine what effects the theory dictates will occur. The purpose of the model is to explain how social phenomena will change based on a set of assumptions about how changes in social phenomena are related. The model is in effect an experimental environment. The problem is that it is very difficult to capture the complexity of the social world because it is very difficult to accurately represent that social world in data. Therefore, social scientists normally use models to generate hypotheses or to attempt to predict how certain policy actions will impact the social world. Modeling, when it is done in the social sciences, tends to happen in conjunction with other empirical descriptions and theory development on the topic under study. All pre-existing research on the topic under study helps inform the model's structure and validates the model. In social science, the results of modeling efforts combine with this other backdrop of information to help provide insights and possibly recommendations. A model is best used as one methodology among many that collectively provide the DoD with an understanding of the social phenomenon they are attempting to study.

COMPARING METHODS

The best social science models are built to answer specific research questions, rather than capture the entire social world. They are developed based on a thorough examination of the social phenomenon in question. The examination of a social phenomenon is most effective when it begins with a literature review, followed by field research to establish an understanding of how it functions and develop hypotheses, followed by field research on a set of cases to test these hypotheses.

| The examination of a social phenomenon is most effective when it begins with a literature review, followed by field research to establish an understanding of how it functions and develop hypotheses, followed by field research on a set of cases to test these hypotheses. |

Once this empirically grounded theory is generated, a model may be useful to experiment with changing different variables that are related to the social phenomenon to see how they might influence the outcomes. But the model is only as good as the research question asked along with the data and theory that support it.

There are some basic trade-offs when choosing a single modeling method. While ethnography provides the most detailed description of a specific social group, the detailed way in which these groups vary makes it difficult to compare across social groups. There is therefore a trade-off between specificity and comparability. The greatest contrast is between ethnographic field research and formal modeling. Theories that compare across cases to develop a general theory for all cases, which are then depicted in a model, by their nature strip away the specific facts of each individual case in order to generalize across cases. The trade-off results from these two methods having different purposes: one is designed to demonstrate a theory, and the other is designed to describe a society or translate a culture. The DoD, therefore, needs to be cognizant that the best approach to these problems may be to use multiple methods.

An example of the utility of multidisciplinary methods of social science research and analysis would be when asked to provide a description of a set of social groups in a weak or failing state. The first step in such a task might be to conduct ethnographic field research in a set of communities from each social group. This would provide an accurate description of the social groups in those specific communities that could be used by the intelligence and operational communities responsible for those areas. The ethnographic description would also identify the main social issues, cultural norms and forms, economic system, and political competitions, among other things, in each community. The deeper understanding of these multiple communities would allow researchers to develop a set of research questions for a larger study of the country. This larger study might include research from smaller country case studies that were developed using the same questions as researchers worked to develop a theory about the causes of behaviors of beliefs in that country. This study might be followed by the development of a random sample survey. A survey based on exploratory ethnography and comparative case studies would allow the DoD to identify the main issues; determine how prevalent those issues are throughout the society; and learn whether they are more or less prevalent based on some other characteristics, like gender, age, income, ethnic identity, and social class. To determine if a theory developed for one weak or failed state applies to all weak or failed states, researchers could study communities across different countries to determine whether the initial hypothesis developed and tested in the first country applies outside it. Once the theory is developed, researchers might then generate a model in which the theory is demonstrated, generating more hypotheses from the model or using it to attempt to predict how social groups might behave in any failed state.

As this example demonstrates, different methods are appropriate for different tasks. All methods have inherent flaws, and since the DoD seeks the most accurate descriptions and explanations of social phenomena possible, it is often best to use multiple methods. The use of multiple methods can compensate for the weaknesses of each method. However, the combination of methods should vary based on the purpose, topic and research environment found in the model in question.

PEER REVIEW OF THE PLAN

It is also important that once a research and analysis plan (such as the one described above) is developed, that the plan is vetted using a set of academic experts from multiple fields. This vetting process is called *peer review*, and in academia, it is the method through which publishers ensure that the books and articles they produce meet high standards. Because the DoD wants its research and analysis to meet high standards, it should allow a group of outside academics from multiple disciplines to review research and analysis plans and provide suggestions on how it might be improved. When using outside academics for peer review, it is important to choose people who cannot personally benefit from the provision of either a positive or a negative review. Further, if the DoD allows outside academics to review a research and analysis plan *before* it is applied, this ensures that the research will be conducted in the most accurate way possible. Additionally, bringing respected members of the academic community in to review a project builds a better and more open relationship between the DoD and social scientists.

PARTICIPANT OBSERVATION

Participant observation is the process of immersing oneself in an unfamiliar society to observe it as a participant. Effective participant observation requires speaking the same language and spending months or even years living in the social group under study. The researcher observes the actions and talk that occur around them and conducts informal interviews with people from the community, some of whom are key informants. Key informants are people in the community who are good observers and can help translate the culture and social structure for the researcher. The field researcher records these observations and discussions in field notes. *Field notes* are detailed chronological logs of what was observed and the details about the discussions with people from the social group. They are written at the end of every research day and make a clear distinction between what happened (data) and what that might mean (analysis). When the researcher returns from the field, they review the notes to identify patterns in the data. These patterns form the basis of the findings of the research project. Participant observation originated in the field of anthropology, but it is now a technique used by social scientists from many disciplines (e.g., sociology, political science, and interdisciplinary area studies).

Security and the openness of society can sometimes impinge on a researcher's ability to get into and integrate with a given society or sub-group. The social sciences have developed a variety of techniques to manage this problem, including interviews with expatriates (characterized by sampling problems, as these individuals are not necessarily representative of the population still in the country), relying on a local network of sources who may or may not know they are providing information to a researcher (which are characterized by researcher training problems), and the rapid assessment process (RAP), which is a mixture of observation and targeted interviews (discussed in more detail in next section).

It is often not possible for DoD personnel to do extended ethnography because of their official capacity or direct involvement in operations. The RAP is a method developed in applied anthropology to get insider insight in a much shorter time frame (several weeks at a time). The goal is still to get the insider's perspective, but it is usually focused on a more narrow set of topics. Eliciting stories—usually in semi-structured interviews—target core members of a community and knowledgeable insiders, not necessarily views of the "average guy." A research team collects and analyzes data, discussing and making course corrections while the research is in progress. The team travels together and conducts interviews together. There are therefore multiple notetakers. At the end of each day, the team discusses what each team member thought was said, what the respondents meant, and the overall relationship of the day's events to research questions. If and when possible, the team should conduct follow-up interviews with the same respondents to clarify their meaning and get any additional information.

THE STANDARD OPERATION PROCEDURE FOR ETHNOGRAPHIC FIELD RESEARCH/PARTICIPANT OBSERVATION IS AS FOLLOWS

• Conduct an exhaustive literature review.
• Learn the local language.
• Move into the community and participate in it.
• Spend months or even years observing talk and actions.
• Have informal discussions with key informants.
• Write field notes every night (separating facts from analysis).
• Constantly assess the possible errors that are reactive, interpretive, or sampling-based.
• Return from the field and search for patterns in the data.

About the Author

Dr. A. Jackson is a political scientist who studies social movements, insurgencies, insurgent organizations, civil wars, and ethnic and sectarian conflict. She holds a PhD from Johns Hopkins University.

NOTES

1. Science is the sum of all knowledge and methods for accumulating more knowledge. It is divided into disciplines that accumulate knowledge of different topics and develop specific sets of methods for studying their topic.

2. Social structure is the set of relationships among groups of persons within a system of groups. Social structure is comprised of enduring patterns of behavior by participants in a social system in relation to one another. (Army FM 3-24) Examples of social structure include tribal systems and other social networks, social classes, ethnic and religious groups. Social structure is often described as the skeleton upon which cultural, political, economic, institutional, security, etc. systems hang.

3. Culture is a system of shared beliefs, values, customs, behaviors, and artifacts that members of a society use to cope with their world and with one another. Culture is an "invisible web of meaning" shared by members of a particular society or group within a society. Culture is patterned, meaning that people in a society live and think in ways forming definite, repeating patterns. Culture includes a system of beliefs about behavior that is acceptable and unacceptable, called a normative system. These norms are generally associated with a set of sanctions (positive and negative) that are used by members of society to encourage individual compliance with the social norms. Culture is learned through the process of enculturation, in which the society teaches individuals the accepted norms of behavior, symbols and their meanings. Culture is shared by members of a society, and it is changeable through social interactions. (FM 3-24) Culture therefore effects the behavior of individuals, while those same individuals are, as a group, generating the culture in question.

4. Politics is the competition of interest for power. This competition can be in the formal political realm of government, or in the informal social realm. For instance, the competition between a pair of tribes for control of grazing land is politics, as is the selection of a new chief for one of the tribes and the relationship of that chief to the formal government. The tools of political competition can include coercion, money, ideology and cooperation with traditional or charismatic authority figures.

5. Economics is the study of the production and consumption of goods and services.

6. A discipline is a loosely structured institution. It is composed of academic departments, grant making foundations, publishers, and academic professional associations. These institutions decide what the standards are for the profession by controlling access to degrees, research grants, publication and jobs. The standards pertain to the purpose and methods of research and analysis, the topic of that research and analysis and ethics.

7. Andrew Abbott. *Methods of Discovery: Heuristics for the Social Sciences.* New York: W.W. Norton and Company, 2004.

8. John Lofland, David Snow, Leon Anderson and Lyn Lofland. *Analyzing Social Settings: A Guide to Qualitative Observation and Analysis.* Belmont, CA: Thomson-Wadsworth, 2006. p. 91

8. Abbott.

9. Lofland, Snow, Anderson and Lofland. p. 91

10. Abbott.
11. The LOC Subject Headings can be found at http://www.loc.gov/catdir/cpso/lcco/

12. Within political science, the field of comparative politics contains many area specialists, as does the field of area studies within international relations.

13. See appendix for detailed description of participant observation.

14. Abbott, p. 19

II. Appreciating the World: A Framework for Doing Socio-Political Analysis

By David P. Oakley

ABSTRACT: This chapter [1] offers a framework consisting of concepts drawn from international relations theory that can be leveraged by practitioners to better appreciate the complex operational environment. The framework complements existing doctrinal concepts and serves as a starting point for practitioners to explore the operational environment's socio-political dynamics. The framework's concepts expose surface area for further exploration, thus enabling a collaborative discussion among practitioners and commanders.

INTRODUCTION

Academic scholarship is essential for practitioners to appreciate the operational environment. Although design doctrine now provides leaders a methodology to help deal with complex problems, it does not provide all the tools necessary for the individual to explore and better understand complexity. Appreciating the operational environment requires practitioners to understand structures, institutions, and narratives—the same political dynamics that are discussed in international relations theory. International relations theories might be focused at the state level, but their explanatory value can shed light on local dynamics within the operational environment. The case study in the analysis section is intended as a demonstration of how the concepts can be applied by the practitioner to understand complex operational environments while supplementing current doctrinal concepts. It is also intended as an easily accessible primer to help the practitioner understand and explore these important concepts.

Although practitioners should embrace an interdisciplinary approach when trying to make sense out of complex operational environments, *international relations theory* is a particularly valuable discipline because it is influenced by other liberal arts disciplines such as history, sociology, anthropology, psychology, and philosophy.[2] The diversity and richness of international relations theory makes it an exceptional platform to introduce practitioners to concepts that explain the socio-political dynamics within the operational environment.

| The diversity and richness of international relations theory makes it an exceptional platform to introduce practitioners to concepts that explain the socio-political dynamics within the operational environment. |

Although traditionally applied to international affairs, its explanatory concepts can be used to better understand socio-political dynamics at any level. A collaborative application of the three primary international relations theories (realism, liberalism, and constructivism) provides an appreciation of the complexity of the world and the role narratives, structures, individuals, and ideas play in shaping reality.[3]

International relations theory, individually pursued or institutionally provided, goes a long way in developing critical and creative thinkers who have a better appreciation for complexity, a richer understanding of local dynamics, and are knowledgeable about America's effect on the operational environment. The richness of international relations theory and the way it lends itself to cross-discipline exploration can help develop thinkers who appreciate the nuances and diversity of the world around them.

Some argue that a purpose behind political science research is to "produce valid inferences about social and political life."[4] These inferences are usually developed through a scientific empirical process that seeks to develop models and theories that provide broad generalizable explanations for political phenomenon. In this regard, political scientists are not as concerned with individual events per se but are more concerned with how several similar events might tell us something about the world at large.[5]

Although there is great utility in these models for the practitioner, the practitioner is usually more concerned with understanding one event or environment (e.g., Iraq, Afghanistan, Russian intentions, or Chinese intentions) and not an entire field of political phenomenon (e.g., civil war or terrorism). With this in mind, the framework of this section is not intended as a framework or process for action but can better be viewed as a framework for understanding. Although merely understanding might not wholly satisfy those who are attempting to uncover some panacea that can be broadly applied, the author posits that it is of utmost importance that we first understand the environment before we attempt to act. As the author, my goal is not to devise a framework that can be applied to solve any problem, but instead to show how certain concepts can be pulled from international relations theory to explore and better understand the operational environment. Finally, my desire is not for practitioners to seize on one set model, but to develop an appreciation and "fox" like knowledge of a variety of disciplines and use them to gain a greater appreciation of the socio-political dynamics within the complex operational environment.

In his 2010 book on combining realism and constructivism to develop a richer understanding of international relations, Dr. Samuel Barkin highlights the difficulty with trying to blend multiple theoretical approaches due to the lack of a "conceptual framework for communication among practitioners."[6] Barkin discuss the paradoxical conflict between not having a framework as a common reference for practitioners being confusing, and how frameworks can often oversimplify or create rigid categorization that detracts from important dialogue between the varying schools. Barkin takes the middle ground between a rigid framework that leaves little flexibility to add and remove concepts when the lack of explanatory value limits their utility in understanding and the other side of the spectrum where there is no conceptual framework.[7]

Instead, Barkin suggests an informal matrix that lays out the compatibility and incompatibility of different paradigms that can be leveraged by academics to encourage collaboration. Although Barkin is speaking to academia, his push for a collaborative approach can inform the practitioner. But, unlike the academic, the practitioner need not worry about keeping the theories separated in a matrix but can draw unabashedly from them all without concern for maintaining theoretical purity. The practitioner should think of these theories and concepts as residing in a toolkit and collaboratively draw from them to improve understanding of the operational environment.

ANALYSIS

This section breaks down analysis into two parts—one developing the framework and one using this framework in a case study to illustrate its usefulness. Inspired by Graham Allison's classic book on the Cuban Missile Crisis, *Essence of Decision*, a framework is built based on key concepts drawn from international relations theory. The key concepts are taken from three principle international relations theories: *liberalism, realism*, and *constructivism.*[8] This framework is designed to consider how, when the different theories are collaboratively applied, they can help the practitioner better understand the operational environment.[9] Each theory has its own section, and each section includes a brief description of the general theory and how it evolved over time and identifies concepts drawn from the theory that are used to build the framework. To keep the general theory descriptions short, extensive footnotes are used to provide more background and offer information about how the other theories might view certain key concepts.[10] After establishing the framework, it is applied to a case study, demonstrating how concepts and ideas from international relations theory can be leveraged to understand political dynamics at all levels. The intent is for the reader to grasp the fundamental theories and collaboratively apply them to better understand socio-political dynamics within the complex operational environment.

LIBERALISM

Liberalism originated as an international relations theory in the early 20th century just prior to World War I and comprises three "strands" of theory.[11] The first strand argues that economic interconnectivity between states will limit war because conflict negatively affects their own prosperity.[12] A contemporary example of this strand is Aaron Friedberg's argument that war between the U.S. and China is unlikely because of economic interests.[13] The second strand argues that democracies are unlikely to go to war because they share similar fundamental values and their leaders' actions are hindered by a voting population.[14] The second strand can be further broken down into a pragmatic argument and an idealistic argument. The pragmatist argues that states spread democracy and democratic principles to primarily protect their own interests. Although pragmatic liberalists believe democracy for the world is a good thing, their primary concern is their nation's interests. Rights and freedoms for other states and people are a beneficial byproduct, but not the main objective. The idealistic argument supports the spread of democracy because they believe it is the right moral and ethical action. These two sub-strands are often difficult to parse out in debates, but they are important to remember.[15] The third and final strand of liberalism argues that international institutions, laws, and protocols help states overcome their selfish behavior and encourage them to work toward common interests for the collective good.[16] This interdependence between states makes war less likely because it runs counter to the interests of all involved.

In 1977, Robert Keohane and Joseph Nye wrote their influential book, *Power and Interdependence*, in which they contrast the *traditionalist* (realist) view that states were still the dominant actor in international affairs with the *modernist* view that corporations and other non-government entities were becoming the dominant players. Keohane and Nye argue that both state and non-state actors were important players in the international arena because they influence each other and develop an interdependent relationship.[17] With the increase in communications and the development of what they refer to as *globalism*, this interdependence only increases.[18] Keohane and Nye's argument about the interdependent relationship between state and non-state entities and the increasing technology that strengthens this interdependence helps account for gaps not necessarily considered focusing on structure alone.

One way to look at liberalism is as the pursuit of a more peaceful natural order within the structure. Where realism speaks of how things are, liberalism offers hope and prescriptions to nudge reality toward a more ideal version. Whether or not mankind is capable of manipulating the "realist reality" and progressing to a better state, liberalist efforts to improve the structure through laws, institutions, and other mechanisms influence the environment and the effect of their ideals must be appreciated.[19] Three key concepts can be identified from this basic understanding of liberalism:

Progress. This human endeavor is about progress and moving forward. At the individual level, people are influenced by a desire to see life improve. At the state level, people want an increase in rights provided to them by the state and then protected by the state. At the intrastate level, progress speaks to the spreading of democratic principles and human rights and away from conflict and war.[20]

Interdependence. Developing dependent links between people through interests and common identity can be one way to connect people together to achieve common objectives within the structure.[21] At the state level, nations look to develop interdependence as a way to limit conflict and progress toward a more stable peace and achieve common objectives.[22] At the local level, individuals look to develop interdependence to achieve common objectives that benefit mutual interests.

Institutions. Institutions can moderate or inflame conflict at all levels.[23] Although the structure is firmly entrenched, institutions and the laws they develop can make conditions within the structure more mutually beneficial or unintentionally worse. A key aspect with institutions is that the "setting-up of certain intersubjectively present institutions channels people unintentionally in certain directions at some later point." Humans form institutions to change behavior within the structure or to further their interests. Sometimes institutions and/or rules that are developed by the institution result in unintended consequences.[24]

Concepts from liberalism can be applied to most human interaction to gain a better understanding of the dynamics involved. Think of a small non-governmental special interest organization that seeks to end ivory poaching in Africa. The group is focused on improving a condition they view negatively, thus progressing to a better state of affairs. They intuitively understand the structure will result in some individuals poaching elephants for personal gain and know that selfish interest can drive individual action. Although they accept the structure, they look for ways to pass laws and establish institutions that will cause the poachers to reevaluate their interests. Understanding there is power in numbers, they develop links with people who hold similar beliefs. The links result in greater interdependence between people from different groups, thus strengthening their ability to influence action and obtain results.

REALISM

Realism in international relations theory developed largely in response to what some perceived as naïve but dangerous liberal arguments.[25] Realism can be broken down into two primary strands, classical realism and neo-classical realism.[26] Classical realists argue that "states, like humans, have a natural desire to dominate each other."[27] The desire for domination means nations seek power in order to pursue their own self-interest. Neo-classical realists switched the focus from the biological nature of man to the structure of the international system. They argued there was no "sovereign" to restrict state actions, limit their interests, or protect them from other states. This "anarchic" condition causes states to pursue their self-interest by "balancing" and "bandwagoning" against other states when it suits their need.[28] For the practitioner, this anarchic environment can exist not just at the state level but also where laws and institutions have broken down or between individuals who have no "sovereign" above them to limit their actions.

Concepts from realism can be used to understand the socio-political dynamics at any level of human interaction. Consider the structure of an organized crime family. There is a structure with set positions (don, capo, consigliore), and individuals operate and behave according to their position within the structure. Less powerful individuals might join forces to balance against the stronger individuals or bandwagon with the most powerful to increase personal power and further their interest. Although there are both formal and informal rules that guide behavior, what really affects behavior is their relative power against the others in the gang.

Realism, at its core, is about dealing with the world "as it is and not how you wish it to be."[29] For a realist, this means dealing with the anarchic structure of the system and understanding that organizations and people pursue self-interest first and foremost. From this basic understanding of realism, three concepts are identified:

Structure. Structure is important at the local, intrastate, and state levels, but for slightly different reasons. At the local level, structure is important in regard to tribal, family, religious, and other social structures because people naturally fall into certain structural roles (e.g., parent, employee, cleric, son, husband). Although these structures are slightly different, they still affect the individuals involved in the structure. At the intrastate level, there are governing and bureaucratic structures that influence the behavior of people and organizations that are part of the structure. At the state level, there are structures that are not as rigorously or formally governed as structures are at other levels.[30] It is important to remember that structure is relatively constant and difficult to manipulate in the short-term.

Interest. Individuals, groups, and states (any organization that has a shared identity) primarily pursue their interest.[32] Success for individuals, groups, or states is defined as preserving current capabilities while strengthening others.[33] Jeffrey Isaac's paper "Beyond the Three Faces of Power: A Realist Critique" is used to differentiate between three types of interest: (1) *subjective interest*: What the individual believes is in his interest or is good for him; (2) *objective interest*: "What really is in the interests, or good, of an agent whether he thinks so or not;" and (3) real interest: The effect or function of identities, "roles," or "social norms," on interests in enduring relationships. "Interests are real because they are causally effective in practice in a sense objective interests are not."

As an example of these three interests, consider the decision-making process many practitioners might go through on a Friday. A practitioner has a subjective interest in leaving early on Friday to play golf. The same practitioner has an objective interest as a professional in attending an educational event that would help nurture them intellectually. The practitioner has a real interest in spending his Friday afternoon working on a presentation due Monday morning.[36]

Power. Once again, Isaac's paper, "The Three Faces of Power" is leveraged to define the concept of power. Power is "the capacity to act possessed by social agents in virtue of the enduring relationship in which they participate."[37] There are two types of power: "power to" and "power over."[38] When most people think of power, they probably think of "power over." "Power over" speaks of a person or organization possessing the strength or capacity that provides the ability to influence the behavior of other actors in accordance with one's own objectives.[39] On the other hand, "power to" is not necessarily about strength in the traditional sense of strong vs. weak, but about the positional power an individual has within a structure and as part of a relationship.[40] An example is the relationship between a coach and their athletes. The coach has the "power to" schedule practices, build the roster, assign positions, and decide amount of playing time. The players have the "power to" decide whether to show up to practices or games. The success of the team depends on both coach and players exercising their "power to" in a manner that benefits the team (players and coach).[41]

CONSTRUCTIVISM

The term *constructivism* in international relations theory was first coined by Nicholas Onuf in the late 1980s.[42] Arguably the most well-known constructivist article was written by Alexander Wendt and entitled "Anarchy is what States Make of It." The article was a counter to realist arguments that the anarchic world forces states into a self-help mentality. Wendt argues that while the concept of anarchy might be true, the notion that all states naturally seek power over other states is incorrect. He argues that all nations react to anarchy differently and before we can determine how an individual state will react, we must know something about how that state's "identity" is formed and "interests" are constructed. A key distinction between constructivism and the other two theories is that liberalism and realism put the emphasis on the structure, while constructivism focuses on the agent. Whereas liberalism and realism treat states as "interest-based," "utility-maximizing" entities constrained by the structure, constructivism focuses more on the individual state and not the structure it operates within.

Constructivism argues that a state's interests can be redefined by the acceptance of "new knowledge and new normative beliefs." Therefore, the embrace of new ideas and redefinition of interests shape state's actions, not a static and unchangeable structure.[43]

Constructivism can easily be applied to the group (local) or individual level. A good example of constructivism is the relationship a parent has with their child. The relationship might begin as one of caregiver/protector (parent) over the young fragile infant. Both the parent and child are defined and shaped by their roles, but these roles are not stagnant; they change through time and intersubjective meaning. The parent, who was once viewed as the protector, evolves into a disciplinarian figure and then later in life to a friend. Through the years, this relationship changes through interaction and is constantly intersubjectively redefined.[44]

From this basic understanding of constructivism, we can identify key concepts, which, unlike our liberalist and realist, are concerned with the unique identity of the actors involved instead of the structure of the system. The constructivism concepts are:

Narratives. Individuals and collective groups of people all have a narrative informed by a shared history and a common identity that affects the way they perceive interests and how they navigate the social world.[45] Narratives are important because they form the lenses through which individuals and groups see and experience the world. Narratives are not static tales or stories. They are dynamic and ongoing "stories of peoplehood" whose final chapter is neither known nor determined.[46]

Social construction. Individuals and groups of people "define their interests in the process of interpreting the social situations in which they are participants."[47] New knowledge or experience can alter how entities (individuals, groups, states) perceive their interests and can change how they view others in the system. "Intersubjective" reality between two entities is framed and reframed through continual interaction.[48] A close ally can become an enemy, or an enemy can become a close ally due to the two intersubjectively redefining their relationship.[49]

Contingency. Narratives and conditions are shaped by contingent human actions that could have turned out differently if other "unanticipated and novel" elements gained prominence.[50] Take Iran, for example. Iran's collective narrative is very much influenced by its 33-year theocracy. If, during the 1979 revolution, other movements came to prominence (e.g., democracy advocates or communists) instead of Ayatollah Khomeini, then Iran's narrative would be different today. Rogers Smith succinctly captures the concept in his book *Stories of Peoplehood*, where he posits "all stories of peoplehood (narratives) are always contingent things that human beings create collectively, through both conflictual and cooperative processes, with some elements that are patterned and predictable, but others that may be novel and unanticipated."[51]

THE CASE STUDY

Now that the framework has been established, it will be applied to a case study to demonstrate its value for the practitioner. The framework, and concepts within, go beyond surface deep analysis of the operational environment and allow exploration of the richness of the agents, structures, institutions, and interests involved. The framework provides a common reference during staff group discussions, but its concepts are rich enough to enhance collaborative dialogue and encourage exploration. The framework is complementary to political, military, economic, social, information, infrastructure, physical environment, time (PMESII-PT) and area, structures, capabilities, organizations, people, and events (ASCOPE), but its concepts take exploration further than either of these models can separately. This framework can be leveraged during mission analysis and as part of the design methodology as a metaphoric tool to mine the operational environment.

The case study leverages a 2005 documentary entitled Gaza: *The Fight for Israel*, which covers a 96-hour period during which Israeli Defense Forces (IDF) and police forces removed Israeli citizens from Gaza settlements in accordance with the government's disengagement plan. The case study uses the framework to explore the socio-political factors involved during the removal of Jewish settlers from the Gush Katif settlement to develop a greater appreciation of the operational environment. The case study focuses largely on Gershon Hacohen, a brigadier general at the time and the IDF operational commander responsible for closing the settlement, and his interaction with settlers. Three key events that occurred during the four-day ordeal will be investigated: (1) IDF entrance into the settlement; (2) Hacohen's meeting with a religious IDF officer; and (3) the "Synagogue Crisis" in which Hacohen negotiated with the elder rabbis.

Although the documentary serves as the foundation for the case study, other available information on the settlement issue will be leveraged from divers views regarding this issue. The resources are drawn from an array of sources to include books, academic journals, newspaper articles, government websites, and policy papers. The Israeli settlement issue has a rich and complex history, and it is not the purpose of this case study to go into the complex nuances of the broader settlement debate. It is necessary to review some of the background to gain a better appreciation of the narratives and varied interests involved. The first section provides background on the broader settlement issue and the Israeli government's decision to enact the disengagement plan. The second section focuses on three events that occurred in Gush Katif during a four-day period in June 2005.

The name "Gaza Strip" (shown in Figure 2.1) was first used during the 1948 Israeli–Arab War and gained wide recognition in the 1949 Israeli–Egyptian Armistice agreement to denote a separate piece of terrain under Egyptian authority.[53] The area that would later become the Gush Katif Settlement in the Gaza Strip was "seized from Egypt" and occupied by Israeli

Figure 2.1. The Gaza Strip. Source: CIA World Factbook at https://www.cia.gov/the-world-factbook/static/maps/WE-map.jpg.

Forces during the Six Days War.[54] Starting in the 1970s, the Israeli government allowed settlers to develop Gush Katif settlements as a means to establish a barrier between the Israeli and Palestinian populations.[55] Although many in the Israeli government supported the occupation of the territory for security purposes, most settlers moved into the area for religious ideological reasons. Believing the area to be Jewish holy land, Zionists established settlements in the area that were not meant for short-term security gains but focused on the long-term incorporation into the Jewish state.[56]

In December 2003, Israeli Prime Minister Ariel Sharon proposed a disengagement plan for the removal of 11 Israeli settlements in Gaza and Northern Samaria. The stated intent of the plan was to "break the current deadlock by removing the too-often lethal friction between Israelis and Palestinians in the Gaza Strip and northern West Bank, thereby improving the situation."[57] Although this was the official position, others have argued that the cost of securing the territory was not worth the money spent and the Israeli government could strengthen their international position by removing the settlements. Removal would ease the international pressure for concessions they viewed as more painful to the Israeli state.[58] The plan passed the Israeli cabinet in late summer of 2004, the Knesset in the fall of 2004, and implementation of the plan began in the summer of 2005.[59]

Gershon Hacohen, a 37-year veteran of the IDF, was assigned the task of commanding a 12,000-member force responsible for overseeing and removing (by force, if necessary) the 8,500 Gush Katif settlers. Hacohen was an interesting selection to lead the removal operation. His parents were founders within the settler movement, and some of his siblings lived in West Bank

Figure 2.2. The West Bank. Source: CIA World Factbook at https://www.cia.gov/the-world-factbook/static/maps/WE-map.jpg.

settlements (shown in Figure 2.2). Hacohen's background and empathetic nature led him to pursue a non-confrontational approach, quietly listening to complaints and offering emotional support. Although focused on the mission, he tried to be as accommodating and sympathetic as possible.

When the IDF and police forces arrived, more than half of the 1,550 registered families had left the settlements in the Gaza Strip. Of the 832 families that remained in the Gaza settlements, approximately 700 were in the Gush Katif settlements.[60] These remaining families were joined by scores of external groups and individuals that came to Gaza to show support for the settlers and to make the government's task more difficult.[61] The settlers were given 48 hours to leave Gush Katif. If they remained after the 48-hour period, the military would physically remove them. The government provided transportation to move belongings and offered compensation money for them to relocate somewhere else in Israel. Sixty percent of the settlers rejected the Israeli governments' monetary and relocation offers.[62]

The confrontation between Hacohen's forces and the settlers began when settlers blocked the entrance to the settlement, impeding the movement of forces and transportation into the area. The environment was tense, with angry protestors intermixed with wailing settlers, creating a potentially combustive situation. The emotional toil was not limited to the settlers. The police and military forces, many who came from settlements, internally struggled between their loyalty to the state and their support for their fellow countrymen and religious brethren. The forces were constantly under barrage from settlers and protestors, who often questioned their compatriots' humanity by comparing them to Nazis, Hitler, and Stalin. Forty-eight hours into the operation, 2,000 individuals remained in the settlement, and Hacohen's forces had to transition from urging and "persuasion" to removing people by force.

The confrontation between Hacohen's forces and the settlers began when settlers blocked the entrance to the settlement, impeding the movement of forces and transportation into the area. The environment was tense, with angry protestors intermixed with wailing settlers, creating a potentially combustive situation. The emotional toil was not limited to the settlers. The police and military forces, many who came from settlements, internally struggled between their loyalty to the state and their support for their fellow countrymen and religious brethren. The forces were constantly under barrage from settlers and protestors, who often questioned their compatriots' humanity by comparing them to Nazis, Hitler, and Stalin. Forty-eight hours into the operation, 2,000 individuals remained in the settlement, and Hacohen's forces had to transition from urging and "persuasion" to removing people by force.

The most potentially violent confrontation between troops, settlers, and protestors was at one of the settlement's synagogues. Approximately 1,000 individuals occupied the synagogue and refused to leave, forcing Hacohen to reach out to some of the rabbinical leadership to call on the protestors to stand-down. Hacohen and the rabbis came to an agreement that protestors would not turn violent if forces entered the synagogue. But the protestors insisted on being physically removed by the units to publicly demonstrate their ire over being compelled to leave the settlement.

Gush Katif was eventually evacuated and the structures torn down (all but the synagogues). Although the operation was successful, it opened a rift between the secular Israeli government and many of its religious citizens. Explore the rift and further unpack the different socio-political dynamics at play in August 2005 by focusing on three key events that occurred during the operation and applying the concepts to achieve better understanding.

EVENT #1: ENTRANCE INTO THE SETTLEMENT

The concepts progress, narrative, contingent, social construction, power, and interest help explore the socio-political dynamics at play during the IDF's entrance into Gush Katif. The concept of *progress* provides an important starting point for exploration of the socio-political dynamics within the operational environment. Although both the settlers and the Israeli government wanted progress, they each defined progress differently. The Israeli government defined progress as either getting closer to an Israeli/Palestinian agreement or creating enough space to alleviate international pressure. The Israeli leadership believed returning Gaza would either move the Israelis and Palestinians closer to an agreement or provide some space for Israel absent international pressure. The settlers' concept of progress was to retain and attain land they believed was Jewish holy land. In the settlers' view, the return of Gaza was retrogression, not progression.

When the IDF arrived at the settlement, they were accompanied by a positive narrative as the "holy" protector of the Jewish state and its population. This narrative was especially strong within the Gaza settlements because of the heightened threat of conflict and the IDFs 30-year role as protector of settlers on the Israeli frontier. When the forces arrived to evacuate the settlers, this narrative was reframed through social construction. Nadia Matar, an Israeli settler, exemplified the reframing of the IDF image when she stated, "we were all educated that the Army was holy. The Army is holy as long as it carries out what it is meant to do…to defend the Israeli people and fight the Arab terrorists."[63] The settlers no longer viewed the forces as legitimate protectors of the Jews, but as a powerful military force uprooting them from their rightful land. The ongoing reframing of the social narrative is heard in the documentary when the settlement population compares the IDF to Hitler's Nazis. The IDF actions reinforced the overarching narrative among the settlers of the Jews as an "uprooted" people, continuously victimized by non-Jews. The interesting aspect of the narrative reaffirmation is the fact the victimizers were now fellow Jews who only a short time prior were viewed as protectors of the faithful.

It is important to remember this narrative was contingent on past "conflictual and cooperative processes." The Israeli government's promotion of settlements and their use of settlers as a policy tool created a narrative of a pro-settlement state whose interests were nearly identical to Zionist interests. If the decision was made in the 1970s to turn over Gaza and the West Bank before settlements were built, the narrative of an Israeli government supporting religious settlements might never have surfaced. The idea of a "settler" might not have the same meaning today (or any relevance at all), within the Israeli narrative.

The IDF arrived with a large force, expecting the massive "scale of the operation" to show the population that resistance was "futile."[64] The use of "power over" to compel the settlers played toward the military's strength and their image as a powerful entity capable of exerting its authority. IDF leadership embraced the concept of "power over," and many believed that negotiating with the settlers only displayed weakness and irresoluteness. For example, when protestors complained that black-clad police forces reminded them of Nazi SS troopers and requested their removal, Maj. Gen. Dan Harel (Hacohen's commanding officer) told Hacohen not to negotiate with them. Harel believed the government had the power; negotiating would only strengthen the settlers' resolve. The comparison of Israeli police to Nazi SS troopers highlighted the strength of the Jewish narrative and the inability of the population to separate from it.

The population also asserted power toward the government. The population had the power to peacefully depart Gush Katif or the power to blockade and delay the government's disengagement plan. Although the government maintained the power over the population, its ability to assert that power was restrained by its own real interests not to take actions that would escalate the situation toward violence and risk weakening the Israeli state.

The population's objective interest was to stand down and peacefully evacuate the settlement. Hacohen's statement about the train travelling down the track and settlers had to choose to step aside or get run over, speaks directly to objective interest.[65] The settlers' subjective interest was to push back against the government and gain enough external support to force the government to postpone or cancel the disengagement plan. The settlers' real interest was a balance between their objective and subjective interests and is related to their split identity.

The settlers had two identities. They were Israeli citizens within a secular, but predominately Jewish, state that was part of the international order. This identity was concerned with the strength of the state and its ability to operate. Any action that weakened the state ran counter to the settlers' secular Israeli identity. Their second identity was one of a Zionist Jew, who believes it is God's will that they occupy all of the holy land. Any action that cedes land goes against God's will and runs counter to this Zionist identity. These two complementary, yet competing, identities created a paradox for the settlers. If they resisted too hard or resorted to violence, they could severely weaken the Israeli state and possibly ignite a civil conflict. Weakening the Israeli state could derail some of their Zionist objectives. This derailment could be more severe than the damage done by turning over Gaza. The settlers' real interest was to display a level of civil disobedience that remained true to their Zionist identity, but not severe enough to irreparably damage both identities.

The use of the framework to explore Event #1 demonstrates the role narratives play in creating and shaping identities. All people have narratives, and as demonstrated above, many people have competing narratives that are often at odds with other narratives. It is important to remember that narratives are not stagnant, but constantly morphing as new experiences socially (re)construct identities. People who were once viewed as protectors could quickly become enemies as new experiences intersubjectively redefine the narrative. It is also important to remember that narratives are contingent on previous "conflictual and cooperative processes." If another group or agent gained prominence, or other unforeseen events evolved, then the narrative would have developed differently.[66]

The framework also shows that power relationships are not only about the "strong do what they can and the weak suffer what they must."[67] Although "power over" is an important element of power, so too is "power to." If "power over" were the only form of power, one could easily calculate outcomes by comparing relative strength. As seen from the case study and known from experience, relatively stronger agents do not always achieve their objectives.

EVENT #2: HACOHEN'S MEETING WITH AN ISRAELI OFFICER/SETTLER

The concepts of interdependence, narrative, structure, interest, and social construction help explore the socio-political dynamics at play during Hacohen's meeting with an Israeli officer/settler. Toward the end of the operation, a young "religious" IDF officer requested a meeting with Hacohen to read a letter he composed about the emotional turmoil he was experiencing. The officer was not only a soldier but also a resident of the Gush Katif settlement who was removed days earlier by fellow soldiers. His identity as a soldier and a settler symbolized the interdependence between the military and the settler movement. The settlers depended on soldiers for more than 30 years to provide the protection they needed to occupy and expand the settlement. Many of the religious soldiers looked toward the settlers as keeping the idea of Zionism alive, while others in the Israeli state embraced secularism.

The story of the soldier/settler confronting Hacohen also depicts the degree to which the narrative of the settler and the narrative of the soldier were intertwined before the settlement removal. The soldier's distress was caused by what he saw as a contradiction within his own narrative(s). Until the day he was ordered to remove his fellow settlers, his roles as a soldier and settler were mutually reinforcing. As a soldier, he viewed himself as protecting his fellow countrymen and helping pursue Israeli interests. As a settler occupying what he believed was Jewish holy land, he viewed himself as pursuing not only Israel's interests but also God's will. The government's decision to remove settlers, and the soldier's subsequent split identity, caused him to socially reconstruct his own narrative. His inability to balance the once supportive, but now distinct, narratives resulted in a personal contradiction that he had to accept.

The concept of structure can be leveraged to help explain the soldier's turmoil. The soldier was a leader within the military structure, a structure that is relatively permanent. He understood that he had certain responsibilities within the structure to include a duty to obey orders and an obligation to serve the nation of Israel. Although his identity as a soldier can explain some of his emotional turmoil, it is focused on the particular individual and not the demands the structure places on him as a military officer. The demands and responsibilities placed on any officer within the structure will be similar, but the fashion and manner each officer emotionally handles the demands can differ greatly.

The soldier provides a unique opportunity to explore the impact of interests on individual decision-making. His split identity meant he had a variety of different interests that were often contradictory. As an officer, his objective interest was to enforce the laws of Israel and remove the settlers. As a settler, his objective interest was to stand down and allow the government to enforce the disengagement plan. His subjective interest was for the state to overturn the disengagement plan and end the resettlement operations. Determining the soldier's real interest is slightly more difficult because his two identities come into confrontation—and his real interest may have shifted during the four-day period. Initially, when there was hope that the government would change course, his real interest was to stand with his fellow settlers to force change (his settler identity was dominant). As the resettlement operation progressed, his real interest was influenced more by his officer identity than his settler identity. When hope receded and despair set in, he realized his real interest was with the military to ensure his own preservation and the security of the state.

The use of the framework to explore Event #2 allowed exploration of the interdependence between the settlers and the IDF. This interdependence was not only present within institutions or groups of people but also within individuals. The religious IDF officer had two separate identities that melded together to form his personal narrative. These two identities were dependent on each other to maintain balance within his narrative. Once these two identities became unbalanced, the officer was forced to reconstruct his narrative based on the introduction of new information and experience.

The framework also showed that identifying interests is not a simple objective process but is difficult and complex. The officer's turmoil shows that defining personal interests is difficult enough; an outside observer trying to determine others' interests is even more complex. Like narratives, interests are dynamic and can change due to obvious external causes or unobservable internal assessments.

EVENT #3: "SYNAGOGUE CRISIS"-NEGOTIATIONS BETWEEN HACOHEN AND THE RABBIS

The concepts of institution, interest, structure, narrative, power, and interdependence help explore the socio-political dynamics at play during the synagogue crisis. On day three of the operation, the IDF faced the unpalatable task of removing 1,000 protestors from a local synagogue. The idea of Israeli forces entering a synagogue, dragging out fellow countrymen, and possibly igniting a violent conflict was not in the interest of the government or the settlement's rabbinical leadership. To avoid a violent confrontation, Hacohen and the senior rabbis negotiated an interesting compromise: The protestors agreed to refrain from violence but insisted on being physically carried out by IDF soldiers as a show of civil disobedience. This seemingly irrelevant show of civil disobedience was of utmost psychological importance if the protestors were to maintain their narrative of struggle and sacrifice for Jewish holy land.[68]

There were multiple narratives being socially constructed and reframed within the synagogue. The IDF forces and the protestors embraced, cried on each other's shoulders, and prayed together. The collective Jewish narrative, which suffered damage throughout the four-day ordeal, once again came together. Despite all their differences over settler removal, the IDF and the protestors identified a common interest in the preservation of the Israeli state. The protestors realized the degree of interdependence between them and the IDF; they needed the IDF to ensure future security and the maintenance of the state, without which the state would cease to exist.

The concept of institution is valuable in deciphering the dynamics at play during the synagogue crisis. The elder rabbis were part of a respected institution in the settlement, and their word and actions probably held more weight than any other settlement constituent. Their status within the structure gave them the power to negotiate and establish informal rules that would be abided by. The protestors in the synagogue also had the power to continue with civil disobedience, resort to violence, or capitulate. What is interesting about the rabbis using their institutional position within the structure to quell violence is that it was Hacohen who requested their involvement.

Hacohen had the power to go into the synagogue with armed troops and violently remove the settlers—and the power to find a more amenable alternative. He turned to an outside institution (elder rabbis) when his institution's (military/government) laws and actions failed to end the protests. This decision served his real interest: to achieve an end to the disorder without creating conditions for a more violent civil uprising. Although Hacohen eloquently described a possible violent conflict when he spoke of a train bearing down on settlers, he realized that his real interest resided in exhausting all possible means to avoid violence. His pursuit of this course allowed him to balance his internal narrative(s) as a son of settlers and a soldier of Israel.

Using the framework to examine Event #3 allows us to explore the role institutions can play on increasing and decreasing violence. It allows us to further explore the power dynamics and interests at play during the synagogue standoff, and, most importantly, it allows us to better appreciate the interdependence between the protesters and the IDF. Although the tension between the two groups was volatile, the protesters realized that preserving their individual group identity meant they could not sever the interdependence between the secular state and the Zionist cause. The Zionist cause is dependent on the strength of the secular state, and, in return, the secular state finds strength in the Zionist cause.

CONCLUSION

The Gush Katif case study demonstrates the value international relations concepts provide the practitioner so they can better understand the operational environment's complexity. It can be used in conjunction with other doctrinal concepts such as PMESII-PT and ASCOPE to assist in design methodology and/or the planning process. The socio-political framework provides practitioners with greater depth to explore complexity within the operational environment. The concepts within the framework serve as launching points to stimulate discussions within planning groups. The practitioner can use the framework during "white-board sessions" to explore socio-political dynamics present at all levels within the environment. These nine concepts can serve as a starting point to initiate discussion in pursuit of understanding. This framework becomes even more powerful if it is supported by an inquisitive mind that seeks out a diverse education to enable greater exploration of socio-political dynamics.

About the Author

Dr. David P. Oakley serves as the Academic Director at the University of South Florida's Global and National Security Institute (GNSI). He served in the U.S. Army as a field artillery officer and strategist, in the Central Intelligence Agency's National Clandestine Service/ Directorate of Operations as a Staff Operations Officer, and as a contractor at the U.S. National Counterterrorism Center. His last assignment in the U.S. Army was as an assistant professor at National Defense University's College of International Security Affairs in Washington, D.C. Prior to joining GNSI, he was an associate professor at Joint Special Operations University in Tampa.

NOTES

1. This chapter is a revised and shortened version of a paper that was written for the Army's Command and General Staff College's Local Dynamics of War Scholars Program in 2012. Although a decade has passed, the military's recurring call for greater critical and creative thinking makes this paper relevant despite the time that has elapsed.

2. For example, constructivism is based on some sociology principles; realism and constructivism have a psychological bent within their paradigm, political philosophy has influenced all three, and history is regularly used by IR theorists to apply their theories. Also, a review of diplomatic history provides a rich understanding of how these different theories shaped each other and help create an American narrative. Constructivism's discussion on narratives is also related to anthropology.

3. James Der Derian, *Critical Practices of International Theory Selected Essays* (New York, NY: Routledge, 2008). Der Derian takes an interdisciplinary approach mixing international relations with other disciplines and theories (such as critical theory) to gain a better appreciation of the world's complexity.

4. Sidney Verba, *Designing Social Inquiry: Scientific Inference in Qualitative Research* (Princeton, NJ: Princeton University Press, 1994), Kindle-106.

5. David E. McNabb, Research Methods for Political Science: Quantitative and Qualitative Approaches (Armonk: M.E. Sharpe Inc., 2010), 48.

6. Barkin, 6.

7. Ibid.

8. Practitioners should appreciate that IR scholarship is much broader than these three schools and the author encourages practitioners to go beyond these schools to better appreciate the operational environment. Although scholarship from these three schools is valuable, there is much more scholarship of value in IR.

9. Graham Allison and Philip Zelikow, *Essence of Decision: Explaining the Cuban Missile Crisis* (Boston, MA: Longman Publishing, 1999), Kindle-106.

10. David Leopold and Mark Stears, eds., "Political Theory and History," *Political Theory: Methods and Approaches* (Oxford: Oxford University Press, 2008), 128.

11. Stephen Walt, "International Relations: One World, Many Theories," *Foreign Policy* no. 110 (1998): 32.

12. Walt. Norman Angell's 1910 book *The Great Illusion* (Seattle, WA: Kindle Digital Services, 2012), 103, argued that intertwined European economic interests meant war would not occur.

13. Friedberg, 12.

14. Democratic Peace is a descendent of Immanuel Kant's Perpetual Peace where he lays out a blueprint for peace among states. See http://www.mtholyoke.edu/acad/ intrel/kant/ka nt1.htm). Sociologist Dean Babst wrote about democratic peace in a 1964 article for the Wisconsin Sociologist in which he argues that elected governments are less likely to go to war with each other because of pressures from the electorate. See http://miles.oppidi.ne t/Babst1964.pdf). See the book by Michael E. Brown, Sean M. Lynn-Jones, and Steven E. Miller, eds. *Debating the Democratic Peace* (Cambridge: MIT Press, 2001) for more information on the ongoing debate.

15. Woodrow Wilson's Fourteen Points was based off both these premises. He believed colonialism should end and nations should have self-determination because it was morally the right thing, but he also believed that by doing so the world would avoid wars in the future.

16. Walt, 32.

17. Robert Keohane and Joseph S. Nye, *Power and Interdependence* (New York: Longman, 2011), 5.

18. Ibid., 225. Keohane and Nye define globalism "as a state of the world involving networks of interdependence at multicontinental distances, linked through flows and influences of capital and goods, information and ideas, people and force, as well as environmentally and biologically relevant substances (such as acid rain or pathogens)."

19. Craig Parsons, *How to Map Arguments in Political Science* (Oxford: Oxford University Press, 2007), Kindle-497-498. For the purposes of this paper, I use Parson's definition of structure "as a sort of obstacle course that is treated analytically as an intersubjectively present, given environment." I interpret this to mean that structures are relatively constant and difficult to change in the short-term.

20. Walter A. McDougall, *Promised Land, Crusader State: The American Encounter with the World since 1776* (Boston, MA: Hougton Mifflin Co., 1997), 120. Although sometimes looked upon as distinct schools of thought, progressivism and liberalism share many similarities to include that the world can progress to a better state of affairs. According to McDougall, "Herbert Croly, founder of the New Republic magazine, defined progressive foreign policy as the pursuit of a perfected American system of states." The belief in changing the world in America's image for the better is a deeply held belief among many throughout American history.

21. Terry O'Callaghan and Martin Griffiths, *International Relations: The Key Concepts* (London: Routledge, 2002), 157. Realism argues that nations pursue self- interest first and foremost. Although they might enter into agreements with other nations, these are temporary and would be violated if one of the nations believed it was no longer in their interest to remain in the alliance.

22. O'Callaghan and Griffiths, 180.

23. Ibid., 181. Realists would argue that as the highest sovereign in the system, nations can largely ignore these rules and institutions if they aren't in their interest, if the reward for ignoring outweighs the cost, and if they are powerful enough.

24. Parsons, 823-828. I take the concept of unintentional consequences and the majority of my Institutions definition from Parsons. FM 3-24 also speaks about institutions but does not speak about the unintentional consequences aspect.

25. Early realist writers argued that liberalism's progress toward perfection was unachievable and could have disastrous repercussions. Reinhold Niebuhr's famous book, *Moral Man and Immoral Society: A Study in Ethics and Politics* (New York, NY: Public Affairs, 2004), argued against the progressive notion that mankind was on an upward trajectory toward lasting peace.

26. Walt, 31.

27. Ibid.

28. Ibid. Neo-classical realism is a developed theory with many nuanced elements.
Within the theory there are defensive realists, offensive realists, among others. The important distinction for our purposes is between the realists who base their theory on human nature/agency (classical) and the realists who base their theories on structure (neo- classical).

29. This is a major concept of Niebuhr's in his book *Moral Man and Immoral Society.*

30. Anarchy is a realist concept that argues there is no sovereign over states and this leads to nations pursuing their own self-interest. In Kenneth Waltz, *Man, the State, and War: A Theoretical Analysis* (New York, NY: Columbia University Press, 1954), 160, he argues "In anarchy there is no automatic harmony" he continues "A state will use force to attain its goals if, after assessing the prospects for success, it values those goals more than it values the pleasures of peace. Because each state is the final judge of its own cause, any state may at any time use force to implement its policies. Because any state may at any time use force, all states must constantly be ready either to counter force with force or to pay the cost of weakness." Liberalism and constructivism do not necessarily deny anarchy but believe that anarchy is "what we make of it" (constructivism) or that institutions and structures can be put into place to ensure anarchy is dealt with in a peaceful and common interest (liberalism).

31. Parsons, 796. FM 3-24 discusses social structures in a similar fashion to the way I discuss them above.

32. Liberalism does not deny self-interest but believes that mutual self-interest can be developed to ensure conflict will not rise again. This mutual self-interest can also be enforced through international institutions and laws. A constructivist would argue that interests are socially constructed and trying to determine what exactly one nation would identify as an "interest" is difficult.

33. Kenneth Waltz, *Theory of International Politics* (New York, NY: McGraw Hill Publishing, 1979), 118.

34. Jeffrey C. Isaac, "Beyond the Three Face of Power: A Realist Critique," *Polity* 20, no. 1 (Autumn 1987): 25-26.

35. Ibid., 26.

36. FM 3-24 discusses interests but focuses more on soldiers understanding population's interests and does not specify different types of interests involved.

37. Isaac, 22.

38. Ibid., 5.

39. O'Callaghan and Griffiths, 253.

40. Isaac, 23.

41. FM 3-24 discusses power, but only in a narrow "power over" context.

42. Robert Jackson and Georg Sorenson, *Introduction to International Relations: Theories and Approaches* (Oxford: Oxford University Press, 2010), 166-168.

43. J. Martin Rochester, Fundamental Principals of International Relations (Boulder, CO: Westview Press), Kindle-630.

44. Chris Ferrero, "Constructivism and US-Iran Relations," http://www.us-iran-relations.com/index.php?p=1_14_Constructivism-and-US-Iran-Relations (accessed 24
April 2012).

45. Rogers M. Smith, *Stories of Peoplehood: The Politics and Morals of Political Membership* (Cambridge, MA: Cambridge University Press, 2003), 9. I believe that Smith's book and constructivist theory both speak about how socially constructed narratives shape and influence people and governments. These narratives are important because (like Smith says) they are how we see ourselves and help shape the lens through which we view others.

46. Ibid. "Stories of peoplehood" is taken from the title of Rogers Smith's book. FM 3-24 mentions narratives and describes them as "means through which ideologies are expressed and absorbed by members of a society." The narrative description above encompasses this but also speaks about individual narratives and counter-narratives. I am concerned with how narratives are framed and reframed through social interaction.

47. O'Callaghan and Griffiths, 51.

48. Ferrero.

49. Realists might argue that a state (can be applied to person or group) changes their alliances because it suits their interests within the structure and it allows them to garner more power.

50. Smith, 212.

51. Ibid.

52. Monica Garnsey, *Gaza: The Fight for Israel*, Documentary, Raw T.V., 2005. The documentary provided all the insight into the event, unless noted otherwise. You can access the documentary at http://www.youtube.com/watch?v=ybRMfwkS-kk (accessed 29 April 2012).

53. Elisha Efrat, *The West Bank and Gaza Strip* (London: Taylor and Francis, 2007), Kindle-3733. Photo in Figure 2.1 from the *CIA World Factbook*, located at https://www.cia.gov/the-world-factbook/countries/gaza-strip/map.

54. Michael Oren, *Six Days of War*, (New York: Rosetta Books, 2004); Shlomo Gazit, *Trapped Fools: Thirty Years of Israeli Policy in the Territories* (Portland, OR: Frank Cass Publishers, 2003), Kindle-638.

55. Gazit, 1516, 1634.

56. Efrat, 3984. Efrat writes, "despite the high importance which right-wing politicians attached to the Qatif block, that regions were characterized by features which it really did not have: it had no Jewish history behind it, and it was not a parsimony, even most Israelis had no special emotional relationship with this corner."

57. Israel's Ministry of Foreign Affairs, "Israel's Disengagement Plan: Renewing the Peace Process," 2005, http://www.mfa.gov.il/NR/rdonlyres/23EFC707-AEBA-4195- BB90-B6BA8AB616FF/0/disengagement2.pdf (accessed 29 April 2012), 13.

58. Efrat, 4093.

59. Israel's Ministry of Foreign Affairs, 5.

60. Efrat, 4302.

61. Garnsey.

62. Ibid.

63. Gaza: *The Fight for Israel*, circa 1:10 min.

64. Gaza: *The Fight for Israel*, circa 3:06 min.

65. Gaza: *The Fight for Israel*, circa 6:50 min.

66. Smith, 212.

67. Robert B. Strassler, ed., *The Landmark Thucydides: A Comprehensive Guide to the Peloponnesian Wars* (New York, NY: Touchstone, 1996), 352.

68. Hacohen acknowledged the protestors' psychological need to maintain their identity of fighting against removal and perceived injustices. Hacohen said, "what is important is the story of the struggle and not the results of the struggle...a way which a human being can keep his obligations without losing his honor."

III. Building Resistance in Georgia

By Brian Petit

ABSTRACT: This section describes applying the principles of national resistance in Georgia within the context of world events in 2023.[1] First, a brief history of Georgia is given, including the recent role of U.S. security assistance. The second half of this section describes how Georgia applied the ideas within the *Resistance Operating Concept*[2] to fashion a national resilience and resistance concept. Next, inhibitors to implementation are described. Finally, the implications for U.S. Special Operations Forces working in this Black Sea region are explored.

INTRODUCTION

Can a tiny country, fully surrounded and partially occupied by Russia, conceptualize, organize, and build a citizen-based national resistance plan? Surprisingly, yes. The country of Georgia has done exactly this. Georgia's achievement, however, is incomplete. Georgia's thorough resilience and resistance strategy remains a *concept* and not a *law*. Without a legislative and legal foundation, the Georgian national resistance approach lacks authoritative status. In this current form, is Georgian national resistance a viable, if unofficial, collective? Or is it an idea stuck and stymied in committee?

EXECUTIVE SUMMARY

- Georgian resistance development shows the power and limitations of the bottom-up model.
- The *Resistance Operating Concept (ROC)* stimulated and galvanized Georgian thinking.
- Georgian Special Operations Forces (SOF) are central to Georgia's national resistance plans.
- Georgia exhibits excellent interagency cooperation on resilience and resistance.
- A concept authorizes and allow agencies to act (build, plan), but with limits.
- A law grants agencies the national authority to implement, compel, and synchronize.
- U.S. SOF must address challenges to improve aid and support to the Georgian resistance.

GEORGIAN HISTORY

Georgia, population four million, is roughly the size of West Virginia.[3] Eighty six percent of its inhabitants are ethnic Georgians.[4] To its north are the towering Caucasus Mountains, where Georgia shares a 556-mile border with Russia. With peaks over 17,000 feet high, the Caucasus Range forms a natural border from the Russian sub-states of Dagestan and Chechnya.[5] Georgia's western border is the Black Sea, fully controlled by the Russian Black Sea fleet. To its south, Georgia shares a border with Turkey (a NATO nation), and Armenia, a client state of Russia. To its southwest is Azerbaijan. It is an understatement to say that this neighborhood is complicated (see Figure 2.3).[6]

Figure 2.3: Map of Georgia. Source: CIA World Factbook, located at www.cia.gov/the-world-factbook..

Georgia's modern history begins in 1919. With the collapse of Czarist Russia in 1918, Vladimir Lenin and the communist Bolsheviks seized power and fought a series of regional wars to retain the Russian hold on the former empire.[7] The Baltics, Poland, and Finland, for example, defeated the Red Army and gained their independence. Georgia, like Ukraine, did not suitably organize and fight, and thus became subjects (republics) within the Union of Soviet Socialist Republic (USSR) in 1922.[8] Upon the collapse of the USSR in 1991, Georgia declared and gained its independence; there was no peace dividend. Georgia fought a series of wars in the early- and mid-1990s with aggravated populations in South Ossetia and Abkhazia as part of broader Georgian civil unrest.[9] Newly independent and rife with infighting, Georgia languished in the 1990s with Russian military forces remaining inside Georgia, both in the contested regions (Abkhazia, South Ossetia), near the capital, Tbilisi, and in coastal Batumi. On its borders, Russian forces never left Armenia, leaving Georgia with Russian forces on its north, south, and west.

To exacerbate this precarious situation, Russia invaded Georgia in August 2008, waging a five-day war against the overwhelmed security forces of Georgia.[10] Strategically, this invasion was a spoiling attack to prevent Georgia from joining NATO and the European Union (EU) and the signal of a new, muscular Russian foreign policy in its near abroad. Russian forces remained in the regions of South Ossetia and Abkhazia,[11] with Georgians citing that 20 percent of their sovereign territory is under occupation.[12] This is where Georgia remains today. Russian occupation forces secure and support the Russian-installed, breakaway governments of Abkhazia and South Ossetia.[13]

SECURITY ASSISTANCE HISTORY

Georgia's current national defense is underpinned by decades of western security assistance. In 2002, in the wake of 9-11, the U.S. initiated the Georgian Train and Equip Program (GTEP).[14] Intent on joining NATO and gaining favor with the west, Georgia was the largest troop contributing nation to Iraq[15] for many years, with nearly a full brigade deployed until the Russian invasion in 2008 reduced those numbers.[16] The latest NATO effort, initiated in 2014 and titled the "Significant NATO-Georgia Package (SNGP),"[17] continues, along with other bilateral efforts, to assist the Georgians with defense reform.[18] For their part, the Georgians have grown a functioning democracy, purged their security sector of Soviet-era leaders, and are educating and training to achieve NATO interoperability. U.S. Army Special Operations Forces (ARSOF) episodically engaged inside Georgia for many years and shifted to a longer dwell "heel-to-toe" presence in January 2020.[19]

BUILDING RESISTANCE

U.S. Special Operations Command Europe (SOCEUR) sponsored a one-week JSOU[20] *National Resistance Course (NRC)* in Tbilisi in October 2020.[21] The Georgians were surprisingly new to the concept of combining whole-of-government with whole-of-society resistance. This newness did not last. The Georgian Special Operations Forces (GSOF), a skilled and well-led regiment, embraced the intellectual and organizational idea of building a Georgian "porcupine" comprehensive defense model. Aided by the *Resistance Operating Concept* and other resistance material translated into the Georgian language, they internalized and localized these ideas in a Georgian context. This was a "bottom-up" process with GSOF as the catalyst and the full participation of the ministries.[22] In a short period, and in the midst of the COVID-19 pandemic, Georgia went from a late adopter to a creative innovator.

Under the guiding hand of the Ministry of Defense, the Georgians developed a national resistance concept.[23] This concept later blossomed into a broader national resilience concept, one that includes all aspects of Georgian society: national guard, first responders, educators, mayors, border control, citizen self-defense groups, and all ministries of the government. The development of this national resilience concept and its fusion into existing strategies, plans, defense reforms, and ministerial roadmaps was initiated in autumn 2020. Two years later, after a series of resistance-focused forums, a Georgian inter-ministry body produced a cohesive, well-socialized resilience and resistance plan.[24] The Georgian national resilience and resistance concept, as of this writing, has not been codified into law. Thus, this is still an ongoing process with a public-facing side and a more discrete policy and plans component that is not detailed here.

GEORGIAN ORGANIZING PRINCIPLES

The Georgian resilience and resistance concept, as understood by the author, is captured below from the unclassified, releasable portions.[25]

- National resistance is organized and led by the legitimate Georgian government.
- Resistance will delay and degrade an enemy invasion and prevent consolidation of control.
- Resistance is a whole-of-government and whole-of-society system.
- Resistance must be organized, authorized, and built pre-crisis to be effective.
- Certain resistance infrastructure should remain compartmented and protected.
- All resilience infrastructure should function with "unity of effort" whether managing a humanitarian crisis, natural disaster, or an enemy invasion.
- All Georgian ministries will support a national resilience concept with an "annex-type" planning document that details out their tasks, requirements, and shortfalls.
- Lead agencies, command-and-control, and key responsibilities: this delineation flows from concepts, policies, ministerial directives, and authoritative laws / legislation.

INHIBITORS TO FULL IMPLEMENTATION

The Georgians have not elevated their national resilience and resistance concept to a law for three reasons. First, the political and legislative adoption of such a measure is considered by some to be provocative and openly anti-Russian. This is not explicitly stated as much as inferred and observed by both Georgians and western observers.[26]

Second, the Georgians are operating a hedging strategy. Politically and economically, Georgia is wedged between a heavy-handed Russia and an open, yet distant, west. Georgia has a mixed score sheet in tying itself to western alliances or markets.[27] Georgia's desire for NATO and EU membership, two decades running, remains more aspirational than probable. Thus, Georgia balances relationships with both Russia and the west, satisfying neither party and suspending itself precariously on this east–west axis.[28] A national resilience and resistance law, it could be argued, disrupts this balance.[29]

Third, the Russia–Ukraine war has proven politically disadvantageous for Georgia.[30] Ukraine–Georgia relations have deteriorated since the Russian invasion of Ukraine in February 2022.[31] At a time when western nations continue to show solidarity for Ukraine and against Russia, the Georgia government—though not the Georgian people—are largely absent from this bloc.[32] Such fractures, internally and internationally, impede the progress of advanced national security concepts and laws such as citizen-centric resistance schemes. Currently, there is no evident political will to sponsor or advance whole-of-society resilience and resistance measures.[33]

IMPLICATION FOR U.S. SPECIAL OPERATIONS FORCES

The current state of Georgian security presents the U.S. Special Operations Forces (SOF) with three challenges and three opportunities.

The first challenge for SOF is to deploy special operations teams to Georgia that are well-versed on the theory, history, and tactical application of *national resistance against a foreign occupier*. This is a difficult task that only gets harder with time. To advise Georgians, in Georgia, against a multi-domain Russian threat, requires skilled and smart operators prepared with curated and tailored education and training for their mission. Showing up with a "good attitude and a good shot" is necessary but not sufficient. The SOF promise of small teams creating outsized impacts requires that Georgians are met on their level and that U.S. teams can perform qualified advisory work both at and above the small-unit level.[34] U.S. SOF must come prepared—on the first day of their deployment, and not their last—to improve Georgian special operations and putative resistance networks, both practically and intellectually.

| U.S. SOF must come prepared—on the first day of their deployment, and not their last—to improve Georgian special operations and putative resistance networks, both practically and intellectually. |

The second challenge for SOF is to devote sufficient resources to Georgian SOF and associated resistance elements so that they may achieve their NATO interoperability goals and to fulfill their resistance writ. This is difficult given that SOCEUR forces are in high demand. SOCEUR must (1) meet the needs of U.S. forces in Europe and the demands of NATO, writ large; (2) respond to the most-threatened NATO nations in East Europe and the Black Sea that border Russia; and (3) meet emerging requirements levied by the NATO-accession of Finland and Sweden. If Georgia slips down the priority list, it will reduce the impact of U.S. SOF as an element of integrated deterrence.

The third challenge for SOF is to navigate the shifting tides of policy, politics, and great power moves. As mentioned above, Georgia is operating a balancing strategy in which Russian and western interests are both courted and frustrated. In this geopolitical environment, U.S. Special Operations are subject to the policy shifts and political forces that can disrupt or suspend persistent engagement strategies.

There are also three opportunities for U.S. Special Operations Forces (SOF) in Georgia. The first opportunity is well understood and is suitably underway. Even without a codified, legal national resistance plan, the Georgian Ministry of Defense and the Georgian special operations forces have the authority to develop concepts, plans, force structures, and capabilities that support resistance against invasion and occupation. U.S. SOF must recognize that this window to build, advise, and assist is open, at least for now. If the Russian invasion of Ukraine taught the west one lesson, it is that these windows are not open indefinitely, and that U.S. SOF, contributing NATO partners, and Georgian SOF must, together, move fast and hard toward capability development. The current planning assumption is that this window is open and will remain so; a better planning assumption is that this window will close, adding urgency and tighter timelines to the mission.

The second opportunity for SOF is to capitalize on the growing role of Georgian SOF as the resistance synchronizer. Georgian SOF inhabit this role informally and without compelling authority. But without question, Georgian SOF have proven in public and private forums to be the acknowledged leaders on resistance development. In the same manner that U.S. SOF is a force integrator and a knowledge broker, Georgian SOF have built a reputation in the Georgian interagency as a reliable brain trust of resistance thinking and concepts integration.

The third opportunity for SOF is to learn from Georgians on their experiences in contesting Russia in both hot and cold wars. Georgians have fought Russians in combined arms, high-intensity conflict; in all manner of political warfare to include misinformation, propaganda, subversion, and sabotage; and are currently contending with an estimated 5,000 Russian occupiers inside of Georgian territory.[35] Another credible, if unofficial source, is the Georgian Legion, a volunteer group of Georgians fighting Russians in Ukraine.[36] It is claimed that Georgians are, per capita, the largest group of volunteer fighters in Ukraine.[37] As U.S. SOF seek to educate themselves on Russian methods, the Georgians continue to be experienced and well-informed partners.

CONCLUSION

Given this situation, can we ascertain if the Georgian national resilience and resistance is a viable, functional collective of plans and propositions? Or is it a concept frustrated in committee, destined to crumple under pressure? We will explore this question in the JSOU Support to Resistance and Resilience (SRR) lessons on national resistance and resilience methods. This section is a primer to investigate this question for Georgia and for similar partners facing Russian aggression. This investigation will further aid in understanding how external supporters, such as U.S. SOF, can assist their partners in crafting regionally appropriate resilience and resistance systems.

About the Author

Brian Petit serves as the CEO of Garfield Syndicate, LLC. He is a former U.S. Army Special Forces Colonel (Green Beret) with 25 years of service in more than 30 countries. He also serves as adjunct faculty at Joint Special Operations University.

NOTES

1. Author visits to Georgia and work with Georgian government officials and U.S. advisors, multiple trips, from October 2020 to February 2022.

2. Special Operations Command Europe and the Swedish Defence Academy, author credit to Otto Fiala, *Resistance Operating Concept*, 2020, published by JSOU Press and the Swedish Defence Academy.

3. CIA factbook. https://www.cia.gov/the-world-factbook/countries/georgia/#geography

4. CIA factbook. https://www.cia.gov/the-world-factbook/countries/georgia/#people-and-society

5. World Atlas. https://www.worldatlas.com/mountains/caucasus-mountains.html

6. Map of Georgia from CIA World Factbook, located at www.cia.gov/the-world-factbook.

7. Victor Sebestyen, *Lenin: The Man, The Dictator, and the Master of Terror*, (New York: Pantheon Books, 2017).

8. Anne Applebaum, *Red Famine: Stalin's War on Ukraine*, (New York: Anchor Books, 2017), 49-95.

9. Donald Rayfield, *Edge of Empires: A History of Georgia*, (London: Reaktion Books, 2012).

10. Ronald D. Asmus, *A Little War that Shook the World: Georgia, Russia, and the Future of the West*, (New York: Palgrave, 2010).

11. Congressional Research Service (CRS), "Georgia: Background and U.S. Policy," updated July 3, 2023, 19, https://crsreports.congress.gov/product/pdf/r/r45307

12. Natia Seskuria, "Russia's 'Silent' Occupation and Georgia's Territorial Integrity," Royal United Services Institute (RUSI), April 18, 2019, https://www.rusi.org/explore-our-research/publications/commentary/russias-silent-occupation-and-georgias-territorial-integrity

13. International Crisis Group (ICG) Georgia, November 15, 2023, https://www.crisisgroup.org/europe-central asia/caucasus/georgia

14. Congressional Research Service (CRS), 19-21. Also see, Civil.ge, "Train and Equip Officially Launched," May 27, 2002, https://civil.ge/archives/101745.

15. Robert E. Hamilton, "Taking Stock of U.S. Military Assistance in Georgia," Foreign Policy Research Institute (FPRI), December 21, 2021, fpri.org, https://www.fpri.org/article/2021/12/taking-stock-of-u-s-military-assistance-in-georgia/

16. Ronald Asmus, 173-175.

17. Ministry of Defense of Georgia, Substantial NATO-Georgia Package (SNGP), website (mod.gov.ge) accessed November 15, 2023, https://mod.gov.ge/en/page/65/substantial-package

18. CRS, 15-21.

19. Special Operations Command Europe (SOCEUR) is the joint command that deploys Army Special Operations Forces. https://www.socom.mil/soceur

20. Joint Special Operations University (JSOU), jsou.edu

21. The author is the director of the JSOU-sponsored National Resistance Course (NRC).

22. The author participated in this process, with both Georgian SOF and Georgian inter-ministry participants.

23. This concept is not a published product; it remains a working document with various versions.

24. This resilience concept, like the resistance concept, remains a working document, sponsored by the Ministry of Defense, and is not releasable as a finished or approved concept.

25. The author description is from the portions released or made public by involved Georgian representatives.

26. CBS, 60 Minutes, Sharyn Alfonsi, "A Quiet Invasion," October 26, 2023, 13:26, https://www.cbsnews.com/video/a-quiet-invasion-sunday-on-60-minutes/; and author travels and research in Georgia, 2020-22.

27. Sasha Stone, Sandro Kevkhishvili, Alexander Kupatadze, et al, Center for European Policy Analysis (CEPA), Report on "Reform and Resistance: Georgia's Path to EU Candidacy," April 26, 2023, cepa.org, https://cepa.org/comprehensive-reports/reform-and-resistance-georgias-path-to-eu-candidacy/

28. Ivan Nechepurenko, "Georgia drops a draft 'foreign agents' law that set off mass protests over parallels to Russia," New York Times, March 10, 2023, https://www.nytimes.com/2023/03/10/world/europe/georgia-foreign-agents-protests.html?searchResultPosition=1

29. Ivan Nechpurenko, "Lavrov accuses NATO of pushing Georgia and Moldova away from Moscow," New York Times, February 2, 2023, https://www.nytimes.com/live/2023/02/02/world/russia-ukraine-news?searchResultPosition=8#lavrov-accuses-nato-of-pushing-georgia-and-moldova-away-from-moscow

30. CRS, 2-18.

31. Francis Fukuyama and Nino Evgenidze, "Russia Is Winning in Georgia: America Needs to Get Tough on Tbilisi," Foreign Affairs, April 6, 2023, https://www.foreignaffairs.com/georgia/russia-united-states-winning-georgia

32. Amy MacKinnon and Jack Detsch, "Georgia Walks a Fine Line After Ukraine Invasion," Foreign Policy, March 11, 2022, https://foreignpolicy.com/2022/03/11/georgia-ukraine-invasion-russia/

33. Arshan Barzani, "Who is Running Georgia?," Politico (politico.eu), April 26, 2023, https://www.politico.eu/article/who-is-running-georgia-tbilisi-ukraine-russia-war-vladimir-putin-sanctions-georgian-dream/

34. Hamilton, FPRI.

35. European Union Monitoring Mission – Georgia (EUMM – Georgia), eumm.eu, accessed November 15, 2023.

36. John Sparks, "Ukraine war: Meet the Georgian Legion joining the fight against Russia's invasion," Sky News, news.sky.com, November 7, 2022, https://news.sky.com/story/ukraine-war-meet-the-georgian-legion-joining-the-fight-against-russias-invasion-12737984

37. Kacper Rekawek, "Foreign Fighters in Ukraine: The Brown-Red Cocktail," Routledge Press, December 30, 2022.

IV. Measuring and Quantifying State Fragility: Why Governments Lose Internal Conflicts and what that Means for Counterinsurgency

By Chris Mason

ABSTRACT: Despite what U.S. Joint doctrine states, counterinsurgency does not increase the legitimacy of, or support for, central governments engaged in internal conflicts. Recent research shows quantifiable degrees of government legitimacy, national identity, and population security are necessary precursors and accurate predictors of a government's ability to outlast a civil uprising. Because the first two predictors—government legitimacy and national identity—can be measured and do not increase during a conflict, the probability of government failure in most cases can be accurately predicted when the conflict starts. Some of these research findings have been discussed previously in *Parameters* journal.[1]

INTRODUCTION

In 2016, the Study of Internal Conflict (SOIC) research program was launched at the Strategic Studies Institute of the U.S. Army War College. The U.S. nation-building effort in Iraq had already collapsed, and the war in Afghanistan was going badly. The purposes of the study were to determine the underlying reasons for state failure in an internal conflict and to solve the paradox of how it was possible to win every tactical engagement in fighting insurgents in places like Vietnam, Iraq, and Afghanistan, yet still fail to achieve state security at the strategic level of war.

The idea behind the study was to set aside the contested issues of counterinsurgency theory and doctrine and to apply statistical analysis to the problem of state failure. Why did most countries fail in combating internal rebellions, while some succeeded? What were the commonalities of success and failure? The research findings developed after several years into a tool for analyzing and predicting outcomes of insurgencies and civil wars—a rubric that senior leaders could use in considering U.S. involvement on one side or the other of any given internal conflict.

CONFLICTS OF STUDY

The first step was to create a list of all internal conflicts since 1945 to be researched and analyzed. The end of World War II was chosen as the starting point for the conflict list because of the overwhelming extent to which World War II has shaped the modern world geographically, politically, militarily, and socially. The postwar world, as it came to be known, is very different from the era before it in almost every way, and the parameters of human behavior and the limits on military actions it established affect almost every dimension of internal conflict. There have been exceptions, such as Saddam Hussein using poison gas on the Kurdish people in Iraq in Hallabja in 1988, for example, but many tactics that might be pursued by a government engaged in an internal rebellion before World War II are, by and large, not possible in the postwar world. To give an example, some of the tactics used by the U.S. in suppressing the Moro Rebellion in the Philippine Islands (1899–1913) could not be used today, so a study of those particular tactics would not get us very far. Fortunately, a number of excellent databases of human conflict are available to researchers, and the SOIC program used the widely referenced conflict list maintained by the Correlates of War group.

Some limits had to be placed on the conflict list because the number of conflicts around the world since World War II is quite large. The first and most obvious filter was "internal conflict vs. international conflict." Wars fought between two countries were excluded, as were wars of colonial liberation, as those conflicts (largely fought in the 1950s) were essentially a form of international conflict fought between the European colonial power and a rebel group seeking independence inside a colonial territory. Although wars of colonial liberation were fought inside one country (the colonial territory), and two groups of citizens of that territory often fought each other (some indigenous group usually found it beneficial to side with the colonial power), but at their conceptual heart, they were wars fought by a European country against a rebel group seeking independence from the European country. The conflicts the SOIC study is interested in are those fought primarily between two groups of citizens of the same country for political power. Outside powers often get involved in internal rebellions, such as the U.S. involvement in the Vietnam War from 1965–1975, for example, and these are included in the study. However, the war being fought in Ukraine as this chapter is written is not considered in the study, even though there are (almost entirely Russian-speaking) citizens of Ukraine fighting their fellow citizens of Ukraine. The war in Ukraine today is primarily an international war between Russia and Ukraine for control of Ukrainian territory.

There are a significant number of internal conflicts worldwide that are not waged by a group of citizens against the government of the country to achieve political power but are simply between two groups of citizens of the country who don't get along with each other. Sometimes these are local territorial disputes fought over local resources, and others are simply blood feuds, like the legendary war that occurred in the U.S. between the Hatfields and the McCoys, two extended families in West Virginia. So, another criteria for being on the SOIC research list (since we want to know why governments fail) is self-evidently that the government of the country had to be a party to the conflict.

This still left an unmanageably large number of internal conflicts to consider, so a filter for the intensity of the conflict was added to the SOIC criteria. Only conflicts since World War II large enough to cause the deaths of at least 1,000 people in any continuous 12-month period of the conflict were included. The Correlates of War database was a good choice for this determination, as it is very detailed and very well documented. The total casualty number includes all deaths attributable to the conflict in question: civilian, guerilla fighters, and government security forces. And of course, the conflict had to be over for at least a year, so a determination could be made on whether the government had won or lost the conflict. If the war is still ongoing, it's obviously not possible yet to assess the outcome.

Applying these criteria left a list of 53 conflicts to consider. This number has fluctuated slightly in the eight years of the study. Most conflicts are pretty clear-cut when evaluating the parameters listed above for inclusion in the study, but a few are ambiguous. A number of conflicts in Chad since 1945, for example, could be interpreted as internal conflicts (in other words, two groups of citizens of Chad fighting each other for political power in Chad), but on closer inspection might also be considered as international conflicts between Chad and Libya, as some of the combatants are members of nomadic or transnational groups that move across the international border with the seasons. So occasionally, upon review, a particular conflict will be moved to the study Annex and be removed from the overall database. In general, however, the list of conflicts has been stable since the beginning of the program.

Of course, the criteria used for inclusion in the study does eliminate some fairly well-known internal conflicts, such as the "Troubles" in Northern Ireland between 1968 and 1998, fought between Irish sympathizers and the British government. There was not a 12-month period during this conflict in which at least 1,000 people died. Because the war was fought in Great Britain, this conflict received a lot of prominent media attention, while conflicts in remote places around the world with far higher death tolls rarely capture world news coverage. (Some reconsideration is being given to including the conflict in Northern Ireland in the SOIC study, simply because it is so well known.)

With this list of internal conflicts to consider, the study sought to determine what factors were common to defeats and what factors were common to government victory—in other words, *correlation.* Correlation is not causation, but for the purposes of creating a predictive model for whether a government would win or lose, it doesn't have to be. For example, for decades, researchers found a consistently high correlation between cigarette smoking and lung cancer. People who smoke cigarettes are much more likely to develop lung cancer than those who don't. That's correlation. It was a far different scientific problem determining that smoking cigarettes *causes* lung cancer. That is much harder to do, and of course the companies that make and sell cigarettes fought hard against this determination. Ultimately, that causal linkage was proven, and warning labels were added to cigarette products. For the SOIC, we are not looking at *causation*, we are simply assessing *correlation.*

Another pratfall to be avoided is encapsulated in the old Latin logic fallacy of *post hoc ergo propter hoc*, or literally "after this, therefore because of this." If you have a cup of coffee at 8 a.m., and it rains at 11 a.m., you drinking that cup of coffee did not cause it to rain. Sequence is also not causation. In other words, the order in which things happen does not necessarily mean the things are related to each other. Frequently in discussions of counterinsurgency, a proponent of one line of action or another will argue that the counterinsurgents did x, and enemy activity fell off. Thus, they argue, doing x caused a reduction in enemy activity. But just like that cup of coffee didn't make it rain, x didn't necessarily cause the reduction in enemy activity: they simply happened in sequence. A classic example of this fallacy is the attribution of a reduction in insurgent activity in Iraq (known as the "Anbar Awakening") to be the result of specific U.S. counterinsurgency tactics in the area. Detailed analysis of the reduction has determined that this argument is *post hoc ergo propter hoc*—the reduction was a result of different factors unrelated to the U.S. tactics.

To give a sense of how the SOIC study was conducted, imagine if those same tactics were used in 53 different conflicts, with the local conditions measured before and after they were applied. One could then calculate correlation over a substantial study size (not just two or three conflicts) and calculate the percentage of cases in which use of a particular set of tactics correlated with a particular measurable outcome. If the correlation was, say, 90 percent (in this example, a hypothetical reduction in enemy activity occurring in 90 percent of all 53 cases where the tactics were employed), one could say there was a definite correlation between the two. There wouldn't be proven *causation*, but there would be shown *correlation*, and enough evidence to suggest a course of action. That is the approach the SOIC has taken for the past eight years.

To be sure, to validate the findings, the SOIC program had to define almost every term used in the study. If different researchers used different definitions of what constitutes "external sanctuary" for a rebel group, for example, then the data for that factor would become meaningless. For the most part, the study used off-the-shelf political science definitions from reliable academic sources rather than spend time creating its own definitions, which might later be contested by reviewers evaluating the findings. Those definitions are included at the end of this section.

FACTORS TO CONSIDER

The next question for the SOIC researchers was "what factors do we consider?" A lot of time was spent in this phase of the research, and a list of more than 60 factors was created. Essentially anything that might conceivably have an impact on the outcome of an internal conflict was added to the list: annual per capita income, infant mortality rates, literacy rates, type of terrain (i.e., jungle terrain, mountain terrain), prevailing religion, access to clean drinking water, type of government in power, and so on. Basically, any political, economic, military, social, or geographic factor that could be measured was put on the list. Then, case by case, researchers determined how often each of these factors correlated with government defeat. (The program began by looking for correlation with government *victory*, but the simple fact is that insurgents usually win, so correlation with victory wasn't deriving much useful data.) For several years, all 53 conflicts were assessed for the correlation of all 60+ impact factors with government defeat. The result was plotted on a scatter graph. See Figure 2.4.

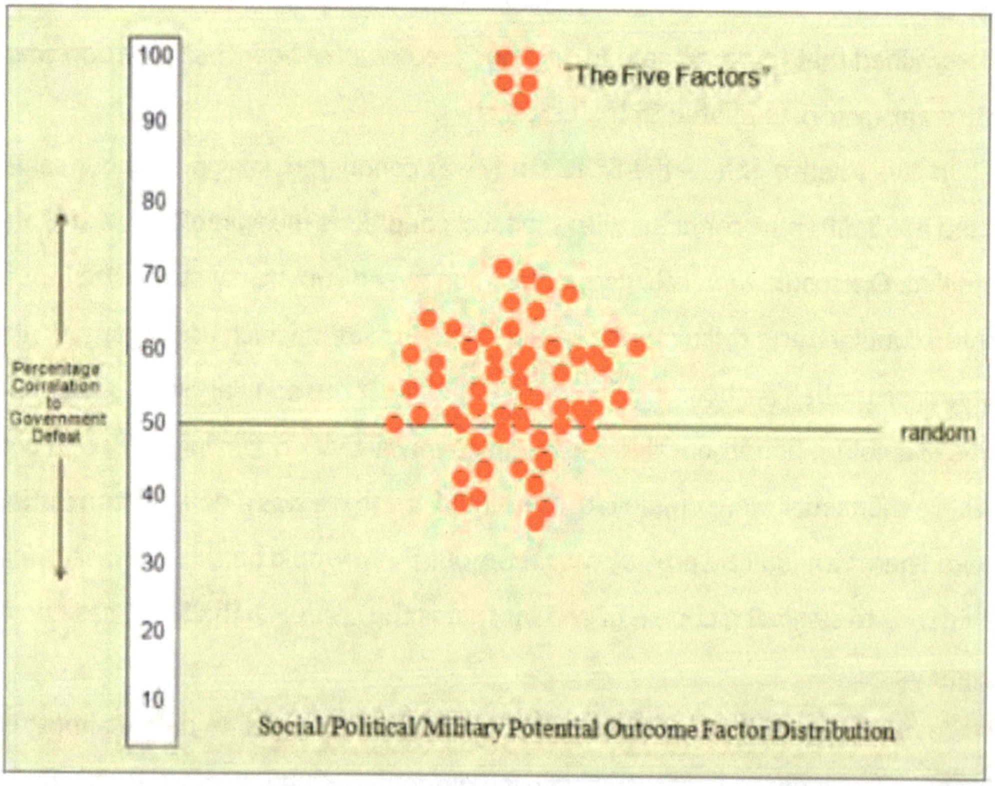

Figure 2.4. Social/Political/Military Potential Outcome Factor Distribution. Source: Author

Almost all 60 factors considered were essentially random in correlation, clustered around the 50 percent mark. (If a factor is present in half of government defeats and not present in half of government defeats, it is exactly random—in effect, a coin toss.) However, five of the assessed conditions on the list emerged as being present individually in more than 90 percent of all government defeats. If a factor was present in, for example, 96 percent of government defeats, this indicates that there is a 96 percent probability of a government losing a conflict when this factor is true. Likewise, four percent of governments waging a counterinsurgency when this factor is present could be expected to prevail against the rebels.

However, whenever more than one of these five factors was present, government failure was 100 percent. In other words, it is possible, but unlikely, for a government to defeat a rebel group if only one factor is against it, but not if more than one is present. The existence of more than one of the five factors in a conflict is invariably fatal for the government. That is, in none of the 53 cases studied did a government survive more than one adverse factor. Again, this is not causation—the SOIC does not assert that the presence of these factors causes government defeat, only that the correlation between these factors and government defeat is extremely high. For predictive purposes, this is sufficient.

THE FIVE FACTORS

So, what are these five "fatal factors"?

1. NATIONAL IDENTITY

In conflicts with less than 85 percent of the total population of a country identified at the national level (i.e., a country that was not a "nation" using the standard political science definition of the word), the government loses 94 percent of the time. In no case among the 53 conflicts studied did the percentage of the population that placed its primary identity at the national level ever increase during the period of the conflict. It has decreased due to poor government policies and actions affecting the civilian population, but it has never increased. Thus, "nation-building" in time of civil conflict is evidently impossible.

2. GOVERNMENT LEGITIMACY

In conflicts where the existing government was considered legitimate by less than 85 percent of the total population, the government lost 96 percent of the time. Critically, in none of the 53 conflicts examined did the percentage of the population that considered the government to be legitimate ever increase during the time period of the conflict. In a number of cases, the legitimacy percentage decreased during the conflict, but it never increased.

3: POPULATION PROTECTION

In only two cases among the 53 postwar conflicts studied (in which the existing government was able to protect less than 85 percent of its population from meaningful contact with insurgents) did the insurgency fail and the government remain in power. As little as 5 percent popular support may be necessary to keep a guerilla movement alive,[2] but the data show that 15 percent is the minimum necessary for victory.

4. EXTERNAL REBEL SANCTUARY

In no case among the 53 conflicts in which the guerillas had and maintained reliable, persistent external sanctuary across a land border into a neighboring country has any government been able to defeat them. External support (from the neighboring country or another country) is not necessary for this factor to pertain. External support, in the form of weapons, money, or other supplies, is often very valuable to an insurgency, but what is measured in this factor is the simple existence of a cross-border sanctuary. The precise definition of "external rebel sanctuary" can be found in the glossary at the end of this section.

5. A PRE-EXISTING ARMY OR SECURITY FORCE

In no case among the 53 conflicts assessed in which there was not a reasonably competent, sustainable security force (i.e., army, police, gendarmerie) in existence at the outbreak of violence has a government been able to form one after the start of the conflict and remain in power. This is a relatively minor factor, because almost every country on earth has a standing security force. However, in cases where there is not—the U.S. wars in Iraq and Afghanistan, for example, it is invariably fatal to the government.

Again, factors 1 and 2 (national identity and legitimacy of governance) have never been found to increase in percentage after an insurgency has started. This has profound ramifications for counterinsurgency doctrine, which in most cases is heavily predicated on measures to increase the government's legitimacy among its people. However, the data show that the best case scenario in counterinsurgency is actually maintaining the *existing* level of legitimacy of the government by avoiding measures that alienate the population. And when this percentage is below 85 percent at the start of the conflict, defeat is virtually inevitable.

A CASE STUDY OF THE FIVE FACTORS: THE VIETNAM WAR

A person wanting to know the probable outcome of a conflict at the beginning of (or during) the conflict can research these five factors and create a chart that will provide a mathematical probability of government defeat. Figure 2.5 illustrates the five factors in the case of South Vietnam.[3]

South Vietnam was ultimately overrun by North Vietnamese military forces in 1975. Scholars agree that South Vietnam never achieved a separate, distinct national identity, despite existing significant regional differences between north and south in terms of culture and personality and a strong strain of anticommunism in the southern population. There might have been clay there to mold into a separate South Vietnamese national identity, as there eventually became a separate South Korean identity. However, this did not happen in Vietnam. The communist north adroitly manipulated a higher, pan-Vietnamese level identity in all its messaging and narratives, often citing heroic historical events and characters. Thus, the question of whether South Vietnam achieved national identity would certainly be "no."

CASE: SOUTH VIETNAM	
NATIONAL IDENTITY	NO
GOVERNMENT LEGITIMACY	NO
POPULATION SECURITY	NO
EXISTING SECURITY FORCES	YES
EXTERNAL SANCTUARY	YES

Figure 2.5: The Five Factors in the Case of South Vietnam. Source: Author

Similarly, the government of South Vietnam likely never approached even 20 percent legitimacy among the people of South Vietnam—at least 80 percent of whom were rural subsistence farmers, craftsmen, or small merchants. The succession of governments in Saigon, which came and went during the war like a revolving door, were composed almost entirely of members of South Vietnam's educated, urban Catholic elites, who sneered at the rural Buddhist population as simple-minded peasants and went out of their way to alienate them with offensive anti-Buddhist policies. The question of government legitimacy in this case is also a "no."

Communist guerillas in the South, who were quasi-independent from the government in the North, were referred to by the epithet "Viet Cong." The North tried to keep them on a short leash, but there remained important differences in political outlook and ideas about Vietnam's future. A small number of Viet Cong were not communist at all, but rather nationalists trying to expel the latest foreign invasion of their country. U.S. strategy in the conflict was largely to try to bring communist forces in the South to decisive battle away from the populated areas, and the Viet Cong exploited this to infiltrate and inculcate the rural villages clustered along the coast. As a result, they largely controlled the rural population, and at no time did U.S. and South Vietnamese security forces ever come close to protecting 85 percent of the South Vietnamese people from contact with Viet Cong guerillas and political cadres. A realistic estimate would be around 40 percent at best, including the urban population of South Vietnam's major cities and towns. So, the question of population security is also a "no."

Regarding the fourth factor in Figure 2.5, existing security forces, an argument could be made both ways. South Vietnam did have an army, navy, and police force at the time U.S. ground forces arrived in 1965. They had been maintaining security, with decreasing success, since the Geneva Peace Accords partitioned the country into North and South in 1955 in what was meant to be a short interval before national elections (which never happened). The South Vietnamese army, the Army of the Republic of Vietnam (ARVN), was riddled with corruption, and unit commanders were chosen by President Diem based on their personal loyalty to him, not their military skills. As bad as it was overall, it still had some good units—the ARVN paratroopers were as good as any, and ARVN Marines and Rangers were good, reliable, and aggressive combat forces. However, the U.S. made several sweeping changes to the structure and composition of the ARVN in exchange for funding almost all of it—major restructuring efforts that repeatedly threw the entire force into confusion. So, in a sense, the U.S. tried to rebuild the engine while the car was driving. Both sides of the argument have a point. But on balance, given the SOIC definition of a" pre-existing, reasonably competent security force which answered to the civilian government," the answer to this question would be "yes."

The final factor, the existence of external sanctuary for the rebel forces, is a resounding "yes," one of the biggest ever in history and comparable to the Taliban's cross-border sanctuary in Pakistan from 2001 to 2021. The North Vietnamese Army, the NVA, and the Viet Cong guerillas of the South had vast sanctuaries in Laos and Cambodia (as shown in Figure 2.6) along thousands of miles of roads and trails from the North into the South—a supply web known as the "Ho Chi Minh Trail." The U.S. dropped twice as many tons of bombs on the Ho Chi Minh Trail as it dropped on Germany and Japan during all of World War II, with virtually no effect at all. And except for two large, short-term (and very controversial) ground incursions into Cambodia and Laos during the war, covert U.S. operations into those countries were limited almost exclusively to intelligence-gathering patrols for targeting in the strategic bombing campaign. So this factor would definitely be against the South Vietnamese government.

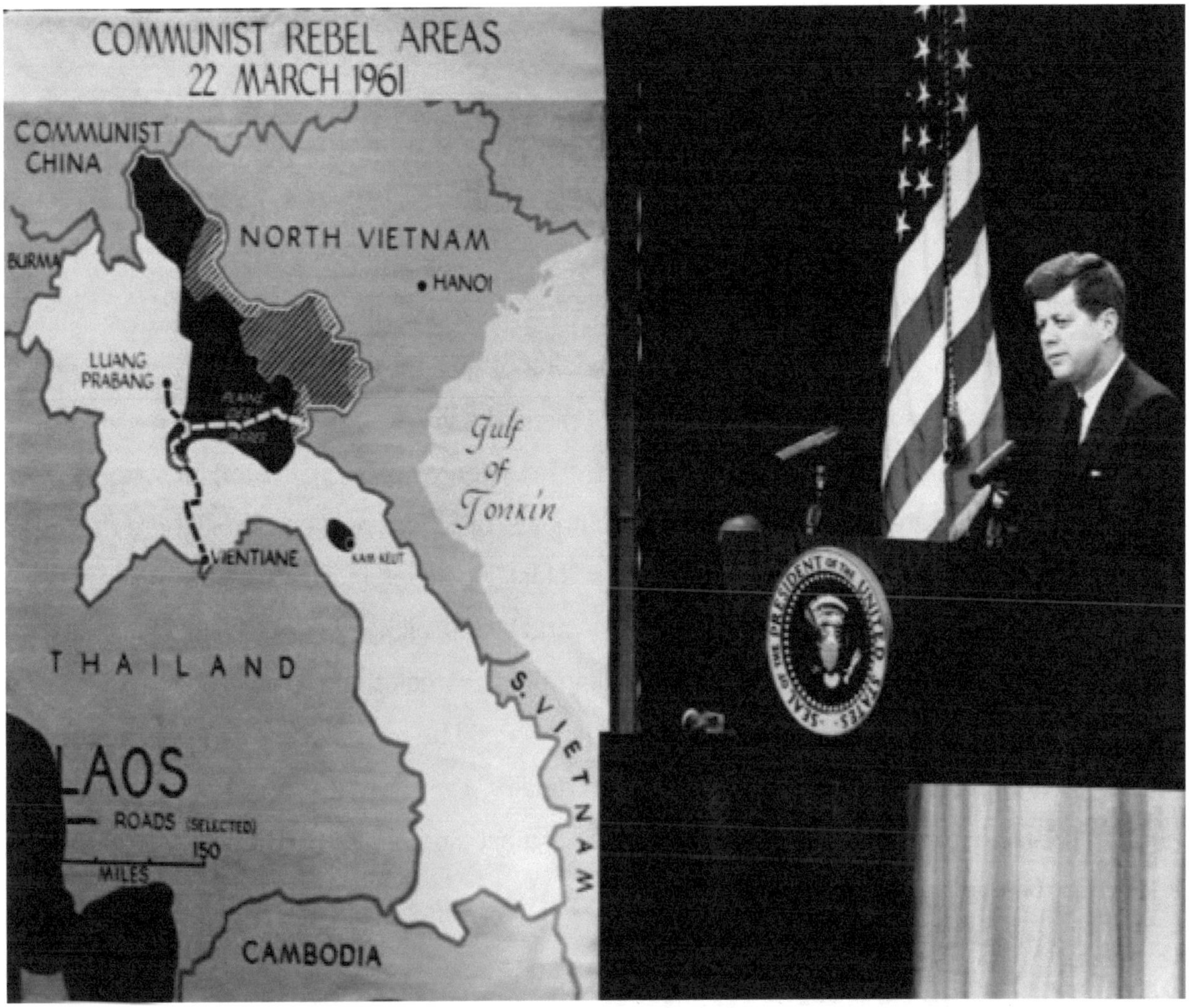

Figure 2.6. President John F. Kennedy and a map of Indochina, 23 March 1961. Source: Central Intelligence Agency/public domain

From assessing these five factors, then, we can see that victory in South Vietnam was never possible. The U.S. tried but could not change the behavior of the South Vietnamese government (another parallel to Afghanistan); the U.S. tried but could not close off a nearly thousand-mile long cross-border sanctuary into Laos and Cambodia; and U.S. military strategy did not prioritize population protection. Using the mathematics of statistical analysis to cross-multiply those morbidity factors, it is evident that the South Vietnamese government literally had a zero percent probability of survival in this insurgency.

This same analytical framework can be applied to any past, present, or future insurgency or civil war. In more than 100 double-blind case studies (where researchers did not have access to previous SOIC analysis of the conflict), it has proven 100 percent accurate at predicting outcomes. When one of the five factors is against the government, failure is between 94 and 100 percent certain (depending on which factor). When more than one of the five factors is against the government, state failure is 100 percent certain. There are no exceptions.

> ***| When more than one of the five factors is against the government, state failure is 100 percent certain. There are no exceptions. |***

AN IMPORTANT CLARIFICATION

There is one important possible misunderstanding of this rubric to clarify: Having all five factors in favor of the government does not mean the government *will* win. It means the government *could* win. Counterinsurgency is hard, even when all structural factors are favorable. There are examples in the SOIC database of 53 postwar conflicts where the government had all five factors in its favor and still lost. However, in almost all the cases of government success against insurgencies since World War II, all five factors in the chart have been "green" for the government. There are a handful of examples where the government overcame having one of the five factors against it. It should be noted when applying the rubric that cross-border sanctuary and/or the lack of a standing army at the start of the conflict is 100 percent fatal. No government has ever succeeded against a rebel movement with one of these two binary (yes or no) factors against it.

RAMIFICATIONS FOR COUNTERINSURGENCY DOCTRINE

During the past eight years, researchers found information that has critical importance to U.S. counterinsurgency doctrine as expressed in Field Manual FM 3-24.2, *Tactics in Counterinsurgency* (April 2009).

When determining whether the 53 countries included in the study met the requirements to be considered a "nation" and to have a "legitimate government," researchers found that these percentages have historically *never* increased during a period of armed conflict. Examining data points available for different periods of the same conflict showed a decline, in numerous cases, in both the percentage of the population expressing a national-level identity and the percentage of the population that believes their current government is legitimate. A steady, unchanging percentage of both factors has also been seen in many cases. What has never been found in any of the 53 conflicts is an increase in either factor.

This has paradigm-changing implications for counterinsurgency strategy (and tactics). Much of the doctrine predicated in FM 3-24.2 is based on the stated purpose of "increasing government legitimacy." Beyond killing guerillas and protecting the population from contact with them, the purpose of the "build" phase of "clear, hold, and build" is to decrease the perceived popular legitimacy of the rebels—and to increase the legitimacy of the existing government. However, the data show unambiguously that no government since 1945 has ever experienced an increase in legitimacy during the time of the conflict. It can go down, and it can stay the same, but it has never gone up. Therefore, the "build" line of action simply does not work. SOIC researchers found several studies on the impact of development programs ("build") during a conflict, which show *no increase* in support for the government in locations where things like hospitals, roads, and schools were built and new government offices were built or expanded. Researchers also found a number of studies showing a decrease in stability (more violence) even during and after construction of infrastructure projects. In addition to less stability, they found decreases in support for the government after their construction, but never an increase. Simply put, the current U.S. counterinsurgency doctrine is wrong because there is no evidence since 1945 of the delivery of development programs increasing government legitimacy. Money is not a weapon system, and development does not help defeat insurgencies.

This concept of what makes a government legitimate in the eyes of the people (e.g., delivering schools, medical clinics, and safe drinking water) is almost uniquely an American idea of where legitimacy comes from. Scholars call this approach "transactional counterinsurgency"—development projects are a transaction (i.e., "we build you a school, and in exchange you support us.") However, different societies have different sources of legitimacy of governance. The people of Jordan, for example, don't think their king is legitimate because he carries out development; they think he's legitimate because he is their king. Development, in and of itself, is a good thing. *But it does not improve a government's chance of defeating an insurgency.* There is simply no evidence to support that contention.

This phenomenon has been discussed in the literature of counterinsurgency. The best theory for why development does not increase support for the government is that for people caught in the middle of an internal conflict, the war is a total war. Making a mistake choosing sides is usually fatal, so civilians avoid choosing sides. In the case of development projects, civilians realize they will get the benefit of development if they support the government (i.e., the government has already decided to build the school). So there is no need to risk supporting the government because they will get the benefit anyway. In the calculus of survival in the middle of an insurgency, supporting the government (or at a higher level of allegiance, believing it to be legitimately in power) because it delivered some goods or services is a non-starter. And the empirical evidence reflects that.

Arguably, development projects in areas that already demonstrably support the government might reinforce that support. If taking a side and supporting the government results in later getting benefits like clean water, irrigation projects that benefit everyone in the village, or the construction of a new school (with a teacher, desks, and supplies), people's existing, already-demonstrated support could be bolstered. *In other words, reward good behavior rather than try to buy loyalty.* Military personnel who deployed to the conflicts in Iraq and Afghanistan know what real (versus spoken) community support looks like: villagers provided useful intelligence on guerilla movements and improvised explosive devices (IEDs), they sent their sons to join the army, enemy-initiated attacks against government forces in the local area decreased, and they displayed pictures of government leaders, for example. And these things could be measured month-to-month. This suggests that if development is to be done, it should be done in already secure areas. In other words, it would make more sense to focus development programs on areas where the metrics of government support are already strong and increasing. Then, theoretically, people in neighboring areas might see life getting better in the developing areas, and decide they, too, want some of those benefits—which could be earned by increasing support for the government over a period of time (say, a year of commitment and improving numbers). But this is beside the point of the SOIC data.

CONCLUSION

Why governments lose to insurgencies is no longer a mystery. It's no longer a theory: The data are in, and it is incontrovertible. It is possible to determine, in advance of a decision to support a foreign government experiencing internal conflict, what the outcome will be. The purpose of the SOIC is to determine *why* governments fail in counterinsurgency, not to determine how to assist insurgents. However, it could be possible to reverse engineer the data in the study to suggest courses of action that could improve the odds for guerilla movements. For example, achieving and maintaining external sanctuary in a neighboring country would logically be a high priority for insurgent leaders. A number of case studies can be utilized considering the rubric of the Five Factors. For more information about the Study of Internal Conflict, including many case studies in the Five Factors format, visit the SOIC webpage at https://www.usf.edu/gnsi/education-programs/study-of-internal-conflict.aspx. See Figure 2.7.

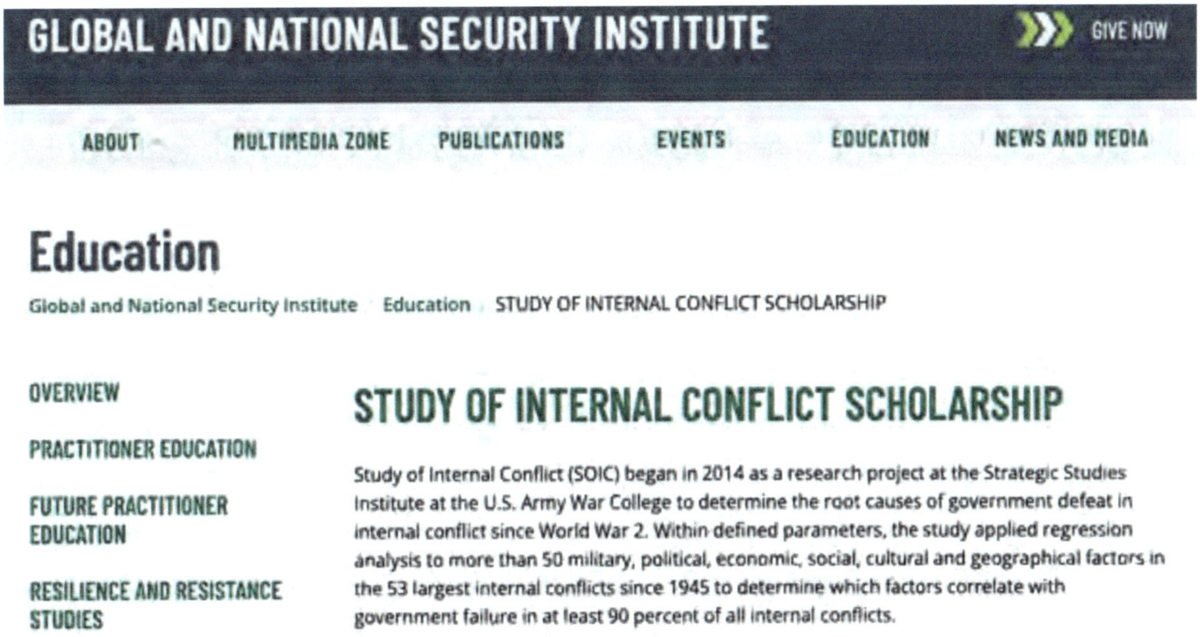

Figure 2.7. Study of Internal Conflict Website. Source: University of South Florida. https://www.usf.edu/gnsi/education-programs/study-of-internal-conflict.aspx.

STUDY OF INTERNAL CONFLICT GLOSSARY OF TERMS AND STUDY DEFINITIONS

SOIC utilizes several terms regarding the five factors analysis, mostly derived from common academic use, but described below to clarify.

External sanctuary: External sanctuary normally comprises a land border with a neighboring country that rebel combatants cross in militarily significant numbers (i.e., platoon-size elements with weapons and equipment) without excessive difficulty or danger. For the destination to be considered sanctuary, rebels must be reasonably certain of safety from detention, repatriation, or pursuit once across the border. Certainty of external *support* in the form of funding, new equipment, training, or new recruits is not required for the cross-border area to be considered sanctuary, but it must be across an international border. During the course of study, determining whether external sanctuary existed was relatively easy in actual practice; there have been no cases in which adjudication was required whether the criteria were met. In other words, all cases have been clear-cut. If sanctuary was available at the start of the conflict, but the government succeeded in sealing off its borders to eliminate the sanctuary during the conflict, the study defines this as a "negative" finding for external sanctuary despite its earlier existence, and it appears as "Yes → No" in the data matrix.

Government victory: A victory entails when the government in power at the start of a conflict or its natural successors (i.e., not a change of government) remains in power 18 months after the end of hostilities, *and* when the territory of the country remains unchanged from the start of the conflict. If the rebels succeeded in bringing down the government or achieved their goal of establishing an independent breakaway country, even if the government remained in power, the government cannot be said to have won the conflict.

Legitimacy of government: The study uses established political science principles in defining legitimacy of government. Princeton University Encyclopedia, for example, defines it as: "the belief that a rule, institution, or leader has the right to govern. It is a judgment by an individual about the rightfulness of a hierarchy between rule or ruler and its subject and about the subordinate's obligations toward the rule or ruler." The level established by the study for analysis is 85 percent of the population holding such a belief.

National identity: The study uses established political science principles in defining national identity as citizens who locate their personal identities at the level of the state and not at a sub-state stratum such as a tribe, clan, ethnicity, linguistic or religious group. The level established by the study for analysis is 85 percent of the population holding such a state or nation-level identity.

Population protection: Although population protection is a bedrock principle of counterinsurgency doctrine, it was necessary to define what the SOIC means by the term so researchers could assess whether a population was in fact being protected. The definition the study uses states that a population is protected if 85 percent or more of the civilian population is effectively sealed off from meaningful contact with the rebel group. Examples of meaningful contact would be printing and distribution of leaflets, pamphlets, or other propaganda material in any significant degree; the persistent ability of a rebel group to make radio broadcasts that reached any significant number of listeners; the ability of rebels to give speeches in public or address groups of more than a few individuals in private; and the ability of rebels to credibly threaten, intimidate, or bribe government officials or to assassinate government workers or leaders. This definition acknowledges that it is not possible to create a completely airtight firewall between civilians and rebels that will prevent 100 percent of rebel propagandists or terrorists from slipping past government security cordons 100 percent of the time. Thus, population protection refers to the reliable, continuous maintenance of a high level of public security that prevents all but rare interaction between the civilian population and rebel elements.

Sustainable, pre-existing security forces: This term refers to an existing security force at the outset of the conflict funded by the government that wears some sort of uniform, observes some sort of hierarchical rank structure in which the soldiers largely obey the orders of their superiors, is capable of transporting itself and sustaining itself in the field, and generally carries out the orders of the head of state or the appointed government official responsible for security. In the case of military rule, the standing security force had to be subordinate to the military element holding state power.

About the Author

Dr. Chris Mason is a Professor of National Security Affairs at the Strategic Studies Institute of the U.S. Army War College. He holds a Master's Degree in Military Studies from the Marine Corps University and a PhD in Military and Central Asian History from George Washington University.

NOTES

1. Chris Mason, "Why COIN Doctrine is Wrong," *Parameters* 51, no. 2 (18 May 2021).

2. Rhyne, Russell, "Victory in Vietnam." *Military Review*, Urbana-Champaign: University of Illinois Press, Volume 50, Part 2, February 1970. p. 39.

3. See, for example, Ian Roxborough's article "Counterinsurgency" *Contexts*, Spring 2007. I also recommend reading Andrew Gawthorpe's "All Counterinsurgency is Local" https://www.tandfonline.com/doi/full/10.1080/09592318.2017.1322330 and Christian Tripodi's "Hidden hands: The failure of population-centric counterinsurgency in Afghanistan 2008-11."

V. Understanding Institutional Resilience

By John H. Mongan

ABSTRACT: This section explores resilience as a concept, as well as this term's applicability to peoples, institutions, cultures, and national governments. In so doing, it raises a series of considerations that personnel operating abroad (either diplomats, development workers, or military personnel) might consider when evaluating an entity's resilience. It uses case studies to explore the topic, including Serbia, Ukraine, and France. It also utilizes a foreign policy and social-political perspective as a foundational view from which other governmental agencies might learn from. In addition to completing a historiography of current publications and policies on the topic of resilience, this section highlights the importance of: (a) identifying the correct entity requiring resilience, (b) the centrality of an institutional narrative, (c) the role of leadership, (d) capabilities and limitations of enhancing partner capacity, and (e) understanding the efficiency trap.

INTRODUCTION

The concept of "resilience"—defined by Google and Oxford Language as "the capacity to withstand or recover quickly from difficulties"[1] —has been gaining prominence in the social sciences for more than a decade as an effort to understand why some social groups, such as governments, companies, civil society organizations, even families and individuals, weather or even thrive under adversity, while other groups struggle or even collapse under those same stresses. Perhaps a repurposing of how the Supreme Court Justice Potter Stewart described pornography during a 1964 case that sought to define obscenity, "we know resilience when we see it."[2] Thus, we know now that Ukraine is more resilient than policymakers believed it to be in spring 2022. And, while many of the people of Afghanistan seem extraordinarily resilient, the Government of the Islamic Republic of Afghanistan clearly was not—or it stopped being so at some point before 2021. We also can conclude from history that resilience has limits; like an individual human, any group, organization, or institution composed of humans will collapse eventually under enough stress.

The essence of many forms of competition, including irregular warfare is to promote one's own resilience, and the resilience of one's partners, while degrading the resilience of one's opponents. *Resistance and resilience* is not a matter of two sides of a coin; an entity must simultaneously build its own resilience while attacking its opponent's resilience. Understanding what makes institutions resilient is central to strategy for both resistance organizations and entities opposing them.

Unfortunately, much of the resilience literature falls into Justice Stewart's trap, either explaining resilience or the lack thereof in an *ex post facto* fashion (i.e., it survived [or not], so it must have been resilient [or not]), or in a mirror-imaging fashion (i.e., it did this thing I do, and is doing well, and I am doing well, therefore we must both be resilient). *Presentism* also plays a very strong role in defining resilience. The Roman, Ottoman, and British empires are gone as political systems, yet Rome, Istanbul, and London thrive, as do myriad cultural and political aspects of those empires worldwide, for better and worse. So, we ask questions like "*were* they resilient?" Or "when did they *stop* being resilient?" Or "*are* they, in fact, resilient through the present day?" And finally, "is the 'they' we are discussing the city, its current polity (Italy/UK/Turkey), its former empire, or its cultural impact?"

This section will raise a series of questions which personnel operating in the field—diplomats, development workers, and military personnel—might consider when evaluating an entity's resilience. The perspective here is the political one of a diplomat. If politics and policy define and shape strategies in diplomacy, defense, and development, then a social and political understanding of resilience needs to underpin other analyses. One immediate implication of this is that a proper understanding of institutional resistance can only come from direct and sustained engagement with the institution. In other words, one cannot gauge resilience effectively from a distance.

COMMON PRACTITIONER MODELS OF ANALYSIS

Two common analytical shorthand methods utilized in the non-governmental community of practice regarding resilience and resistance are worthy of consideration: SWOT and Pillars of Support.[3] Both have great utility but also limitations, in understanding resilience. SWOT (strengths, weaknesses, opportunities, and threats) is a very perspective-dependent form of analysis. Strength to do what? Opportunity to achieve what? An entity must already have a strong sense of self and purpose to evaluate SWOT effectively. Pillars of Support, meanwhile, is a very government-centric form of analysis that is hard to apply to non-governmental entities, which might be better off applying the venerable Porter's Five Forces business school template or Jim Collins's Hedgehog Concept.[4] One possible illustration of Pillars of Support is shown in Figure 2.8.[5]

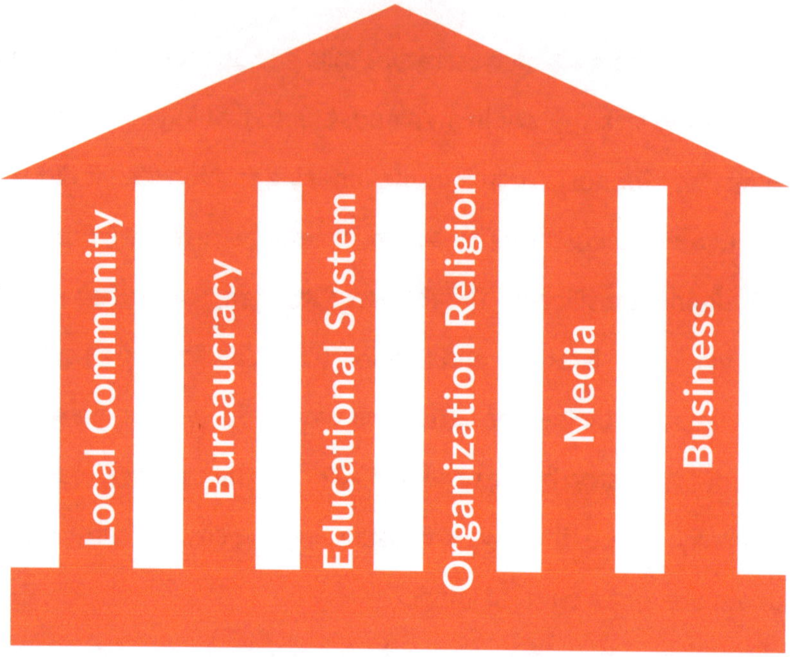

Figure 2.8: Pillars of Support concept example. Source: Author

In recent years, governments have also gotten involved in attempting to understand resilience. NATO has approached the challenge as a means to understand how some smaller countries—notably, new members like Finland and Sweden—deter Russian aggression, as well as how some smaller, weaker countries—like those in the Balkans—could both become more internally stable and less vulnerable to outside influences from malign actors. NATO evaluates resilience by seven "baseline requirements:"[6]

- assured continuity of government and critical government services;
- resilient energy supplies;
- ability to deal effectively with uncontrolled movement of people;
- resilient food and water resources;
- ability to deal with mass casualties;
- resilient civil communications systems;
- resilient civil transportation systems.

It should be obvious from this list of requirements that NATO's understanding of resilience is the understanding of a defensive alliance of states trying to understand the civil defense capacities of states. It almost certainly is useful for that purpose, but it is much less informative for understanding non-state groups, movements, or even particularly politically unstable states. For example, in 1986, when it hosted the Winter Olympics in Sarajevo, the Federal Republic of Yugoslavia would have scored high in all these categories. Five years later, it would not. But nothing that transpired in Yugoslavia did so because of changes to its baseline requirements.

The U.S. Agency for International Development (USAID) created its own Resilience Policy in 2012, revising and updating it in 2022.[7] In its original meaning for USAID, resilience was focused on ensuring food security in areas affected by climate change. USAID even organized a Bureau for Resilience, Environment, and Food Security.[8] By 2022, USAID had decided "the challenges facing the world require USAID to take a broader approach to strengthening resilience that is inclusive of all geographies and technical sectors and well-being outcomes." The closest USAID's policy comes to defining or understanding resilience itself is in two different parts of the document: first in describing four "Sources of Resilience:"

- systems, structures, and governance
- assets and natural resources
- individuals' skills
- inclusive relationships/networks[10]

The policy later describes "Five Conditions of Collective Impact for Resilience:"

- a common agenda
- shared measurement
- mutually reinforcing activities
- continuous communication
- backbone support—taken to mean a dedicated supporting entity like a USAID-implementing partner[11]

NATO and USAID sources both contain a tautology and a contradiction in that "systems, structures, and governance" only promote resilience *if they are resilient*. While it certainly is true that groups with few assets and natural resources are less resilient, there also remain many cases where abundant natural resources have actually weakened institutional resilience. Resources can become both a curse or a blessing in fostering resilience. The Five Conditions in the USAID policy, meanwhile, could be applied as best practices to any development activity and seem designed more to shape programming to support institutional resilience than to understand institutional resilience itself. And can we really say an entity is resilient if its resilience relies on backbone support?

It may be perfectly legitimate to assert that all development work, by definition, promotes resilience if done properly. However, if so, that adds little to one's understanding of resilience itself, or to the resilience of a particular institution. If the USAID Policy is an improvement over NATO's technocratic understanding of resilience, it is in the articulation of its "common agenda," a phrase the Policy uses repeatedly but spends little time dwelling on. This is the closest either framework comes to discussing politics, policy, and culture.

The Joint Special Operations University's *Resistance Operating Concept* dedicates its entire Chapter 1 to "Resilience as a Foundation for Resistance," touching on "National Identity," "Psychological Preparation," "Knowledge and Reduction of Vulnerabilities," and "Identification/ Preparation Against Threats."[12] However, most of the chapter is devoted to questions of planning and organization rather than trying to understand resilience itself. Even its Appendix G, "Assessing Resilience," provides a framework for… assessing foreign malign influence.[13] This is quite valuable but is an assessment of *threats* to an entity's resilience, not the resilience of the entity itself.

There are characteristics of resilient institutions that run deeper than the factors considered in these analytical models, all of which have their own utility. These characteristics are not easily quantified or assessed but are observable to an astute practitioner. Some are relatively simple for an outside practitioner to assist; others simply are up to the entity and are not as susceptible to outside influence. Understanding these characteristics is important to evaluating an entity's resilience, as well as how to enhance or degrade that resilience. These characteristics include: (a) narrative (with a key corollary, openness); (b) leadership; and (c) capacity (with a key corollary, efficiency).

DEFINING THE ENTITY

Before going further, however, one critical caveat: *one must very precisely define the entity whose resilience one is trying to understand.* Misunderstanding the entity will lead to failure of analysis, perhaps disastrously.

| one must very precisely define the entity whose resilience one is trying to understand. Misunderstanding the entity will lead to failure of analysis, perhaps disastrously. |

Take Kosovo. It is hard to call Kosovo a resilient country. Twenty years after Western intervention, Western troops remain central to its stability. Ten percent of the population—Kosovo Serbs—utterly reject the government. It is one of the poorest countries in Europe, and many other states in the world still do not recognize it as a sovereign country, including its largest neighbor, Serbia. This is not a prescription for resilience.

Yet anyone who knows Kosovo knows its two core constituents—Kosovar Albanians, now commonly called "Kosovans," and Kosovar Serbs, who reject "Kosovan" as a term—are incredibly resilient. For all their internal political divisions, Kosovans point to a common history of centuries of repression at Ottoman and Serbian hands, a war of liberation, and a struggle to build a new state. Meanwhile Kosovar Serbs have maintained a common front of staunch opposition to all of that for over 20 years, and it would be naïve to believe the strength and depth of their defiance is purely based on external support from Belgrade. Ironically, the resilience of the entities saps the resilience of the whole.

Hence the importance of great clarity when identifying the particular entity whose resilience one is assessing. Assessments of Kosovo's resilience likely will yield very different results than assessments of its core constituencies or particular institutions. It is vital to such an assessment that the assessor and the subjects share a common definition of the entity under discussion.

THE NARRATIVE

The overriding determinant of an entity's resilience is a common narrative among its members, akin to USAID's Common Agenda or the *Resistance Operating Concept's* national identity and psychological preparation. For purposes of this discussion, narrative means considerably more than an entity's messaging or strategic communications. Here we mean the shared story, stated or intuited, that binds peoples together into an entity. The story can be as straightforward as: "we are related by blood," to "we work together to make money," to the more abstract ideals of corps, corporations, states, and international institutions—Benedict Anderson's famous "imagined communities" of people who share fellow-feeling without even knowing each other, even to the extent of risking life to help each other.[14]

Whatever their origins, entities with strong shared senses of belonging, unity, and purpose are more likely to cooperate, collaborate, and adapt to changes, regardless of whether the entity is a regiment, civil society organization, tribe, ministry, or religious group. Entities whose members lack a strong shared sense of purpose are unlikely to summon the ability to collaborate in the face of stresses and will come apart quickly.

France's Third Republic in 1940 is an excellent example. Riven by decades of infighting among entrenched and corrupted political parties, liberal parties wavered over war measures against Germany while the communists and rightists all but actively supported Germany. *Meilleur Hitler comme Blum* was a common phrase on French streets.[15] In the face of opening shocks, France folded even though its material capacity for further resistance remained quite strong.

At times a purely negative narrative—one focused less around group unity and more around opposition to an outside foe—will suffice. French Resistance leader Henri Frenay noted that Charles de Gaulle was an acceptable leader to most of the Resistance precisely *because* he was a general with no political baggage, letting every group rally to him purely as a leader in fighting against Germany without prejudging the future politics of a liberated postwar France, a topic too divisive for the groups to rally around.[16]

Negative narratives have their limits, however. Between wildly-divergent visions of a future Syria and pervasive distrust due to decades of living in a police state, the Syrian opposition never succeeded in unifying around an ideal greater than overthrowing the hated Assad regime, with disastrous results. At no time were opposition groups able to present a plausible alternative governing vision sufficient to move a majority of Syrians to support them actively.[17] American revolutionaries declaring independence recognized it was insufficient to simply say "we don't like King George because he has done all these bad things to us." Instead, it became necessary to state affirmatively that Americans deserved, and would build, a government devoted to protecting inalienable rights to life, liberty, and the pursuit of happiness.

Diversity and inclusion enhance a narrative's power. All things being equal, the narrative that appeals to the most people will prevail over a narrower one. Yugoslav resistance in World War Two reflects this lesson: the ostensible leader of Yugoslavia's resistance was career Royal Yugoslav Army officer Draza Mihajlovic. A Serb Royalist, Mihajlovic did not just represent resistance to the occupiers—he represented the eventual return of the Royal government after liberation.

Unfortunately for Mihajlovic, the Serb Royal Yugoslav government, which had been an authoritarian ethno-centric government for 20 years, enjoyed little popular support among most of its subjects, who were not Serbs. As hated as the German and Italian occupiers were, few people would risk their fierce reprisals just to bring back an unpopular monarchy. Only a relatively few devoted Serbs joined Mihajlovic's Chetnik movement.

Josip Broz Tito's Communist partisans did not have this problem. The average Yugoslav peasant had no idea what communism was (and likely did not care), but they could see the partisans were led by a Croatian-Slovene with a diverse, multi-ethnic leadership structure, promising that whatever a Communist Yugoslavia would be, it would not be dominated by Serbs. This vision ultimately proved much more attractive than Mihajlovic's and led, inevitably, to the Partisans' ultimate consolidation of power.

The British and later the Americans sent missions to engage and work with both Chetniks and partisans, but here again narrative determined the destinies of the two movements.[18] Committed to restoring Yugoslavia's pre-war order, Mihajlovic and the Chetniks were horrified by Axis atrocities, which concentrated their ire on leading members of society, and in response scaled back their resistance operations. The partisans, envisioning a new social order post-war, were untroubled by Axis retaliation against pre-war elites whom they had little use for. Rather than back down, they used reprisals as a recruiting tool. Some Chetniks even focused on fighting the partisans alongside the Germans because communism seemed a greater threat to their post-war vision.

For the British and Americans, the priority was to support whichever movement was doing the most fighting against the Germans. They abandoned the Chetniks and shifted all their support to the partisans, whose narrative ultimately was more appealing both to most Yugoslav citizens and the Allies, even though the Allies were hardly pro-communist.[19]

From the perspective of external actors working to support or oppose an institution, it should be clear that narrative is the hardest thing to impact, even as it is the most important thing the institution is grounded on. When opposing an institution, great caution must be taken in confronting its narrative head-on. As odious as the narrative may be to the outsider, it plays an important role in this institution's sense of self and its members' values system, and efforts to change or challenge it require the utmost circumspection and patience, to say nothing of a superior counter narrative, articulated in a convincing way in both expression and action. We must also recognize that material conditions have little or no impact on the power of a narrative unless that narrative is grounded mainly in profit, as with a business. People committed to a belief system will commit suicide, march barefoot in the snow, or starve to maintain it.

| People committed to a belief system will commit suicide, march barefoot in the snow, or starve to maintain it. |

Finally, we should see the importance of *engaging* (vice simply *observing*) an entity if we are to understand its narrative to the degree needed to affect it.

More useful approaches for external actors might include broadening a narrative's diversity and/or exacerbating any internal contradictions in a rival narrative. De Gaulle's aforementioned success in presenting a broad front of French resistance is an excellent example of the former; America's use of the Helsinki Accords—and its use by courageous dissidents—to highlight the inconsistencies of the Soviet Union in its proclaimed and actual treatment of its citizens and satellites—its narrative of the benefits of communism—is an excellent example of the latter.[20]

A useful tool for the external actor might be to evaluate and support the entity's capacity to plan. A plan is an excellent articulation of an entity's narrative—even when plans "don't survive first contact," they provide the blueprint by which people can continue to support the entity even if cut off from it or when conditions change. Well-facilitated planning can adjust, elevate, and diversify narratives in ways that enhance an entity's resilience dramatically. U.S. support to Syrian civilian opposition planning efforts helped turn these local councils from protest organizers into effective governing bodies of liberated areas as long as the armed opposition was able to hold off the Syrian military. Nevertheless, these efforts still failed in building a narrative strong enough to unify enough Syrians against Assad.[21]

Great caution must be taken by the outside actor to remain a facilitator and not be dragged into a leading role in planning. Doing so risks becoming astro-turfing—the creation of an entity that appears grassroots when it is actually run by an outside group. When the supported entity is not making its own plan and narrative, it is simply falling in on the outsider's narrative because it may seem easier. This may make the entity a useful auxiliary to the external actor's efforts but does nothing to promote the entity's resilience.

LEADERSHIP

Narratives bring an entity together, but it takes leadership to move that entity in a coherent direction. No institution is more resilient than its own leadership structure and the personalities of those occupying its key leading positions. Resilience is about more than absorbing shock. It is about improvement after that shock, which requires capacity to execute vision that only effective leadership can provide.

Ori Brafman's and Rod Eckstrom's famous-to-the-point-of-cliché *The Starfish and the Spider* is one of the best-known works in this field, describing the give-and-take between management structures capable of authoritative direction (spiders) and those able to absorb shock via their diffusion (starfish).[22] Historically, outside actors, and particularly the U.S. , have tended to prefer working with spiders. It is much easier to work with one trusted point of contact who can deliver on commitments reliably than it is to try to find and build consensus among more amorphous and divided bodies, even if they are more democratic and less susceptible to the particular shock of decapitation. American diplomatic history relates a sad litany of strong men the U.S. has supported. Concurrently, Roger Myerson notes how British colonial officials *specifically instructed* to work with local institutions still tended to end up centralizing those institutions beyond recognition because of their natural instinct to work with a few preferred individuals.[23]

There is no question that history is full of institutions whose resilience was grounded in the dynamism of one person: Washington, Churchill, De Gaulle, Gandhi, Tito, Ho, Magsaysay, Mandela, or Zelenskyy. These are only prominent examples from the last hundred years. That begs the question of "does this mean that leadership of a resilient institution requires a 'great person?'"

Maybe. It might be better to say a resilient institution might require a great person at a particular moment in time, like Churchill and De Gaulle in summer 1940, or Zelenskyy the night he recorded his video from downtown Kyiv showing Ukrainians and the world he had not fled. However, one might argue that a key defining characteristic of a great person is *bringing great people around them.* Churchill and De Gaulle may have been essential in 1940, but by 1941–42 the talent they had assembled was enough to see their countries through the war (though it is less clear what a De Gaulle-less postwar France might have become). Zelenskyy included most of his senior leadership team in his video. Ho Chi Minh's North Vietnam had enough talented leadership to press on to ultimate victory in 1975 despite his death in 1969. George Washington probably was essential in 1776, but by 1778 the talent of (still loyal) Benedict Arnold, Nathanael Greene, and Henry Knox could have taken over the Continental Army, to say nothing of the astonishing political talent bench in the Continental Congress. By contrast, Tito somehow failed to sustain (or simply outlived) a talent bench willing and able to sustain his vision of Yugoslavia after his demise. Meanwhile Magsaysay died before he could consolidate his reforms in the Philippines.

Leadership is a quality an outsider has little or no control over. Occasionally, an outsider may play a role in talent-scouting a particular individual or group or provide some critical advice or coaching, but the historical record of success with such efforts is mixed at best. Again, we see that efforts at coaching require direct and personal engagement—yet this has its own risks. While some leaders have benefited from the impression of external support, others have been sabotaged by the perception of independence. Indeed, the most compliant or friendly appearing or strongly supported leader may end up being the least effective. Hamid Karzai was regarded as the most broadly acceptable consensus leader of a new Afghan government, but in hindsight it is clear he may have been so acceptable to so many deeply opposed factions precisely because he was not strong enough to jeopardize anyone's vital interests. Likewise, Ngo Dinh Diem may have been the most capable anti-communist leader in Vietnam, but at some point, he chose to defend his own position and power as opposed to building a broader-based anti-communist movement. Eventually the U.S. was reduced to desperately seeking alternatives to the very people it had promoted and elevated in the first place.[24]

The resilient leader and the one who gets along best with the U.S. will not always be the same. Roosevelt's and Churchill's detestation of De Gaulle is well-known, and neither man desired a Communist government in post-war Yugoslavia. Certainly, few people would choose a 40-year-old comedian to lead their country or ally in war. Yet outside efforts to find an alternate leader for an entity are highly unlikely to result in greater resilience. Either a capable leadership structure will emerge, or it will not. The external actor can often do little but engage, search, and advise in finding the best leader. More importantly, in many cases, when external forces have imposed their will on deciding another population's leader—like in the case of Karzai in Afghanistan or Diem in South Vietnam—the results have been disastrous.

CAPACITY

Cohesive and well-led movements still can witness their resilience fail in cases where human security needs collapse. Britain's resilience could stave off the *Luftwaffe's* bombers, but nearly cracked from the *Kriegsmarine's* U-boats. Any organization benefits from greater capacity, measured both in resources as well as in intellect.

Capacity is the easiest thing for an external actor to support.

| Capacity is the easiest thing for an external actor to support. |

They are limited in what they can do to engineer narratives or leaders, but they can "train, advise, and assist" just about anyone with just about anything. The U.S. has not always made the right choices in what it provides, from trying to develop the armies of South Vietnam and

Afghanistan as conventional forces to trying to train the Ukrainians in its airpower-dependent form of maneuver warfare. The U.S. has had a tendency to support partners in its mirror image rather than with what is best for "meeting them where they are" (as social workers put it). There is also the risk that too much additional material capacity creates new resources to fight over, or that one might concentrate one's effort with one partner at the expense of another, creating tensions that inadvertently sap the resilience of the group one is trying to support.

Yet, there remains plenty of success stories in external support to intrastate resiliency. From France's help to the U.S. during its revolution, to Allied support to European and Asian resistance movements in WWII, to more modern external support to non-violent resistance movements and even simple "democracy promotion" efforts with nascent political parties, there are plenty of examples of outside resources, training, and advice playing a significant role in building—or breaking—institutional resistance.

The lessons of historical successes and failures of external support to institutional resilience collectively illustrate the primacy of narrative and leadership over capacity. Those institutions with appealing narrative and capable leadership benefit from additional intellectual and resource capacity provided by external supporters. Institutions with weaker narratives and leadership do not. One might even speculate that recent and unfortunate efforts to mirror-image one's partner to that of its external supporter may have directly negatively impacted the partners' own lack of effective narrative and leadership (i.e., the Afghan national government). Unable to articulate its own vision for external partners to support, some organizations have simply followed the ideas of its external supporter for lack of better ideas. This rarely leads to long-term success.

THE EFFICIENCY TRAP

In this section, factors for understanding resilience have been largely in the context of supporting resistance movements or governments opposing them. The factors discussed here apply to understanding the resilience of other institutions as well, from civil society groups to private businesses large and small. From restaurants to charities and organizations with effective narratives, leadership, and capacity continue to perform well, generating results and/or profits, even under great adversity such as the COVID-19 pandemic. Interplay with private sector organizations will be important to the resilience of any institution operating in a free-market economy, because even most government organizations depend on some level of private sector resourcing, such as the military–industrial complex.

It stands to reason that private entities in free-market capitalist systems would be highly resilient given the right narrative, leadership, and capacity. After all, in a free market, entrepreneurs should be able to adapt rapidly to changes and create new innovations in response based on the profit incentives that stand to follow from those innovations. Innovation and adaptability certainly do support resilience. However, free-market capitalism also contains a key vulnerability: efficiency.

Much of the profit derived in the private sector comes from the ruthless pursuit of efficiencies (i.e., minimizing expenses to maximize profits). Unfortunately, inefficiencies all too often equate to inadequate capacity. In other words, minimal personnel, with minimal training, and minimal resources to do the job generate the most profit.

Governments are not immune to this problem. It is well-known how much the military values professional military education for its officers. Consequently, the services are funded generously to have a large number of "training float" officers over-and-above the minimal required number so that some always are available for training and educational opportunities. Civilian agencies like the State Department and USAID are criticized for not investing so much in educational opportunities for their workforces, but they are barely funded for the number of positions they need for their regular work—a veritable model of efficiency that certainly does nothing to promote institutional resilience.

Promoting resilience in, with, or among private-sector entities will involve planning and preparation to step in to support essential profit-driven institutions that cannot be expected to have necessary reserves for a crisis because holding such reserves would be wasteful and cut into profits. Regulatory approaches must require some private entities to maintain certain key reserves, or they must be incentivized to do so, or they must be regulated more like a public utility.

CONCLUSION

It is uncertain which institutions will prove resilient under pressure or crisis and which will not. However, the practitioner can be confident that an entity's resilience is most connected to those things that analysis might be least able to provide—particularly its leadership and its fundamental narrative.

The inability to change or evolve indigenous institutions proves a challenge when faced with the ambitions of foreign policy objectives. Governments tend to try promoting (or degrading) the resilience of entities based on how they relate to policy objectives that promote national interests. Thus, there typically exists attempts to support entities that, in the long run, simply are not sustainable. This is often conducted in the hope that such efforts will build capacity and that train, advise, and assist will overcome weaknesses of narrative or leadership—or even in the hope that those key factors will arise spontaneously if an external sponsor just builds capacity long enough. This type of approach to circumvent innate indigenous weakness in leadership and narrative may work for short-term or transactional ventures but is not the basis for long-term resilience.

It is a practitioner's responsibility to make policymakers aware of the limitations of capacity-building and to provide the frankest-possible assessment of the narrative and leadership of the group(s) whose interests might coincide with the external sponsor. This is not an impossible task. The British officers who worked in Yugoslavia were successful both at understanding the difference in resilience between the Chetniks and the Partisans and at persuading their political leaders to adjust policy accordingly.

Therefore, there is no hard-and-fast formula for resilience as a quality in people or in institutions, but if there is one thing gleaned from experience and study, it is that the qualities of resilience are those that are best—perhaps only—found through direct engagement with, and a deep understanding of, an institution and the people within it.

About the Author

John H. Mongan is a senior stabilization advisor currently serving in the State Department's Bureau of Conflict and Stabilization. The views expressed here are his own and do not represent those of the U.S. Department of State.

NOTES

1. https://www.google.com/search?q=resilience&oq=resilience&aqs=chrome..69i57j0i433i512j0i131i433i512j0i-512j0i131i433i512j0i433i512l2j46i175i199i512i654j0i512j0i131i433i512.1861j0j7&sourceid=chrome&ie=UTF-8

2. See concurrence of Justice Potter Stewart in *Jacobellis v. Ohio*, 1964. Although the justices were able to agree the object of the case was not "obscene" in the sense of Ohio's obscenity laws at question, they were unable to articulate a legal understanding of what *would* constitute "obscene" to guide future artists, lawyers, or legislators. Stewart got as far as equating "obscenity" to "hard-core pornography," but could not define that term beyond his own unhelpful aphorism.

3. For a useful review of SWOT, see https://www.mindtools.com/amtbj63/swot-analysis. For Pillars of Support, https://www.nonviolent-conflict.org/wp-content/uploads/2019/02/Pillars-of-Support-PDF-English.pdf.

4. For Porter, see https://www.cascade.app/blog/porters-5-forces. For Collins, https://www.jimcollins.com/books/g2g-ss.html.

5. Example taken from Srdja Popovic, Slobodan Djinovic, Andrej Milivojevic, Hardy Merriman, and Ivan Marovic, *Canvas Core Curriculum: A Guide to Effective Nonviolent Struggle*, (Serbia: Centre for Applied Nonviolent Actions and Strategies, 2007), 33.

6. https://www.nato.int/docu/review/articles/2019/02/27/resilience-the-first-line-of-defence/index.html.

7. https://www.usaid.gov/sites/default/files/2022-12/Resilience-Policy-Revision-Jan-2023.pdf.

8. https://www.usaid.gov/about-us/organization/bureau-resilience-environment-and-food-security.

9. USAID 2022 Resilience Policy Revision, p. 8. https://www.usaid.gov/sites/default/files/2022-12/Resilience-Policy-Revision-Jan-2023.pdf.

10. Ibid., p. 9

11. Policy, p. 19

12. Fiala, Otto C. *Resistance Operating Concept.* Tampa: Joint Special Operations University Press, 2020, pp. 7-10.

13. *Resistance Operating Concept*, 199-202.

14. Anderson, Benedict, *Imagined Communities: Reflections on the Origin and Spread of Nationalism.* London: Verso, 1983.

15. https://www.britannica.com/biography/Leon-Blum.

16. Frenay, Henri, translation by Dan Hofstadter. *The Night Will End.* New York: McGraw-Hill, 1976. P. 89

17. The author personally oversaw operations to support to the Syrian Opposition from 2012-14.

18. The outstanding work on the British-American missions to Yugoslavia in WWII is Fitzroy Maclean's *Eastern Approaches.*

19. Thayer, Charles. Guerrilla. New York: Harper and Row, 1963. Pp. 51-65.

20. Fried, Daniel. *Fifty Years Later, the Helsinki Process Stands as a Turning Point for Human Rights in Europe.* Washington, D.C.: The Atlantic Council, 2023. https://www.atlanticcouncil.org/blogs/new-atlanticist/csce-fifty-years-human-rights-europe/

21. Author's experience.

22. Brafman, Ori and Beckstrom, Rod. *The Starfish and the Spider*. New York: Portfolio, 2006.

23. Myerson, Roger. *Stabilization Lessons From the British Empire*. Austin: Texas National Security Review, Vol. 6, Iss. 1, 35-50. https://tnsr.org/2022/11/stabilization-lessons-from-the-british-empire/.

24. Works abound about U.S. support to Karzai in Afghanistan and Diem in Vietnam, but two of the best first-hand recountings are: Dobbins, James. *Foreign Service*. Santa Monica: The RAND Corporation, 2017, pp. 232-260; and Halberstam, David. *The Best and the Brightest*. New York: Random House, 1972.

CHAPTER 3: SOCIOLOGY

I. Social Movements: What Causes Resistance to Succeed or Fail?

By A. Jackson

ABSTRACT: This section explains what causes social movements to succeed or fail and applies those findings to examining one successful social movement and one failed Iranian social movement. It describes the social movement that overthrew the Iranian Shah in 1979, which resulted in an Iranian Islamic Republic, and explains why it succeeded. It then examines why Iran's Women, Life, Freedom protests in 2022 and 2023 have not succeeded in overthrowing the Islamic regime in Iran.

INTRODUCTION

From time to time, ordinary people rise up against governments that are adversaries or allies of the United States. Whether they succeed or fail has a great impact on U.S. and international security. It is therefore critical that we answer these questions: What makes a social group resilient enough to resist a tyrannical government? And what makes a social group willing and able to resist an armed organization that wants to violently impose its will on them? In the answers to these questions lies the ability to assess whether a population is resilient enough to resist an overt or covert incursion and the ability to provide insight into what factors and processes an outside actor, like the U.S. , might seek to effect in order to assist a movement that resists an authoritarian adversary.

Political scientists and sociologists have studied social movements in the U.S. and Western Europe (including, but not limited to, the civil rights movement, the free speech movement, the environmental movement, and the anti-war movement) and have drawn some conclusions about why they succeed or fail to change society or government policy within liberal democracies. The findings of these studies contain some of the answers to the questions posed above. However, less is known about what causes social movements against authoritarian governments or invading forces to succeed or fail. This section seeks to describe what is known about the factors and processes that cause a social movement to succeed or fail; uses an example of one successful and one failed social movement in Iran to demonstrate the important impact successful or failed social movements have on U.S. and international security; and illustrates what caused one to succeed and the other to fail.

ONE MOVEMENT THAT CHANGED THE WORLD, AND ONE THAT FAILED TO DO SO

On 9 January 1978, students at the seminary in Qom, Iran, merchants from the town's traditional *bazaar*, and university students gathered to protest an article written by an agent of the Shah, which claimed exiled Ayatollah Ruhollah Khomeini was an agent of British colonials and a mad Indian poet bent on selling Iran out to communists and colonial interests. Government security forces used live ammunition to suppress the protests, killing and wounding hundreds. This protest was the spark that lit the dry tinder under a society that had experienced rapid top-down social, economic, cultural, and political change under the Shah's White Revolution.[1] The White Revolution broke down the traditional norms of behavior and authority structures, generating multiple competing visions of Iran's future.

Violence from Qom spread to Tabriz in early January of 1978. "The suppression of each [demonstration] was marked by a mourning ceremony every 40 days, thus maintaining the momentum. This was the beginning of the real mass mobilization, organized by the clergy but in alliance with the religious educated young. ... On 8 September, a large number of people were killed in a half million person demonstration in Jaleh Square, Tehran,[2] and the opposition campaign gathered pace, culminating in a huge demonstration on 'Ashura[3]" that brought an estimated 17 million protestors to the streets of towns and cities across Iran.[4] Soldiers began to refuse to shoot protesters, some units within the Army mutinied, and the regime began to fail.[5] Mohammad Reza Shah Pahlavi, *Shahanshah* of Iran, defeated and mortally ill with leukemia, departed Iran on 16 January 1979. While he left behind a regency, the collapse of the army spelled the end of the Pahlavi dynasty. By the time Ayatollah Khomeini returned from his exile on 1 February, the state had effectively ceased to function, and millions greeted him as the leader of their revolution.[6]

Khomeini was a symbol of resistance to the Shah for many different groups in Iranian society. Khomeini and his followers, including the roughly 500 senior clerics he had trained while teaching in Qom in the 1950s and some portion of the 12,000 religious students who had attended his lectures in Qom,[7] combined with traditional merchants from Iran's bazaars and the urban poor, sought to create a hierarchical, authoritarian theocracy, the *wilayat al faqih*, in which Khomeini would use the state and his followers to violently enforce his interpretation of Islamic law. These three groups organized for the clear purpose of establishing an authoritarian, hierarchical theocracy. While they had to accommodate the desires of other groups within the protest movement, they were able over time to establish a state that enforces sharia on its citizens and seeks to export its model to other societies.

The social movement that overthrew the Iranian Shah in 1979 resulted in an Iranian Islamic Republic that surveils and punishes its citizens for violating rules few of them chose; is bent on the destruction of western power in the Middle East; actively destabilizes neighboring countries using a series of proxy political parties, militias, and gangs; seeks the eradication of Israel; and seeks to export its authoritarian Islamic theocratic model of government.

| The social movement that overthrew the Iranian Shah in 1979 resulted in an Iranian Islamic Republic that surveils and punishes its citizens for violating rules few of them chose; is bent on the destruction of western power in the Middle East; actively destabilizes neighboring countries using a series of proxy political parties, militias, and gangs; seeks the eradication of Israel; and seeks to export its authoritarian Islamic theocratic model of government. |

The impact of the successful anti-Shah social movement, and Khomeini's ability to harness it for his own ends on U.S., European, Iranian, Middle Eastern and South Asian security has been immense.

Since 2009, a series of social movements within Iran have sought to end the regime that was brought to power in 1979. All these movements have failed to overthrow the regime of the Supreme Leader. The most recent of these began in September 2022 and slowed in the winter of 2023, but continues at a low level today.

On 13 September 2022, morality police in Tehran arrested a young Iranian Kurdish woman named Mahsa Amini because they determined that the head scarf and robe she was wearing did not meet their standards for Islamic *hijab*. The morality police pulled her from the car her brother was driving and placed her in a morality patrol van, where they appear to have severely beaten her before taking her to a "re-education center" on Vozara Street, where she collapsed. Amini was taken to the hospital, where she died on September 16. Pictures of Amini's bruised and swollen face and the tubes inserted into her 22-year-old, comatose body spread on social media throughout Iran. Amini's death stuck a chord across Iran because nearly everyone, "felt that Amini could have been them or at least a female member of their family" who was arrested, beaten, humiliated, and killed for the way she wore her headscarf.[8] Tens of thousands of ordinary women tore off their headscarves and burned them in public, along with dumpsters, police cars, and stations. Within the first weekend, the protesters burned down 167 police stations.[9] The protesters were not part of large-scale organizations, and they lacked a leader or leaders to organize their collective actions. But, for the first time, protests against the Islamic Republic of Iran "gained broad public support among rich and poor, urban and rural, young and old, female and male."[10] Throughout Iran, day after day, thousands of protesters gathered and chanted, "death to the dictator," "I will kill whoever killed my sister," and "women, life, freedom." They did so in the face of machine gun, shotgun, and pistol fire; tear gas; and severe beatings, imprisonment, and public executions.[11]

The protestors suffered more than these near-term threats because of Iran's brutal and effective authoritarian regime. Iran is a country where local gangs of morality police, a mass political organization of those who would defend the regime called "the popular uprising" or *Basij*, intelligence operatives and their agents, police, closed circuit cameras and other electronic surveillance monitor citizens' behavior. They seek to identify people who break the regime's rules, focusing particularly on those seeking to overthrow the government.[12] When the regime finds people who have violated their rules, they arrest, imprison, and sometimes torture and execute them. Despite these enormous threats, tens of thousands of people protested over many months. Between 16 September 2022 and October of 2023, Iranians had conducted 4,816 protests. By February 2024, they had conducted 5,793. The regime's crackdown was brutal. It killed 641 protesters, including 80 minors; arrested 23,739 people; and executed eight by publicly hanging them from cranes.[13] By the middle of the winter of 2023–2024, the protests had slowed, but they have never completely stopped. And many women continue to refuse to wear *hijab*, persisting in individual, non-violent protest against coercive theocracy, demonstrating a continued seething anger that outweighs the very real fear of violent retribution by the state.

On 11 February 2024, the 45th anniversary of the Islamic Revolution, hundreds of thousands of Iranians gathered to chant, "death to America!" This is a common occurrence at many state organized rallies in Iran.[14] If the Woman, Life, Freedom protests had succeeded in overthrowing the Iranian regime, they would have eliminated an adversary that wishes harm upon the U.S. and its allies. If they had succeeded in overthrowing the Iranian regime, the set of Iranian proxies across the region, including Khatib Hezbollah and HAMAS, would have disintegrated without the resources, training, and guidance of Iran's Quds Force. Instead, threats to shipping as well as U.S. and allied naval vessels, in the Red Sea from Iran's proxy militia, the Houthis, which Iran has trained and armed with missile and unmanned vehicle technology, persisted. The failure of these protests to overthrow the Iranian regime had enormous impacts on regional, U.S., and international security. So, why did the Iranian Revolution succeed and the Women, Life, Freedom protests fail?

WHAT WE KNOW ABOUT WHY SOCIAL MOVEMENTS SUCCEED OR FAIL

Political scientists and sociologists have long studied social movements. Their work provides some knowledge of what causes ordinary people to organize themselves to change something in society and what causes them to succeed or fail.

Social movements are precipitated by catalytic events (like the death of Mahsa Amini in 2022, or similarly, that of African American Emmett Till in 1955) and broad social processes like urbanization, economic change, and international political realignments. These political events and social forces undermine the assumptions members of a regime—and those whose interests it does not serve—use to guide their behavior.[15] This shift in political opportunities "facilitate(s) increased political activism on the part of excluded groups by seriously undermining the stability of the entire political system or by increasing the political leverage of a single insurgent group."[16] When political opportunities present themselves, all sufficiently organized groups contest the status quo.[17]

But not all insurgent groups succeed in their quest to change the status quo. The members of an insurgent group must have a set of shared grievances—but grievances alone do not generate a social movement. These same people must have a reduced fear of retribution for organizing to act collectively against the regime. Fear and grievance are not objective facts. They are shared beliefs about what members of a group are willing to tolerate, and how much retribution they are willing to risk, to eliminate their grievances. Increases in the level of grievances and the fear of retribution provide the opportunity for members of a group to take collective action to change the status quo.[18]

To succeed, insurgent social groups must additionally be rooted in social structures[19] that can motivate action by their members. Membership by individuals in families, friendship groups, religious or ethnic organizations, unions, sports teams, and other social structures where members care about what their fellow members think of them, and where people share the same increased grievances and decreased fear of retribution, is a necessary component of any successful social movement.[20]

But a naturally occurring, untargeted social structure is insufficient to sustain collective action beyond a set of riots or sporadic protests. Sustained protest that succeeds in generating and consolidating change requires organizations dedicated to that change.[21] Effective organizations take resources like personnel, money, goods, and services and converts them into sustained collective action.[22] An effective organization shapes grievances, generates and reinforces shared narratives about what people should do and why, and aims the resulting energy at actions that will effectively generate change.

| An effective organization shapes grievances, generates and reinforces shared narratives about what people should do and why, and aims the resulting energy at actions that will effectively generate change. |

An effective social movement requires political entrepreneurs who can perform these same functions and lead organizations to do so. These political entrepreneurs can inspire their followers to take risks through their personal authority or their actions. Political entrepreneurs and organizations successfully recruit people to join and take action as part of a social movement by framing desired actions in the values and beliefs of the people they seek to motivate. Successful framing connects the grievances of the people the organization seeks to motivate to the collective action the organization wants them to undertake.[23] One way to do so is to identify the modes of behavior or end states of existence the members of the group in question values.[24] Having identified these values, the leader or organization should describe how the goals of the social movement align with the values of the group.[25] Another way to motivate members of a social group to engage in collective action is to identify, amplify, and transform specific sets of beliefs they have about the object of the action the organization or leader wants them to take. In particular, leaders and organizations should shape group beliefs about: (1) the seriousness of the problem, issue, or grievance in question;[26] (2) the causes of the problem and/or the people who are to blame for it;[27] (3) the nature of antagonists who can act as unambiguous coordinating symbols to galvanize action,[28] emphasizing the ways in which they are beyond the span of sympathy;[29] (4) the high probability of change if they engage in collective action;[30] and (5) the necessity and propriety of rising up to act as a group.[31]

| Another way to motivate members of a social group to engage in collective action is to identify, amplify, and transform specific sets of beliefs they have about the object of the action the organization or leader wants them to take. In particular, leaders and organizations should shape group beliefs about: (1) the seriousness of the problem, issue, or grievance in question; (2) the causes of the problem and/or the people who are to blame for it; (3) the nature of antagonists who can act as unambiguous coordinating symbols to galvanize action, emphasizing the ways in which they are beyond the span of sympathy; (4) the high probability of change if they engage in collective action; and (5) the necessity and propriety of rising up to act as a group. |

If these messages are to motivate members of the group in question to engage in collective action, they must be framed in narratives, stories, about what people should do under particular circumstances. Strong social structures often have powerful shared narratives that motivate collective action and are accompanied by a toolbox of symbols.[32] An excellent example of such a shared Shi'ite narrative is the story of the Martyrdom of Hussein in the year 680. The evocative, emotional recitation and reenactment of the bravery and painful death of Hussein motivates Shia to fight against corrupt governments that rule in the name of Islam in order to acquire wealth and power. In the story, Hussein fights against an oppressive government even though he knew he would be martyred. The story of Hussein provides a lever skilled political entrepreneurs can use to mobilize Shia to collective action against tyrannical governments. The martyrdom narrative comes with a set of rituals and symbols that reinforce its power to mobilize collective action by Shia populations.

Even if a social group has strong social structures, shared narratives about collective action, increased shared grievances, and decreased fear of retribution leveraged by effective organizations and nimble political entrepreneurs, its collective action may be defeated by the regime's social control apparatus.[33] If the regime can collect information about people violating its rules and brutally and publicly punish them, it can prevent the social movement from achieving its goals of changing the regime.[34] In order for the regime to be able to do so, it requires personnel it can motivate to undertake the risky task of collecting information about people organizing collective action to overthrow or coerce the regime into change. Further, the regime requires more resources than the social movement can organize to resist them.

Just as it matters what the regime does in attempting to repress a social movement, the specific actions the social movement takes matter. A social movement organization facing a violently repressive government should attempt to protect its personnel and leaders from regime retribution. The social movement organization should simultaneously focus its collective action on defeating the regime's ability to collect information about social movement participants, along with the regime's ability to violently punish those people. The most important task the social movement organization can undertake to protect its people from retribution is to disguise their identities and locations so the regime cannot target them. The social movement can then undertake an effort to target regime information collection equipment and patrols. It can also endeavor to convince members of the regime security forces to refrain from engaging in information collection about the social movement and from perpetrating violence against people engaging in collective action.

In conclusion, a social movement can succeed if it is built out of social groups with strong social structures, shared narratives about collective action, increased shared grievances, and decreased fear of retribution leveraged by effective organizations and nimble political entrepreneurs who communicate them through shared narratives about collective action.[35]

| A social movement can succeed if it is built out of social groups with strong social structures, shared narratives about collective action, increased shared grievances, and decreased fear of retribution leveraged by effective organizations and nimble political entrepreneurs who communicate them through shared narratives about collective action. |

Such a movement will succeed only if its leaders can frame the movement's goals, and the actions it wants people to take, in the values and beliefs of social groups it wants to act on its behalf. Specifically, the social movement organization must be able to convince members of social groups the seriousness of the grievance(s) they seek to address;[36] identify the causes of the problem and/or the people who are to blame for it;[37] describe the unsympathetic nature of antagonists against whom they should act;[38] determine the high probability of success if they engage in collective action;[39] and evaluate the necessity and propriety of rising up to act as a group.[40]

The social movement organization should attack the mechanism the regime uses to repress populations, i.e., the collection of information about the location and identity of members and leaders of the social movement and violent attacks on them. It should protect information about its members, while also attacking the equipment and personnel the regime uses to collect information about their identities and locations and to violently target them. In some cases, the social movement organization may be able to convince members of the regime's security services to simply not perform these tasks, which will cause the regime to lose control over the population. But even if the social group that serves as the basis of the movement has all the characteristics required for success and its leaders and organization direct its members to do all the right things, it may lose to the regime if the regime can motivate its personnel to do all the right things and has more resources than the social movement can organize.[41]

The next two sections will use the causal framework described above to explain how and why the Iranian Revolution succeeded in overthrowing the Pahlavi regime and subsequently why the Women, Life, Freedom movement failed in its goal of overthrowing the Iranian regime.

WHY WAS THE SOCIAL MOVEMENT FOR THE OVERTHROW OF THE PAHLAVI REGIME SO EFFECTIVE?

Mohammad Reza Shah Pahlavi, Shahanshah of Iran, sought to change Iran—to create a modern, westernized, urban, secular society run by a monarch legitimized by the splendor of the pre-Islamic Persian Empire. His "White Revolution" was an effort to use the authoritarian mechanisms of the state to remake Iranian society in the image of Western modernity. He did so using a great deal of money from Iran's sale of oil. The Shah's efforts were augmented by aid and advice from the U.S. and other European allies, who assisted him in creating a large state structure.

Mohammad Reza Shah Pahlavi Shahanshah Aryamehr at the 2,500 year anniversary of the Persian Empire at its ancient capital, Pasagard, in October 1971. Source: Wiki Commons

In particular, the Shah conducted coercive land reforms in which he seized the property of the great *khans*, the leaders of the patrilineal clans whose members tilled the soil of lands owned by the khans. While most of the khans resided in Tehran and traveled to their native lands to manage their affairs, their traditional authority and relationships with their farmers formed the tie between Iran's parliament and the rural population. Newly socially untethered peasants, many of whom suffered in the transition to land ownership on plots that could not generate enough revenue, flowed into Iran's booming cities searching for a new form of social organization. The traditional businessmen of Iran's bazaars and the manufacturers of the handicrafts they sold railed against the losses they faced as the Shah sought to supplant their businesses with foreign imports. Inflation, mismanagement, and the contrast between the wealth of the upper classes and abject poverty among migrants to the cities generated many grievances among the urban poor, who no longer had political representation through their khans in parliament. The khans were also enraged by their loss of land and power within their tribes.

A newly urban generation of middle- and upper-class young people attended universities where they imbibed knowledge of many foreign ideologies, including nationalism, democracy, capitalism, communism, and Islamism. They emerged fighting with one another over competing visions of Iran's future. When they expressed these views, they were often arrested by the notoriously brutal SAVAK, the Shah's Secret Police, and imprisoned and tortured at Evind Prison.[42]

When these young idealists arrived there, they encountered young clerical activists[43] whose teacher and object of emulation, Ayatollah Khomeini, railed against the corruption, ineptitude, and anti-Islamic stance of the Shah and his White Revolution. The Shah's White Revolution sought to diminish the role of all clerics in Iranian society, generating a powerful group of enemies. The White Revolution sought in general to eliminate the traditional elite classes of Iran and to indoctrinate and provide basic health and education services for the urban and rural poor. These activities were accompanied by a propaganda campaign aimed at "modernizing" Iranian society by making it more like European societies.

Having unmoored Iran's society, polity, and economy, the Shah relied upon censorship, secret police, imprisonment, torture, propaganda, and single-party rule to hold his opponents at bay. He centralized power in his own hands, becoming the decision maker on all important issues. The Shah was diagnosed with leukemia in 1974, and the disease gradually destroyed his ability to react decisively to challenges.[44]

The Shah's nemesis was Ayatollah Ruhollah Khomeini. Khomeini believed that Iran's society suffered from corruption by foreign powers who had diminished the role of Islam. He believed that society should be led by government that would implement the *sharia*, thereby creating a just and Islamic society. Khomeini believed leadership of such an Islamic state should rest in the hands of those clerics, who, like himself, carried the title of *mujtahid* because they had studied the *sharia* so extensively they had been granted the right to interpret it. Khomeini believed that clerics who had the selflessness and discipline to seek oneness with Allah and knowledge of his laws would act for the good of the people. This government would implement the *sharia*, protect the weak, and exclude foreign influences. The government should be strong enough to enforce and implement the *sharia*. Khomeini believed that to achieve such a government, a revolutionary vanguard led by clerics would be required.[45]

Ayatollah Ruhollah Khomeini, Speaking to Clerical Students at the Feyziyeh School in Qom June 3, 1963. Source: emam.com/ Wiki Commons

In 1962, the Shah sought to enfranchise women and open local councils to non-Muslims. Khomeini organized a campaign of criticism by clerics in their Friday sermons and provided support and organization to the leading bazaar merchants to generate popular protest against the proposed law. The offending law was subsequently repealed.[46] Khomeini had used his top clerical students to build a vanguard composed of the traditional bazaar merchants, the urban poor, and clerics and clerical students who followed him.[47]

In response to the resistance he was experiencing from Khomeini, in January of 1963 the Shah proposed a referendum be held on his entire White Revolution. The program included the redistribution of land from landlords and the government to peasants, the privatization of factories, the right of women to vote and stand as candidates for parliament and local government, the nationalization of natural resources, a literacy corps for remote towns and villages, health care and other public services, and an industrial profit-sharing scheme.[48]

Khomeini and other clerics organized a large protest in Tehran against the White Revolution, which they saw as an effort to gravely weaken Islam. The 1963 protests against the White Revolution followed a program that would become very familiar by the end of the Iranian Revolution. Clerics delivered political sermons aimed at the Shah's government. *Bazaars* were closed, and the leaders of the *bazaars* organized protests.[49]

The foot soldiers in the 1963 protests were composed of the vast populations of newly urban poor who had been displaced from their traditional social structures by the White Revolution's land reforms and agricultural policies. These families had moved to the cities where they embraced Islam as a social structure that would provide them with a set of values and norms and a community that shared them. For them, Islam replaced the community, values, and norms they had lost when they moved away from their traditional villages. In the confusing new environment of the cities, Islam provided new migrants with a community and a set of beliefs with which to make sense of the world and a set of norms they could follow to please their new community and improve their chances of entering Paradise on Judgement Day. Khomeini's religious followers reached out to this community to build a political base to oppose the Shah's rule.[50]

A series of provocations and counter provocations between the Shah and his security forces and Khomeini and his vanguard ensued. The confrontation came to a head on 'Ashura in 1963 when Khomeini delivered a sermon in Qom attacking the Shah, Israel, and the United States.[51] Standing next to the shrine of Prophet Mohammad's only child, Fatima, Khomeini asked whether the Shah wanted to destroy Islam and its clerics, just like the Umayyads had sought to destroy the family of the Prophet by martyring Fatima's son, Hussein, and his relatives on 'Ashura. Khomeini compared the corruption of the Shah taking money from the poor to build himself palaces and fill his foreign bank accounts with the poverty of the clerics of Qom. He advised the Shah not to make the same mistakes that led his father to be removed from power by serving foreign rulers, in this case the U.S. and Israel. He mocked the White Revolution, saying, "What do you mean, a White Revolution? Why do you try to deceive the people so? Why do you threaten the people so?"[52] It was the most insulting speech made by anyone to date against the Shah.[53]

A series of protests against the Shah followed Khomeini's sermon, and a group of Khomeini's supporters camped outside his house to protect him from the SAVAK. Two days later, hundreds of members of the SAVAK surrounded Khomeini's house in Qom. Khomeini emerged and submitted himself to them to prevent the SAVAK from harming his followers.[54] Rioting ensued in Tehran and Qom and the Shah declared martial law. After Khomeini spent 10 months in prison, and confident that the clergy had been defeated, the Shah released him.[55]

In Autumn of 1964, Iran's parliament secretly granted immunity to members of the U.S. military from criminal or civil prosecution. News of the law was leaked to one of Khomeini's students. On the birthday of the Prophet's daughter Fatimah, pilgrims came to Qom to hear the sermons of prominent clergymen. Khomeini delivered a lengthy political sermon lambasting the Iranian government for having sold Iran to America. He inveighed against the White Revolution and against imperialism and the leaders who had allowed it. He accused the members of parliament of treason against Iran and the Koran. Khomeini had become the first teaching mujtahid to become a political leader. He had done so by seizing on popular anger over colonial encroachment on Iran's sovereignty, which he had tied to the sovereignty of Islam. In response, the Shah had Khomeini exiled from Iran on 4 November 1964.[56] But Khomeini had built his vanguard of clerics, students, the leaders of the traditional bazaars, and the urban poor, which maintained itself within Iran in his absence and communicated with him regularly.[57] In 1977, the leaders of this vanguard sought to unite all militant clerics and Muslim activists in a countrywide organization, which became known as the Society of Militant Clergy. It began to print political Islamic books about opposing and fighting tyrants and colonialists and distributing tapes and statements from Khomeini. The organizations' branches were based out of mosques and formed the basis of what later became the revolutionary committees that organized the protests of 1978–1979 and consolidated control in the wake of the collapse of the Iranian government.[58]

In September of 1978, on the Eid festival at the end of Ramadan, Khomeini's vanguard generated a two-day demonstration in Tehran that attracted an estimated half-million people. On what became known as Black Friday, the government declared martial law and security forces began shooting protesters.[59] Opposition spread throughout the country. Khomeini's clerics spread his cassettes and statements and gave sermons deep in the countryside. Demonstrations, marches and mourning for the martyrs of the movement were a regular occurrence throughout rural Iran, which had never been politically mobilized before.[60]

Mass Demonstrations of People Protesting Against the Shah on 11 December 1978 at College Bridge. Source: Wiki Commons

Worse still for the Pahlavi regime, the commitment of people within the state to its preservation began to falter. The Shah himself, now stricken with advanced leukemia, was indecisive and lacked the will to lead.[61]

Throughout this period, Khomeini's statements inveighed against the Shah, his secularism, and his reliance on foreigners to maintain his power.[62] He emphasized the principles of Islam and the need for clerics to play a supervisory role. He did not, however, mention his plan to install a government run by the clergy that would implement and enforce the sharia.[63] He called for a government that would bring social justice to the people through Islam and defend Islam from foreign and secular encroachment.[64]

On 9 and 10 December 1978, during the 'Ashura commemorations of the martyrdom of Hussein by an un-Islamic, corrupt government, Khomeini's vanguard generated peaceful marches throughout the country of an estimated 17 million people (2–3 million in Tehran alone) who demanded the removal of the Shah and the return of Khomeini. The following day, Khomeini delivered a message from exile, thanking the participants for their peaceful march. He enjoined foreign states to take notice of what he termed a referendum on the Shah.[65] At this point, security forces began to refuse to fire on or arrest protesters and units in the Army mutinied. The state had begun to collapse under the weight of the protests.[66]

The size of the protests resulted not only from the effectiveness of Khomeini's vanguard but also from the widespread grievances against the Shah and his White Revolution. Political repression and the pervasive presence of the secret police, lack of freedom of speech and the press, and the lack of political mechanism for communicating grievance to the Shah, who had centralized power to implement change in his own hands, led to a glut of grievances. The grievances were multifold because of the Shah's top-down, rapid efforts to change Iran's economy and society. Land reform and poorly implemented agricultural policies led to the displacement of the rural population into the cities where they found few jobs, houses, or services.[67] These neglected migrants gravitated to activist, Khomeini-related clerics who advocated social justice through Islam. They also joined Islamic associations that provided them with a social and welfare framework. These mosques and associations were the basis of Khomeini's vanguard of protest committees.[68] Sudden policy shifts, like the introduction of machine-made bread and shoes, led to deep job losses by the traditional guilds in Iran's cities, and outrage among the populace grew as it lost access to its preferred products in favor of modern, western goods.[69] Problems across the economy were compounded by inflation coupled with recession after the rise in oil prices in 1973. As the poor got poorer, the gap in wealth between the urban rich and the urban poor grew. All the while, a flood of foreign advisors, businessmen and goods coursed through Iran's cities, deepening the feeling that Iran was losing its culture to imperial powers.[70] This popular perception was inconveniently coupled with the naked corruption of the Pahlavi regime and its dependence on the United States.[71] At the same time, the Pahlavi regime went to war with the powerful clerical class, which included not only the activists of Khomeini's

| The size of the protests resulted not only from the effectiveness of Khomeini's vanguard but also from the widespread grievances against the Shah and his White Revolution. |

vanguard but also the more traditional quietists who did not seek to overthrow the regime. The Shah's White Revolution was modernizing and secularizing, and he saw the clerical class as anachronistic. By the late 1970s he was in open conflict with them as he tried to usurp their roles and minimize their impact on society.[72] In so doing, the Shah had made enemies of Khomeini's religious competitors.

Since ancient times, Iranians have believed in the concept of *farr*, or divine grace. Those favored by the gods (or currently God) enjoy success, health, and prosperity.[73] If a ruler becomes unjust, he will lose farr and fall from power.[74] A repetitive series of ancient tales about farr is ever present in modern Iran because of the place of the *Shahname*, or *Book of Kings*, in Iranian culture. This epic poem describes the rise and fall of Iranian kings from the first king at the dawn of time through the 10th century Sassanian Shahs. One of the central stories in the Shahname is of a tyrant called Zahhak, who rules unjustly for 1,000 years until a child named Fereydun is born with god's blessing—with farr. Fereydun's farr allows him to escape death against all odds and perform supernatural feats, while Zahhak's loss of farr causes him to suffer and lose power.[75] "In this work, rulers are said to have the right to succession because they possess farr; their rule is both legitimate and just for the same reason; and they lose farr and therefore their legitimacy when they become unjust. Rebellion in such cases is then justified, and if successful, the rebel leader is deemed to have farr, [and] he becomes the legitimate successor and leader."[76] In this context, the Shah's advanced, terminal leukemia and his disappearance from public view in 1978, coupled with hapless policies and increasingly repressive and unjust actions by his secret police,[77] appeared to the Iranian people like a loss of farr. For many ordinary Iranians, the Shah's increasing weakness and injustice indicated he needed to be overthrown.

As a result of widespread grievances resulting from the Shah's own efforts at radical change, coupled with his apparent weakness and shared Iranian beliefs about authority, the crowd of 17 million demanding the removal of the Shah and replacing him with Khomeini included secular nationalists, khans, small farmers, members of the middle class (which the Shah had built), university professors and students, leftist Islamists, oil workers, and communists. To them, Khomeini was a symbol of righteous resistance to a Shah they saw as corrupt, oppressive, un-Islamic, or, at a minimum, beholden to foreign powers.[78]

Khomeini provided a stark contrast to the Shah. He had built a following based in part on the authority he derived from his adherence to the disciplined, ascetic life of a cleric and his willingness to put himself in harm's way to stand up for Islam and Iran in the face of foreign domination and the Shah's security apparatus. As a careful political entrepreneur, Khomeini tailored his message to take aim at the Shah and his foreign ally, the United States. To hold all of these groups together, he placed almost no emphasis on what would come after the movement removed the Shah from power. Each of these anti-Shah groups was therefore free to assume it would be able to achieve its desired political goals. As a result of his ability to obfuscate his ultimate goals, Khomeini was able to hold together a coalition of fractious organizations and groups.

Once the Shah was gone, Khomeini returned to Iran. The protest committees that had been built out of the network of mosques and Islamic associations, out of the traditional bazaar guilds and the urban poor—and led by Khomeini's former clerical students—formed the basis of an effective organization that could implement his vision for an Islamic state. In the end, they were the only group organized enough to fill the vacuum left by the departure of the Shah. Once in place, Khomeini and his followers set about systematically eliminating or diminishing the power of the competing organizations and groups described in the earlier. While this process took some time, Khomeini's social movement succeeded in achieving its goals of establishing a government run by clerics that implements and enforces the sharia and fights against foreign cultural influences that impinge on Islam.

If we consider the framework from social movement theory, we find that that the anti-Shah social movement brought together social groups with strong social structures and shared narratives about collective action. Each of these social groups had increased grievances against the Pahlavi regime that resulted from the broad social changes the regime had generated through its top-down White Revolution and the massive expansion of the state and its security apparatus. Ayatollah Khomeini and his vanguard of former clerical students, traditional bazaar guilds, and urban migrants sought to use this political opportunity to generate a social movement to remove the Shah and replace him with an Islamic state. The government they envisioned would be led by mujtahids who would selflessly seek to apply their knowledge of the sharia to generate social justice by implementing and enforcing it through an Islamic state.

When speaking with his vanguard, Khomeini framed the collective action he required from them in their shared values and beliefs. However, he simultaneously communicated to outside groups, who did not share these goals, using shared narratives, particularly the martyrdom of Hussein, to motivate collective action against a corrupt and un-Islamic government. Khomeini, as a political entrepreneur, was able to provide an example of fearless resistance to the regime, as did members of his vanguard. He was also able to carefully frame his messages to fit within the values of those members of his protest coalition who would not have signed up for a hierarchical Islamic theocracy. For them, he framed collective action as a way to remove the Shah and as a fight against corruption, incompetence, and foreign colonial domination. He pointed their grievances at the Shah and characterized him and his entourage as beyond sympathy. He convinced them of the necessity of acting and provided them with a model of protest. Khomeini also framed the need to act against the Shah in the most powerful shared narrative in Iran, the martyrdom of Hussein, which demands that Shia fight against oppressive, unjust, un-Islamic governments. He did not inform them that his intention was to establish an authoritarian theocracy that would impose the sharia upon them.

Khomeini's vanguard was built out of a set of insular, opaque social organizations (the Shia clerical class and the bazaar guilds) that naturally denied information about their efforts to organize large-scale protests to the government until those protests had reached such a size that the security forces were unable or unwilling to punish the leaders. This natural tendency toward secrecy protected the movement from the secret police.

In its efforts, Khomeini's vanguard's success was facilitated by the Pahlavi regime's serious missteps. The regime generated massive social changes that created the political opportunity Khomeini and his followers seized upon. The regime's efforts to change Iranian society also destroyed the traditional clerical and aristocratic social structures that historically buttressed Iranian monarchs. Further, the regime's inability to motivate its personnel to crack down on protestors, coupled with the Shah's frailty, poor judgement, and excessively centralized decision-making, greatly facilitated the success of the social movement to overthrow the Shah.

The success of Khomeini's vision of an Islamic state that implements sharia through a cleric-led, strong, centralized state results from the organization he built out of a narrow set of social groups. The vanguard of Khomeini's revolution was able to capitalize on the Shah's departure because of shared beliefs about the goals of their organization and the collective actions they needed to take to reach those goals.

WHY DID THE WOMEN, LIFE, FREEDOM PROTESTS FAIL TO OVERTHROW THE IRANIAN REGIME

Due to the authoritarian nature of the Iranian Islamic regime, and the recent nature of the protest movement in Iran, there is significantly less information about these protests than there is about the Iranian Revolution of 1978–1979. But when we consider the social movement theory framework, we can see some of the reasons why the Women, Life, Freedom movement was not successful in its goal of eliminating the authoritarian theocratic regime it sought to destroy.

Like the Pahlavi regime before it, Iran's Islamic regime has generated vast social changes while providing only limited mechanisms for addressing grievances. Sanctions resulting from the Islamic regime's nuclear program and proxy confrontation with the U.S., Israel, United Arab Emirates, and other states has impoverished Iranian society. Simultaneously, the Iranian regime's leaders' wealth and power has expanded. The security state has vastly expanded its surveillance and violent enforcement of sharia upon citizens. The Iranian election in 2021 was the first time the Islamic Republic blatantly prevented any candidates to run for the presidency of Iran who offered any hope of reform.[79] Therefore, these grievances, among many others, now exist without any foreseeable prospect for change within the government system. But grievances alone do not generate effective social movements.

Despite the powerful security and intelligence organizations of the Iranian state, Mahsa Amini's arrest and subsequent death at the hands of the morality police liberated Iranians from their fear of retribution by very publicly demonstrating the total vulnerability—and possibility—of any Iranian to be murdered at the hands of the morality police, regardless of whether they obey the rules of the sharia. This realization resulted in spontaneous protests and unusually violent and risky collective action, like burning down police stations or women filming themselves burning their headscarves. The communal anger at the regime boiled into a set of narratives that resonate broadly with beliefs and values across Iranian society, but the protests lack a shared narrative about collective action (like the martyrdom of Hussein).

Women, Life, and Freedom Protests in Iran on 25 September 2022. Source: Taymaz Valley/WikiCommons

It is clear that the protesters were drawn from across Iran's ethnic groups, from rural and urban communities, and from the entire span of Iranian society—with the exception of the urban poor, who were affiliated with clerics tied to the regime. But the protests appear to be generated by individual or small friendship groups rather than larger social structures that can motivate prolonged, concerted collective action. There does not appear to be a single organization or political entrepreneur shaping the collective actions of the protesters in Iran and targeting them at relevant outcomes. If there were such an organization, it could focus group collective action on more useful and less dangerous activities, like reducing basij patrols' ability to enter communities to collect information about people organizing to resist the regime. There is also no political leader or leaders urging people to join the protests and providing instruction for what protesters should do to resist.

About the Author

Dr. A. Jackson is a political scientist who studies social movements, insurgencies, insurgent organizations, civil wars, and ethnic and sectarian conflict. She holds a PhD from Johns Hopkins University.

NOTES

1. The White Revolution was a long-term policy program the Shah pursued throughout the 1960s and 1970s. Its purpose was revolutionary transformation of Iran's society into a modern, prosperous, secular nationalist society.

2. Vanessa Martin. *Creating an Islamic State: Khomeini and the Making of a New Iran* (I.B. Tauris, 2003),151-52.

3. 'Ashura is a Shia religious commemoration of the martyrdom of Hussein, the grandson of the Prophet Mohammad, who died in 680 AD. Shia believe that he died because he refused to submit to a vastly militarily superior Umayyad Caliph who Hussein and his followers, who became the Shia, believed ruled corruptly and un-Islamically in the name of Islam. Hussein's martyrdom occupies 40 days a year of social activity in Shia communities, where people describe in detail and re-enact the martyrdom of Hussein in order to feel the pain Hussein suffered in order to fight against a corrupt, un-Islamic government.

4. Martin, *Creating an Islamic State*, 149; Baqer Moin. *Khomeini: Life of the Ayatollah* (I.B. Tauris, 2009), 196.

5 . Martin, *Creating an Islamic State*, 149; Moin. *Khomeini: Life of the Ayatollah*, 196.

6. Martin, *Creating an Islamic State*, 149; Moin. *Khomeini: Life of the Ayatollah*, 196.

7. Said Amir Arjomand. *The Turban for the Crown: The Islamic Revolution in Iran* (Oxford University Press, 1988), 98

8. Mahsa Rouhi. "Woman, Life, Freedom in Iran" *Survival: Global Politics and Strategy, 66, no. 6* (December 2022): 193

9. Hooman Majd. "The View from Here: As Protests Consume Iran, the Specter of the Shah's Toppling Haunts the Ayatollahs." *Airmail*, October 1, 2022; Bill Chappell and Joe Hernandez. "Why Iranian Women Are Burning their *Hijabs* because of the Death of Mahsa Amini." National Public Radio, September 21, 2022.

10. Majd, "The View from Here."

11. Human Rights Watch. "Iran Events 2023." *World Report 2023.* New York.

12. Said Golkar. *Captive Society: The Basij Militia and Social Control in Iran.* (Columbia University Press, 2015), 44-47, 86-87, 92-95.

13. Mark Dubowitz. "Mapping Protests in Iran." Foundation for the Defense of Democracies. 11 February 2024. Associated Press. "Iran Acknowledges it Has Detained Tens of Thousands in Recent Protests." February 5, 2023. Number of executions from statement by United Nations' Special Rapporteur on Human Rights in the Islamic Republic of Iran.

14. Reuters. "Iran Anniversary Marchers Chant, 'Death to Israel' Amid Regional Tensions Over Gaza." 11 February 2024.

15. Doug McAdam. *Political Process and the Development of the Black Insurgency*, 1930-1970. (University of Chicago, 1982), 41

16. McAdam, *Political Process*, 42.

17. McAdam, *Political Process*, 42.

18. Sydney G. Tarrow. *Power in Movement: Social Movements and Contentious Politics.* (Cambridge University Press, 1994).

19. A social structure is a set of values and norms in which members of a social group believe. In a strong social structure, members of the group care about what other members of the group think of them and fear that other members of the group will think badly of them if they do not comply with the group members' shared beliefs about what group members should or should not do. These shared beliefs are called norms of behavior. See John Levi Martin. *Social Structures.* Princeton: Princeton University Press, 2009. ,5-8.

20. Tarrow. *Power in Movement*; McAdam, *Political Process*, 44-46.

21. As referenced by McAdam, *Political Process*, 44; Anthony Oberschall. *Social Conflict and Social Movements.* (Prentice-Hall, 1973), 119

22. Doug McAdam. "Revisiting the US Civil Rights Movement: Toward a More Synthetic Understanding of the Origins of Contention" in Jeff Goodwin and James M. Jasper, eds. *Rethinking Social Movements, Structure, Meaning and Emotion.* (Rowman & Littlefield, 2004), 203-4.

23. David A. Snow, E. Burke Rochford, Jr., Steven K. Worden and Robert D. Benford. "Frame Alignment Processes, Micromobilization, and Movement Participation." *American Sociological Review* 51, no. 4 (August 1986): 468-9.

24. Milton Rokeach. *The Nature of Human Values.* (The Free Press, 1973), 2

25. Snow et al., 469-470.

26. As referenced by Snow et al, 470. (See William A. Gamson, Bruce Fireman and Steven Rytina. *Encounters with Unjust Authority.* (Dorsey, 1982); Doug McAdam. *Political Process*; Frances Fox Piven and Richard A. Cloward. *Poor People's Movements.* (Vintage Books, 1977); Ralph H. Turner. "The Theme of Contemporary Social Movements." *British Journal of Sociology*, 20 (1969):

27. As referenced by Snow et al., 470. (See Myra Marx Ferree, Frederick D. Miller. "Mobilization and Meaning: Toward an Integration of Social Movements." Sociological Inquiry 55 (1985): 38-51; Piven and Cloward. *Poor People's Movements*; Louis A. Zurcher, Jr., David A. Snow. "Collective Behavior: Social Movements." in Morris Rosenberg and Ralph H. Turner, eds. *Social Psychology, Sociological Perspectives.* (Basic Books, 1981): 447-82.

28. As referenced by Snow et al., 470; (See Tamotsu Shibutani. "On the Personification of Adversaries" in Tamotsu Shibutani, ed. *Human Nature and Collective Behavior: Papers in Honor of Herbert Blumer.* (Prentice Hall, 1970): 223-33. Ralph H. Turner, Lewis M. Killian. Collective Behavior. (Prentice Hall, 1972).

29. As referenced by Snow et al., 470. (See Lewis A. Coser "The Visibility of Evil" *Journal of Social Issues* 20 (1969): 101-109.

30. As referenced by Snow et al, 470. (See Bert Klandermans. "The Expected Number of Participants, the Effectiveness of Collective Action and the Willingness to Participate: The Free Riders Dilemma Reconsidered" Paper Presented at the American Sociological Association, Detroit, 1983. Bert Klandermans. "Mobilization and Participation: Social Psychological Expansions of Resource Mobilization Theory." *American Sociological Review*, 29, 1984. ,583-600. Anthony Oberschall. "Loosely Structured Collective Conflicts: A Theory and Application" in Lewis Kreisberg, ed. *Research in Social Movements, Conflict and Change.* (JAI Press, 1980), 45-88; Mancur Olson. *The Logic of Collective Action: Public Goods and the Theory of Groups.* (Harvard University Press, 1965); Piven and Cloward. *Poor People's Movements.*

31. As referenced by Snow et al., 470. (See Gamson et al., *Encounters with Unjust Authority*; Pamela Oliver. "Rewards and Punishments as Selective Incentives for Collective Action: Theoretical Investigations," *American Journal of Sociology*, 84 (1984): 1356-1375; Piven and Cloward. *Poor People's Movements.*

32. Tarrow, *Power in Movement*, 10.

33. McAdam, *Political Process*, 56-58

34. Stathis N. Kalyvas. *The Logic of Violence in Civil War.* Cambridge: Cambridge University Press, 2007. ,173-183.

35. Tarrow, *Power in Movement*, 10. McAdam, Political Process, 39-58.

36. As referenced by Snow, et al., 470. (See Gamson et al. *Encounters with Unjust Authority*; McAdam, *Political Process*; Piven and Cloward. Poor People's Movements; Ralph H. Turner. "The Theme of Contemporary Social Movements." *British Journal of Sociology*, 20: 390-405, 1969.)

37. As referenced by Snow, Rochford, Worden and Benford, 1986, ,470. (See Myra Marx Ferree, Frederick D. Miller. "Mobilization and Meaning: Toward an Integration of Social Movements." *Sociological Inquiry*, 55 (1985): 38-51; Piven and Cloward. *Poor People's Movements*; Louis A. Zurcher, Jr., David A. Snow. "Collective Behavior: Social Movements." in Morris Rosenberg and Ralph H. Turner, ed. *Social Psychology, Sociological Perspectives.* (Basic Books, 1981): 447-482

38. As referenced by Snow, et al., 470. (See Tamotsu Shibutani. "On the Personification of Adversaries" in Tamotsu Shibutani, ed. *Human Nature and Collective Behavior: Papers in Honor of Herbert Blumer.* Englewood Cliffs, NJ: Prentice Hall, 1970. ,223-33. Ralph H. Turner, Lewis M. Killian. *Collective Behavior.* Englewood Cliffs, NJ: Prentice Hall, 1972.). Also as referenced by Snow et al., 470. (See Lewis A. Coser "The Visibility of Evil" *Journal of Social Issues*, 20, 1969. ,101-9)

39. As referenced by Snow, et al., 470. (See Bert Klandermans. "The Expected Number of Participants, the Effectiveness of Collective Action and the Willingness to Participate: The Free Riders Dilemma Reconsidered" Paper Presented at the American Sociological Association, Detroit, 1983; Bert Klandermans. "Mobilization and Participation: Social Psychological Expansions of Resource Mobilization Theory." *American Sociological Review* 29 (1984): 583-600; Anthony Oberschall. "Loosely Structured Collective Conflicts: A Theory and Application" in Lewis Kreisberg, ed. Research in Social Movements, Conflict and Change. (JAI Press, 1980): 45-88. Mancur Olson. The Logic of Collective Action: Public Goods and the Theory of Groups. (Harvard University Press, 1965); Piven and Cloward. Poor People's Movements.

40. As referenced by Snow, et al., 470. (See Gamson et al., *Encounters with Unjust Authority*; Pamela Oliver. "Rewards and Punishments as Selective Incentives for Collective Action: Theoretical Investigations." *American Journal of Sociology* 84, (1984): 1356-1375; Piven and Cloward. Poor People's Movements.

41. McAdam, *Political Process*, 39-58.

42. Abbas Milani. *Tales of Two Cities: A Persian Memoir.* (Mage Publishers, 1996).

43. Milani, Tales of Two Cities.

44. Milani, Tales of Two Cities.

45. Martin, *Creating an Islamic State*, 103-124.

46. Moin. *Khomeini: Life of the Ayatollah*, 78-81.

47. Moin. *Khomeini: Life of the Ayatollah*, 80-85.

48. Moin. *Khomeini: Life of the Ayatollah*, 83.

49. Moin. *Khomeini: Life of the Ayatollah*, 83-86.

50. Arjomand, *The Turban for the Crown*, 98.

51 . Arjomand, *The Turban for the Crown*, 198.

52. Ayatollah Ruhollah Komeini, 'Ashura Sermon at the Feyziyeh School in Qom, Iran. 3 June 1963.

53. Arjomand, *The Turban for the Crown*, 198-200.

54. Moin. *Khomeini: Life of the Ayatollah*, 106-108.

55. Moin. *Khomeini: Life of the Ayatollah*, 109-119.

56. Moin. *Khomeini: Life of the Ayatollah*, 121-127.

57. Moin. *Khomeini: Life of the Ayatollah*, 175-181; Martin, *Creating an Islamic State*, 145.

58. Moin. *Khomeini: Life of the Ayatollah*, 179-181.

59. Moin. *Khomeini: Life of the Ayatollah*, 188.

60. Moin. *Khomeini: Life of the Ayatollah*, 194.

61. Arjomand, *The Turban for the Crown*, 114-119.

62. Martin, *Creating an Islamic State*, 125-126.

63. Moin. *Khomeini: Life of the Ayatollah*, 195.

64. Martin, *Creating an Islamic State*, 125-126.

65. Moin. *Khomeini: Life of the Ayatollah*, 196.

66. Moin. *Khomeini: Life of the Ayatollah*, 196.

67. Martin, *Creating an Islamic State*, 147-49; Arjoamand, *The Turban for the Crown*, 91.

68. Arjomand, *The Turban for the Crown*, 91-94.

69. Arjomand, *The Turban for the Crown*, 105-107.

70. Martin, *Creating an Islamic State*, 148; Arjomand, *The Turban for the Crown*, 105-106.

71. Martin, *Creating an Islamic State*, 147.

72. Arjomand, *The Turban for the Crown*, 117-118.

73. Richard Foltz. *Religions of Iran: From Prehistory to the Present.* (One World Publications,2013): 15; Homa Katouzian. *The Persians: Ancient, Medieval and Modern Iran.* (Yale University Press 2010): 5, 20-27, 396-398.

74. Katouzian, 5.

75. Foreword and Introduction by Azar Nafisi of Abolqasem Ferdowsi. *Shahname.* (Penguin Classics, 2016); Katouzian, 20-27.

76. Katouzian, 396.

77. Arjomand, *The Turban for the Crown*, 114-119.

78. Arjomand, *The Turban for the Crown*, 114-119.

79. Rouhi, 92.

II. The Arab Spring as a Case Study

By David R. DiOrio, PhD

ABSTRACT: This section introduces the application of a human-centric model to study a series of resistance movements and national resilience actions during the Arab Spring uprisings that spread across North Africa to the Middle East in 2011. The case explores the shocking self-immolation of a young Mohamed Bouazizi in response to Tunisian government corruption and repression and the resulting domino effect of protest and revolution across the Arab world. This section will explore the specific situations in Tunisia, Libya, Egypt, Yemen, and Syria. Each country's case exhibits unique social movement characteristics along the *resistance* continuum and distinct government attributes that ascertain *resilience*. The essay introduces the root causes of the movements, the key factors that determined political and social outcomes, and the potential application of lethal and non-lethal hybrid warfare tools. The national security practitioner's ability to understand the social dynamic of the Arab Spring and identify the relevant factors that inspired resistance or fostered resilience may facilitate the application of all elements of national power—*diplomatic, information, military, and economic*—to advance our national interests. Military planners and operators may study the Arab Spring to develop appropriate courses of action with partner nations to defeat our adversaries and achieve our mutually aligned goals. U.S. military engagement requires special consideration to avoid unintended political repercussions or prevent a protracted conflict disproportional to the intended outcome. The multi-ethnic, religious, and multi-faceted social nature of Arab Spring requires a comprehensive analysis of the operational environment to develop effective *Integrated Country Strategies* and associated military *Theater Campaign Plans* to achieve our national objectives.

INTRODUCTION

The Arab Spring uprisings provide national security professionals an insightful opportunity to understand and influence contemporary social movement dynamics. The movement encompassed a wave of social protests with a significant escalation of aggression spread across the Arab world. The strategic landscape changed at an accelerating rate with increasing political instability, leading to extensive violence and bloodshed. The present-day political instability that emanates from the uprisings enabled terrorism, emboldens strategic competitors, and may limit our access to the global commons. The Arab Spring has not run its course.[1] In the affected countries, the social grievances that triggered the uprisings—widespread corruption, political oppression, poverty, lack of freedom of speech or the internet, deficient women's rights, and high youth unemployment—are just as bad now as they were 10 years ago.[2] A myriad of threat actors, inspired by the Arab Spring, continue to work in concert to exert political influence and challenge us militarily. State-sponsored Iranian or Arab proxies, criminal organizations, and ideologically inspired militants often combine efforts to advance their mutual interests. Outside power brokers—China, Russia, and Iran—leveraged the social unrest and government destabilization to supplant Western influence. They strive to neutralize political opposition and strengthen their control in regions central to their national interests. Exploring the Arab Spring offers several practical case studies to comprehend and respond to socially or ideologically motivated militancy enabled by our global competitors.

The 2022 National Security Strategy describes the global strategic environment as malign powers seeking overly authoritarian governance with a revisionist foreign policy by preparing, inspiring, or waging conflict by traditional or irregular means and leveraging technology.[3] Renegade militias, inspired by social depravity or ideological disenfranchisement and often supported by malign foreign powers, were on full display during the Arab Spring—with significant and persistent present-day ramifications.[4] Encouraged by an Islamic ideology that professes equity and justice, people of diverse nations and cultures demanded governance reforms to address widespread social and political grievances.[5] Internal or regional political factions and external state-sponsored support to national resistance created a fury of conflict and bloodshed that did little to alleviate social depravity and locked many regions into a cycle of enduring and devastating violence.[6] Millions of people were displaced by the conflict, particularly in Syria, Yemen, and Libya, where the social movement evolved into a Civil War.[7] Although each Arab Spring uprising had unique attributes, common provocations and similar structural or contingency factors drove distinct political results. Using a pragmatic model, a broad analysis of the relevant factors may generate a more comprehensive grand strategy and effectual operational design to achieve desired political, economic, and social outcomes.

MODEL FOR ANALYSIS – RESILIENCE AND RESISTANCE

The Arab Spring uprisings were an escalating series of peaceful protests that turned violent against the governing establishment.[8] Adverse social conditions or political corruption precipitated the protests.[9] The situation started as an isolated personal indignation that became an advanced social movement. In each case, violence escalated with a disproportional government reaction—a *resilience* method common to authoritarian regimes.[10] As the governing authority challenged people's freedoms, honor, and dignity with a violent response, the perceptions of government corruption and dysfunction mounted.[11]

Social unrest then reached a tipping point when the masses likely moved toward fierce *resistance* to reform governmental practices and restore fundamental human rights. Individuals or groups that participated may have resisted for different reasons and had different perspectives on discreet political or social injustices. However, the ability of diverse groups to collectively unify based on common grievances or ideology proved formidable and dangerous. A prolonged escalation provided an opportunity to organize and was likely to lead to a fully developed insurgency or revolution to achieve a complete restructuring and redistribution of power.[12]

Robert Burrell and John Collison developed a resilience and resistance framework based on defense analyst Gordon McCormick's model for an insurgency that characterizes the resistance continuum.[13] Resistance represents opposition to governance and occurs across a range from legal and peaceful protest to violent belligerency, although no two resistance movements are the same.[14] Resilience embodies the ability of the prevailing governance system to endure a resistance movement categorized by "Pillars of Stability" —security, economy, infrastructure, governance, and rule of law.[15] Resistance and resilience confront each other while seeking domestic and international support.[16] External involvement in either element has significant legal ramifications depending on the existing partner-nation relations or the international recognition of the social movement.[17] The Arab Spring uprisings demonstrated resistance and resilience activities across the continuum. The Arab Spring ebbed and flowed in the dynamic and chaotic security environment. In some countries, the government authorities made concessions, and the movement lost momentum and normalized to a relatively stable political situation. Other cases culminated in a full-blown insurgency or civil war with a protracted and brutal state of belligerency. A prolonged conflict challenges the existing system and creates a security situation that is difficult to control. Countries in a persistent conflict become dependent upon external support and are vulnerable to internal power actors with specific agendas. The chaos created by a social movement provides a means for change but does not always steer the direction of change.

STRUCTURAL, CONTINGENCY, AND PROXIMATE FACTORS

The affected governments, including some long-standing U.S. partner nations, responded to the Arab Spring protests differently. The precipitous spread and intensification of the protest rallies demonstrated the power of resistance movements and exposed the strengths and weaknesses of government resilience measures. Some governments embraced appeasement through concessions and policy reform, while others resorted to violence and imprisonment without due process. The political outcomes varied as much as the government reaction—some positive, others detrimental, and some temporary, with others enduring. Government authorities may assess resistance and resilience and adapt or reform policy to pacify the activists and re-stabilize the social environment. Resistance and resilience are complicated human endeavors influenced and shaped by structural, contingency, and proximate factors.[18] Leadership and planners must understand these factors before considering an appropriate reaction and response. Some analysts may link the elements to a social movement's cause or inspiration. However, the movement timing may be a matter of spontaneity related to a dramatic event or series of events that spark an uprising.[19] While there are many collective factors, the Arab Spring uprisings were profoundly different, and it is essential to look at each country within its context.[20] Countries across the Middle East and North Africa (MENA) had unique national issues, social conditions, levels of government trust, and expressed freedoms. These factors were sometimes shared but often unique to each country.[21]

Regaining popular confidence and achieving political stability requires understanding the social environment and relevant factors to craft an appropriate government response. *Structural factors* (pre-existing systemic or structural conditions), *contingency factors* (evolving political and social situations), and *proximate factors* (internal and external catalysts for inspiration) provide critical insights into the social movement dynamics.[22]

| Structural factors (pre-existing systemic or structural conditions), contingency factors (evolving political and social situations), and proximate factors (internal and external catalysts for inspiration) provide critical insights into the social movement dynamics. |

Before the outbreak of violence, government responses to these factors may serve to preempt bloodshed and avoid destabilization. In contrast, failing to respond to these conditions adequately may escalate social unrest.[23] The social movement can organize, unify, and accelerate if the government response is lethargic or indolent. Conversely, if the government response is overbearing and violent, the action becomes personal, and vengeance may energize a protracted and bloody conflict. Such an escalation presents a challenging governance problem because regaining popular confidence or "winning hearts and minds" after government security forces turn from protector to aggressor is not likely and usually results in long-term instability.

Analyzing the contributing social movement factors in the proper context is crucial. Structural factors are generally pre-existing political, cultural, or ideological conditions that define perceptions and drive behavior based on individual belief systems or expectations of personal freedoms. Ideological legitimacy, religious jurisprudence, tribal traditions, women's rights, and autocratic governance may all be structural factors.[24] These systemic conditions may or may not be changed in the short term to achieve desired outcomes. However, to avert escalation to violence, leadership may manage or address issues to stabilize situations. Contingency factors are conditions aggravated by ineffective governance and typically circumscribe political and social conditions that decline over time. Political corruption, deteriorating economic conditions, poor living standards, access to healthcare, and educational opportunities are potential contingency factors.[25] The people may unleash social unrest when contingency conditions reach a tipping point. Protesters often link the condition to dysfunction within the existing governance structure, and they demand reform. Once unleashed, proximate factors may aggravate or result in an assertive spread of the movement. Social media, the abusive role of the military, and detrimental foreign interference are important proximate factors. These factors often arouse a cumulative effect that may quickly inspire a social movement or just as promptly subdue the movement. Any one or combination of these factors may be the root cause of dissent and provide the basis for a persistent struggle.

In Tunisia, the Arab Spring revealed the latent potential of a relatively isolated malignant social condition. Government injustice exposed in a Tunisian marketplace provided the spark that ignited a raging fire of social unrest that spread across MENA. The isolated protests broadened regionally, invigorated by common contingency factors and fueled by proximate factors that unleashed a determined social movement. Perhaps governmental policy actions or reforms to address the mal-condition that started the unrest could have been quelled in the opening protest. It is now clear that the overbearing and unjust government response was on full display for the world, especially the Arab world, to witness. Social media triggered an awakening of social discontent that found support from diverse peoples in a variety of places who sympathized with a 26-year-old market vendor abused by local Tunisian police. A more comprehensive understanding of the contingency, structural, and proximate factors may have provided sufficient insight to forestall the impending movement.

ARAB SPRING SOCIAL MOVEMENT FACTORS

Pre-existing structural factors can create enduring perceptions and foster a foundation of dissent that may fester for generations. Colonialization of the MENA during the 18th and 19th centuries became the nexus for structural problems. Colonial power abuses, subjugation, domination, and oppression of the legacy governance systems are common root causes of

MENA dissent.[26] Strong European powers successfully capped the fermenting dispute, particularly after World War I, when the last great caliphate—the Ottoman Empire—collapsed and fractured into segments regulated by the West. However, colonial dominance and the ability to suppress the underlying dissent softened after World War II, leaving the Islamic world in a quandary concerning the best way to govern.[27] Under the quasi-colonial authority and even after achieving independence, Muslims in the post-colonial era generally experienced a stagnant or declining standard of living with a growing inability to select their preferred governance construct or pursue their desired spiritual belief system. The ever-increasing rejection of a Westphalian nation-state model with a separation of religion and state and a perceived unequal common law jurisprudence became a foundational grievance. The debate about maintaining the nation-state model with Western common law versus reestablishing an Islamic caliphate with sharia jurisprudence—or perhaps a combination of both—remains unsettled.

A common breach of faith between the government and the people involves a breakdown in the "authoritarian bargain" aggravated by several factors.[28] At the time of the Arab Spring, MENA governance structures, authoritarian or semi-democratic, were generally led by a ruling elite class presiding over a subjugated majority experiencing a high poverty rate and a dysfunctional middle class growing unhappy with government oppression.[29] The consolidation of power in the hands of a few and denial of fundamental human rights, notably to voice dissent or demonstrate their anger, exposed a complete breakdown in the social covenant between the government and the people. The acute contrast between the plight of ordinary people and the indulgences of the corrupt class fed public rage.[30] Alleviating the resentment, acceptance, and tolerance of government inadequacy is often achieved through promises of social services, employment, and subsidies in return for political support.[31] Fulfilling the expectations for such promises is problematic, especially during global economic declines or significant population explosions, as seen in the last few decades when subsidies and social programs became unsustainable.[32] Broken promises during the Arab Spring exposed government ineptitude—authorities incapable of stimulating the economy or implementing meaningful reforms to provide opportunities for the younger generations.[33] Public distrust and anger due to the lack of jobs, denial of rights, corruption, and inequality stirred a consciousness to restore national and individual dignity.[34] In the Islamic world, poor social conditions and personal suffering are often spiritually linked to Allah's displeasure. Thus, many Muslims across the MENA believed they had a societal and moral obligation to rectify the situation by placing a more legitimate and righteous authority into power—the foundation of an enduring social movement.

Several factors aggravated the emerging social movement after the outbreak of protests. Social media and information networks significantly increased awareness of cruelties and maltreatment.[35] Social media contributed to a "demonstration effect," propelling and expanding the Arab Spring swiftly across the Arab world.[36] Afflicted populations witnessed events in real-time, and protesters across the MENA were inspired to follow suit.[37] The state often responded with human rights abuses, heavy-handed violence by security forces, and misguided policy blunders that galvanized the movement.[38] In some cases, the armed forces played a decisive role, particularly in the early stages in Tunisia and Egypt. The Army opted to side with the people, forcing leaders to step down. In Libya, by contrast, the Army fractured along opposition and regime lines, instigating civil war.[39] In Syria, the nucleus of the Army stayed loyal to President Assad, resulting in a prolonged conflict.[40] Regular and irregular forces became targets of opportunity for external influences emanating from Iran, Russia, and China. Indeed, the West leveraged existing military relationships to intervene by providing direct and indirect military assistance or humanitarian support. The leading Western objective was to champion potential democratic opportunities to guide change and reform. Whether for noble or nefarious agendas, the injection of external influence often simply prolonged the conflict. The Arab Spring social movement may have had limited endurance at the onset of the protests. However, outside interventions from global power brokers facilitated a more violent and prolonged proxy campaign.

Ideology may not have precipitated the Arab Spring, but the chaos enabled sectarian opportunists. Radical Islamic groups with formidable military capabilities and political influence soon propagandized an ideological agenda.[41] The movement rejuvenated fledging extreme resistance activities partly because the collapse in order and absence of governance allowed for more fighting.[42] Reformist Islamist groups such as the Muslim Brotherhood emerged from exile to espouse a new political path that guaranteed positive governance reforms and better social conditions that resonated with the general population. Militant Islamism proxies sensed a favorable opportunity to advance, given a general sense of apathy and anticipated retreat signaled by the West. Jihadi Islamist groups such as Al-Qaeda and the Islamic State cleverly used social media to propagandize their ideology and recruit militant forces from all facets of society into their ranks.[44] These groups played on Islamist propaganda that defines the West as a decadent enemy. Islamists champion a new and morally righteous path to gain favor with Allah. Al-Qaeda articulated a long view where Islamic states—absent Western influence—would gradually supplant existing nation-states and unite to form a grand caliphate exercising sharia jurisprudence. The Islamic State strategy was more aggressive, whereby a caliphate governance system backed by a formidable militant force would gain an immediate foothold among a sympathetic population and then spread into an expanding area of control reminiscent of the glory days of the Rashidun (founding) caliphs. Absent a viable alternate governance system, both Islamist militant approaches have achieved surprising support—albeit tacit—from Quietist moderates who desire reform but would rather divorce their faith from politics. The Arab Spring occurred when the strategic environment warranted change.

> ***| The Arab Spring occurred when the strategic environment warranted change. |***

The governing authorities across the MENA failed to recognize the warning signs of growing social unrest at the local level and a surge in global Jihadi militancy across the broader region.

PRACTICAL APPLICATION – THE ARAB SPRING MOVEMENT

Political corruption and social depravity sparked an uprising in Tunisia in December 2010. The shocking events appeared in real-time at the speed of social media and the internet. Social unrest—directly inspired by events in Tunisia—erupted in Morocco, Libya, Egypt, Syria, Kuwait, Bahrain, Saudi Arabia, the UAE, and Yemen.[45] In later years, the remnants of upheaval spread to the Western Sahara, Algeria, Sudan, Jordan, Oman, Iraq, and Lebanon.[46] Figure 3.1 depicts the scope of the Arab Spring movement. Surprisingly, millions embraced the wave of resistance, but it only produced meager political concessions and virtually no real social change.[47] Most affected governments responded with violent and brutal security force reactions. Some authoritarian elites fled when the crowds demanded democracy or democratic institutions, creating power vacuums that produced even more chaos and instability. Nevertheless, democratic reforms and processes did not last, except in Tunisia.[48] Most governments could not reconcile the masses' grievances, often resulting in regime change or a protracted civil war. The idea of a true democratic republic within the MENA remains elusive as regimes trade one authoritarian regime with another.

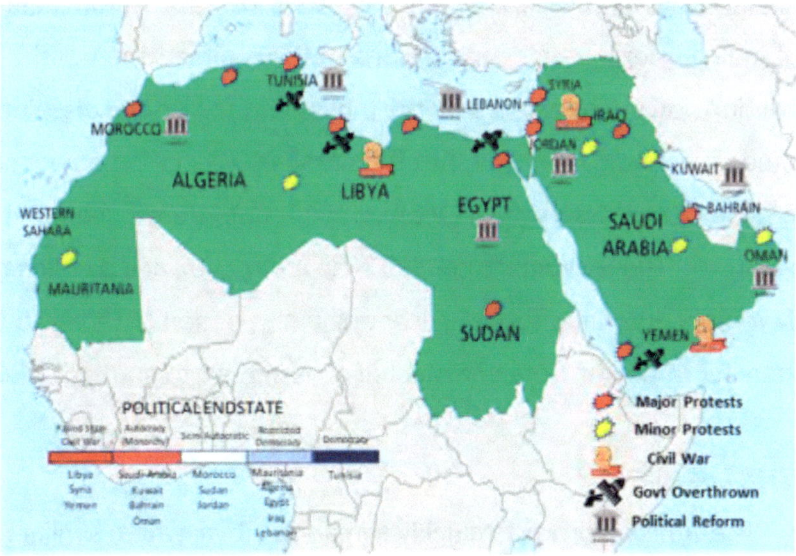

Figure 3.1. Arab Spring movement across Middle East and North Africa. Source: WikiCommons/updated by author

THE TUNISIA CASE

In 2010, Tunisia was a parliamentary state but authoritarian in practice. Personal freedoms were restricted, especially the freedoms of speech and the press.[49] Social depravity was prevalent, with a notable lack of employment and rampant inflation.[50] Courts administered justice unequally and unfairly—a small cadre of anointed judicial elites repressed trained legal professionals. Secret police forces frequently exercised brutality in quelling government dissension. The first demonstrations took place on 17 December 2010, incited by the self-immolation of Mohamed Bouazizi, a 26-year-old local vendor who was enraged when government authorities coerced bribes and seized his merchandise when he opposed.[51] He immersed himself in gasoline and set himself on fire. The social movement, designated the "Jasmine Revolution," swiftly expanded throughout the region.[52] Tunisian constabulary and military used unwarranted force to quell the rising dissent, resulting in extensive bloodshed.[53]

Tunisian President Ben Ali's weak and corrupt government offered some political concessions, but protests continued and eventually overwhelmed the nation's security forces, forcing a regime change. The crowds compelled Ali to resign and flee in January 2011.[54] Islamist party factions (Ennahda) competed with Secularist party factions (Nidaa Tuones) for governance power. The government held free elections and crafted a new constitution with Secularists maintaining tenuous control, but unrelenting human rights violations created an unstable political environment that currently persists.[55] The Jasmine Revolution exhibits an interesting social movement case. A significant grassroots protest movement initially put aside long-standing internal sectarian tensions to focus directly on political change and social reform. The friction and chaos created by the movement forced the dissolution of the ruling party and drove modest economic reforms, but it unleashed a sectarian power struggle. The people elected a Constituent Assembly in October 2011, but internal sectarian tensions paralyzed social reform.[56] The movement needed a clear direction and a charismatic leader to achieve a more favorable outcome. Tunisia became the first nation of the Arab Spring to endure an ineffectual transfer of power from one elected government to another.

THE LIBYA CASE

The Tunisian protests quickly spread to Libya, where similar repression and social depravity existed. In February 2011, protests against Muammar Qaddafi's corrupt regime escalated into the "Libyan Revolution of Dignity."[57] Qaddafi death squads killed hundreds, compelling a NATO air campaign—Operation ODYSSEY DAWN—that enforced a no-fly zone and targeted abusive Libyan security forces.[58] Qaddafi's military campaign against the rebels was brutal, targeting protesters with teams of snipers and squashing mass rallies with tanks and helicopter gunships, killing innocent families. NATO stepped up operations to protect civilians with Operation UNIFIED PROTECTOR.[59] NATO intervention shifted the military balance to a rebel faction that eventually gained control of Tripoli and, subsequently, the government in a bloody last stand.[60] Qaddafi was ultimately tracked down and killed in October 2011, and his political faction, the Libyan Arab Jamahiriya, was dissolved.[61] An interim Libyan Transitional National Council assumed control and attempted to impose a new Islamic State founded in sharia jurisprudence. However, historical tribal disputes undermined the weak council, fracturing the country and escalating the conflict to civil war.[62] Numerous Islamist militias organized along tribal lines emerged in response, often supported by foreign powers. The Islamists seized an opportunity to advance ideological agendas amid the chaos. One such militia, Ansar al-Sharia, attacked the U.S. consulate in Benghazi in September 2012—killing the U.S. Ambassador—to diminish U.S. influence and commission a new government that supported Islamic sharia law.[63]

The mass protests of 2011 devolved into a decade-long civil war along tribal, cultural, and religious lines. The situation created a security dilemma for the United Nations and NATO to decide which political camp to support, and there was no apparent option favorable to the West. In 2014, the Libyan House of Representatives, composed of Loyalists located in Tobruk, was at odds with Islamists within the Governing National Council in Tripoli.[64] The former Libyan National Army fractured and aligned with both camps. Salafists supported by al-Qaeda and the Muslim Brotherhood formed a National Salvation Government to dissolve all legacy institutions and reform into a new sectarian caliphate embracing sharia law. The injection of foreign state and non-state influence created a perfect storm of violence that raged for another seven years. With the conflict nearing culmination and outright exhaustion, a weak unitary government emerged in 2021.[65] A fragile peace is currently at hand, supported by the United Nations. However, the lack of competent government authority has hindered properly codifying a functioning democratic governance system. The dysfunctional National Transitional Council needs to address the lack of essential services and widespread social depravity that triggered the initial movement.

THE EGYPT CASE

The Arab Spring astonishedly impacted one of the U.S.' most robust and reliable regional allies—Egypt. In January 2011, on a national holiday to honor police, a long-standing anti-government political faction took advantage of the spreading social unrest by organizing protests throughout Egypt.[66] Egyptians from diverse cultures and backgrounds precipitated a significant "demonstration effect" where strangers leveraged common grievances to unify against a familiar foe—the corrupt government.[67] The Mubarak regime was a ruling elite that built immense fortunes for themselves at the expense of the ordinary class. There was a long history of rigged elections, political corruption, and nepotism. Mubarak named himself President for life and maintained dictatorial powers brought on by a State of Emergency declaration that gave the President unlimited authority with little or no basis. Thousands of protestors marched on Tahrir Square in Cairo to finally end his tyranny.[68]

An overly aggressive police force supported by the regular military began killing indiscriminately.[69] The government imposed a military curfew and shut down all internet and cellular communications. The bloody government response unleashed a massive social movement against President Mubarak and Prime Minister Nazif.[70] Experienced police and military were unwilling to support the slaughter of innocents and began to sympathize with the people by changing sides and abandoning their posts. Even Mubarak's long-standing relationship with the U.S. dissolved remarkably fast when the U.S. President could not be associated with such a brutal dictator.[71] Mubarak's feeble attempt to replace some ministers and form a new government was too late, and the new authorities jailed him, his sons, and other corrupt ministers for life. The Supreme Council of the Armed Forces assumed control, suspended the Egyptian Constitution, and directed new elections and constitutional reforms to limit presidential powers.[72]

The Egyptian ideological factions began vying for power amid the turmoil. The elections of June 2012 produced surprising results when the Muslim Brotherhood, led by exiled or imprisoned dissidents, assumed control. The Muslim Brotherhood Islamists embraced socialism and promised social change, but the new President Morsi displayed the same autocratic excesses of his predecessor. The implementation of strict sharia jurisprudence was rejected outright by a people that demanded representative government, fair and equal justice, and a quality of life consistent with the West.[73] The Egyptian military, considered the only remaining trusted and competent institution, stepped in again, and the people elected Field Marshall Abdel Fattah el-Sisi to power in May 2014.[74] The new government was slow to reform, and essential services remained well below modern standards, but it stopped the bloodshed, and the political environment became relatively stable.[75]

The Arab Spring movement in Egypt did not meet the people's expectations for various reasons.[76] There needed to be an explicit agreement on the proper governance model and acceptable legal jurisprudence. The protesters yearned for democratic reforms and social justice but the movement was disorganized with competing messages. The most prominent opposition movement came from the Islamists and later the Secular elites who hijacked the movement, to advance their power base.[77] The two sides never resolved the debate between adopting an Islamic state imposing sharia versus a democratic nation–state exercising common law. Mubarak's abusive security forces and corrupt legal system were gone, but a competent replacement remains elusive. Identity politics and religious ideology confused the message, and a capable leader has yet to emerge. Horrific human rights violations during the protests amplified distrust in all government institutions, including the revered Egyptian military. Governmental authority today is maintained only by a strong force of arms to subdue the constant revolutionary undercurrent.[78]

THE YEMEN CASE

Yemen has a historical cultural divide between the North and South that is both tribal and ideological (Zaydi Shi'a vs. Sunni). The British maintained a tenuous cap on tribal tensions until 1962 when an Arab nationalist movement compelled independence. A civil war along tribal fault lines ensued until exhaustion. The war resulted in the emergence of a weak unitary government led by Ali Abdullah Saleh in 1990.[79] Saleh exercised autocratic rule marked by political corruption. Police brutality, rampant unemployment, an abysmal education system, and an enduring economic crisis sullied his regime. The Arab Spring inspired a "Day of Rage" when the people called for Saleh to resign in March 2011.[80] The incident precipitated a fracture along the traditional tribal fault lines, and several governates separated from the national government. Iran immediately took notice of the instability and chaos that erupted on the southern flank of their competitive sphere of influence. The Islamic Revolutionary Guard Corps (IRGC), with their special operations Quds Forces (QF), began supporting the Zaydi Shi'a Houthis (Ansar Allah movement) with training and weaponry.[81] In response, Saudi Arabia, supported by the Gulf Cooperation Council (GCC), stepped in to mediate and resource the Sunni factions, and an irregular warfare campaign along ideological lines ensued.[82]

Protesters in Aden, Al Mansoora during the Arab Spring 2011. Source: AlMahr/ WikiCommons

The leadership in Yemen collapsed, and remnant forces aligned between the old guard with President Saleh and his Vice President Hadi.[83] The U.S. position followed the Saudi preference to encourage political reforms and human rights under any legitimate Arab leader chosen by the people—and to ensure that Iranian influence did not gain a foothold. The Saudis and GCC chose Hadi, who formed a new National Unity Government in February 2012. The Hadi government struggled to hold religious and political factions together as social conditions deteriorated. In 2014, the Zaydi Shia-inspired Houthi tribe supported by Iran precipitated an insurgency and seized the capital at Sana'a during the "Glorious Revolution" of 2014.[84] The Houthis made fighting corruption and social reform the centerpiece of their struggle, but they lacked international recognition and countrywide legitimacy.[85] The Houthis formed a new government led by a revolutionary committee and tribal leader, Abdul-Malik al-Houthi, but failed to gain majority support.[86] They called for a North–South partition, eventually fracturing the Yemeni political structure that destabilized the country. A full-blown civil war erupted in March 2015.

The Saudi-backed Hadi government became engaged in a significant conflict with the Houthis, supported by Shia militants and proxies such as Hezbollah, resourced by Iran, and directly supported by the IRGC-QF.[87] Al-Qaeda (AQAP) and the Islamic State in Yemen (ISIY) leveraged the chaos and exploited propaganda to insert their followers into the conflict. The U.S. and UK viewed the potential spillover of international terrorism to be the greatest threat and initiated a concerted campaign to target the AQAP and ISIY factions. The Saudis, concerned that the conflict could threaten Mecca and Medina, launched an aggressive air campaign that caused significant non-combatant loss of life and sympathy for the Houthi cause.[88] The United Arab Emirates and GCC distanced themselves from the indiscriminate Saudi campaign by promoting a third political solution—the Southern Transition Council—with regional control along the coastal commercial centers vital to GCC global commerce. In 2022, the civil war centered around three political factions: The Houthi Supreme Political Council in the West, the Presidential Leadership Council (successor to the Hadi regime) in the North, and the Southern Transitional Council separatists along the Southern coast. The multi-front internal conflict along ideological fault lines supported by external power brokers became "wars within a war." The Omanis and the United Nations intervened to broker a political solution to end the escalating human carnage. The Omanis failed, but the UN brokered a fragile ceasefire. Today, the Houthis claim victory, a remarkable accomplishment for a militia group with no Navy or Air Force.[89] However, an enduring political solution remains elusive, social conditions continue to deteriorate, and sporadic violence continues, suppressed only by the international effort to impede external resources from fueling the war.[90] The conflict has yet to culminate, and the outcome remains uncertain.

THE SYRIA CASE

In Syria, the Arab Spring broke out in earnest in March 2011, triggered by the arrest and torture of teenage boys posting anti-government graffiti.[91] President Assad ordered an aggressive crackdown against the protesters, and the Syrian Army quickly responded with fervent violence and bloodshed.[92] When evidence of chemical and nerve agents surfaced, the U.S. and the United Nations called for Assad to step down.[93] Anti-government political factions organized along ideological lines and supported by formidable militias saw an opportunity to rise in rebellion. By 2012, the Alawite Shi'a Arab Republic (Assad regime Ba'athists), supported by Hezbollah and the IRGC-QF, engaged in an all-out ideological civil war with the Sunni factions of the al-Nusra Front (al-Qaeda affiliate), Free-Syrian Army (FSA, backed by Turkey and the GCC), Islamic State (ISIS) and the independent Kurds (YPG). Internal dissent erupted within the Sunni factions when the radical ISIS faction came into direct conflict with the more moderate FSA, creating a conflict within a conflict.[94] The inter-rebel fight escalated as alliances shifted within the Islamist militants. Russia entered the fray to support Assad by conducting limited air strikes on moderate rebel factions and by providing air defense systems.[95] The U.S., France, and the UK faced a security dilemma with no clear faction to support while innocent civilian casualties mounted and a mass refugee crisis ensued.[96] The West resorted to a limited campaign against the abusive Syrian Army to protect innocents while monitoring the activities of the warring groups.

| Internal dissent erupted within the Sunni factions when the radical ISIS faction came into direct conflict with the more moderate FSA, creating a conflict within a conflict. |

The Islamic State quickly expanded and enjoyed significant freedom of maneuver in eastern Syria and western Iraq during the chaos of the civil war. An American-led coalition, "Operation INHERENT RESOLVE," supported the Syrian Democratic Forces (primarily Kurdish and Arab militias) directly against the Islamic State and the al-Nusra front—two factions that threatened the stability of the region and were capable of inspiring a new wave of international terrorism.[97]

The Kurds enthusiastically and competently supported the Western initiative in return for the possibility of achieving international recognition and support for an independent state upon the cessation of hostilities. Kurdish support from the West alarmed Turkish leadership, who were outraged that their NATO allies would support their long-standing revolutionary rivals—the Kurds (PKK). Russia and Egypt eagerly supported Assad to gain favored commercial trading status and military basing rights. The social movement and internal political fracturing combined with the external power competition created a melee of violence as the number of Syrian refugees swelled. Radical Islamist militants actively recruit young Syrian fighters within the refugee camps, and they return to the fight.

Today, the Syrian Civil War rages on, ravaging the country. The Syrian gross domestic product has plummeted, and the social depravity is extensive.[98] The United Nations estimates that more than 600,000 Syrians have died (~1.5 percent of the population), and more than 14 million Syrian displaced persons and refugees have swarmed neighboring countries and Europe.[99] The U.S. eventually defeated ISIS, and the coalition withdrew. Russia then renewed its partnership with the Kurds and anxiously filled the leadership void created as the Allies decreased their presence. The region remains an active fault line between Iranian Shi'a and Arab Sunni factions competing for domination along the fertile crescent. The Israelis anxiously monitor the advancing wave of radical Shi'a ideology that they consider to be an existential threat. Assad's power base is currently secure with Russian and Egyptian patronage. The social movement and promise for social change that sparked the civil war seem lost among the carnage. Syria represents the far right of the resistance continuum—an example of a social movement without direction that escalated to a tenacious belligerency and created unmitigated devastation.

CONCLUSION

The Arab Spring's success in achieving meaningful political reform, enhancing personal freedoms, and improving living standards did not meet expectations. In cases where governments displayed formidable resilience, the cost of retribution from government security forces overcame the advantages of political reform. In most cases, the uprisings lacked a charismatic leader to rally the masses and achieve a unity of effort or provide a legitimate government alternative. Government anti-protest and counterinsurgency methods proved particularly effective when some select—albeit not comprehensive—reform measures were sufficient to appease the agitators. Governments with relatively weak resilience devolved into civil war along ideological fault lines, as is the case today in Syria and Yemen. Foreign influence and interference prolonged these conflicts by enabling proxy forces and injecting

military resources to gain a strategic advantage in the broader hegemonic competition. The U.S. championed pro-democratic initiatives and provided some assistance to long-standing partners such as Saudi Arabia and the GCC, Egypt, and partners of opportunity such as the Kurds, but engaged cautiously to allow the movements to convey a natural course.
cautiously to allow the movements to convey a natural course.

The impact of the Arab Spring is not over and has arguably inspired a new round of hegemonic competition within the MENA. To advance American interests, our national policymakers and military planners must understand the social dynamics and repercussions of the movement. The joint operational planning process should consider and accommodate the complexities and nuances of the intense population-centric struggle along the resistance continuum. The unpredictable social dynamic of the Arab Spring movement, combined with capricious foreign influence, resides in the gray zone and is inherent to contemporary hybrid warfare. The Arab Spring displayed the complexities of a globalized society with information flow and real-time images that traveled at the speed of the internet. The involvement or intercession of malign foreign states or non-state actors becomes even more problematic. Planners must address foreign intervention—either calculated in advance to entice or imposed afterward to fuel the unrest—to prevent undermining a legitimate government authority from achieving an enduring settlement. The continued interference from Russia, China, and Iran to displace Western influence and advance their worldview will significantly prolong the violence, infringe personal freedoms, and threaten global lines of communication. The multi-faceted conflicts inspired by diverse ideological, cultural, and social conditions and buttressed by international competitors present a complex security dilemma for the West. The task for our leadership and military planners is daunting. Each country's situation requires a comprehensive operational environment analysis to develop effective Integrated Country Strategies and associated military Theater Campaign Plans to achieve our national objectives.

About the Author

Dr. David R. DiOrio (Captain, U.S. Navy retired) is a National Security Professional with a Doctor of Philosophy degree in Public Policy and Administration from Walden University. He served as the Deputy Director at the Joint Forces Staff College of National Defense University and is currently Adjunct.

NOTES

1. Matt Bradley, "10 Years After the Arab Spring, Autocratic Regimes Hold the Upper Hand," NBC News, December 19, 2020, https://www.nbcnews.com/news/world/10-years-after-arab-spring-autocratic-regimes-hold-upper-hand-n1251710, accessed August 6, 2023.

2. Kali Robinson and Will Merrow, "The Arab Spring at Ten Years: What's the Legacy of Uprisings?' Council on Foreign Relations, December 3, 2020, see Standard of Living, Youth Unemployment, Internet Freedom, Women's Empowerment, https://www.cfr.org/article/arab-spring-ten-years-whats-legacy-uprisings, accessed September 15, 2023.

3. The White House, *National Security Strategy*, 2022. (Washington, DC: Government Printing Office): 8.

4. Elfaith A. Salam, "The Arab Spring: Its Origins, Evolution, and Consequences – Four Years On," 127–128. Research Gate, Review Article 23, no. 1 (2015): 121–122, https://www.researchgate.net/publication/284181844, accessed October 28, 2023.

5. Kat Eghdamian, "What was the role of religion in the Arab Spring?," *Open Democracy* (July 28, 2014), https://www.opendemocracy.net/en/north-africa-west-asia/what-was-role-of-religion-in-arab-spring/, accessed November 1, 2023.

6. Bradley, "10 Years After the Arab Spring," see Conclusion.

7. Robinson and Merrow, "The Arab Spring at Ten Years?' see Displacement.

8. Salam, "The Arab Spring," 122.

9. Iffat Idris, "Analysis of The Arab Spring" Governance – Social Development – Humanitarian – Conflict (GSDRC) Applied Knowledge Services, Research Report (August 4, 2016): see Overview, https://gsdrc.org/publications/analysis-of-the-arab-spring/, accessed September 17, 2023.

10. Nick Schifrin and Ali Rogin, "10 Years After the Arab Spring: Democracy Remains Elusive in Egypt." PBS News Hour, February 11, 2023, see the video. Found at https://wwwsa.pbs.org/newshour/show/ten-years-after-the-arab-spring-democracy-remains-elusive-in-egypt, accessed October 27, 2023.

11. Schifrin and Rogin, "10 Years After the Arab Spring."

12. Schifrin and Rogin, "10 Years After the Arab Spring.".

13. Robert Burrell and John Collison, "The Resilience and Resistance Model: An Overview," *Resilience and Resistance: Interdisciplinary Lessons in Irregular Warfare, Competition, and Deterrence* (JSOU Press, 2025).

14. Burrell and Collison, "The Resistance and Resistance Model."

15. Burrell and Collison, "The Resistance and Resistance Model."

16. Burrell and Collison, "The Resistance and Resistance Model."

17. Burrell and Collison, "The Resistance and Resistance Model."

18. Idris, "Analysis of The Arab Spring," 1-3.

19. Idris, "Analysis of The Arab Spring," 1-3.

20. Idris, "Analysis of The Arab Spring," 1-3.

21. Idris, "Analysis of The Arab Spring," 1-3.

22. Idris, "Analysis of The Arab Spring," 1-3.

23. Idris, "Analysis of The Arab Spring," 1-3.

24. Idris, "Analysis of The Arab Spring," 1-3.

25. Idris, "Analysis of The Arab Spring," 1-3.

26. Salam, Salam, "The Arab Spring," 124.

27. Salam, Salam, "The Arab Spring," 124–127.

28. Idris, "Analysis of The Arab Spring," 1-2.

29. Elena Ianchovichina and Hair-Anh Dang, "Middle-class Dynamics and the Arab Spring," Bank Blogs, May 16, 2016, https://blogs.worldbank.org/arabvoices/middle-class-dynamics-and-arab-spring, accessed October 22, 2023.

30. Idris, "Analysis of The Arab Spring," 2.

31. Idris, "Analysis of The Arab Spring," 2.

32. Idris, "Analysis of The Arab Spring," 3.

33. Idris, "Analysis of The Arab Spring," 2.

34. Idris, "Analysis of The Arab Spring," 2.

35. Idris, "Analysis of The Arab Spring," 2–3.

36. Idris, "Analysis of The Arab Spring," 2.

37. Idris, "Analysis of The Arab Spring," 2.

38. Idris, "Analysis of The Arab Spring," 2.

39. Idris, "Analysis of The Arab Spring," 2.

40. Idris, "Analysis of The Arab Spring," 2.

41. Jon B. Alterman, "Radicalism Four Years into the Arab Spring," Center for Strategic and International Studies, December 15, 2014. Found at https://www.csis.org/analysis/radicalism-four-years-arab-spring, accessed October 24, 2023.

42. Alterman, "Radicalism Four Years into the Arab Spring," para. 3.

43. Alia Chughtai, "Timeline: How the Arab Spring Unfolded," *Aljazeera News* (January 14, 2021), see Egypt, https://www.aljazeera.com/news/2021/1/14/arab-spring-ten-years-on, accessed September 22, 2023.

44. Salam, "The Arab Spring," 135.

45. Michael Safi, Antonio Voce, Frank Hulley-Jones, and Lydia McMullan, "How the Arab Spring Engulfed the Middle East – and Changed the World," See the Guardian Interactive Video (January 25, 2021), https://www.theguardian.com/world/ng-interactive/2021/jan/25/how-the-arab-spring-unfolded-a-visualisation, accessed October 20, 2023.

46. Safi et al, "How the Arab Spring Engulfed the Middle East."

47. Robinson et al., "The Arab Spring at Ten Years: What's the Legacy of Uprisings?" see Standard of Living.

48. Robinson et al., "The Arab Spring at Ten Years: What's the Legacy of Uprisings?" para. 2.

49. Robinson et al., "The Arab Spring at Ten Years: What's the Legacy of Uprisings?" see Freedom of Speech.

50. Robinson et al., "The Arab Spring at Ten Years: What's the Legacy of Uprisings?" see Standard of Living and Youth Unemployment.

51. Safi et al. "How the Arab Spring Engulfed the Middle East," see Tunisia.

52. Chughtai, "Timeline: How the Arab Spring Unfolded," see Tunisia.

53. "Arab Spring Pro-Democracy Protests," Britannica Online, October 17, 2023, see para. Tunisia, https://www.britannica.com/event/Arab-Spring (Accessed October 6, 2023)

54. "Arab Spring Pro-Democracy Protests,"

55. "Arab Spring Pro-Democracy Protests,"

56. "Arab Spring Pro-Democracy Protests,"

57. Safi et al., "How the Arab Spring engulfed the Middle East," see Libya.

58. Chughtai, "Timeline: How the Arab Spring Unfolded," see Libya.

59. Chughtai, "Timeline: How the Arab Spring Unfolded," see Libya.

60. "Arab Spring Pro-Democracy Protests," see Libya.

61. "Arab Spring Pro-Democracy Protests," see Libya.

62. "Arab Spring Pro-Democracy Protests," see Libya.

63. Chughtai, "Timeline: How the Arab Spring Unfolded," see Libya.

64. Chughtai, "Timeline: How the Arab Spring Unfolded," see Libya.

65. Chughtai, "Timeline: How the Arab Spring Unfolded," see Libya.

66. Safi et al., "How the Arab Spring engulfed the Middle East – and Changed the World," see Egypt.

67. Schifrin and Rogin, "10 Years After the Arab Spring: Democracy Remains Elusive in Egypt." See video.

68. Schifrin and Rogin, "10 Years After the Arab Spring: Democracy Remains Elusive in Egypt." See video.

69. Schifrin and Rogin, "10 Years After the Arab Spring: Democracy Remains Elusive in Egypt." See video.

70. Schifrin and Rogin, "10 Years After the Arab Spring: Democracy Remains Elusive in Egypt." See video.

71. Schifrin and Rogin, "10 Years After the Arab Spring: Democracy Remains Elusive in Egypt." See video.

72. Schifrin and Rogin, "10 Years After the Arab Spring: Democracy Remains Elusive in Egypt." See video.

73. Schifrin and Rogin, "10 Years After the Arab Spring: Democracy Remains Elusive in Egypt." See video.

74. Schifrin and Rogin, "10 Years After the Arab Spring: Democracy Remains Elusive in Egypt." See video.

75. Schifrin and Rogin, "10 Years After the Arab Spring: Democracy Remains Elusive in Egypt." See video.

76. Kali Robinson and Will Merrow, "The Arab Spring at Ten Years: What's the Legacy of Uprisings?" see Egypt.

77. Schifrin and Rogan, "10 Years After the Arab Spring," see the video.

78. Schifrin and Rogan, "10 Years After the Arab Spring," see the video.

79. Bruce Riedel, "The Houthis have won in Yemen: What Next?" Brookings Institute (February 1, 2022): para. 5-8, https://www.brookings.edu/articles/the-houthis-have-won-in-yemen-what-next/, accessed September 16, 2023.

80. Safi et al., "How the Arab Spring engulfed the Middle East," see Yemen.

81. Seth G. Jones, Jared Thompson, Danielle Ngo, Joseph Bermudez, and Brian McSorley, "The Iranian and Houthi War Against Saudi Arabia," (December 21, 2021): see Military Capabilities, https://www.csis.org/analysis/iranian-and-houthi-war-against-saudi-arabia, accessed October 25, 2023.

82. Jones et al., "The Iranian and Houthi War Against Saudi Arabia," see Irregular Warfare Campaign.

83. Jones et al., "The Iranian and Houthi War Against Saudi Arabia," see the Houthi Movement.

84. Jones et al., "The Iranian and Houthi War Against Saudi Arabia," see the Houthi Movement.

85. Jones et al., "The Iranian and Houthi War Against Saudi Arabia," see the Houthi Movement.

86. Jones et al., "The Iranian and Houthi War Against Saudi Arabia," see the Houthi Movement.

87. Riedel, "The Houthis have won in Yemen: What Next?" See What's Next section.

88. Jones et al., "The Iranian and Houthi War Against Saudi Arabia," see Irregular Warfare Campaign.

89. Riedel, "The Houthis have won in Yemen: What Next?" See What's Next section.

90. Jones et al., "The Iranian and Houthi War Against Saudi Arabia," see Irregular Warfare Campaign.

91. "Twelve Years-On from the Beginning of Syria's War," *Al-Jazeera News* (March 15, 2023): see How It Started, https://www.aljazeera.com/news/2023/3/15/twelve-years-on-from-the-beginning-of-syrias-war, accessed October 26, 2023.

92. Mona Yacoubian, "Syria Timeline: Since the Uprising Against Assad," United States Institute for Peace (January 1, 2021): para. 1-6, https://www.usip.org/syria-timeline-uprising-against-assad, accessed September 26, 2023.

93. "Twelve Years-On from the Beginning of Syria's War," see the United States.

94. Yacoubian, "Syria Timeline," para. 3-4.

95. Yacoubian, "Syria Timeline," para. 6.

96. "Twelve Years-On from the Beginning of Syria's War," para 2.

97. Safi et al., "How the Arab Spring engulfed the Middle East," see Syria.

98. "Twelve Years-On from the Beginning of Syria's War," para. 3.

99. United Nations, UN News Report: Syria in Numbers (January 2022), https://news.un.org/en/focus/syria, accessed October 30, 2023.

CHAPTER 4: HISTORY

I. Support to Resistance: Case Study of Cuba, 1959–1961

By Robert S. Burrell

ABSTRACT: While the 1961 U.S. invasion of Cuba, dubbed the Bay of Pigs Invasion, has received much public scrutiny, academic scholarship on this topic lagged behind for many years. One of the obstacles to scholastic investigation derived from the classification of primary material, which remained unattainable to academia for decades. Consequently, first accounts of the Bay of Pigs Invasion initially emerged under a cloud of secrecy and with some confusion. In contrast, this section principally relies on primary sources from the Central Intelligence Agency (CIA) using some material that was declassified in 2016. It argues that when the CIA sponsored an indigenous resistance on the island that proved unachievable, they grew to favor an amphibious invasion made up of a proxy land force of Cuban exiles. This important historical case study in unconventional warfare, contrived and executed in an era of strategic competition, stands apart as a failure but with applicable lessons in the modern era. Note: this section was previously published in *Inter Populum: Journal of Irregular Warfare and Special Operations.*[1]

OVERVIEW

In the early 1960s, the U.S. backed several Cuban resistance movements in an effort to oust the communist dictator Fidel Castro. The culminating event of this support resulted in what is commonly known as the Bay of Pigs Invasion—an event that stands apart as a disastrous failure due particularly to the poor strategic concept adopted by the CIA. The CIA's approach eventually evolved into a conventional land force made up of Cuban exiles who would then attempt to reinsert themselves into Cuba via an amphibious invasion. When the plan was executed, Castro responded quickly—with 10 times the number of troops—which killed or captured the entire CIA-backed force. As a case study on U.S. support to resistance, this section shows how and why the secret war in Cuba developed and eventually failed. It principally utilizes declassified sources.

BACKGROUND

The U.S.' relationship with Cuba became inextricably linked following the Spanish–American War in the late 19th century. From 1895 to 1898, Cuba fought its third war for independence from Spain. Watching its small neighbor struggle in the fight against European imperialism, the U.S. enthusiastically intervened, and the Spanish–American War began. A few months later, at the treaty of Paris in 1898, Spain relinquished all rights to the territory.

After two years of U.S. military occupation, in 1901 the U.S. Congress signed the Platt Amendment, which created an unequal treaty between the U.S. and Cuba. The U.S. withdrew its Army garrison, but the amendment allowed for U.S. dominance over Cuban policies and markets. Through the Platt Amendment and later treaties, the U.S. maintained its authority over Cuban sovereignty until 1934.

For 60 years, the Cuban economy was dominated by American retail, banking, oil, food, and trade companies. Meanwhile, Cuban income was derived primarily from tourism and sugar exports. A growing number of Cubans held animosity toward the U.S. , who they saw as interfering with their sovereignty and dominating the economy. With the rise of Marxist ideology during the Cold War, Cuba grew ripe for revolution.

FULGENCIO BATISTA

Fulgencio Batista came to power in 1933 and dominated Cuban politics for two decades. He previously served as Chief of Staff of the Cuban armed force and held two terms as president, from 1940–1944 and (after a successful coup d'etat) from 1952–1958. Batista's primary external support came from the U.S. —the U.S. government, U.S. business, and the American mafia. In accordance with the Truman doctrine and as a strategy to contain communism, the CIA assisted Batista in building a secret police force designed to eliminate communist movements. Batista reigned through fear and retribution, and a number of resistance movements sought his overthrow. On New Year's Eve 1958, as his legitimacy plummeted, Batista fled the country and his government collapsed.[2]

FIDEL CASTRO

Fidel Castro was a lawyer and politician in Havana. He formed a group called "the Movement" in 1952, which published an underground newspaper called *El Acusador*. A revolutionary with socialist ideology, Castro, nevertheless, avoided an alliance with the growing communist party in Cuba—the Popular Socialist Party. The CIA and the U.S. State Department feared Castro was communist and began investigating his movements from 1948 onward.[3]

Fidel Castro circa 1950s. Source: Public domain

On 26 July 1953, Castro, with his brother Raul and 150 armed comrades, attacked an army barracks in Santiago in an overtly iconic attempt to free political prisoners. They failed miserably. Half of his force was killed during the fighting and most of the rest executed. Castro was publicly tried as a criminal, but, due to domestic concerns and the popularity of what he had attempted, he only served two years in prison. Additionally, Castro's public trial served to increase his popularity among dissidents. His followers became known as the 26 July Movement (or M-26-7).

Directly after his release from prison in 1955, Castro left Cuba and established a headquarters in Mexico.[4] Three years later, he returned with 80 highly motivated supporters. Most of Castro's force did not survive initial contacts with Batista's army. He then embarked on a guerilla warfare campaign—attacking Batista's government at weak points followed by a highly successful media campaign. While his resistance only consisted of a small armed component, Castro simultaneously allied himself with vibrant underground movements in urban centers. The declared aim of Castro and M-26-7 was put in simple terms—revolution—for which the many Cuban resistance groups could agree.

By the CIA's estimates, most Cubans supported a revolution by 1957. Batista was no longer able restore order to the mounting unrest. Opposition included (a) rebels supporting the previous president, Carlos Prio; (b) Castro's guerillas; (c) domestic political opposition parties; (d) workers' unions; and (e) even some in Batista's military.[5] By August, the capital in Havana was rife with public demonstrations, strikes, and publicized terrorist attacks.[6] After only a short time in power, Batista had proven himself very unpopular with many segments of the population. By default, Castro absorbed widespread popularity through resistance to Batista. Castro also wielded two strong subordinates who proved themselves central to his revolution. One was his brother Raul. The other was Ernesto Che Guevara, a Marxist revolutionary from Argentina.

Castro's revolution succeeded in 1959, and he triumphantly marched into Havana unopposed. Afterward, he consolidated his power and control as a dictator. One of Castro's immediate goals was the nationalization of foreign-owned business, principally those owned by persons in the United States. The U.S. government responded with sanctions and denunciated Castro's regime. Castro reciprocated by establishing diplomatic relations with both the Soviet Union and communist China. To the U.S. government's alarm, the Soviet Union became an overt external sponsor to Cuba both economically and militarily.

DIASPORA

Once Castro consolidated power into a dictatorship, he immediately alienated many other revolutionaries, some of whom opposed communism and desired democracy and continued relations with the United States. Others wanted socialism but not a dictatorship. As summarized by one scholar, "a number of leaders who shared power with the 26th of July in Castro's early governments held democratic, liberal principles; many of them had fought dictatorship in Cuba even before Batista."[7] In 1959 and 1960, Cuba realized a large exodus of dissidents. A diaspora composed of more than 100,000 influential Cubans settled primarily in Miami, Florida.

Hundreds of political movements evolved within the diaspora with divergent aims against the Castro regime. Their ambitions ranged from regaining privileges (like family estates) to complete regime change. Cuban exiles included right-wing supporters of Batista as well as democratic enthusiasts who opposed both Castro and Batista. However, none of the dissidents had enough charismatic leadership to unify resistance to Castro abroad, let alone at home.[8] There were hundreds of exiled groups, but several primary groups emerged that would be important to the eventual Bay of Pigs Invasion. These groups would tentatively unify under the Cuban Revolutionary Council.[9]

CUBAN REVOLUTIONARY COUNCIL

The Cuban Revolutionary Council (or *Consejo Revolucionario de Cuba*) was formed in March of 1961, just three weeks before the invasion. Its formation was highly influenced by the desire of U.S. President John F. Kennedy to demonstrate united opposition to Castro. Consequently, this organization represented a confederation of various resistance movements. Some were pro-Batista and sought to reinstate the Batista regime. José Miró Cardona, the former prime minister of Cuba, emerged as the group's central leader. While Cardona was marked as the provisional president in the event of Castro's demise, these disparate groups had several other objectives, many of which were incongruent with one another.

PEOPLE'S REVOLUTIONARY MOVEMENT

The People's Revolutionary Movement (or *Movimento Revolucionario del Pueblo*) consisted of former 26 July Movement members who had become dissatisfied with Castro after he assumed power. It formed in May 1960 under the leadership of Manuel Ray, a former Castro government official and Minister of Public Works, who disagreed with the increasing communist ideology of the Cuban government. Ray eventually joined the Cuban Revolutionary Council just prior to the invasion.

DEMOCRATIC REVOLUTIONARY FRONT

The Democratic Revolutionary Front (or *Frente Revolucionario Democratico*) was a resistance group led by Miró Cardona. This organization initially headquartered in Mexico but eventually founded a chapter in New Orleans. It consisted of five major anti-Castro groups, and opposition to Castro may have been its singular identifying theme. U.S. statesman Charles Thayer would describe it as a "highly artificial organization without any genuine political solidarity and the tendency to fall apart at the slightest provocation."[10] In October 1961, the Democratic Revolutionary Front finally merged with the Cuban Revolutionary Council.

MOVEMENT FOR REVOLUTIONARY RENEWAL

As Cuban resistance attempted to centralize, the Democratic Revolutionary Front aligned with the People's Revolutionary Movement and the two formed the Movement for Revolutionary Renewal (or *Movimento De Recuperacion Revolucionaria*). The leader of the Movement for Revolutionary Renewal was Manuel Francisco Artime, a former Castro supporter. As it received CIA sponsorship, the Movement for Revolutionary Renewal took on an armed component structure and would form the basis for Brigade 2506 that would invade Cuba.

BRIGADE 2506

Brigade 2506, sometimes referred to as the Cuban Brigade or the Blindado Battalion, evolved into the armed component of resistance. Recruitment for the Cuban Brigade began in

May of 1960 from anti-Castro organizations like the Democratic Revolutionary Front. It consisted of "students, workers, former Castro supporters, former Army personnel, professionals, the rich, the poor, and the middle class."[11] The number 2506 represented the secret identifier for Carlos Rafael Santana Estevez, a member who died in a training accident in Guatemala in September 1960.

Flag of the Cuban Brigade or Brigade 2506. Source: MrPenguin20/WikiCommons

At first, the CIA sent 40 men from Brigade 2506 to train on a small island in the Caribbean. Initially, the training focused on guerrilla warfare. Then, in the summer of 1960, the CIA began airlifting brigade members to training camps in Panama and eventually to Guatemala. Training was supervised by American professionals, and the volunteers received excellent equipment.[12] By November 1960, the plan changed entirely from guerilla warfare to a conventional invasion. This transformation may have occurred due to the large numbers of available Cubans recruited by the Brigade.[13] As the Brigade grew, political leadership remained with Manuel Artime. Artime also led the Movement for Revolutionary Renewal organization. Military command fell to José Alfredo Pérez San Román, commonly referred to as Pepe.

Recruitment was overt. The brigade made public statements about their undertakings, which, in hindsight, appear incongruent with the secrecy required for the upcoming invasion. On 6 February 1961, Artime held a news conference in his house in Miami. He stated that the U.S. was training Brigade 2506, made up of 1,400 to 1,500 members. He also confirmed that the U.S. Army, Navy, and Air Force were conducting the training in various locations within the United States.[14] In short, it was no secret to Castro that the U.S. was planning an invasion.

There were about 375 officers in Brigade 2506. Each received about 13 weeks of training, which initially included guerilla warfare. Additionally, many of these officers already had guerilla warfare experience from their fights against Batista. As time progressed, the CIA attempted to train Brigade 2506 to operate as a conventional force and "wean" the officers from their "marked inclination to guerilla operations."[15] Table 4.1 indicates the general organization of the Brigade prior to the invasion.

TABLE 4.1. BRIGADE 2506 GROUND, SEA, AND AIR FORCES

UNIT	PERSONNEL	WEAPONS
Headquarters and Service Company	156	Rifles
Heavy Weapons Company	114	50 caliber MGs 81 mm mortars 4.2" mortars 75 mm recoilless rifles Flamethrowers 2.5 ton trucks
(5) Infantry Companies	175 each	30 caliber MGs 60 mm mortars 57mm recoilless rifles Browning Automatic rifles ¼ ton trucks
Airborne Infantry Company	177	Browning Automatic rifles Rifles
Tank Platoon	24	M41 Tanks
Boat Section	36	Unknown
Intelligence/Recon	68	Rifles
AIR FORCES		
15 x B-26 light bombers		
10 x C-54 transports		
5 x C-46 transports		
SEA FORCES		
(2) Landing Craft Infantry		11 x 50 caliber MG each 2 x 75 mm rifles each
(3) Landing Craft Utility		2 x 50 caliber MG each
(4) Landing Craft, Vehicle, Personnel		50 caliber MG each
(7) Chartered commercial freighters		3 x 50 caliber MG each

Source: Author

The closer the invasion became, the more the CIA advocated for a conventional defeat of Castro in pitched battle. In the five months of training leading up to the operation (between November 1960 and March 1961), conventional tactics were prioritized over all else. The concept of operations included insertion by sea and air, constitution of a land force, and then a conventional fight to defeat Castro's army. By one CIA estimate, "at no time did the brigade once organized receive training to fight as a guerrilla force."[17] However, if the desired direct confrontation were to result in defeat, one secondary course of action included dispersion into the hills and countryside including the Escambrays, Pinar del Rio, and the Oriente. Nevertheless, the CIA offered little to no training on the secondary option.

UNCONVENTIONAL WARFARE

President Dwight Eisenhower's Anti-Castro Program started 17 March 1960 and was eventually codenamed JMATE.[18] The principal U.S. agent supporting Cuban resistance remained the CIA. Due to Castro's effective intelligence efforts, the CIA kept the operations on close hold, leaving a number of important players, including large segments of the Department of State and the Department of Defense and even the American Embassy in Havana, out of the planning.[19]

To fund its efforts, the CIA requested monies in 1960 and 1961 for five principal activities. It requested $950,000 to support opposition elements in the Cuban diaspora. It requested $1,100,000 for radio broadcast operations directed against Castro. It asked for $600,000 for publications and press. It asked for $250,000 for intelligence collection. And, finally, it requested $1,500,000 for maritime and air training and equipment to enable a paramilitary force.[21]

The CIA established a network of U.S. bases and stations to support Brigade 2506 activities, including a forward operating base in Miami, Florida at Opa Locka Naval Air Station. The

| The CIA's "Anti-Castro Program" had four main lines of effort.
1. Create unified opposition to the Castro regime within the Cuban diaspora.
2. Create opposition to Castro in the indigenous Cuban population by using psychological operations and mass propaganda.
3. Create resistance on the island of those who can work in concert with the resistance in the diaspora.
4. Form an armed component of resistance, a paramilitary force, made up of members of the diaspora. |

airfield served as a storage point for arms and equipment and was the port of embarkation for personnel flights to Central America. It also operated a field and facilities in the Florida Keys due to its proximity to Cuba. Eglin Air Force Base in Florida supported logistics flights to Central America. Additionally, Vieques, Puerto Rico, provided the original training base for maritime activities.[22]

SUPPORT TO RESISTANCE

The CIA's initial plans for JMATE mirrored the Jedburgh program utilized in France during World War II. It contemplated the training of small units composed of two to three persons: (1) a communications expert, (2) a guerilla warfare trainer/advisor, and (3) possibly a native Cuban from the diaspora. Such a plan, conceived as Jedburgh, should come as no surprise. The CIA agent in charge of training during this period was Gerry Droller (code name Frank Bender), a former member of the French resistance in World War II.[23]

From September 1960 to April of 1961, the CIA began inserting agents into Cuba. Seventy personnel deployed in small teams, including 19 radio operators. All but two of the radio operators succeeded in establishing communications following insertion.[24] However, in the following weeks and months, most were killed or captured.

Materiel support to resistance had uneven results. Air delivery of support was generally unsuccessful. As a result of poorly trained Cuban pilots, only four of 27 missions achieved their objectives. In comparison, supply runs via sea fared better. Boats from Miami to Cuba delivered 40 tons of arms and equipment.[25]

The most prominent resistance element in Cuba identified by the CIA was a 600-1,000 person force operating in the Escambray Mountains of Las Villas Provence. However, the CIA never made direct radio contact with these guerillas and could only relay information through underground organizations in Havana. Due to Castro's well-organized intelligence apparatus in Cuba and Miami, the armed component in the Escambray Mountains never succeeded in gaining popular support—and apparently operated only for about six months.[26]

The CIA supported a rather extensive sabotage campaign. Much of this was industrial disruption including sugar cane fields, warehouses, refineries, railroads, and power stations. The targets were selected to disrupt Castro's support and cause confusion, but, by the CIA's own admission, they had no substantial direct impact on the regime.[27]

INVASION BY PROXY FORCES

The real nail in the coffin in the CIA's support to resistance lay in a couple factors. First, it maintained little direct interaction with indigenous resistance, a situation exacerbated by Castro's effective counterintelligence efforts. Second, the failure of aerial resupply made effective support to armed resistance elements impossible. With little perceived progress in support to indigenous resistance, the Jedburgh approach was generally shelved in favor of an amphibious assault by a proxy army. Planners began to fixate on counting the number of troops, tanks, vehicles, and seacraft required to establish a foothold on the island.[28] By November 1960, the plan gelled as a "conventional amphibious assault force of least 1,500 infantrymen."[29]

In 1961, the CIA established a base of operations in Guatemala as a staging point for the invasion. It also utilized Puerto Cabezas in Nicaragua. Invading from Central America would maintain the covert sponsorship of the United States. Brigade 2506 stood up its forces for the invasion from bases in Guatemala and Nicaragua from 25 March to 7 April.

Guatemala had officially severed relations with Castro, and President Miguel Ydigoras Fuentes offered the CIA use of his nation to conduct psychological operations, establish training areas, and operate a staging base. Ydigoras' government was quite unstable, and he desired U.S. support to shore up his domestic popularity. Castro may have been sponsoring resistance to Ydigoras at the time. For the president's support, Ydigoras received a number of payouts in loans and aid as well as arms and equipment. The CIA planned to use Brigade 2506 to back his regime if needed and did fly air sorties in support of the Guatemalan army during a revolt in late 1960.[30]

The CIA chose an airstrip in Guatemala near the city of Retalhuleo for renovation and made it large enough to accommodate C-54 aircraft. By 30 September 1960, the U.S. engineers had extended the airstrip to 5,000 feet. The B-26 and C-46 aircraft utilized in JMATE were sold to the Guatemalans as a cover; they turned over these aircraft to Brigade 2506. Within the JMATE program, JMTRAV made up the ground training program, while airfield operations were named JMADD.[31]

Cuban ground forces in Brigade 2506 received some training by U.S. Army Special Forces. Initially, this occurred at Fort Randolph in the Panama Canal zone. Training generally consisted of communications and sabotage. Later, in December 1960, the training moved to Guatemala, at which point 21 agents and five staff personnel had joined the JMATE project. At this point, the CIA requested 38 Special Forces personnel to assist in training requirements. The U.S. Army was concerned about the legal status of its Soldiers and desired Guatemala to sign a Status of Forces Agreement, which it eventually did. CIA agents provided a large part of the paramilitary training, but Special Forces Soldiers conducted training of the brigade and trained a "Guatemalan Battalion" to "deter the revolutionary activities against the Ydigoras Government."[32]

The CIA maintained considerable interest in roping in Guatemala as a regional sponsor of the Cuban resistance. It wanted both Guatemalan pilots and soldiers integrated within Brigade 2506. The CIA also desired recruitment of foreign mercenaries to supplement the Cuban ranks, including "Germans, Greeks, and Turks."[33]

The U.S. State Department vehemently opposed these ideas, and they never took root. In fact, the State Department opposed the invasion of Cuba throughout the planning, primarily based on the potentially negative repercussions to the U.S. reputation internationally.[34]

Castro was well aware of the CIA activities in Guatemala. He likely had operatives within the diaspora, or at least was collecting intelligence from them. He requested support from the Organization of American States to inspect Guatemala and ensure it was not being used as a staging area for a future invasion. They agreed but never found any evidence.[35] The CIA's concerns about maintaining secrecy around the future invasion likely hampered its ability to share information with many important U.S. partners.

The tentative diplomatic situation in Guatemala gave rise to the idea of using Nicaragua as a second staging area. The CIA negotiated with President Luis Somoza Debayle. Somoza was an unpopular dictator, so the CIA kept negotiations clandestine to maintain the U.S. image within the international community. Eventually, in late 1960, Somoza agreed to the use of the airport at Puerto Cabezas. Additionally, he allowed Brigade 2506 airplanes to use Nicaraguan Air Force insignias to mask their origin. CIA operations in Nicaragua became known as JMTIDE.

> ***The tentative diplomatic situation in Guatemala gave rise to the idea of using Nicaragua as a second staging area.***

THE INVASION

In early 1961, the CIA briefed the newly inaugurated President John F. Kennedy on JMATE and the upcoming Cuban invasion. To establish air and maritime control, the U.S. Navy had been tasked to deploy a large amphibious force near Cuba under the guise of a training exercise. In reality, this armada was forward postured to support Brigade 2506. The CIA envisioned the U.S. amphibious component as directly supporting the invasion, with direct use of military capabilities if required.

While Kennedy gave the go ahead to proceed with JMATE, he subsequently grew increasingly concerned about the ability of the U.S. to maintain a covert status. The upcoming invasion increasingly relied on U.S. military support, both in terms of air strikes and resupply. As Kennedy's concern over international implications grew, he would become increasingly reluctant to offer direct U.S. support, even for support that the CIA had planned on. Figure 4.1 portrays the region and operational situations regarding the invasion.[36]

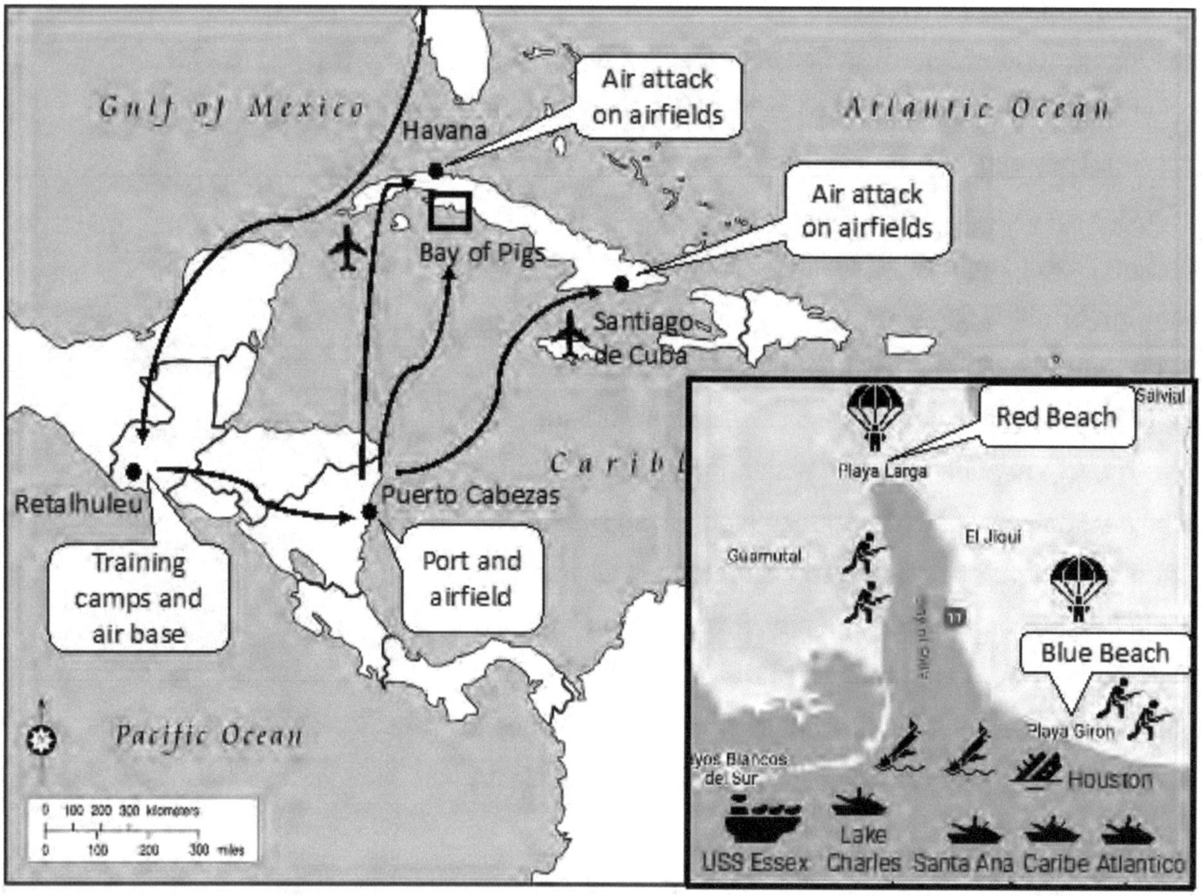

Figure 4.1: Brigade 2506's Invasion of Cuba at the Bay of Pigs, 15-20 April 1961 (Map Data @ 2024 Google)

On 15 April 1961, the invasion began with air strikes flown by Cuban pilots in CIA procured planes. Eight aircraft left from Puerto Cabezas, Nicaragua to drop bombs on airfields in Cuba at Santiago and Ciudad Libertad (16 were originally planned, but this figure was scrubbed by Kennedy). These air strikes damaged a number of Cuban planes, but not enough to impede Castro's air force from gaining air supremacy during the invasion.

On 17 April, Brigade 2506 launched four battalions in amphibious assaults on Red Beach (Playa Larga) and Blue Beach (Playa Giron). Simultaneously, a fifth battalion of paratroopers seized key terrain at Jocuma and San Miguel do Pita. Once Brigade 2506 established the beachhead and the airfield was secured, its forward momentum stalled, and it established a defensive posture. Accounts differ, but Brigade 2506 insists that the CIA told them to halt and wait for U.S. air and naval support.

For certain, the U.S. had promised the brigade resupply via air and clandestine maritime aircraft. These did not arrive. While the aircraft carrier USS *Essex* and accompanying task force had deployed to support the invasion force with tactical air support, Kennedy decided not to use overt U.S. military force, and the USS *Essex*, as well as other aircraft in the U.S., were told to stand down. With no air cover, two support ships, the *Houston* and the *Rio Escondido* were sunk by Cuban planes. Castro maintained air supremacy over the invasion area as well as artillery fires, for which Brigade 2506 had no response. Meanwhile, the brigade increasingly ran short on ammunition, a situation exacerbated in the following days.

In a short three days, Castro had mobilized thousands of soldiers in an aggressive counterattack, a force that included members of his Cuban Revolutionary Army Forces as well as the National Revolutionary Militia. In the preceding years, the Soviet Union had poured 40,000 tons of military equipment into Castro's regime, which he used to organize his army and militia forces.[37] Total strength of the attacking forces was overwhelming and included men, tanks, bombers, jets, and artillery. Running short on ammunition and with no U.S. support forthcoming, Brigade 2506 surrendered on 20 April. Casualties included 118 killed, 360 wounded, and 1,202 captured.

Of the captured personnel, a number were executed. Others were put on trial—Castro put 1,189 soldiers from Brigade 2506 on trial and sentenced them to imprisonment for 30 years. Twenty months later, the U.S. paid $53 million in food and medicine for their release and return to Florida.

AFTERMATH

David had defeated Goliath. The botched invasion of Cuba provided Castro with more popular support at home and abroad than ever. The invasion also ensured Castro's cooperation with the Soviet Union and precipitated the Cuban Missile Crisis in 1962. Many of the members of the Cuban Brigade blamed the U.S. for the disaster—no air support came, no resupply occurred, and no promised reinforcements arrived. Additionally, some of the U.S-provided equipment failed to work properly, including the landing boats and tanks.[39] The CIA became especially critical of the Kennedy administration for not providing air cover to the brigade. As summarized by case officer E. Howard Hunt, the Cuban exiles "were betrayed by America."[40]

The blame placed on Kennedy remains misplaced. He inherited a plan with many shortcomings. Castro successfully carried out a popular revolution in Cuba and ousting him would require a similar grassroots effort. Castro's dictatorship alienated many Cubans, and the opportunity certainly existed for a counterrevolution.[41] However, the title "Bay of Pigs Invasion" implies everything that was wrong with this operation. *Bahia de Cochinos* is a geographic reference, and the term "invasion" entails a conventional military maneuver. Successful resistance has little to do with either geographic terrain or military maneuver. Instead, resistance competes over human terrain instead of geography; the population, not armies, remains the center of gravity.

| Successful resistance has little to do with either geographic terrain or military maneuver. Instead, resistance competes over human terrain instead of geography; the population, not armies, remains the center of gravity. |

The U.S. failed to develop support to indigenous resistance in Cuba. Instead, it defaulted to its own proclivity for conventional warfare. By the time Brigade 2506 arrived in Cuba, it no longer resembled the organization needed to inspire a Cuban uprising. Armed Cuban exiles had become a U.S. surrogate force, rather than an indigenous resistance.

The CIA had grossly underestimated Castro's army and militia, including their strength, capabilities, and morale. It conducted little psychological operations on the population to influence their loyalties. In fact, the CIA appears to have had little access, knowledge, or placement within the Cuban population. Consequently, optimistic assessments that the Cubans would rise up to oppose Castro following the invasion did not have any basis in evidence. An additional criticism of the CIA, which has merit, is that the Agency could not handle an operation of this size and magnitude and simply became overwhelmed by events.[42]

About the Author

Dr. Robert S. Burrell is a resilience and resistance interdisciplinary scholar using data-driven and human-centric methodologies to analyze intrastate conflict ranging from nonviolent protest through belligerency. He is a Senior Research Fellow at the Global and National Security Institute of the University of South Florida. From 2020–2024, he taught irregular warfare at Joint Special Operations University and was the editor in chief of special operations doctrine from 2011–2014.

NOTES

1. Robert S. Burrell, "The Secret War in Cuba: The Bay of Pigs Invasion," *InterPopulum: The Journal of Irregular Warfare and Special Operations*, Winter 2023.

2. Jack B. Pfeiffer, *The Official History of the Bay of Pigs Invasion: Volume III, Evolution of CIA's Anti-Castro Policies*, 1959-January 1961, (Central Intelligence Agency, 1979). Found in CIA Archives. Between 1979 and 1984, Central Intelligence Agency (CIA) staff historian Jack Pfeiffer prepared five volumes of the Agency's Official History of the Bay of Pigs Operation. The titles of the five volumes were *Air Operations, March 1960-April 1961; Participation in the Conduct of Foreign Policy; Evolution of CIA's Anti-Castro Policies, 1951-January 1961; The Taylor Committee Investigation of the Bay of Pigs*; and *CIA's Internal Investigation of the Bay of Pigs.* Another historical account from the CIA was written by J. Hawkins, *Record of Paramilitary Action Against the Castro Government of Cuba, 17 March 1960 – May 1961*, (Clandestine Services History, 5 May 1961). Hawkins' history was sanitized and declassified in 1997.

3. Pfeiffer, *The Official History of the Bay of Pigs Invasion: Volume III*, 1.

4. Michael Dunne, "Perfect Failure: The USA, Cuba and the Bay of Pigs, 1961," *Political Quarterly*, vol. 82, no. 3, July 2011, pp. 448–58.

5. Memorandum for the Director of the Central Intelligence Agency, "The Situation in Cuba," *Central Intelligence Agency*, 13 June 1957. Found in CIA Archives.

6. *Central Intelligence Agency Bulletin*, 6 August 1957, 5. Found in CIA Archives.

7. Victor Andres Traiy, "Castro's Agenda Prompted the Formation of the Anti-Castro Militia," found in Haynes Johnson, *The Bay of Pigs: The Leaders' Story of Brigade 2506*, (Greenhaven Press, 2004), 18.

8. Charles W. Thayer, *Guerilla*, (Harper and Row: New York, 1963), 16.

9. Dunne, "Perfect Failure," 452-453.

10. Thayer, *Guerrilla*, 17.

11. Traiy, "Castro's Agenda," 23.

12. Manuel Panabaz, "The U.S. Betrayal of the Cuban Militia Caused Heavy Losses," found in Haynes Johnson, *The Bay of Pigs: The Leaders' Story of Brigade 2506*, (Greenhaven Press, 2004), 45.

13. "The Real Story of the Bay of Pigs," *U.S. News and World Report*, 7 July 1963. Accessed from CIA archives.

14. "U.S. to Train Members of Cuba Brigade," *AP Press*, 6 February 1961. Found in CIA Archives.

15. "What Briefing, if Any, Was Given the Brigade or the Brigade's Staff on Going Guerilla," *Central Intelligence Agency Memorandum*, 22 May 1961. Found in CIA Archives.

16. Information in this table drawn from Hawkings, 25-26. I've made some assumptions about which weapons Brigade 2506 assigned to individual companies.

17. "What Briefing, if Any, Was Given the Brigade or the Brigade's Staff on Going Guerilla," *Central Intelligence Agency Memorandum*, 22 May 1961. Found in CIA Archives.

18. Jack B. Pfeiffer, *Official History of the Bay of Pigs Operations: Volume II, Participation in the Conduct of Foreign Policy*, (Central Intelligence Agency, 1979), footnote on page 4.

19. Dunne, "Perfect Failure," 453.

20. Jack B. Pfeiffer, *Official History of the Bay of Pigs Operations: Volume I, Air Operations, March 1960 – April 1961*, (Central Intelligence Agency, 1979), 408-410.

21. Pfeiffer, *Official History of the Bay of Pigs Operations: Volume I*, 417.

22. Hawkins, *Record of Paramilitary Action*, 2-4.

23. Panabaz, "The U.S. Betrayal of the Cuban Militia," 46.

24. Hawkins, *Record of Paramilitary Action*, 7-9.

25. Hawkins, *Record of Paramilitary Action*, 7-9.

26. Hawkins, *Record of Paramilitary Action*, 7-9.

27. Hawkins, *Record of Paramilitary Action*, 7-9.

28. Pfeiffer, *The Official History of the Bay of Pigs Invasion: Volume III*, 143-144.

29. Pfeiffer, *The Official History of the Bay of Pigs Invasion: Volume III*, 147.

30. Pfeiffer, *Official History of the Bay of Pigs Operations: Volume II*, 35-38.

31. Pfeiffer, *Official History of the Bay of Pigs Operations: Volume II*, 1-28.

32. Pfeiffer, *Official History of the Bay of Pigs Operations: Volume II*, 93.

33. Pfeiffer, *The Official History of the Bay of Pigs Invasion: Volume III*, 118.

34. Haynes Johnson, *The Bay of Pigs: The Leaders' Story of Brigade 2506*, (Greenhaven Press, 2004), 11.

35. Pfeiffer, *Official History of the Bay of Pigs Operations: Volume II*, 1-28.

36. Map created by author. For a similar map, see Norman Friedman, "Debacle at Bahía de Cochinos," Naval History Magazine, April 2021. Also see Maria C. Werlau, "The Bay of Pigs – 50 Years Later," SFPPR News and Analysis, 20 July 2011. Graphic from Werlau originates from the Miami Herald.

37. Hawkins, *Record of Paramilitary Action*, 6.

38. Johnson, *The Bay of Pigs*, 11-13.

39. "The Real Story of the Bay of Pigs," *U.S. News and World Report*, 7 July 1963. Accessed from CIA archives.

40. Matthew C. Quinn, "Fiasco of 1961 Regretted by New, Young President," *Washington Times*, 17 April 1986.

41. Thayer, *Guerilla*, 15-16. Also see Chesly Manly, "Donovan Acts as Plan Fails at Bay of Pigs," *Chicago Tribune*, 11 January 1966.

42. Lyman Kirkpatrick, "The CIA Could Not Adequately Handle the Bay of Pigs Operation," found in Haynes Johnson, *The Bay of Pigs: The Leaders' Story of Brigade 2506*, (Greenhaven Press, 2004), 71-74.

II. Support to Resilience: Case Study of Laos, 1954–1975

By Robert S. Burrell

ABSTRACT: During the Vietnam War, the U.S. State Department and the CIA oversaw the largest sponsorship of a foreign paramilitary force in U.S. history. The primary objective of Operation MOMENTUM was the defense of the Royal Laos Government from an aggressive communist takeover. Indigenous communist belligerents included the local Pathet Lao and the neighboring Viet Minh from North Vietnam. In an era of great power competition, Communist China and the Soviet Union energetically supported these Indochinese communist movements. Correspondingly, the U.S. and its ally, the Kingdom of Thailand, eventually committed to arm, train, and equip a minority ethnic group in Laos, the Hmong. This case study provides important lessons in the use of irregular warfare utilized during an age of strategic competition. Most of this research draws upon declassified sources from the CIA Archives.

OVERVIEW

During the Vietnam War in the 1960s and 1970s, the U.S. State Department and the CIA sponsored resilience activities in the Kingdom of Laos. Many scholars and journalists refer to this as the Secret War, although it was not so secret. Generally, the intent of the U.S. effort was to harden and protect the population from the violent overthrow of communists. Laos faced the threat of a communist takeover from inside the country as well as the projection of military forces from Communist North Vietnam into Laos. Simultaneously, Laos served as a transit route for North Vietnamese Army (NVA) troops and supplies between north and south Vietnam. Dubbed the "Ho Chi Minh Trail," this road became a frequent target for the U.S. Air Force, providing Laos the infamous trophy as the most bombed nation ever.[1] Laos also served as the stage for the largest CIA paramilitary operation ever conducted. For all these reasons, Laos provides an excellent case study of U.S. support to resilience during a period of great power competition with the Soviet Union and Communist China.

BACKGROUND

The struggle in Vietnam obviously influenced America's policy in Laos. French Indochina (consisting of the current nations of Cambodia, Vietnam, and Laos) had been in a continual state of war from at least 1939 onward.[2] The constitutional monarchy emerged in 1953 but never realized enough time to solidify its legitimacy for self-rule. Additionally, the Laos people remained disunified with separate ethno-linguistic groups as well as religious and political ideologies. Essentially, there remained little national identity to adequately counter internal unrest. Weak governance coupled with Cold War competition made Laos ripe for insurgency. Various resilience or resistance groups found sponsorships in neighboring countries like North Vietnam, Thailand, and China but also with Cold War competitors like the Soviet Union and the United States.

The U.S. policy pertaining to Laos began as early as 1946. During the First Indochina War, France attempted to reassert its colonial claims in Southeast Asia. Initially, the U.S. opposed European colonialism, instead advocating for a new international order built on freedom and democracy. Nevertheless, President Dwight Eisenhower's policy in 1950 of containing communism marked a change in priorities. Eisenhower put forward the "domino theory" and argued that communism had to be fought globally to prevent the collapse of democracies one at a time. To keep France democratic and to fight communism in Europe and abroad, the U.S. began funding France's efforts in Indochina. In 1951, military aid to France equated to $425 million; in 1952 it was $520 million. From 1950 through the French defeat at the Battle of Dien Bien Phu in 1954, the U.S. provided billions of dollars to fund French efforts in defeating communist resistance, primarily in Vietnam but also in Cambodia and Laos.[3]

Most Americans associate U.S. activities in Southeast Asia with the Vietnam War, and the U.S. intervention in Laos was certainly connected to those efforts. Following its defeat in the First Indochina War, France withdrew from Southeast Asia. By agreement in the 1954 Geneva Conference, four new states were agreed upon: North Vietnam, South Vietnam, Cambodia, and Laos.[4] To counter Thailand's ambitions of regaining territories from the French, the Kingdom of Laos had obtained full independence from France a year earlier in 1953. Meanwhile, the U.S. supported the creation of the new nation of South Vietnam, which took shape in 1955 as the Republic of Vietnam.

Supporting the advancement of the Republic of Vietnam necessitated standing up and training the Army of the Republic of Vietnam (ARVN). Initially, this effort included U.S. advisors and equipment. For a decade, ARVN attempted to combat the NVA, but from 1965 onward the U.S. gradually increased U.S. ground forces to take part in direct actions. However, less well known is the conflict in Vietnam much prior to 1965 that escalated throughout Southeast Asia—commonly referred to as the Second Indochina War. As an example of the regional connotations, the U.S. Air Force began operating from the Kingdom of Thailand in 1960 to support the ARVN. At about the same time, the CIA began support to resilience operations against communist influence in Laos.[5]

Historians refer to the complex and violent period in Laos between 1959 and 1975 as the Laotian Civil War. On one side, the Laos monarchy was backed by a national Royal Lao Army (a composite force of various ethnic groups), which was supported externally by partnerships with the U.S. , South Vietnam, and Thailand. On the opposing side, the Communist *Pathet Lao* (both a political party and a military force) was heavily influenced and sponsored by North Vietnam and ultimately the Soviet Union and Communist China.[6] Importantly, neither the Royal Lao Army nor the Pathet Lao had the inherent force to defeat each other or to unify the country. Within this context, the stage was set for a proxy war with external nations attempting to settle the conflict, and ultimately, the future of Laos.

SUPPORT TO LAOS RESILIENCE

The Viet Minh Front (the Vietnamese communists) had begun sponsoring a communist resistance in Laos as early as 1950—the Pathet Lao. In actuality, the Pathet Lao developed as a surrogate for the Viet Mihn to resist French occupation. Following the French defeat at Dien Bien Phu, the Pathet Lao realized international recognition at the Geneva Conference of 1954, which gave it de facto autonomy in the two most northern provinces bordering Vietnam—Sam Neua (also called *Huaphanh*) and Phongsaly. However, in 1954, the Pathet Lao already controlled all of Huaphanh, most of Xiengkhuan, and parts of the Luangprang and PhongSaly provinces. More importantly, the Pathet Lao had sanctuary in and support from North Vietnam. The central Lao government remained just as disunified as the population, with three opposing political parties: (1) a party that supported the constitutional monarchy, (2) a communist political party and paramilitary force, (3) and a neutralist party attempting to bring about internal peace. Fed up with discussions and compromise, the Pathet Lao in 1959 embarked on a guerrilla warfare campaign against the Royal Lao Government to seize control over the country through force. Both the NVA and the Soviet Union served as vital partners to the Pathet Lao.[7] See Figure 4.2.

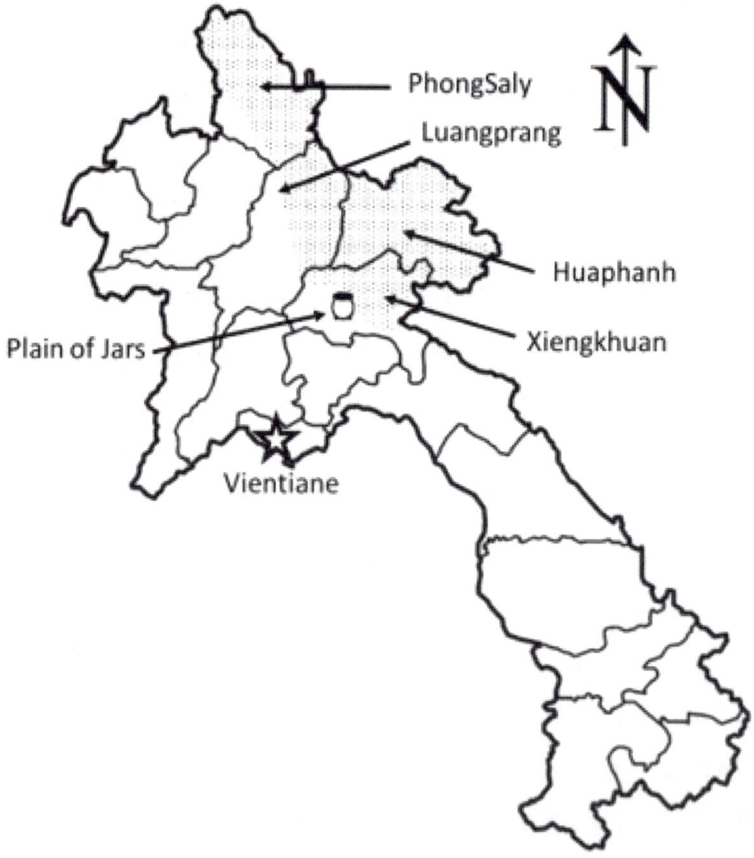

Figure 4.2. Map of Pathet Laos progress in 1954. Source: Author

The Pathet Lao commenced military operations against the forces supporting the constitutional monarchy from bases in the east. As the situation unfolded, the Pathet Lao and their NVA allies would expand to occupy large swaths of the country, eventually including all the Northeast and most of the panhandle bordering southern Vietnam. The territorial linchpin between the Pathet Lao forces and the capital of Vientiane remained the contested province of Xiengkuan and the Plain of Jars in particular.[8] The name Plain of Jars (*Plaine des Jarres* in French) derives from the thousands of large stone jars scattered for miles in this region. Their ancient origins remain a mystery. In geographic terms, the plain consisted of a circular plateau measuring 500 square miles in total area "with an average elevation of 3,500 feet."[9]

In 1961, as threat of the Pathet Lao grew, the CIA approached a Royal Lao Army officer named Vang Pao to act as an American proxy in Laos. Ethnically, Vang came from the hill peoples (*Hmong*) who lived in pockets of central and northern Laos, including the hills near the Plain of Jars. Ethno-linguistically, Laos has three main populations: Lao, Khmer, and Hmong. Of the three, Hmong represents about only 10 percent of the population. As a minority group, the Hmong had been fighting against the influence of other major populations in Southeast Asia, including the Lao, Khmer, and Vietnamese, for centuries. As an outsider group constantly fighting for autonomy and equality, the Hmong were easily influenced into accepting support from an external power, with the hope of achieving parity with neighboring ethnicities.

By mutual agreement at the 1962 Geneva Accords (signed by China, the Soviet Union, North Vietnam, South Vietnam, the Kingdom of Laos, and the U.S.), all parties agreed to stop directly interfering in the internal politics of the Kingdom of Laos. However, at least 8,000 NVA troops remained inside the country, shed their uniforms, and disguised themselves as Pathet Lao. In 1963, in response to the persistent NVA threat, Kennedy reauthorized the covert program to support the Kingdom of Laos. The entire operation ran under the authority of the U.S. Ambassador. The CIA dubbed this effort to arm, train, and equip the Hmong Operation MOMENTUM. This activity also received support from the Kingdom of Thailand who deployed forces into Laos called Thai Police Aerial Reinforcement Units (PARU).[10] Operation MOMENTUM should not be confused with Operation PINCUSHION, which was a similar activity run by U.S. Army Special Forces in Southern Laos to arm, train, and equip hill tribes of Lao Theung ethnicity (commonly referred to as *Montagnard* by the French).

Thailand's role in supporting Lao resilience with covert actions proved critical to the aims of U.S. but obviously served the Thai Kingdom's self-interest. By 1907, Thailand had ceded the territories of Laos to the aggressive imperial ambitions of France. In the 1960s, the now-independent Royal Lao Government continued to serve Thailand's interests as a buffer state to communist threats in China and Vietnam. Recognizing the need for its own inherent special operations forces to address internal irregular threats, Thailand agreed for the U.S. to train the PARU, who were regarded as the kingdom's "best and most capable" troops.[11] Ultimately, Thailand deployed these troops to support the resilience of the Royal Lao Government. Prior to the 1962 Geneva Accords, the PARU trained five battalions of the Royal Lao Army and at least 3,000 soldiers. However, while Thailand supported the Royal Lao Government, it found the amalgamated Lao National Army incapable of countering communist forces. Consequently, Thailand agreed with the U.S. that Vang Pao and the Hmong were the best option for thwarting communist efforts in Laos—an option that both Thailand and the U.S. would need to conduct covertly considering the 1962 agreements.[12] See Figure 4.3.

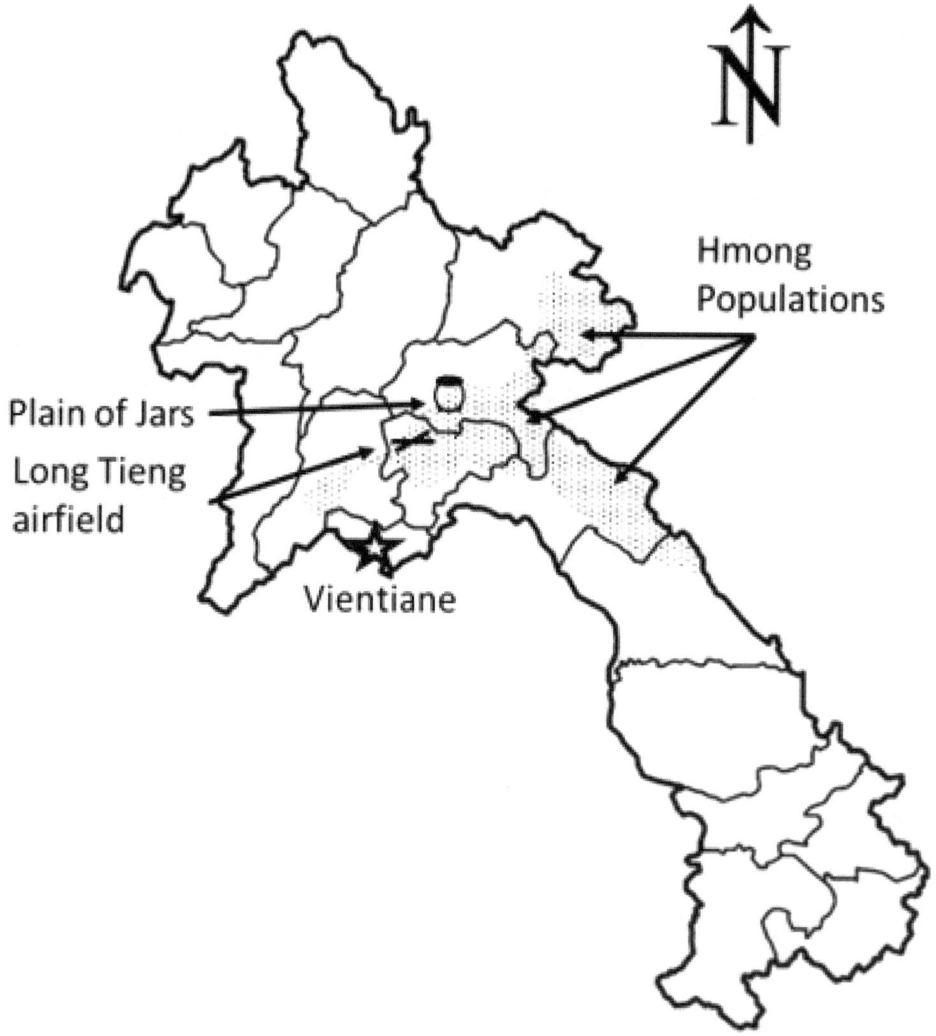

Figure 4.3. Map of Hmong Ethnolinguistic Population. Source: Author. Data derived from Center for Southeast Asian Studies, http://www.seasite.niu.edu/lao/Lao_maps/language_family.htm.
Note: Provinces include modern boundaries.

The Hmong population looked on the possibility of U.S. sponsorship with great favor. Hmong primarily consisted of poor, illiterate, subsistence farmers, and the prospects of improving their lives proved enticing. The charismatic Vang Pao held high ambitions for the Hmong well beyond simply militarization. He envisioned an independent Hmong homeland with higher education opportunities and the types of modern jobs realized in urban centers. To spread his message, Vang operated a radio station in Thailand that broadcasted across the border into Laos.

Figure 4.4: Photo of Vang Pao. Source: Image in the public domain[14]

He then provided radios to the villages, which allowed him to influence the entire population with his vision of independence, modernization, and opportunity. Compelling and energetic, Vang took on legendary proportions for Hmong communities as a hero akin to George Washington. See Figure 4.4. Vang agreed to American payments for Hmong soldiers, and this allowed the village militias to work full time and form a standing military force free from farming requirements. These soldiers were organized into Special Guerilla Units (SGU). Thai soldiers conducted most of the training for SGUs, which helped to lower the signature of U.S. participation.[13]

The CIA ran Operation MOMENTUM from Bangkok, Thailand, adjacent to the Air America parking ramp on Udorn Royal Thai Air Force Base. The Thai named the organization supporting covert operation in Laos *Headquarters 333*. The Thai headquarters was located adjacent to the CIA command center to synchronize intelligence and operations. Lt. Col. Vitoon Yasawad managed these Thai covert activities.[15] Initially, the Thai fielded 13 PARU teams in Laos—about 100 men in total. In concept, one five-man PARU team could support an infantry battalion. PARUs provided training in light weapons, mortars, and explosives, but they also had a communication specialist and a medical technician. Additionally, some PARU spoke Laotian and negated the need for translators. The U.S. dubbed its ground element in Laos the *4802nd Joint Liaison Detachment*. The American footprint remained small and included an Operation Detachment Alpha team of nine Special Forces soldiers, nine CIA case officers, and two more case officers in the capital at Vientiane. These numbers fluctuated over time based on the diplomatic situation, as well as the military one.[16]

Air America was the CIA code word for an essential activity that supported Operation MOMENTUM. These covert flights transported food and weapons to the Hmong via a CIA-constructed airfield at Long Tieng. The hilly region near Long Tieng (and the Plain of Jars) contained the largest concentration of the Hmong population in Laos. Long Tieng also served as Vang Pao's headquarters, and flights transported personnel to and from Thailand. Air America leveraged (24) twin-engine transport planes, (24) short-takeoff-and-landing (STOL) aircraft, and (30) helicopters. More than 300 pilots and support personnel operated Air America. Through a training program called *Water Pump*, many Hmong pilots trained at Udorn Air Force Base in Thailand. Thai pilots also trained in the Water Pump program and flew combat missions in Laos, an operation dubbed *Firefly*.[17] Rice served as an excellent payment to the Hmong for service rendered, and it comprised a major portion of air transportation requirements.[18] Helicopters also provided valuable insertion, extraction, and medical evacuation platforms for Hmong and Thai soldiers.[19]

From 1963 onward, both the U.S. and Thailand expanded covert operations in Laos to counter the NVA. Each time the Hmong won territory, the North Vietnamese would increase the stakes by counterattacking with more and more NVA troops. In 1964, Thailand began inserting its own ground forces into Laos, starting with an artillery battery. The CIA code-named this activity *Sierra Romeo*.[20] Nevertheless, American and Thai covert support never matched the same volume as what the NVA more overtly offered to the Pathet Lao. By 1968, an estimated 35,000 NVA had crossed the border into Northern Laos, basically replacing the Pathet Lao and forcing them into a supporting role. Simultaneously, the Royal Lao Army became overrun by the NVA in the south and was rendered generally ineffective for the remainder of the conflict. As the NVA gained momentum with conventional army operations, the Hmong remained a light infantry force of around 20,000 soldiers.[31] See Figure 4.5.

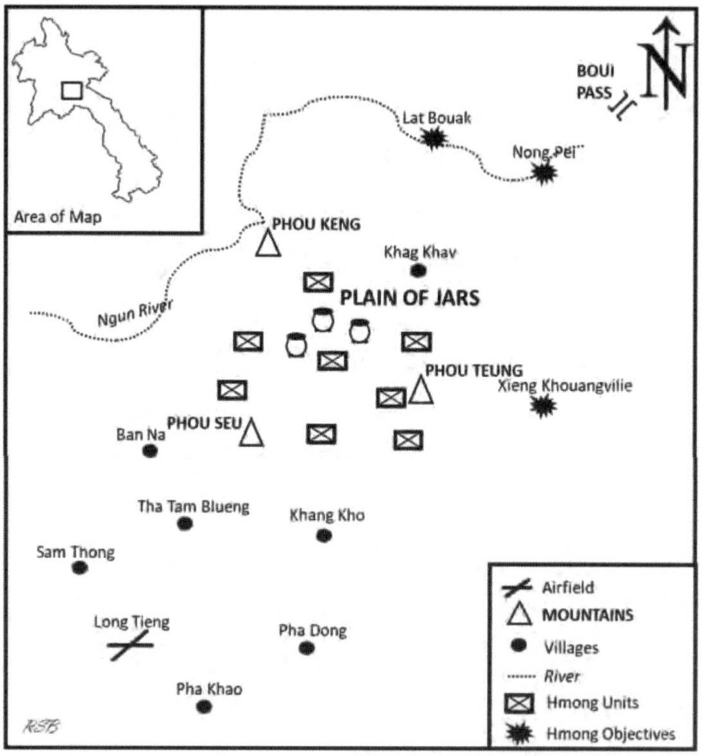

Figure 4.5. Map of the Plain of Jars and Hmong offensive, July 1971. Source: Author. Information derived from Al Haig, "Washington Special Action Group Paper," Central Intelligence Agency, 15 July 1971. Found in CIA Archives.

Amazingly and against the odds, Vang retook the Plain of Jars from the NVA and Pathet Lao in 1969, surprising many, including the CIA.[22] Despite this victory, from this point onward, the Hmong would never be able to reach an equivalent number to rival their NVA opponent.[23] The NVA simply regrouped, increased reserves, and prepared more offensives each dry season. Consequently, the NVA gradually inched closer and closer to Long Tieng. Regardless of the David versus Goliath odds, Vang committed to standing toe-to-toe against the communists as they advanced.

The vital airstrip at Long Tieng simultaneously had become Vang's center of gravity as well as his critical vulnerability. To protect the airfield, by 1970, Vang's tactics had changed from guerilla operations to direct conventional battles—contests primarily fought in the Plain of Jars region. While outnumbered, the Hmong were backed by the U.S. Air Force, including B-52 bombers. Yet even this advantage had a limited lifespan and remained dependent on the deteriorating political situation in Vietnam. Additionally, the American home front was by this time losing confidence and interest in Southeast Asia.[24] Meanwhile, Thailand continued to support the Hmong, even sending a 500-person regimental combat team to defend Long Tieng in 1970.[25]

After 1970, the Royal Loa Government everywhere began eroding under Ho Chi Mien pressure. By 1971, the U.S. State Department concluded that as many as 80,000 NVA troops were "directed against the neutral government of Laos," perhaps with as many as half of these directed against the Hmong in the north.[26] The battles for Skyline Ridge, which served as the final geographic ridge to defend the airfield at Long Tieng, took place just west of the strategically important Plain of Jars between 1970 and 1972. The ridgeline had nine helicopter landing zones to insert and extract defensive forces. At these nine points, the Hmong and NVA engaged in bitter battles for all the stakes.[27] These clashes on Skyline Ridge would set the stage for the final chapter of Operation MOMENTUM.

As the North Vietnamese increased their presence in Laos, the U.S. simultaneously decreased its footprint in Southern Vietnam. U.S. appetite waned for escalation to counter the NVA. By 1971, it halved the number of B-52 bombers stationed in Thailand and withdrew them stateside. In October 1971, Congress capped funding for CIA operations in Laos.[28] The last geographical feature for holding the airfield at Long Tieng, Skyline Ridge, served as the decisive battlefield for the future of Laos. Despite valiant efforts, Hmong losses in the previous years and their inability to field an army the size or composition of the NVA sealed their fate. By 1972, Vang could only field a defensive force of about 3,000 soldiers at Long Tieng, and he was constantly outnumbered.[29] See Figure 4.6. Recognizing the inevitable defeat of the Royal Lao Government in repulsing communists in Laos, the objectives for CIA sponsorship of the Hmong moving into the 1970s was as a diversionary effort to tie up NVA troops from defeating South Vietnam.[30]

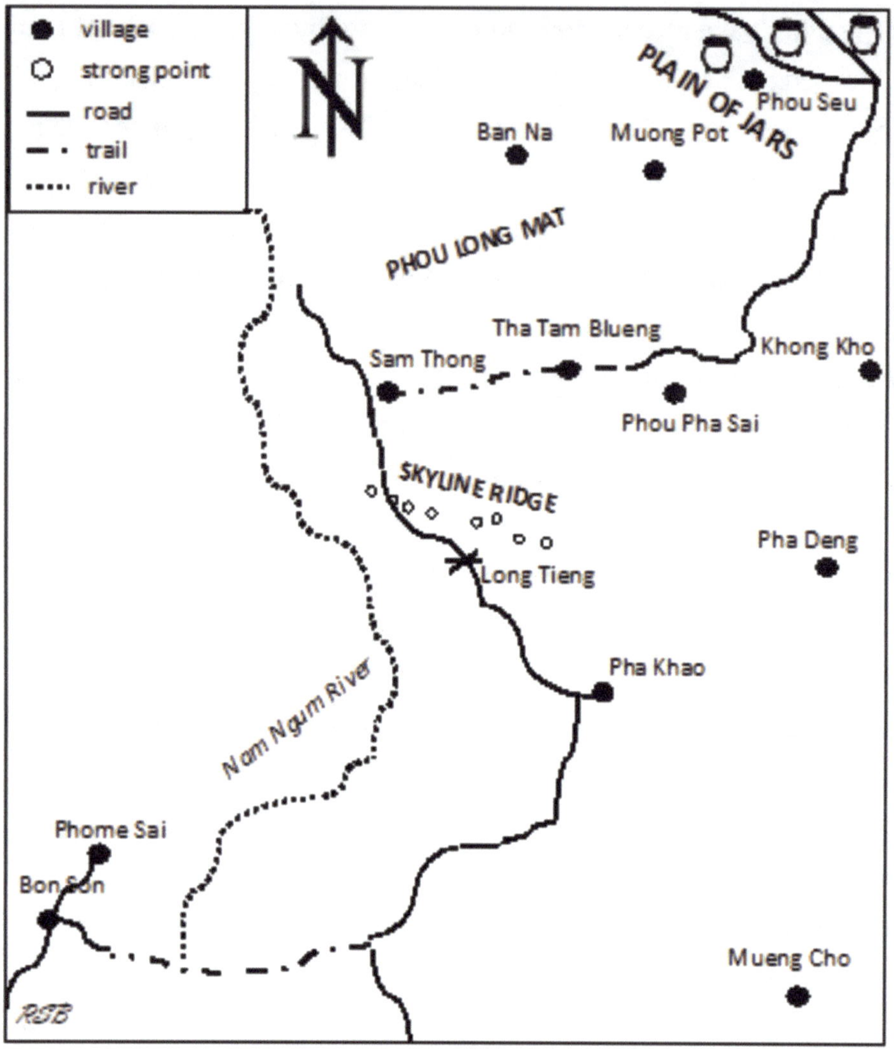

Figure 4.6. Battle for Skyline Ridge, 1972. Source: Illustration by author. Data from Richard T. Kennedy and John Holdridge.

From December 1970 to February 1973, Thailand attempted to stem the tide of the NVA advance through the Plain of Jars through direct intervention. It gradually introduced an expeditionary force called *Seua Pran.* At its peak, Seua Pran consisted of 23,000 soldiers. This impressive force did protect the Hmong and, in particular, saved the crucial airbase at Long Tieng. However, with the U.S. scaling back its covert sponsorship and with the ever-increasing NVA presence, Thailand realized it could not win the fight against communists by itself, particularly when the Royal Lao Army remained ineffective, and the U.S. was withdrawing. In 1973, Thailand decided to withdraw, with the last Seua Pran troops departing in May 1974.[31]

The fall of the Hmong mirrored the timeline of the fall of Saigon. Through February 1975, the NVA seized the remaining hills built to defend the CIA-built airfield at Long Tieng. The loss of Long Tieng and the throughput of CIA war material made defeat of the Hmong inevitable. In May 1975, Air America attempted to extract as many Hmong as possible. About 2,500 key leaders, officers, and their families left via U.S. aircraft, but the vast majority, as many as 50,000 guerillas and their families, were left behind to face the wrath of the communists and their ethnic enemies.[32]

CONCLUSION

Probably the foremost lesson about irregular warfare and great power competition from Operation MOMENTUM is regarding potential populations for partnerships. The Hmong comprised a minority group that was largely disenfranchised. Minority groups can prove very effective partners as they are eager to achieve agency and control over more dominant neighboring groups. Consequently, the Hmong found U.S. sponsorship quite attractive. With a population of only 250,000 to 300,000 people, the Hmong fielded a very effective and completely indigenous army of 20,000. While the force was clearly remarkable, the question remains if creating this conventional Hmong Army—instead of reinforcing the Hmong's inherent guerilla warfare heritage—was the best approach.

Another important lesson from the Secret War relates to leveraging regional partners when conducting irregular warfare in global competition. Thailand proved an essential ally in support of U.S. objectives in Laos. They provided a sanctuary for U.S. airpower and even a radio station for the Hmong. Also, Thai soldiers provided nearly all the training for the Hmong militias. Supporting the conflict covertly in violation of the 1962 Geneva Accords, Thai soldiers in the PARU offered a much lower signature than U.S. troops. Additionally, the Royal Thai government had a vested interest in the future of Indochina, much more so than the American public, and could support the type of multi-decade activities required to win. Primarily through partnership with Thailand, the U.S. was able to support a 20,000-soldier army with only a few dozen Americans. Eventually, Thailand offered direct support with infantry, artillery, and air support. Unfortunately, by this point, the U.S. had decided to withdraw its support to Laotian resilience.

The tragic tale of the Hmong following U.S. withdrawal from Southeast Asia remains equally unfortunate to that of the Montagnard and the South Vietnamese. The immediate cost of war was high. For example, the Hmong followed a tradition in funeral ceremonies of the fallen. The bodies were normally returned to their village of origin for burial, during which the elders would pass the rifle of the dead comrade to the next most logical male candidate. In this way, Hmong were conscripted. This process maintained a steady stream of replacements for battlefield losses. However, by 1970, the Hmong villages could no longer maintain this tradition without offering up young male children, instead of men, as replacements. Consequently, the recruitment and weaponization of child soldiers became a terrible component of the conflict. The estimated number of total Hmong deaths due to combat during this period varies but was probably no less than 30,000, which was 10 percent of the total population. The Hmong had supported the defeated French in the 1950s and now the defeated Americans in the

1970s. On a campaign of revenge after 1975, communist reprisals on the Hmong verged on extermination.[33] This led to a large portion of the Hmong fleeing Cambodia. Laos served as a microcosm of a greater war between communism and the free world.

| Laos served as a microcosm of a greater war between communism and the free world. |

America's rivals, Communist China and the Soviet Union, provided essential support to North Vietnam's efforts in Indochina. After 20 years of warfare, the U.S. lost more than the conflict in Vietnam: South Vietnam fell to the NVA on 30 April 1975, and Cambodia fell to the communist Khmer Rouge in the same month. The Kingdom of Laos fell to the Pathet Lao in December 1975. Overall, communist resistance movements seized the entirety of French Indochina. From this perspective, Eisenhower's Domino Theory appears substantiated. The First and Second Indochina Wars remain an important case study for future strategic competition—and perhaps the Secret War holds the most relevance.

About the Author

Dr. Robert S. Burrell is a resilience and resistance interdisciplinary scholar using data-driven and human-centric methodologies to analyze intrastate conflict ranging from nonviolent protest through belligerency. He is a Senior Research Fellow at the Global and National Security Institute of the University of South Florida. From 2020-2024, he taught irregular warfare at Joint Special Operations University and was the editor in chief of special operations doctrine from 2011-2014.

NOTES

1. For an explanation on the Ho Chi Minh Trail, see memorandum Central Intelligence Agency, "The Ho Chi Minh Trail," 30 August 1961. Also see Directorate of Intelligence, "Intelligence Memorandum: The Ho Chi Minh Trail, March 1971. CIA archives. For information on the bombing campaign in Laos, see Victory B. Anthony and Richard R. Sexton, *The War in Northern Laos*, 1954-1975 (Center for Air Force History: United States Air Force in Southeast Asia, 1993).

2. See Central Intelligence Agency, "International Boundaries of Indochina," March 1965. Found in CIA archives.

3. See Yu Insun and Choi Dong-Ju, "A Re-examination of America's Indochina Policy During the French Presence — Arms Transfer for Diplomatic Leverage and the Role of the Catholics," *Journal of International and Area Studies* 6, no. 2 (December 1999): 75-90. Information on funding from pages 81-82.

4. Richard H. Immerman, "The United States and the Geneva Conference of 1954: A New Look," *Diplomatic History* 14, no. 1 (January 1, 1990): 43–66.

5. For a good summary of America's support to the Republic of Vietnam, see Leslie H. Gelb, "The Essential Domino: American Politics and Vietnam," *Foreign Affairs*, April 1972. Also see Gelb's article, "Vietnam: The System Worked," *Foreign Policy*, Summer 1971.

6. According to the CIA, the "command structure" of the Pathet Lao "parallels the civilian apparatus." See Directorate of Intelligence, "Intelligence Memorandum: The Laos Communists," 12 March 1973, 8. Found in CIA archives.

7. Central Intelligence Agency, "Communist Advance Toward Power in Laos," December 1958. Also see, Central Intelligence Agency, "Background on Laos for Points Paper," 2 March 1970. Also, see the Geneva Agreements, "Agreement on the Cessation of Hostilities in Viet-Nam," 20 July 1954. Also see, Central Intelligence Agency, "National Intelligence Estimate 63.3/1-55: Probable Developments in Laos to July 1956," 20 Dec 1955.

8. Central Intelligence Agency, "Background on Laos for Points Paper," 2 March 1970.

9. William M. Leary, "The CIA and the 'Secret War' in Laos: The Battle of Skyline Ridge," *The Journal of Military History*, Issue 59, 1995, 507.

10. Leary, "The CIA and the 'Secret War,' in Laos," 505-506.

11. Sutayut Osornprasop, "Amidst the Heat of the Cold War in Asia: Thailand and the American Secret War in Indochina (1960 –74)," *Cold War History*, August 2007, 355.

12. Osornprasop, "Amidst the Heat of the Cold War in Asia, " 349-371.

13. Leary, "The CIA and the 'Secret War,' in Laos," 510-513.

14. Photo in the public domain and originally taken by J. Vinton Lawrence in the service of the U.S. Government. See Dan Olson, "Hmong Community, Military Leader Vang Pao Dies at 81," MPRNEWS, found at https://www.mprnews.org/story/2011/01/06/vang-pao-dies-hmong-laos, accessed on 2 May 2024. This image also previously in wikicommons. See Vang Pao, "Unfinished Business," *The Economist*, 13 January 2011. Found at https://www.economist.com/blogs/asiaview/2011/01/vang_pao, accessed on 2 May 2024.

15. Osornprasop, "Amidst the Heat of the Cold War in Asia," 360.

16. Leary, "The CIA and the 'Secret War' in Laos," 510-513. Also see Osornprasop "Amidst the Heat of the Cold War in Asia."

17. Osornprasop, "Amidst the Heat of the Cold War in Asia," 362.

18. In 1962, the U.S. Air Force air lifted 100 tons of rice seed to the Hmong. See, Thomas L. Ahern, Jr., *Undercover Armies: CIA and Surrogate Warfare in Laos: 1961-1973*, (Washington DC: Central Intelligence Agency, 2006), 145.

19. For more information on Air America, see William M. Leary, "CIA Air Operations in Laws, 1955-1974: Supporting the 'Secret War,'" *Center for the Study of Intelligence*, 13 June 2007. Found at https://web.archive.org/web/20070613112032/https://www.cia.gov/library/center-for-the-study-of-intelligence/csi-publications/csi-studies/studies/winter99-00/art7.html.

20. Osornprasop, "Amidst the Heat of the Cold War in Asia," 363.

21. Leary, "The CIA and the 'Secret War' in Laos," 506-510.

22. Henry A. Kissinger, "CIA Memorandum on Vang Pao's Victory in Laos," 23 October 1969. CIA Archives.

23. Some primary documents refer to the Hmong by a Chinese term called Meo or Miao. However, the term Meo, which certainly was commonly used, is looked upon by the Hmong as a derogatory slur.

24. Leary, "The CIA and the 'Secret War,' in Laos," 506-510.

25. Osornprasop, "Amidst the Heat of the Cold War in Asia," 364-365.

26. Department of State, "Statement of the United States Government Position on Foreign Forces in Laos," No. 171, 9 August 1971. Found in CIA archives.

27. Richard T. Kennedy and John Holdridge, "Laos Sitrep as of 1800, 20 March Laos Time," 29 March 1972. Found in CIA Archives.

28. "Senate Votes Ceiling on CIA Funds in Laos," *Washington Star*, 5 October 1971.

29. Jack Foisie, "Mortar Fire Hits Laos Major as He Tells of the Toll at Key Base," *Los Angeles Times*, 20 January 1972.

30. For more on the Battle of Skyline Ridge, see Leary, "The CIA and the 'Secret War' in Laos," 505-517.

31. Osornprasop, "Amidst the Heat of the Cold War in Asia," 366-367,

32. For the airlift evacuation at Long Tieng, see Ahern, 515-517.

33. According to one CIA study, the Laos communist government used chemical weapons attacks on Hmong populations from 1976-1979. See Directorate of Operations, "Incidents of Poison Gas Attacks Against Resistance Elements in Louang Probang, Xieng Khouang and Savannakhet Provinces," 21 March 1980. Found in CIA archives.

III. Support to Resistance: Case Study of Tibet, 1951–1975

By Robert S. Burrell

ABSTRACT: Relatively unknown to most Americans, the U.S. sponsored Tibetan resistance to Chinese occupation for two decades during the Cold War. This section principally utilizes recently released primary sources from the CIA archives. It provides the strategic context regarding the Chinese occupation of Tibet in international affairs, particularly the crucial role of India, the U.S. , the Republic of China, and the government in exile—the Dalai Lama. It describes the armed component of Tibetan resistance and how it progressed through four phases from 1955 to 1975. It also details the methods of external support provided by the U.S. through unconventional warfare. Finally, it provides the lessons learned—particularly applicable to strategic competition in a polarized world. "Tibet. Shangri-La. Theocratic state. Roof of the world." - Roger E. McCarthy, 1997[1]

| "Tibet. Shangri-La. Theocratic state. Roof of the world." - Roger E. McCarthy, 1997. |

OVERVIEW

In 1949, Mao Zedong (a founding member of the Chinese Communist Party in 1921) unexpectedly defeated the Chinese nationalist movement under Chiang Kai-shek and officially established the People's Republic of China (PRC). Communist control over mainland China represented a complete reversal of fortunes in only a few years. The Nationalist Chinese regime had been a key ally to the U.S. during World War II. Now, just four years later, the situation in Asia had dramatically changed from an optimistic view to a pessimistic one highlighted by the threat of communist expansion. From 1945–1949, the U.S. had generally supported the ambitions of the Nationalist Party to reestablish boundaries associated with the Qing Dynasty (territory lost by China over the past century). In fact, when it drafted its Constitution in 1947, the *Kuomintang* (the party of Nationalist China) recognized the contested territories of Taiwan, Mongolia, Tibet, and the South China Sea as part of China (see Figure 4.7).[3]

Figure 4.7. Map of China and lost territories, 1911. Source: Wikipedia Commons

Flush with victory, Mao now vied for dominion over those same lost Qing Dynasty territories–essentially embracing a shared Kuomintang's vision of *Zhongguo* (greater China).[4] Meanwhile, the U.S. responded by contesting any further communist expansion in Asia. In 1950, the PRC abruptly invaded Tibet, stimulating a maelstrom of opposing overt, clandestine, and covert actions by the U.S. . Utilizing primarily declassified sources, this section focuses on the activities of the U.S. State Department and the CIA to contest the PRC's occupation of Tibet for two decades from 1950–1972, a task addressed initially through diplomacy and later by supporting resistance activities.

BACKGROUND

For more than 20 years, Tibet served as one of the first battlefields of great power competition between the PRC and the U.S. . U.S. strategy would include information operations, unconventional warfare, and intelligence operations. As early as April of 1949, the State Department recommended treating Tibet as an independent nation, specifically to counter the PRC.[5]

> ***| As early as April of 1949, the State Department recommended treating Tibet as an independent nation, specifically to counter the PRC. |***

In 1950, Chinese troops invaded Tibet; the following year, they declared its annexation. Over the next decade, the PRC instituted major road construction projects with emphasis on Tibet, as, prior to "communist occupation, there were no motorable roads."[6] This highway project embodied an urgent national security issue that would allow for the projection and maintenance of the needed numbers of People's Liberation Army (PLA) troops to occupy Tibet and its contested border with India. Called the "Silver Dollar Road," the PRC paid local Tibetans in lucrative silver coins to work this expansive project, which simultaneously served as an excellent public relations measure. In domestic terms, the road also connected the capital of Lhasa with Beijing.[7] Also of importance, there were several prominent and influential families in Tibet who leveraged the opportunity to work with the Communist Chinese Party for their own personal gain.[8]

From the start of the invasion, the State Department advised the Dalai Lama to leave the country and establish a government-in-exile.[9] The Dalai Lama, Tenzin Gyatso, stalled and attempted to negotiate with China. To make peace, the Dalai Lama's administrators eventually signed the 17-point protocol agreement.[10] Afterward, Tenzin left for Beijing to meet with Mao and remained in Beijing for a year to learn about the communist reforms to be imposed on Tibet. Meanwhile, China initiated a social reengineering program to overthrow the Tibetan feudal systems and destroy its ubiquitous religion, which simultaneously formed the basis for Tibetan governance. The assault on religion proved a particularly insightful act of aggression, as the PRC had promised that "the old systems and customs" would not change and that the thousands of monasteries and other religious property would be protected.[11] Instead, the PRC implemented programs for communist education of children, burned Buddhist archives, and forced the implementation of community farming.[12] The latter resulted in mass starvation, a situation exacerbated by the requirements to feed large numbers of occupation forces. As stated by ex-CIA officer and author Roger McCarthy, "the Chinese made countless mistakes in trying to implement Chinese farming practices in Tibet, most of which turned into disasters."[13] The forced development and suppression of the people over the next decade resulted in what some have described as genocidal numbers of deaths.[14]

From the mid-1950s onward, the U.S. worked closely with India in its efforts to support Tibetan resistance. The CIA quickly realized this opportunity, evaluating that from 1955–1958, "resistance has hardened and grown despite Chinese countermeasures that include military force as well as partial withdrawal of Chinese cadres and postponement of 'reforms' and other programs leading toward socialization."[15] Both India and Pakistan provided sanctuary for CIA covert activities. However, India served in a more open role in protesting China's invasion internationally. Additionally, India and China contested their borders with each other, including the border of Tibet. They also competed in terms of military, cultural, and regional influence. The dispute over the Sino-Indian border would erupt into open hostilities in 1962.

Another strategic player in support of the Tibetan resistance was Chiang Kai-Shek's National Chinese forces in Taiwan. The Generalissimo continued to oppose Communist China and viewed the fight in Tibet as another theater of operations. The Republic of China sought to raise its own army in Tibet under Chiang Kai-Shek's authority. The vast distance and lines of logistics from Taiwan did not support this idea, and the U.S. remained resistant to nationalist Chinese partnership in its support to Tibetan resistance activities. In the U.S. assessment, while it supported Chiang Kai Shek's ambitions to defeat Chinese Communism, many of his ideas appeared overly optimistic. Nevertheless, the Republic of China maintained its seat in the

United Nations Security Council until 1971 and consequently held considerable influence. In 1959, the CIA relented to support a coalition approach to supporting Tibetan resistance with the Republic of China effort labelled *ST Whale*, but it did not prove very successful.[16]

Eventually, the PRC's attempts to communize the Tibetans resulted in open rebellion, initially by the Khamba tribes in Eastern Tibet in 1956.[17] China's subsequent destruction of thousands of Buddhist monasteries further outraged the population.[18] Again, in 1959, a second and larger rebellion made up of nearly 90,000 people erupted throughout the eastern region, during which the Chinese killed about 2,000 citizens. To regain public control, China attempted to capture the Dalai Lama, first by deception and then by force. Consequently, the Dalai Lama finally decided to leave for India. One of the Dalai Lama's concerns revolved around the safety of his family, particularly his older brother Norbu, who had been promised sanctuary by the United States. The Dalai Lama's departure from Tibet proved critical in delegitimizing Communist China's occupation, and he continues to serve in that role today. To reinforce the effort of delegitimizing China, the U.S. introduced UN resolutions in 1959, 1961, and 1965 that condemned Chinese human rights violations, but these initiatives fell short of requesting recognition of Tibet as an independent state.[19]

TIBETAN RESISTANCE

The Tibetan insurgency occurred in four phases over a period of two decades from 1955 to 1975.[20] During the incubation phase from 1951–1954, the Chinese occupational force consisted of about 50,000 troops under general Chang Kuo-hua. The PLA established a Tibetan Military District, but the limited numbers of troops proved insufficient to suppress the population. One of the early resistance movements dubbed itself the *People's Party* and represented a loosely organized confederation of anti-communist groups. Two of the early leaders included monks Lhopto Rimpochhe and Chhandzo. Other members in the Lhasa province included traders and people with trade skills, estimated at about 3,000 members in 1952. Eastern Tibet also realized resistance in the Batang, Ch'angtu, and Kham areas. In Kham, local Tibetans viewed communists as enemies of Buddhism, arising to sporadic and isolated violence against limited numbers of PLA soldiers.[21] However, resistance during the incubation period of 1951–1954 mainly consisted of nonviolent action like tearing down communist posters, local vendors refusing to sell foodstuffs, and Tibetan officials not providing information requested (like biographic information).[22] Protests over food security also persisted during this period.[23]

THE FIRST PHASE OF TIBETAN RESISTANCE

The first phase of armed resistance, dubbed the *Kanting Rebellion*, occurred in the province of Kham in 1955 and 1956. Over the next five years, a large PLA troop presence decreased available commodities for locals and intensified food prices. Additionally, the continuous occupation of the capital by PLA soldiers and communist party agents (likely outnumbering the local inhabitants of 80,000) aggravated the atmosphere of oppression and constrained resources.[24] Secretly, Tibetan resistance networks met in and operated out of places of worship. The Chinese quickly realized that monasteries were the epicenter of resistance, so the PLA demolished several of them, including the complete destruction of a vast religious complex at Litang. This assault on Buddhism and national monuments sowed the seeds for future grievance. In rural areas, particularly in the eastern province of Kham, sizable resistance elements also emerged, with as many as 6,000 armed rebels.[25] "With swords and rifles the Tibetans attacked the Chinese posts in Lithang, Bathang, Kansu, Champdo, Derge, Po, and countless other places."[26] One leader and financier, Andrutsang Gompo Tashi, used his personal wealth and influence to mobilize Khampa resistance from 1956 onward.[27] In summation, the rebellions of 1955 and 1956 appear to have had success: in June 1957, Beijing decided to withdraw significant numbers of PLA soldiers and workers, at least 25,000, believing that it would appease widespread resistance.[28]

Andrutsang Gompo Tashi, an early leader and financier of resistance. Source: WikiCommons/public domain

It was at this point that the CIA began to develop a limited plan to support Tibetan armed resistance, yet another indicator of success by the rebellions of 1955 and 1956. Initially, based on Tibetan geography, the CIA believed armed resistance had favorable odds. The terrain and the distances involved posed "serious obstacles to Chinese Communist operations."[29] The vast and desolate plateaus in the north, with high mountains and deep river valleys, posed barriers to conventional forces. These imposing geographic features, with elevations as high as 13,000 to 15,000 feet in both the north and the east, isolated populations into pockets and intensified clannish loyalties that instinctively resisted centralized authority.[30] Additionally, Tibet borders three Himalayan states that could serve as sanctuary, including Nepal and the semi-independent Indian states of Bhutan, and Sikkin.[31] To vie with this geography, the PRC instituted a road building program to tie these regions together and provide access for military forces. See Figure 4.8.[32]

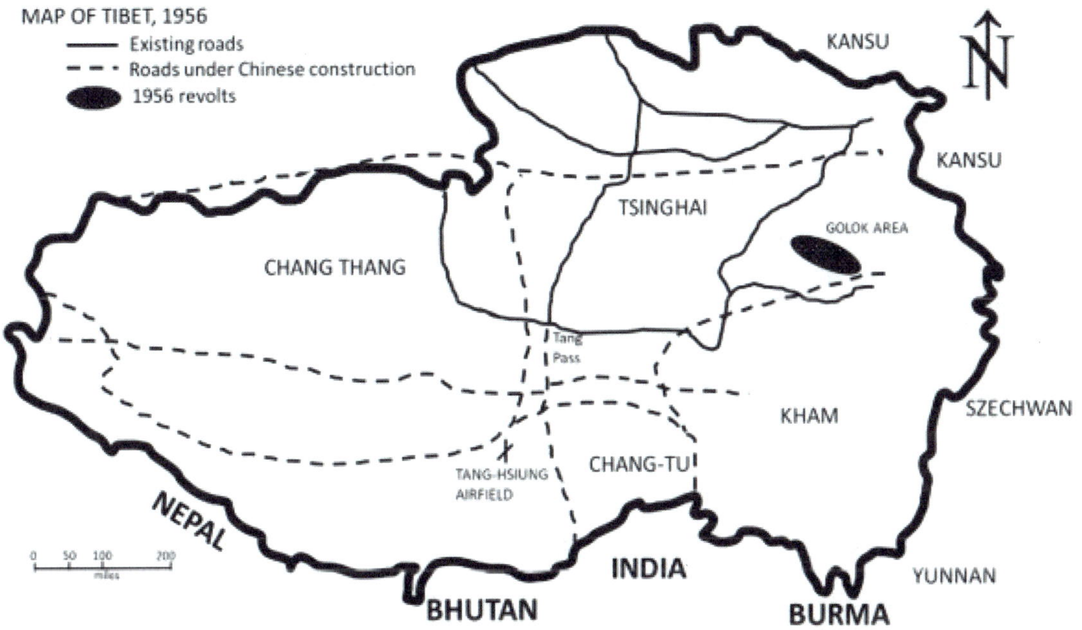

Figure 4.8. Map of Tibetan resistance and Chinese road projects in 1956. Source: Author, based on map from CIA archives.

THE SECOND PHASE OF TIBETAN RESISTANCE

The second phase, from 1957–1959, witnessed a shift to urban-based resistance movements in Lokha (or Shannon), just south of the capital of Lhasa and near the border with Bhutan and India. Lokha was one of the spiritual centers of Tibet, and the native inhabits consisted of Khamba tribes.[33] The communists had established communist associations in Lokha, which, unfortunately for them, became nodes for resistance cells. Outbreaks of violent rebellion sparked in 1957 in eastern Tibet along the border with Szechwan.[34] The final catalyst occurred on 10 March 1959, which led to a major revolt in the capital due to fear that the Chinese were planning to kidnap the Dalai Lama.[35]

Thousands of demonstrators surrounded the Dalai Lama and escorted him outside Lhasa. Subsequently, he departed for India on 17 March while resistance grew exponentially.[36] Simultaneously, the Chinese Communist Party "were furious when they learned the Dalai Lama had escaped" and was given asylum in India.[37] On 28 March, the Premier of China, Zhou

Enlai, issued the dissolution of the former Tibetan government (headed by the Dalai Lama) and replaced it with a preparatory committee of the Tibet Autonomous Region.[38] This act enflamed the political chaos, and the extensive violence garnered international attention. One group of resistance members, which the CIA estimated as "well-organized," overcame a local PLA outpost of 80 soldiers and marked the walls with posters that declared the "independent kingdom of Tibet."[39] At least 61,000 PLA troops deployed to various areas to suppress rebellion.[40] The Chinese Communist Party admitted to a substantial Tibetan resistance, but then downplayed the numbers to 20,000 (the actual size remains likely much higher, in the hundreds of thousands).[41] As adeptly argued by historian Mark Kramer, "the CIA's involvement in the Tibetan resistance began on only a limited scale in 1956–1957, but U.S. covert aid to the Tibetan guerrilla increased significantly" following the March 1959 revolts.[42]

At this point in 1959, the armed component of resistance had finally achieved cohesion and dubbed itself Chushi Gangdruk (sometimes written as Chushi Gangdrug), which means four rivers, six ranges—the land encompassing the people of Kham. It also embraced the title National Volunteer Defense Army (NVDA), perhaps to appeal to Tibetan peoples other than the Kham. In this early period, funding for the NVDA primarily derived from the Andrutsang family.[43] In the fall of 1958, at a Chinese garrison in Tsetang south of Lhasa, 2,000 resistance fighters attacked 3,000 PLA troops and caused hundreds of Chinese casualties.[44] Even in Chinese sources, the extent of the insurgency was pervasive. During 1958 and 1959, the Chinese claim to have fought at least 47 engagements with the NVDA and to have killed 2,000 insurgents.[45]

Following the departure of the Dalai Lama in 1959, the Chinese abandoned its cautious policies toward Tibetan integration, instead adopting one of "all-out repression." In March of 1959, Premier Zhou effectively ordered martial law, which dissolved the local government and replaced it with the communist Preparatory Committee.[46] In his telegram to the United Nations in September, the Dalai Lama explained the situation this way: "every attempt has been made to destroy our religion and culture...thousands of monasteries have been razed to the ground... life and property are no longer safe and Lhasa, the capital of the State, is now a dead city."[47] The expansive Chinese counterinsurgency effort led to a lull in armed resistance for about two years.[48] As the Chinese communists ramped up oppression, the CIA correspondingly advocated for increased armed support to resistance.[49]

| As the Chinese communists ramped up oppression, the CIA correspondingly advocated for increased armed support to resistance. |

Some influential Americans, like scientist Howard D. Grayber, agreed with this approach, believing that "U.S. action could force negotiation with the Communist Chinese (with India as intermediary) leading to...Chinese withdrawal from Tibet."[50]

THE THIRD PHASE OF TIBETAN RESISTANCE

Tibetan monks surrender weapons to PLA soldiers in April 1959. Source: AFP/Getty Images

In 1960 and 1961, the third phase of Tibetan resistance manifested on the Tsang Plain around the towns of Sigatse and Gyantse.[51] The high peaks of this region provided ready access for smuggling of arms via land routes from Nepal. China's logistical support for the large numbers of PLA soldiers proved a critical vulnerability. The mountainous terrain also assisted in raids and ambushes along the highways, particularly the Tibet-Xinjiang Highway, which runs east to west along the border with Nepal. In 1960, at least 128,000 PLA forces deployed to Tibet, requiring 514 tons of food and supplies daily—a rate of consumption unsustainable without local procurement.[52] The population of Tibet was 1.27 million, and the PLA added another 10 percent to this number.[53] The deployment of the PLA troops created an enormous logistical challenge to support (particularly food), but the effect of an increased military presence on population security and containment of the insurgents yielded positive results.[53] Additionally, Chinese use of airpower in the Tsang Plain, with little available cover and concealment, proved a critical factor, and it essentially put an end to armed resistance in this region—but only after significant Tibetan casualties as guerillas refused to disperse. Despite tactical setbacks, resistance appears to have paid some dividends during this period. In 1961, the Chinese Communist Party acknowledged it had committed "too many errors" and postponed the "socialization of Tibet" (meaning its many reforms to the economy, education, and religion) for a four-year grace period.

THE FOURTH PHASE

The fourth, final, and longest phase of resistance occurred from 1961–1974 and was carried out by what might be deemed an American proxy force located on the Mustang plateau in nearby Nepal. Attempting not to provoke China, India would not support a sanctuary for armed Tibetan resistance; this included their satellite states of Bhutan and Sikkim. While Nepal recognized Chinese claims to Tibet in 1956, the Mustang plateau jutted outward into their neighbor and shared historical, cultural, and religious ties to Tibet. Consequently, despite Nepal's overtures to China, the resistance found sanctuary granted at the local level by Tibetan peoples living on Mustang.[57] See Figure 4.9.[58]

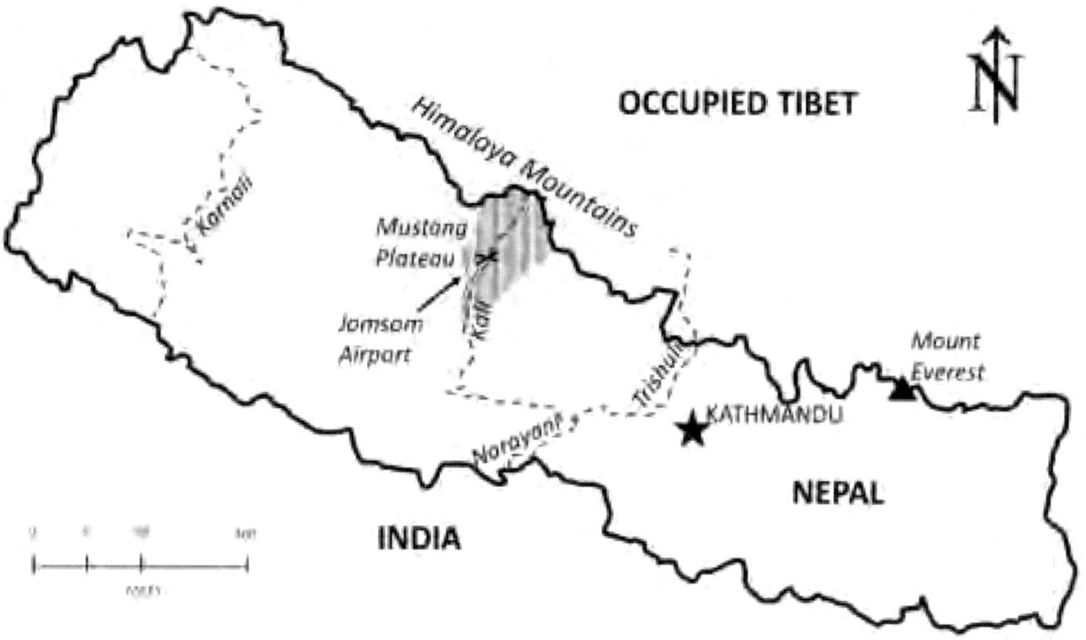

Figure 4.9: Map of Nepal and the Mustang plateau. Source: Author

To ensure external support, the CIA constructed an airfield in the early 1960s to serve as a logistics hub.[59] The CIA's "Tibetan Activity" had three components: (1) political actions, (2) propaganda, and (3) paramilitary activity. The CIA's main purpose was "to keep the political concept of an autonomous Tibet alive within Tibet and among foreign nations, principally India."[60] The paramilitary force consisted of about 2,000 fighters, 800 of whom the CIA had equipped via airdrop in January 1961.[61] In the following years, many more exiled Tibetans "jammed into Indian trains or came by foot to enlist and to demand arms."[62] Additionally, by 1963, the CIA had brought 133 Tibetans to the U.S. to train for support to the three lines of effort in the Tibetan Activity.[63]

As they had promised, the Chinese Communist Party renewed their socialization program for Tibet in 1965 and formally established the Tibet Autonomous Region in September.[64] Governance was implemented through the Party Committee for Tibet, made up of Han Chinese.[65] In response, as many as 6,000 guerillas developed a new headquarters at Mustang. The high plateau served ideally as a base of operations. During this period, the CIA even offered leadership courses from Cornell University for resistance leaders (at least three Tibetans went into this program).[66] Additionally, the U.S. provided education in governance, including guided tours in Washington D.C. as well as the United Nations.[67]

The decade of rebellion supported by the CIA during this period, from 1956–1966, realized several successes, but the resistance forces gradually lost control of indigenous territories in eastern and southern Tibet. By 1968, the PLA brought in larger forces and generally quelled the Khampa populations, leaving only remnants of resistance at Mustang plateau for another five years. By 1973, the CIA had decided to transition away from armed support to Tibetan resistance, particularly one which had been relegated to the role of an American proxy.[68] Finally, in 1974, China put diplomatic pressure on Nepal to clear out the remaining insurgents on Mustang plateau.[69] Nepal labeled them as "bandits."[70] Even the Dalai Lama "urged the soldiers to surrender."[71] Finally, a congenial meeting between U.S. President Gerald Ford and Mao Zedong in 1975 officially put an end to any hope for U.S. support for Tibetan armed resistance.[72]

One more point that needs emphasis in regard to Tibetan resistance is the international outreach conducted by the Dalai Lama. Tenzin effectively reached out to Buddhist communities, including those in India, Japan, and Thailand. The Indian connection, in particular, proved effective in garnering support for Tibetan freedom from a wide-ranging religious diaspora, but never, unfortunately, in a large enough amount to reverse Chinese occupation.

UNCONVENTIONAL WARFARE

Once organized, the CIA established three main operations to support Tibetan resistance: ST Circus (sometimes called Shadow Circus) to arm, train, and equip the ground component; ST Barnum for airlift support; and ST Bailey for the information operation campaign.[73] By way of ST Circus, the CIA determined that the best way to train indigenous operatives was to "establish a secret training base," then recruit the most capable warfighters from the diaspora in India, transport them to the U.S. for education in guerilla warfare, and then reinsert them back into Tibet later.[74] Starting in 1956, the first location used for training was the island of Saipan in the Pacific.[75] This consisted of general infantry classes (pertaining to marksmanship, explosives, and machine guns) and lasted about five months.[76] The next location, used from 1958 onward, was at Camp Hale in Colorado. Camp Hale, at elevations over 10,000 feet, better mimicked the conditions in the Himalayas. The training evolved into guerilla warfare tactics.[77]

In addition to training the Tibetans in infantry skills, the CIA provided them with parachute training at a jump school at Camp Hale, using American 'smoke jumpers' as instructors. From 1959 to 1960, the CIA trained "approximately 260" resistance fighters at Camp Hale.[78] Once trained, these operatives returned to Tibet (normally via parachute), would meet up with resistance movements, and would then coordinate with the CIA through radios to receive U.S. arms and equipment via paradrop. Between 1959 and 1960, the CIA inserted trained Tibetan operatives back into their country in several batches.[79] Almost immediately, the Tibetan guerillas, backed by U.S. arms, equipment, and now prepared operatives, began kinetic operations against the Chinese.[80] After 1960, guerillas could train and equip in the denied area at Mustang plateau, making paradrop of personnel unnecessary. Still, in 1963, the CIA trained another 135 Tibetans at Camp Hale for insertion back into Mustang.[81]

The other two CIA operations in Tibet included ST Barnum (an airlift/delivery operation) and ST Bailey (an information campaign). The CIA placed the primary airport and logistics hub for aircraft supporting ST Barnum at one of the highest elevation runways in the world, an austere airport at Kurmitola in East Pakistan (modern day Bangladesh).[82] From here, the agency's Civil Air Transport (CAT) utilized C-118s to provide airdrops on designated and precoordinated drop zones (although B-17s had been utilized for blind drops in 1957).[83] Eventually, the operation incorporated C-130s, which performed better in high altitudes.[84] To ensure deniability, a number of foreign pilots were considered, included, and utilized to pilot some aircraft, including Polish and Czech dissidents.[85] Between 1959 and 1961 alone, at least 30 sorties, "each carrying an average of 25,000 pounds of arms, ammunition, medical supplies, and food" made airdrops to guerillas.[86] Some of the armaments included Enfield rifles, M-1 rifles, 60mm mortars, 80mm mortars, 57mm recoilless rifles, 75mm recoilless rifles, bazookas, grenades, and .30 caliber machine guns.[87]

Regarding ST Bailey, it's important to remember that the U.S. information campaign was designed for several audiences. One audience was within Tibet. This was the struggle for hearts and minds within the country. Another audience was for the Tibetan diaspora, and the third was an attempt to gain international support. The CIA established Tibet Houses in New York City, New Delhi, and Geneva. These served as "unofficial representation for the Dalai Lama to maintain the concept of a separate Tibetan political identity."[88] Another example of information operations includes a documentary film called *Unconquerable Tibet: Anti Communist Revolt, 1959.*[89] This film emphasized the peaceful nature of Tibetans and their Buddhist society, which, it shows, was devastated by greedy communists who simply wanted to impose slave labor and exploit Tibetans despite centuries of Tibetan self-rule.

As an example of expenditures to finance these three programs, the CIA spent $1,735,000 in 1964.[90] This included:

- Supporting 2,100 guerillas based in Nepal ($500,000)
- Providing a subsidy for the Dalai Lama ($180,000)
- Training paramilitary forces at an undisclosed location ($225,000)
- Training expenses in Colorado ($400,000)
- Funding Tibet Houses ($75,000)
- Providing covert air transportation between Colorado and India ($185,000)
- Developing an education program for 20 junior Tibetan officers ($45,000)
- Miscellaneous operating expenses ($125,000)

All these activities were coordinated with the State Department and other U.S. Government agencies. With remarkable efficiency, this very modest example of an annual budget was able to substantially disrupt the PRC's occupation of Tibet for a decade.

LESSONS

Scholarship on the Tibetan resistance to Communist China 1950–1975 is long overdue and in need of renewed emphasis. As stated by historian Robert McMahon, "crisis and conflict in Tibet had a powerful impact on the Cold War in Asia and beyond. The Tibet problem significantly affected the diplomatic relationships among five different countries—the U.S. , the People's Republic of China (PRC), the Soviet Union, India, and Pakistan—in manifold ways."[91] This case study regarding irregular forms of resistance in an occupied state has significant value not only in terms of international relations but also for future policymakers and strategies applicable to an era of great power competition.

| This case study regarding irregular forms of resistance in an occupied state has significant value not only in terms of international relations but also for future policymakers and strategies applicable to an era of great power competition. |

In this Cold War scenario, the U.S. offered both overt and covert support to the Tibetan resistance with mixed results, which generated some valuable lessons.

It remains unclear whether the CIA or the State Department views on supporting Tibetan resistance were anything more than a disruption effort targeted on thwarting communist expansion.[92] In 1957, when the CIA reinserted its first batch of six operatives into Tibet, the State Department attempted to provide legitimacy to the Tibetan resistance by convincing the Dalai Lama to promote violent opposition. An advocate for nonviolence, the Dalai Lama never took this step, which may have proven a critical factor in the limited aims of the guerillas during the next two decades.[93] Perhaps equally to blame for a limited application of unconventional warfare was the opinion that "both communist bloc and free world media output" regarding Tibet commanded little attention in the world, which included the United States.[94]

Due to its small scale, CIA support to paramilitary operations in Tibet served as an excellent economy-of-force option to contest the PRC. Diplomatically, not a single member of the United Nations recognized Tibet as an independent country. Despite this, the U.S. was able to fund resistance to occupation and even garner international attention to human rights abuses. While the operation remained covert, it served larger strategic purposes of disrupting communism, and it did so with very little resources. In fact, the Tibetan resistance imposed severe challenges for the Chinese to resolve, and the scars and tensions in Nepal remain to this day. Additionally, as in all cases of a sponsored resistance, the U.S. needed ready access to the region. In this case, the U.S. alliance with Pakistan and its use of a rural airbase proved pivotal. Keeping the signature low and keeping the operation covert was also critical in maintaining access.

One of the lessons learned in this case study is the importance of high-quality intelligence when conducting unconventional warfare. Particularly at the beginning of CIA operations, the U.S. had very little information on what was happening in Tibet. This meant they had an incomplete picture of the PLA in terms of size, organization, and activities. Further, the CIA did not understand or have communications with Tibetan groups that protested or rebelled. This lack of situational awareness hampered the CIA's ability to plan support for or coordinate with resistance movements. Such human networks and preparation of the environment needed to be completed years prior to be effective. One of the primary reasons for ST Circus included gathering intelligence from inside the country for export back to the CIA. In this respect, the operatives were helping the U.S. by providing information—more so than vice versa. In what is now known as the "Blue Satchel Raid" in October 1961, the Tibetan resistance ambushed a Chinese convoy. The intelligence haul included many insights on low PLA morale, the impossibilities of invading Taiwan, and PRC tensions with the Soviet Union.[95]

China's annexation of Tibet led to continuous tensions on the borders with India and eventually a war in 1962. As early as 1952, the PLA demonstrated intentions to move into Bhutan, Nepal, and Sikkim.[96] In 1954, the Chinese demanded: (1) that India show no interest in Tibet, (2) that India not object to Chinese border forts in Tibet, (3) that India "eradicate illegal activities of foreign agents working on the Indian side of the border" (an obvious reference to covert actions by the U.S. and Republic of China), and (4) that India not support any question of Tibetan sovereignty via the United Nations.[97] By 1962, hostilities between China and India led to a short war, but tensions continued for decades, with each country conducting exercises and mobilizations on the border regions, and this tension remains.[98] Essentially, the loss of a buffer state, combined with irregular forms of warfare and a border dispute, exacerbate competition. From a realism standpoint, conflict between China and India strengthened the U.S. position of confronting Chinese aggression and even inspired partnered activities. America maintained diplomatic coordination, communication, and logistics activities with India during the totality of its support to Tibetan resistance. Essentially, increased tensions between India and China furthered and expanded up capabilities to support Tibetan resistance.

About the Author

Dr. Robert S. Burrell is a resilience and resistance interdisciplinary scholar using data-driven and human-centric methodologies to analyze intrastate conflict ranging from nonviolent protest through belligerency. He is a Senior Research Fellow at the Global and National Security Institute of the University of South Florida. From 2020–2024, he taught irregular warfare at Joint Special Operations University and was the editor in chief of special operations doctrine from 2011–2014.

NOTES

1. Roger E. McCarthy, *Tears of the Lotus: Accounts of Tibetan Resistance to the Chinese Invasion*, 1950-1962, (Jefferson, North Carolina: McFarland and Company, 1997), 1

2. In the 1949 San Francisco treaty between Japan and Nationalist China, Japan relinquished claims to Taiwan. For more information, see Zhai Xin, "Security Securing Taiwan: Separating the Two Sides of the Taiwan Strait in Japan's China Policy," *Asian Studies Review*, Vol 46 No. 1, 2022, 93-109.

3. Lindsay Maizland, "Why China-Taiwan Relations Are So Tense," *Council on Foreign Relations*, 18 Apr 2023. Found at https://www.cfr.org/backgrounder/china-taiwan-relations-tension-us-policy-biden, accessed on 9 Aug 2023. Also see the Kuomintang constitution of 1947. One online example can be found at https://law.moj.gov.tw/ENG/LawClass/LawAll.aspx?pcode=A0000001, accessed on 9 August 2023.

4. Zhongguo, a name for China, utilized during the Qing period and prevalent in communist China today.

5. See Central Intelligence Agency archives, Central Intelligence Agency, "Covert Action in High Altitudes," *Tibetan Operations*, 17-19.

6. Central Intelligence Agency, *Tibet and China*, 27 Apr. 1959, 16.

7. Central Intelligence Agency, *Tibet and China*, 16-18.

8. Central Intelligence Agency, *Tibet and China*, 16-18.

9. Central Intelligence Agency archives, Central Intelligence Agency, "Covert Action at High Altitudes," 1 Jan. 1960.

10. See Central Intelligence Agency archives, Central Intelligence Agency, "The 'Liberation' of Tibet," 16 Jun. 1951. The Dalai Lama also agreed with the Sino-Tibetan agreement of May 1951, see Central Intelligence Agency archive, *Current Intelligence Review*, "Implications of Chinese Communist Control of Tibet," 21 Nov. 1951.

11. See Central Intelligence Agency archives, Central Intelligence Agency, "Popular Petition to the Chinese Communists and Tibetan Authorities," 20 Mar. 1953. Also see "Conversation from Mao Zedong's Audience with the Tibetan Tribute Mission," *Wilson Center Digital Archive*, 8 Oct. 1951, available at https://digitalarchive.wilsoncenter.org/document/conversation-mao-zedongs-audience-tibetan-tribute-mission, accessed 23 Aug. 2023.

12. For information on communist education, see Central Intelligence Agency archives, Central Intelligence Agency, "Chinese Communists Economic and Cultural Activities, Easter Tibet," 24 Dec. 1953.

13. McCarthy, *Tears of the Lotus*, 74.

14. Some sources estimate as much as one million of the six million Tibetan inhabitants died due to Chinese occupation. However, most sources, including the CIA, estimated Tibet's population as around 1.25 million in total. For the estimates of one million deaths due to starvation, wee Yeshe Choesang, "China is Guilty of Mass Genocide Against 1.2 Million People in Tibet," *Tibet Post International*, 12 June 2019, found online at https://www.thetibetpost.com/en/outlook/opinions-and-columns/4643-china-is-guilty-of-mass-genocide-against-12-million-people-of-tibet. Accessed on 24 Aug. 2023. Also see McCarthy, *Tears of the Lotus, 2.*

15. Central Intelligence Agency archives, Central Intelligence Agency, "Geographic Intelligence Memorandum: Resistance in Tibet," 21 July 1958.

16. For more information on ST Whale, see Kenneth Conboy and James Morrison, *The CIA's Secret War in Tibet*, (University of Kansas Press, 2002), 103, 105, 112, and 120. Also see Kenneth Conboy, "CIA Paramilitary Operations in Tebet, 1957-1975," (Warwick, England: Helion & Company, 2022), 26-18.

17. Central Intelligence Agency archives, "Notes for DCI Briefing of Senate foreign Relations Committee on 28 April 1959," 21 April 1959. For more information on the destruction of monasteries, see McCarthy, *Tears of the Lotus*, 177.

18. Tenzing Sonam and Ritu Sarin, *The Shadow Circus: The CIA in Tibet*, (White Crane Films, 1998).

19. For the United Nations resolutions, see Central Intelligence Agency archives, Central Intelligence Agency, "Covert Action in High Altitudes," *Tibetan Operations*, 17-29.

20. This essay uses the same phases of resistance as described in IDR Research Team, "Studies in Low-Intensity Conflict: The Tibetan rebellion," *Indian Defense Review*, 1988, found online at http://www.indiandefencereview.com/studies-in-low-intensity-conflict-the-tibetan-rebellion/2/, accessed 24 Aug. 2023.

21. Central Intelligence Agency archives, Central Intelligence Agency, "Anti-Communist Activities, Tibet," 31 Dec. 1952.

22. Central Intelligence Agency archives, Central Intelligence Agency, "Chinese Communist Occupation of Tibet," 5 Jan. 1952. Also see, "Anti-Communist Activities in Lhasa," 20 Nov. 1951.

23. Central Intelligence Agency archives. Central Intelligence Agency, "Anti-Communist Demonstration in Tibet Suppressed by Force," *Current Intelligence Bulletin*, 8 May 1952.

24. For the population estimate in Lhasa, see Central Intelligence Agency archives, "Notes for the DCI Briefing of Senate Foreign Relations Committee on 28 April 1959," 27 Apr. 1959. For agents from the communist party, see "Chinese Communist Party Workers Sent to Tibet," 2 Feb. 1953. Also see "Dispatch of Political Cadres to Tibet," 8 Aug. 1953.

25. McCarthy, *Tears of the Lotus*, 105.

26. McCarthy, *Tears of the Lotus*, 104.

27. Andrutsang Gompo Tashi wrote his own memoire of his struggle. See Gompo Tashi Andrugtsang, *Four Rivers, Six Ranges: A True Account of Khampa Resistance to Chinese in Tibet* (Dharamsala: Information Office of His Holiness the Dalai Lama, 1973).

28. Central Intelligence Agency Archives, Central Intelligence Agency, "Peiping Plans Large Withdrawals from Tibet," *Current Intelligence Bulletin*, 19 June 1957. According to McCarthy, the number of PLA troops in Tibet in 1956 was 150,000. See McCarthy, *Tears of the Lotus*, 114.

29. "Geographic Intelligence Memorandum: Resistance in Tibet."

30. "Geographic Intelligence Memorandum: Resistance in Tibet."

31. Central Intelligence Agency Archives, Central Intelligence Agency, "Notes for DCI Briefing of Senate Foreign Relations Committee," 27 April 1959.

32. Map by author and created utilizing one found in the Central Intelligence Agency archives, Central Intelligence Agency, "Geographic Intelligence Memorandum: Resistance in Tibet," 21 July 1958. Also see the map before page 1 in Michel, Peissel, *Cavaliers of Kham: The Secret War in Tibet*, (London: Heinemann, 1972).

33. Central Intelligence Agency archives, Central Intelligence Agency, "Notes for the DCI Briefing of Senate Foreign Relations Committee on 28 April 1959," 27 Apr. 1959.

34. Central Intelligence Agency archives, Central Intelligence Agency, "Peiping Moves Against New Trouble in Tibet Border Region," *Current Intelligence Bulletin*, 17 Aug 1957.

35. Central Intelligence Agency, "Peiping Moves Against New Trouble."

36. See Central Intelligence Agency archives, Central Intelligence Agency, "National Security Council Briefing," 28 March 1959.

37. David Wise, *The Politics of Lying: Government Deception, Secrecy and Power*, (New York, Random House, 1973), 170.

38. "Tibetan Serfs Emancipation Day," *Embassy of the People's Republic of China in the Republic of Namibia*, 28 Mar. 2009, found online at http://na.china-embassy.gov.cn/eng/sgxw/200903/t20090331_6717930.htm#:~:text=Onpercent2028percent20Marchpercent2Cpercent2019595percent2Cpercent20Zhou,thepercent20newpercent20Tibetanpercent20localpercent20government, accessed on 5 Sep. 2023.

39. Central Intelligence Agency archives, Central Intelligence Agency "Notes for the DCI Briefing of Senate Foreign Relations Committee on 28 April 1959," 27 Apr. 1959.

40. Central Intelligence Agency, "Notes for the DCI Briefing."

41. Central Intelligence Agency archives, Central Intelligence Agency, *Central Intelligence Agency Bulletin*, 30 March 1959, 1.

42. Kramer, Mark. "Great-Power Rivalries, Tibetan Guerrilla Resistance, and the Cold War in South Asia." *Journal of Cold War Studies*, vol. 8, no. 3, Summer 2006, 10.

43. McCarthy, *Tears of the Lotus*, 142.

44. McCarthy, *Tears of the Lotus*, 164.

45. "Studies in Low-Intensity Conflict: The Tibetan rebellion."

46. "Studies in Low-Intensity Conflict: The Tibetan rebellion."

47. McCarthy, *Tears of the Lotus*, 211.

48. Central Intelligence Agency archives, *Central Intelligence Agency Bulletin*, 30 March 1959, 1.

49. Howard D. Grayber was a well-known astrophysicist and prolific author. He was also a vocal advocate for U.S. international policy with opinions on the 2nd Indochina War, as well as the Chinese invasion of Tibet. Central Intelligence Agency archives, "Letter from Howard D. Grayber to John Foster Dulles," 25 Mar. 1959.

50. See Central Intelligence Agency archives, "Letter from John Foster Dulles to Howard D. Grayber," 22 Apr. 1959.

51. See "Studies in Low-Intensity Conflict: The Tibetan Rebellion."

52. Central Intelligence Agency archives, Central Intelligence Agency, "Impact of the Tibetan Campaign on the Economy of Communist China," 1 Feb. 1960. One year later, PLA forces appear to have been slightly lower with 106,900 soldiers. See Central Intelligence Agency archives, Central Intelligence Agency, "Impact of Continued Tibetan Resistance on the Economy of Communist China," 5 Jun. 1961.

53. Central Intelligence Agency, "Impact of the Tibetan campaign."

54. Central Intelligence Agency, "Impact of the Tibetan campaign."

55. For one example, see McCarthy, *Tears of the Lotus*, 229.

56. Central Intelligence Agency archives, *Central Intelligence Bulletin*, 9 Mar. 1961, 6. Some documents site that the Chinese announcement to stall socialization efforts began in December, 1961. See Central Intelligence Agency archives, Central Intelligence Agency, "Transition in Tibet," *Geographic Intelligence Report*, 5 Jun. 1961, 3.

57. Conboy and Morrison, 145-146. Also see McCarthy, *Tears of the Lotus*, 234.

58. Map by author. I have assumed that the position of the CIA built airfield was at present day Jomsom Airport, which has been supporting civil aviation since at least 1970. For instance, a Nepalese Royal Flight crashed at Jomsom in 1970. See "Crash of Havilland Canada DHC-6 Otter 100 on 27 February 1970," *Aviation Safety Network*, found online at https://aviation-safety.net/database/record.php?id=19700227-0, accessed on 24 Aug. 2023.

59. For information on the airport established by the CIA at Jomsom, see "Studies in Low-Intensity Conflict: The Tibetan rebellion."

60. Central Intelligence Agency archives, Central Intelligence Agency, "Review of Tibetan Operations," 9 January 1964.

61. In 1960, the footprint of armed fighters at Mustang numbered 1,800. See Central Intelligence Agency archives, Central Intelligence Agency, "Covert Action in High Altitudes," *Tibetan Operations*, 21.

62. Central Intelligence Agency, "Covert Action in High Altitudes," 21.

63. "Review of Tibetan Operations," 9 January 1964. One CIA history sites this as occurring in 1957. See Central Intelligence Agency archives, Central Intelligence Agency, "Covert Action in High Altitudes," *Tibetan Operations*, 21.

64. Central Intelligence Agency archives, *Central Intelligence Bulletin*, 25 Aug. 1965, 7.

65. Central Intelligence Agency archives, *Central Intelligence Bulletin*, 25 Aug. 1965, 7.

66. "Review of Tibetan Operations," 9 January 1964.

67. McCarthy, *Tears of the Lotus*, 5.

68. Jack Anderson, "CIA-Inspired Tibet Raids Wind Down," *Washington Post*, 14 Apr. 1973, B11.

69. Carole McGranahan, "Tibet's Cold War: The CIA and the Chushi Gangdrug Resistance, 1956-1974," *Journal of Cold War Studies*, Vol. 8, No. 3, 2006: 124.

70. "CIA-Inspired Tibet Raids Wind Down."

71. "CIA-Inspired Tibet Raids Wind Down."

72. The actual termination of the resistance activities on Mustang around 1974 proved tragic. In 1969, the CIA forced the removal of the Tibetan commanding officer Baba Yeshe in favor of Gyado Wangdu, whom the CIA deemed was more military-minded and aggressive. Later, Baba Yeshe, apparently out of spite and self-interest, conspired with the Nepalese government to weed out the resistance on Mustang by providing detailed information on the NVDA in 1974. See McCarthy, *Tears of the Lotus*, 246-247.

73. ST likely derives from the two-letter CIA country identifier used for East Asia, including Tibet. See Steve Ferenzi, "Proxy Warfare on the Roof of the World: Great Power Competition Lessons from Tibet," *InterAgency Journal*, Vol. 11, No. 1, 2020: 99. Also see Tenzing Sonam and Ritu Sarin. Ferenzi argues that ST Circus was later subsumed under the codename ST Barnum but describe the same activity, an argument differing how I have described them as separate functional activities.

74. Wise, *The Politics of Lying*, 169.

75. *The Shadow Circus: The CIA in Tibet.*

76. *The Shadow Circus: The CIA in Tibet.* For a much more detailed description of the training on Saipan, see Conboy, 11-16.

77. John Kenneth Knaus, "Official Policies and Covert Programs: The U.S. State Department, the CIA, and the Tibetan Resistance," *Journal of Cold War Studies*, Vol. 5, No. 3, 2003: 69-71. Also see Conboy, 21-25.

78. Central Intelligence Agency archives, Central Intelligence Agency, "Covert Action in High Altitudes," *Tibetan Operations*, 20.

79. *The Shadow Circus: The CIA in Tibet.*

80. *The Shadow Circus: The CIA in Tibet.* For context, see McGranahan, 102-130. For information on Camp Hale, see "C.I.A. Trained Tibetans in Colorado," *New York Times*, 19 April 1973. Also see David Wise, *The Politics of Lying: Government Deception, Secrecy, and Power*, (Random House, 1973). Also see, David Wise, "The Politics of Lying: Government Deception, Secrecy, and Power," interview by WFMT Radio, 25 May 1973, available online at https://studsterkel.wfmt.com/programs/david-wise-discusses-his-book-politics-lying-government-deception-secrecy-and-power, accessed 15 Aug. 2023. You can also read David Wise's book, *The Politics of Lying*, 166-167. Also see L. Fletcher Prouty, "Colorado to Koko Nor: The Amazing Tree Story of the CIA's Secret War Against Red China," *Denver Post Empire Magazine*, 6 Feb. 1972.

81. See Central Intelligence Agency archives, Central Intelligence Agency, "Covert Action in High Altitudes," *Tibetan Operations*, 25. For more information on ST Barnum, also see chapter 7 of Kenneth Conboy and James Morrison, *The CIA's Secret War in Tibet*, (University of Kansas Press, 2002).

82. Kenneth Conboy and James Morrison, 75-76. Also see Conboy, 13-14. Also see Bruce Riedel, "JFK's Forgotten Crisis with Bruce Reidel," *International Spy Museum*, 26 April 2016. Found at https://www.youtube.com/watch?v=5jXwaVPn3_Y, accessed on 12 Sep. 2023. Also see Leary, 64.

83. McCarthy, *Tears of the Lotus*, 242.

84. McCarthy, *Tears of the Lotus*, 242-243.

85. See William M. Leary, "Secret Mission to Tibet," *Air and Space*, Vol. 12, No. 1 (December. 1997/January 1998), 64.

86. Knaus, "Official Policies and Covert Programs," 70-71. Also see McCarthy, *Tears of the Lotus*, 243.

87. McCarthy, *Tears of the Lotus*, 242-243.

88. "Review of Tibetan Operations," 9 January 1964.

89. *Unconquerable Tibet: Anti-Communism Revolt*, 1959, unk. director, release circa 1959. A copy can be found on YouTube at https://www.youtube.com/watch?v=g5sCLUy2r7M, accessed on 23 Aug. 2023.

90. "Review of Tibetan Operations," 9 January 1964.

91. Robert J. McMahon, "U.S. Policy toward South Asia and Tibet during the Early Cold War," *Journal of Cold War Studies*, Vol. 8, No. 3, Summer 2006, 131.

92. *The Shadow Circus: The CIA in Tibet.*

93. Knaus, "Official Policies and Covert Programs," 69.

94. Central Intelligence Agency archives, Central Intelligence Agency, "Tibet: Forthcoming Events of Propaganda Significance," *Bi-Weekly Propaganda Guidance*, 9 October 1961.

95. For more discussion on the importance of this intelligence, see Carole McGranahan, "Tibet's Cold War: The CIA and the Chusi Gangdrug Resistance, 1956-1974," *Journal of Cold War Studies*, Vol. 8, No. 3, 2006, 119-120. Also see *The Shadow Circus: The CIA in Tibet.* Also see McCarthy, *Tears of the Lotus*, 234-235.

96. Central Intelligence Agency archives, "Chinese and Nepalese Communist Activities Along the Indian Boarder, August – October 1952," 4 Mar. 1953. Also see "Chinese Communist Troops on Northern Border of Bhutan," 2 Feb. 1953.

97. India did not agree to any of these conditions. See Central Intelligence Agency archives, "Sino-Indian Treaty of 29 April 1954 on Tibet," 12 Oct. 1954.

98. For information on the Sino-India War, see Central Intelligence Agency archives, Geographic Intelligence Memorandum, "The Sino-Indian Border Dispute in the North East Frontier Agency."

| 259 |

IV. Support to Resilience: Case Study of Uribe's Colombia (2002–2006)

By Thomas A. Marks

ABSTRACT: In this section, Dr. Thomas A. Marks details the insurgency in Colombia and the approach taken by the Colombian government to address it over an extended period. Additionally, he provides analytics on what the U.S. has offered to support partnership with Colombia. Ultimately, Marks describes the efforts to defeat and delegitimatize the FARC as a model to follow. Lastly, he provides lessons learned for future consideration with similar insurgent threats. This essay was previously published in 2007 as "A Model Counterinsurgency: Uribe's Colombia (2002-2006) vs FARC" in *Military Review*.[1]

OVERVIEW

Little is heard of U.S. involvement in counterinsurgency in Colombia. That which does appear is often inaccurate and ideologically skewed. Yet progress in America's "number three war" has been significant and appears all the more impressive given the increasing difficulties experienced in Iraq and Afghanistan. What is noteworthy is that the approach being used is "classic counterinsurgency." In this, there is considerable irony, because many of the significant aspects of the campaign were developed and implemented by American-educated leaders who were assisted, both directly and indirectly, by Americans. That the Colombians have improved upon the original foundation makes examination of the case all the more compelling and urgent.

BACKGROUND

Upon taking office in August 2002, President Alvaro Uribe Velez of Colombia was faced with a difficult strategic situation that required a fresh approach. This was forthcoming in a new document, the Democratic Security and Defense Policy, which radically reoriented the state's posture from negotiating with to confronting its principal security challenge: an insurgency inextricably linked to the narcotics trade and other criminal activity.

Although multifaceted in its dimensions, the new policy effectively assigned the cutting-edge role to the Colombian armed forces, most prominently to the dominant service, the army. It required the forces to pursue counterinsurgency aggressively against a well-funded, entrenched adversary within a complex international environment decidedly unsympathetic to internal war campaigns. Regardless, the armed forces performed in impressive fashion.

These same armed forces had already set the stage for the shift in policy by pursuing a reform movement that had enabled them to conduct more aggressive operations even as Uribe's predecessor, President Andres Pastrana (1998–2002), had unsuccessfully sought a negotiated settlement with the main insurgent group, *Fuerzas Armadas Revolucionarias de Colombia*, the Revolutionary Armed Forces of Colombia (FARC), and to a lesser extent with the distant second group, *Ejercito de Liberacion Nacional*, or National Liberation Army (ELN). Continued combat was necessary because neither FARC nor ELN altered its military posture during negotiations. To the contrary, FARC used Bogota's provision of what was supposed to be demilitarized space, the Zona de Despeje (or Area de Distension), to facilitate an intensification of the conflict via main force warfare while it continued to conduct terror and guerrilla actions. Thus, Colombia's counterinsurgency approach during the Pastrana years was not the result of deliberation and consultation within the government, but of an uneasy, unstated compromise, as Pastrana and his intimates negotiated with a duplicitous insurgent leadership on one hand, while on the other, they confronted the security force's growing unwillingness to accept the administration's increasingly discredited strategic calculus. When, in the face of all evidence to the contrary, Pastrana attempted to push through a second Zona, this one for the ELN, he faced a virtual popular revolt in the designated area. Cutting his losses prior to the first round of that year's presidential elections, Pastrana ordered the military in February 2002 to reoccupy the original Zona.

Maj. Gen. Carlos Ospina, IV Division commanding general, holds an AKM assault rifle captured while fighting in eastern Colombia in 2000. Ospina, at the time Colombia's most combat decorated officer and its acknowledged brain trust, went on to become the four-star commanding general of both the army and the joint forces. Source: Author photo.

SITUATION PRIOR TO URIBE'S ELECTION

Lack of government leadership during the Pastrana years had left security matters to the army (Ejerctio Nacional, or COLAR); navy, of which the marines were a part; and air force. The state, in other words, did not engage in counterinsurgency. This meant that although annual military plans included a basic civic action component, they were necessarily incomplete. That this did not prove disastrous stemmed from the nature of the major security threat, FARC (ELN was essentially a law-and-order concern).

Committed ideologically to Marxism-Leninism, FARC had increasingly drifted to a vaguely defined "Bolivarian" populism that had little appeal in Colombia. Polls consistently found the movement had minimal popular support or even sympathy. Its efforts at armed propaganda had fallen off to nothing after a mid-1980s high, and it was increasingly corrupted by reliance for funding upon criminal activity—drugs, kidnapping, and extortion (in that order, perhaps $250 million total). Consequently, its approach to insurgency—modeled after the "people's war" doctrine of the Vietnamese variant, filtered through, in particular, the *Farabundo Marti National Liberation Front* (FMLN) of El Salvador—had become a perversion of the original and had more in common with *focismo* of Che Guevara than Maoist armed political action built upon mass mobilization.

Mobile brigade soldiers pause during operations in 2000. Source: Author photo. Mobile brigades were the all-volunteer reaction element of the Colombian counterinsurgency. Photo illustrates the mixed-race nature of Colombia's manpower.

FARC's reliance upon the normal apparatus necessary to support armed campaigning—base areas and mobility corridors—resulted in a dual center of gravity vulnerable to Colombian military attack: the insurgent units themselves and their sources of sustenance. Allowing for the low numbers organized in a nationwide support base (frequently inspired by terror), the armed units basically comprised the movement.

FARC's vulnerabilities had been recognized by the new military leadership that emerged following Pastrana's inauguration. They had crafted their approach to neutralize FARC's strategy even as they instituted a far-reaching and comprehensive military reform process that affected everything from recruiting (a largely draftee COLAR became one-third volunteer, with key units essentially 100 percent "professionals"), to military schooling, to assignment policies, to structure, to operational art. The result was a reclaiming of the strategic initiative by the time of Uribe's advent.

Military reform was central to all that occurred during the Pastrana years.

| Military reform was central to all that occurred during the Pastrana years. |

A combination of internal dislocation caused by the growing drug trade, U.S. efforts to "punish" Colombia during the Samper administration (1994–1998) for inadequate "cooperation" in counternarcotics efforts, and mediocre senior military leadership had all combined to cripple a sound military. Reform, primarily a COLAR project, touched upon virtually every aspect of the institution but focused mainly on revitalizing the military education system, turning lessons learned into operational and organizational modifications, and developing sound NCO leadership to enhance small unit performance. Simultaneously, greater attention was paid to human rights instruction, information warfare, and joint and special operations.

The profound institutional and strategic shifts outlined above occurred as the U.S. , in the aftermath of 9/11, altered the approach of the Clinton years (1992–2000) and dropped the artificial barrier that had separated counternarcotics from counterinsurgency. This was critical because, during the Clinton administrations, the war had been artificially divided in accordance with the demands of American domestic politics. Washington was compelled to focus upon counternarcotics to the virtual exclusion of counterinsurgency. Only where counterinsurgency objectives could be subsumed within counternarcotics action was U.S. aid allowed to assist in the security campaign.

Consequently, the U.S. contribution to Plan Colombia, a multifaceted effort to identify Colombia's critical areas for action to facilitate national revitalization, was structured wholly to support counternarcotics (for projects and allocations; see Table 4.2). Its centerpiece was an American-funded, equipped, and trained counternarcotics brigade manned by COLAR personnel but dedicated entirely, for legal reasons contained in the implementing legislation, to support of eradication. The brigade was severely limited in its operational and geographic scope, even though it had several times the number of helicopters in the entire COLAR aviation inventory.

TABLE 4.2. U.S. ALLOCATIONS OF $1,318.6 BILLION FOR PLAN COLOMBIA

LINE OF EFFORT	FUNDING IN MILLIONS
Support for Efforts in Southern Columbia	$416.9
Support for Interdiction (including Forward Operating Locals)	$378.1
Support for Columbia National Police	$115.6
Alternative/Economic Development	$106
Human Rights/Judicial Reform	$122
Regional Support	$180

Source: Author. Monetary figures derived from Colombian military briefing.

Of greater consequence than the lack of fully relevant support was the battlefield fragmentation and distortion—the disruption to unity of effort—that the U.S. strategy entailed. Committed to assistance in the only fashion politically viable, and in an America forced to focus upon the supply side of its own drug problem, U.S. officials, forces, and individuals tended to embrace the flawed logic that Colombia's problem was narcotics, with the security battle merely a by-product. Insurgent reality was stood on its head.

American urgings that Colombian armed action focus upon a narcotics center of gravity were rejected by the military's leaders (often in conflict with the Pastrana administration). As far as they were concerned, U.S. input during this period was appreciated, but tangential to the real issue, counterinsurgency. Committed to area domination by regular (largely draftee) brigades and divisions, with strike forces organic to each of these units, COLAR would deploy but limited additional forces to augment the counternarcotics brigade. The focus of the internal war, in its estimation, had to be the population, 95 to 96 percent of which lived outside the drug-producing zones of the *llanos*, or eastern savannah.

Ironically, even the eventual drop in the bar between counternarcotics and what came to be labeled CT (counterterrorism) assistance, did not change this situation. Although U.S. funding was impressive in raw figures (see Table 4.3), it was still overwhelmingly committed to a counternarcotics campaign driven by its own internal measures (most prominently, hectares of narcotics fields eliminated).[2] Controversial due to its reliance upon aerial spraying, the eradication effort incorporated a variety of other components, from air and riverine interdiction to alternative development, but its actual impact upon insurgent operational capabilities proved difficult to measure.

TABLE 4.3. U.S. FINANCIAL ASSISTANCE TO COLUMBIA, 1997-2005

	1998	1999	2000	2001	2002	2003	2004	2005
INC/ACI	57	200.1	686.4	48	243.5	412	313	313
INC/ACI*	.5	5.75	208	0	134	168	150	150
FMF	0	.44	.02	4.49	0	93	98.45	99.2
IMET	0.863	.92	.9	1.04	1.165	1.676	1.7	1.7
ATA	0	0	0	0	25	3.28	.2	3.92
506	41.1	58	0	0	0	0	0	0
1004	11.78	35.89	68.71	150	84.9	136	110.2	110.2
1033	2.17	13.45	7.23	22.3	4	13.2	13.2	13.2
TOTAL	**113.4**	**314.6**	**971.3**	**225.8**	**492.6**	**827.2**	**686.8**	**691.2**

Source: Author. Information as briefed by the U.S. State Department.
*Note: Figures are in millions of dollars. ATA = anti-terrorism assistance; FMF = foreign military financing; IMET = international military education and training; INC/ACI = International. Not included: excess defense articles ($10.1 million total) and economic support funds ($7 million total). *funding for counter-drug economic and social aid*

Also clouding the picture were periodicals of record in the U.S. that tended to lump overall U.S. aid figures into "support for the Colombian military," thus reviving a Vietnam-era stereotype of a hapless ARVN held together by American money and "advisors." Nothing could have been further from reality in Colombia. The bulk of U.S. funding to date has gone mainly to the counternarcotics effort (e.g., 85 percent of the 2005 figure in Table 4.3), with only incidental impact (from this source) upon the Colombian forces. The funding that has gone directly to the Colombian military has been important, especially as dispersed through the actions and programs of the highly regarded military assistance mission, but during the Pastrana years, Colombia's armed forces were quite on their own in both their operations and their reforms.

Colombia's basic military framework for waging counterinsurgency was created by the geographical assignment of the 5 COLAR divisions (18 brigades) and a joint task force, with a division-strength national reaction force.[5] Of its 145,000 troops, COLAR had some 20,000 in volunteer counterguerrilla units organic to its brigades and divisions. Altogether, the volunteer units amounted to 47 counterguerrilla battalions (*batallones contraguerrillas*, or BCG) and 3 mobile brigades (*brigades moviles*, or BRIM) each composed of 4 BCG, for a total of approximately 59 BCG (each with approximately 40 percent of the manning of a line battalion, but with additional machine guns and mortars).

The regular formations that comprised the rest of COLAR were overwhelmingly draftee. Domination of local areas was the linchpin of the counterinsurgent effort, and a variety of imaginative solutions were tried to maintain state presence in affected areas. Essentially, the draftee regular units were used in area domination and local operations, the BCG and BRIM to strike at targets of opportunity. Specific missions that required specific skills, such as guarding critical infrastructure or operating in urban areas, were carried out by dedicated assets, as were special operations.

But in the absence of local forces, which had fallen afoul of constitutional court restrictions and thus disbanded, it was difficult to consolidate gains. As areas were retaken, they could not be garrisoned with home guards. Instead, regular units rotated in and out in a perpetual shell game designed to keep FARC off balance (to a lesser extent ELN; only FARC operated with main forces).

Further complicating the situation, a legal framework that did not respond to the needs of internal war meant that all action was carried out under the provisions of peacetime civilian law. The Pastrana administration passed no emergency or anti/counterterrorist legislation of any sort. This sometimes placed soldiers in absurd situations, particularly since the police were not available to accompany operations, being preoccupied with their own efforts to survive. Half a dozen times, for instance, towns and their police garrisons found themselves attacked by FARC forces using homemade but nonetheless potent armor.

Faced with such an array of challenges, it was a credit to the power of the military reform movement and the improvements made by its leadership that the strategic initiative had been regained by mid-2002. This occurred because the reform movement in the dominant service, COLAR, was driven by personalities who evinced an understanding of counterinsurgency and Colombia's unique circumstances. Thus, they were able, despite the state's lack of strategic involvement, to arrest the negative trends that had emerged with growing force as early as the Samper administration.

Most importantly, the reform leadership defeated FARC's attempt to transition to main-force warfare (i.e., mobile or maneuver warfare, stage two in the people's war framework). Using the Zona as the staging ground for attacks by "strategic columns" comprised of multiple battalion-strength units, PARC found itself bested by the commanding general (CG) of IV Division, Major General Carlos Alberto Ospina Ovalle, who worked intimately with his superior the CG of COLAR (*Comandante de! Ejercito*), General Jorge Enrique Mora Rangel, and CG of Joint Command (*Comando General de las Fuerzes Militares*), General Fernando Tapias Stahelin.[4]

This trio dominated operational planning throughout the Pastrana years, with Mora eventually taking Tapias' place upon the latter's retirement. Ospina, after serving as the CG of IV Division, became COLAR Director of Operations under Mora; then Inspector General (IG) Joint Command, under Tapias, who used the IG principally as a combat inspectorate; and, finally, CG COLAR (with full general rank) when Mora moved up upon Uribe's inauguration. When Mora himself retired in November 2003, Ospina became CG Joint Command.

What these officers shared was a correct understanding of Colombia's war and a well-developed approach to institutional transformation and strategy realized in operational art. Mora and Ospina were noted for their close working relationship and the general esteem they were held in throughout the armed forces. Both had proven themselves tactically time and again as they advanced through the junior ranks, then operationally and strategically as more senior commanders.

Ospina was apparently the most combat-decorated officer in COLAR at the time he became its CG in addition to being universally regarded as COLAR's "brain trust" with a deep knowledge of insurgency and counterinsurgency. Working together under Tapias, Mora and Ospina fashioned highly effective COLAR annual campaign plans that forced FARC onto the defensive. Their correct appreciation of the situation, though, could not be translated into a true national counterinsurgency until Uribe's election.

URIBE'S DEMOCRATIC SECURITY AND DEFENSE POLICY

A third-party candidate who won an unprecedented first-round victory in May 2002, Uribe introduced a dynamic style to security affairs that prominently included producing, early in his administration and with U.S. encouragement, the aforementioned Democratic Security and Defense Policy (officially released in June 2003). Unlike the Plan Colombia of the

Pastrana-Clinton years (written with U.S. input), which had been a virtual catalog of national ills with proposed solutions beyond Bogota 's ability to operationalize or fund, the new policy was intended to be a course of action. As such, it was built upon a fairly basic syllogism:

- Lack of personal security is at the root of Colombia's social, economic, and political ills.
- This lack of personal security stems from the state's absence from large swaths of the national territory.
- Therefore, all elements of national power need to be directed toward ending this lack of national integration.

Addressing this assessment was the policy itself, its thrust stated directly: "Security is not regarded primarily as the security of the State, nor as the security of the citizen without the assistance of the State. Rather, it is the protection of the citizen and democracy by the State with the solidarity and cooperation of the whole of society. This is, in short, a policy for the protection of the population." According to the policy, citizens and the stability of the country were threatened by an explosive combination of "terrorism; the illegal drugs trade; illicit finance; traffic of arms, ammunition, and explosives; kidnapping and extortion; and homicide." The hitherto intractable nature of Colombia's security conundrum stemmed from the interlocking nature of these threats.

It was this dynamic at which Uribe's plan was aimed. If one course of action stands out as central to the whole, it is "consolidating control of national territory," the indispensable element of any counterinsurgency.

| If one course of action stands out as central to the whole, it is "consolidating control of national territory," the indispensable element of any counterinsurgency. |

The plan details a "cycle of recovery" that evokes images of the approach used in successful counterinsurgencies in Thailand, the Philippines, and Peru, and it outlines precisely the strategic approach to be used:

- "The Government will gradually restore State presence and the authority of State institutions, starting in strategically important areas.
- "Once the Armed Forces and the National Police have reestablished control over an area, units comprising professional soldiers, campesino soldiers [i.e., local forces], and National Police Carabineros [police field force] will maintain security and protect the civilian population. This will enable state organizations and criminal investigation authorities to work in the area.

- Once a basic level of security has been established, the State will embark upon a policy of territorial consolidation, reestablishing the normal operation of the justice system, strengthening local democracy, meeting the most urgent needs of the population, broadening state services, and initiating medium-to-long term projects aimed at creating sustainable development."[6]

Necessarily, since Colombia's plan calls for nothing less than waging internal war against a hydra-headed threat, the security forces undertake the most prominent and difficult tasks. Although responsibilities are outlined for all state bodies, it is the security forces that are to provide the shield behind which restoration of legitimate government writ takes place. Under the Ministry of Defense (*Ministerio de Defensa Nacional*, or MDN) the security forces prepared their own plans to implement the Democratic Security and Defense Policy.[7] Both the military's Joint Command and the national police (*Policia Nacional*, or CNP) were subordinate to MDN and used as their guide the strategic document drawn up by Defense Minister Marta Lucia Ramirez de Rincon and her staff after consideration of the Uribe policy. Their product was issued as a four-year vision applicable to the entire Uribe presidency. COLAR's objectives were, for all practical purposes, those of the Joint Command.

The central elements remained "protection of the population" and "elimination of the illegal drugs trade in Colombia," to be accomplished through the application of national will, resources, and power. As the premier element of national power in the internal war at hand, the military clarified its role further in a "general military strategy" issued by CG Joint Command, General Mora. This is still the key document regarding the application of military action to support the president's Democratic Security and Defense Policy's counterinsurgency approach.

IMPLEMENTING URIBE'S PLAN

With the framework established, implementation followed. In this, the military was far ahead of other state elements since it had already gone through dramatic change during the Pastrana years. So far reaching were the military reforms that, in many respects, the armed forces presented Uribe with a new tool upon his taking office. The key had been a continuity of exceptional leadership able to reorient, under difficult operational and material conditions, the military's warfighting posture.

Central to this reorientation was the inculcation in the officer corps of greater professional knowledge concerning not only the operational and tactical mechanics of internal war but also the strategic knowledge of insurgent approaches and aims. It was here that Mora's faith in Ospina's understanding of counterinsurgency paid off.

Ospina was adamant that seeing the insurgents as merely narcotics traffickers or criminals or terrorists obscured the deadly symbiosis that drove the conflict. Whatever it engaged in tactically, whether terror or the drug trade, FARC was a revolutionary movement that sought to implement a people's war as its operational form, to include focusing upon the rural areas to surround the urban areas.

Hence, for the security forces, the strategic and operational threat had remained relatively constant in nature, regardless of increasing insurgent (especially FARC) involvement in the drug trade and other criminal activity. The insurgents sought to dominate local areas, eliminating through terror those who persisted in their opposition. Guerrilla action targeted the police and smaller military units, with task-organized columns (*columnas*) appearing as main forces whenever a target invited. Other, nonviolent elements of the FARC people's war approach—mass line, united front, political warfare, and international action—remained anemic to the point of irrelevance, leaving the "violence" line of operation the only real issue.

As noted previously, when Uribe took office, the military had already spent nearly four years developing an effective counterinsurgency approach specifically applicable to Colombia. The strategy recognized the need to dominate local areas by providing a security umbrella under which the normal functions of the state could be exercised. The operational vehicle for carrying out the effort was to place a "grid" over the target area, with specific forces carrying out specific missions, all coordinated in such manner as to stifle insurgent activity. The immediate problem was that there had not been enough units or enough funding.

Counterinsurgency is manpower and resource intensive. Uribe sought to provide both assets to a military leadership that was already out of the starting gate. Not only did he raise the military's general funding level, but, in a dramatic gesture of commitment, he also asked Congress to levy a one-time war tax for a substantial expansion of actual forces, primarily COLAR (which in mid-2004 reached a strength of some 202,000). The tax brought in approximately $670 million, which was allocated to *Plan de Choque* 2002-2006 (Plan Shock), a phased scheme to substantially increase the specialized COLAR forces needed to make the grid viable.

Units of all types were integrated into the force structure according to plans predating Uribe, but hitherto unfunded: new BCG and BRIM were added, with every division getting its own organic BRIM (IV Division received two; COLAR-wide, there are now at least 17 BRIM, up from the previous three) and others going to the general reserve (if all formations are considered, there are now roughly 100 BCG, up from the Pastrana total of 59); urban special forces (joining "rural" special forces, the traditional mode of operation); special transportation network protection units (*Plan Meteoro*, or Plan Meteor); high-mountain battalions specifically situated and equipped to block insurgent mobility corridors through hitherto inaccessible heights; strengthened infrastructure protection units (PEEV, from *Plan Energetico y Vial*, or Energy and Road Plan); and local forces (*So/dados de mi Pueblo*, "Home Guards") to provide security, particularly for rural urban centers.[8]

At the same time and from the same funding, some individual soldier effectiveness was to be improved by conveying draftee slots to volunteers at the rate of 10,000 per year—an expensive undertaking since it costs approximately 10 times more to field a volunteer.

All components were related to each other. The standing up of local-forces platoons, for instance, although initially intended to enhance the population's security, was soon found to produce a much greater information flow to the forces, which enabled more accurate and intense employment of regular and strike units. Greater activity in an area forced the insurgents, especially the leaders, to move, presenting targets for the upgraded special operations capability. Loss of leaders led to surrenders, which psychological warfare units exploited with a variety of innovative programs from rallies to radio broadcasts. Fewer insurgents meant greater freedom of movement, and special units secured the transportation arteries, just as they did the critical infrastructure. Business picked up, the economy improved, and kidnappings and murders dropped substantially.

If there was one element in the grid that provided the missing link, it was the deployment of local forces. These were indispensable to establishing a state presence in affected areas and neatly sidestepped legal objections (and fierce opposition from international human rights organizations) by utilizing a forgotten law, discovered still on the books, that allowed a portion of the national draft levy to opt for service in hometown defense units.

These 40-man units were constituted as regular platoons assigned to complement regular battalions stationed nearby. They were trained, armed, and equipped as regular soldiers, officered by regulars, and fielded systematically according to Plan de Choque funding. Soon, they were present in more than 600 locations selected according to the Joint Command campaign plan. Most were COLAR assets, although a number were run by the marines, mainly in a special "mini-divisional zone" assigned to the marines, south of navy headquarters in Cartagena on the Caribbean coast.

Local forces had more impact because the police, responding to the same need for government presence if security was to be guaranteed, systematically established a presence in every municipio (county) in the country. Those areas from which they had been driven, or that historically had been considered too dangerous for police presence, were manned by police field forces, the Carabineros, under regular CNP jurisdiction. The Carabineros functioned in units of the same size and type as the COLAR local forces, but they were more mobile and often better armed. Where necessary, they constructed fort-like police stations to project state presence. Backing them up was a highly trained reaction force.

Incorporation of police involvement into the grid highlighted a further development: the increasingly joint and interagency nature of Colombian operations. Although the military services had always answered to CG Joint Command, they had previously functioned together more as a matter of courtesy than command. This had not posed any insuperable problems, particularly given COLAR's dominance, but it was not the ideal way to conduct counterinsurgency, where unity of command is crucial. It was especially the case that the CNP, under Pastrana, was not integrated at the national level in any of the counterinsurgency planning. This ended under Uribe.

Within the military itself, a clear trend toward greater "jointness"—which had emerged under Tapias as CG Joint Command and matured under Mora (and Uribe)—blossomed under Ospina. Plans to implement "joint operational commands" in place of the exclusively COLAR divisional areas met with fierce resistance in parochial circles but were being pushed through by late 2004.

This transformation alone would be enough to produce a measure of turmoil within the military. Even the existence of the integrated *Fuerza de Tarea Conjunta* (Joint Task Force), controlled by CG Joint Command and operating in FARC's traditional base complexes in the east, generated disquiet in some circles—particularly as it became clear that it was a model of what is to come. If these plans are pushed through, the individual services would become more like "service providers" in the U.S. sense, while CG Joint Command will exercise operational control of joint forces that resemble U.S. combatant commands (e.g., Southern Command, which supports Colombia's effort). Such a development would be entirely logical for waging counterinsurgency but represent a sea change in the way Colombian services have historically functioned.

Integration extended beyond the military. Other government agencies were directed to participate. The state's involvement brought a new closeness to integrated efforts that hitherto had normally depended upon interpersonal relations in areas of operation. In particular, law enforcement and judicial authorities became an important part of operations. This provided government forces with enhanced flexibility because the police and officials could engage in actions not legally devolved to the armed forces (e.g., the right to search).

Operationally, the guiding document was the Joint Command's multi-year *Plan Patriota* (Plan Patriot), which prioritized areas of insurgent activity according to FARC's dispositions and activities and outlined sub-plans for the group's neutralization. FARC's demise was to be achieved via the tested technique of "'holding" in "strategic maintenance areas," where the situation was already considered in hand, while concentrating forces in "strategic operational areas" where insurgents still operated freely. The first such operational area was Cundinamarca, the state surrounding Bogota, which throughout 2003 was systematically cleared of major insurgent presence. So complete was the effort that FARC assessments outlined a disaster of the first magnitude, even as the security forces "moved on" to the insurgent base complexes in the east, especially in the area of the former Zona.

"Moved on," of course, meant only a concentration of forces for the purpose of conducting the continuous operations, unlimited in time but directed at a particular space, that the Joint Command has termed *masa dispersa* (dispersed mass).[9] These were conducted under tight operational security. Once Cundinamarca was cleared, *Fuerza de Tarea Conjunta* assumed priority of effort and systematically combed the "strategic rearguard," as FARC termed its decades-old base complexes, restoring government presence and popular freedom of movement and livelihood. A particular chore was to deal with the numerous and widespread unmarked minefields FARC had emplaced.

CHALLENGE OF ASSESSING COUNTERINSURGENCY PROGRESS

Uribe was able to deliver the state commitment, strategic framework, and enhanced resources that propelled takeoff. While he provided the dynamic leadership, the Defense Ministry's job was to offer further guidance but, in particular, to engage in matters of policy that allowed the military forces to exist and operate. A confusion of roles—a desire to lead the military rather than manage it—led to the replacement of Defense Minister Ramirez in November 2003. Ramirez had clashed repeatedly with the military leadership. CG Joint Command Jorge Mora also stepped down.

The Minister and CG were replaced, respectively, by Jorge Alberto Uribe Echevarria and Carlos Ospina. Moving into the CG COLAR position was the COLAR Director of Operations, Major General Martin Orlando Carreno Sandoval. Mora had planned to step down in December in any case, so the transition was smooth. Minister Uribe adopted a more careful style than his predecessor, and there were no significant changes in the 2004 planning and policy guidance: the military was left to lead the implementation of the counterinsurgency. In this, however, Carreno did not inspire the support necessary to keep his position more than a year. He was replaced in November 2004 by the *Fuerza de Tarea Conjunta* commander, Major General Reinaldo Castellanos Trujillo. Subsequently, Minister Uribe himself, weary of criticism in congress, stepped down and was replaced by Camilo Ospina Bernal. Castellanos, however, was himself replaced only a year later by Major General Mario Montoya.[10] Ultimately, in the second Uribe administration, both Carlos Ospina and Minister Ospina stepped down, and Ospina's deputy, Lieutenant General Freddy Padilla, became CG Joint Forces. Juan Manuel Santos became Minister of Defense.

Mobile brigade soldiers in 2001 operations. Ultimately, the mobile brigades (or BRIM) were made up of some 100 counterguerrilla battalions (BCG), with approximately half the strength of line battalions but with a heavier complement of firepower. Source: Photo by Author

Such personnel upheaval notwithstanding, military support for the Democratic Security and Defense Policy proceeded in near textbook fashion. Politically, the danger was that Colombia would become distracted, as it was by the debate that surfaced about Uribe's then still low-key effort to be allowed to run for a second term, which required constitutional amendment. To oppose a second term for Uribe all but demanded that his first-term record be attacked. The attacks, however, did not involve direct assault on the security forces; rather, they argued that too much effort was being placed upon security, and that "social matters" were just as important. The precise point of Uribe's approach, however, was that the second was not possible without the first.

Nevertheless, what emerged was a FARC response that sought to strike at the counterinsurgents' will to persevere. If Colombia's operational implementation of its plan had been successful where the U.S. had stumbled in Iraq and Afghanistan—the Colombians successively dominating areas and restoring government writ—this did not prevent critics at home and abroad from attacking Bogota 's approach. Their criticism allowed FARC to appear much stronger than it was. Insurgent tactical assaults were given strategic consequence with spin. This spin came not from FARC, but from the president's political enemies and from the media's often dubious reporting. The result was that FARC's minor tactics, inconsequential in and of themselves, stood a chance of generating strategic reversal for the state.

It could be argued that this is the very stuff of insurgency, where every action is intended to have a political consequence. True as far as it goes, the observation misses the point that, in today's international environment, what insurgents and terrorists do is in one sense irrelevant: few citizens accept their proffered agendas. But their actions provide ammunition for political attacks occasioned by the normal infighting inherent to democratic politics. Rather than targeting their intended mass base, the insurgents try to cut corners by attacking the will of their enemies. This is what happened in Colombia.

| in today's international environment, what insurgents and terrorists do is in one sense irrelevant: few citizens accept their proffered agendas. But their actions provide ammunition for political attacks occasioned by the normal infighting inherent to democratic politics. Rather than targeting their intended mass base, the insurgents try to cut corners by attacking the will of their enemies. |

As it was, Uribe was able to adroitly fend off the attacks even while successfully overseeing and completing an arduous process of constitutional amendment and reelection that culminated in an unprecedented second term in office (beginning August 2006) after another first-round victory in the presidential vote. Uribe's win ensured that operational implementation of his strategic framework would continue. This was significant because the approach, as discussed above, was both correct and sustainable—thereby satisfying two of the three requirements of successful counterinsurgency.

What the political controversy highlighted was a little-understood element in successful counterinsurgency. With a correct and sustainable approach in place, the counterinsurgent "plays for the breaks," those shifts in the internal or external situation that work against the insurgent and favor the state.[11] Such play normally requires an extended period of time and leads to a protracted war. This long timeframe makes it difficult for democracies to sustain counterinsurgency campaigns, particularly in the present world environment where there is little agreement upon strategic ends and means, much less operational and tactical concerns. Yet it does not in any way obviate the reality that there is no other option.

How then was the state to think about the tremendous progress it had made in Uribe's first term? What future steps would allow Colombia not only to assess sustainability but also to continue its success?

What drives any assessment is the nature of the situation on the ground as it can be measured. Efforts to judge counterinsurgency progress in Colombia have produced a variety of statistics. These have been used to support both proponents of Colombia's Democratic Security and Defense Policy's efficacy and opponents who question, if not the approach as a whole, certain of its emphases and components.

Statistics, in other words, are a double-edged sword:

- First, there is the political reality: efforts to arrive at metrics for assessing the progress of an approach, although absolutely necessary, take on meaning only as they are interpreted by an audience. All parties to the Colombian political debate, for example, agree that by any metric utilized (e.g., a decline in kidnapping and murder), there has been demonstrable (even stunning) progress toward normalcy. Yet there is little agreement as to what normalcy, as an end state, should actually look like.
- Second, there is the empirical reality that the causes behind insurgency cannot be statistically explained. Hence, to measure counterinsurgency progress by gauging how much the country has moved toward a notional state of normalcy is like looking at annual percentage increases in the gross domestic product (GDP) without actually being able to measure the GDP itself. "Progress," then, ends up being a state of popular mind, a belief by the populace (and its leaders) that the situation is improving.

In the matter of statistics, a combination of quantitative and qualitative indicators has given rise in Colombia to the judgment that progress is being made. This does not mean, however, that merely advocating more of the same is the prescription for further action so much as staying the course.

Demo Democratic Security and Defense Policy was built upon acceptance by the political authorities of the Uribe administration position that the Gordian knot in Colombia's security impasse is FARC. Only FARC continues to seek state power while simultaneously demonstrating the capacity to negate state armed capacity. ELN, the "other" insurgent group, is a nuisance, while the vigilante *Autodefensas Unida Colombia* (AUC, or United Self-Defense Groups of Colombia), the so-called paramilitaries, have historically been a consequence of lack of state presence. As the state has expanded its control, the AUC has been willing to strike demobilization deals. ELN has likewise indicated a desire to open a peace process. In contrast, negotiations with FARC have not proved successful, so only armed action by the state remains. The desired goal is reincorporation of FARC into the political process, but it is recognized that incentive must be created by armed action.

Compelling FARC to undertake a course of action necessarily involves neutralizing its ability to remain viable. Thus, the intent of the government's counterinsurgency grid is to attack FARC's ability to recruit, sustain itself, move, and initiate actions. Domination of populated areas such as Cundinamarca prepared the way for operations against FARC's strategic rearguard in the former Zona and other southern areas. These operations continued due to the sheer size of the counter-state FARC constructed over four decades. The forces committed to these and other priority efforts have not been robbed from established counterinsurgency areas (effectively, the army's divisional zones), but deployed from new assets. Their actions are sustainable virtually indefinitely.

That the government's operations have made life more difficult for FARC is unquestionable. But just how difficult is the query that cannot be answered definitively. The least reliable way to judge results is to match FARC casualties with the organization's order of battle. The top figure of some 17,000 combatants (reached during the Pastrana administration) is now put at below 13,000, with most counts claiming that AUC combatants at the time of their demobilization actually outnumbered their FARC rivals (ELN was perhaps a fifth the size of FARC). It is not that these numbers are necessarily wrong; rather, it is unlikely that they mean much given the realities of an insurgent movement operating with a minimal but adequate support base and funding generated outside any popular base.

During the Mora and Ospina tenures, the need to count insurgent casualties was not driven by the Colombian military, which made a concerted effort to stay away from the "Vietnam body-count trap." Instead, the political authorities (many of whom have business backgrounds) and the press felt it necessary to give the public the numerical equivalent of sound bites that elevated quantitative measures to heights the military itself did not subscribe to. The military's approach was clear if one inspected its internal documents. These gave pride of place not to body count, but to measures of FARC's initiative and armed capacity (such as the ability to initiate major attacks).

Not only do the military's metrics contrast sharply with the indicators favored by the political authorities and the press, but they also serve to highlight the abuse of statistics that became a routine part of the present political debate surrounding President Uribe's desire to earn a second term. Critics of Uribe and the Democratic Security and Defense Policy approach regularly claimed to possess data showing an explosion of FARC incidents and initiative, but their position was not backed by realities on the ground. What must ultimately drive any assessment is the nature of the incidents being counted. The military knows this and has incorporated such an approach into its own analysis. Nature can involve anything from size to context.

An insurgent group such as FARC, forced from mobile warfare back to guerrilla and terror actions, of necessity needs to up the ante. This FARC attempted to do by cultivating an association with the Provisional Irish Republican Army (PIRA), which sent some two dozen training teams into FARC areas before the pipeline was effectively shut down in 2001. FARC efforts to utilize a variety of PIRA terror techniques, rarely or never seen in Colombia—ranging from the precise placement of bombs to inflict maximum structural damage to the use of secondary explosions to wreak havoc upon crews responding to incidents—were all designed to inflict maximum casualties and generate maximum terror. That they failed to do so left FARC with the one option it has now pursued: pinprick attacks that can produce tactical heat but lack strategic fire.

In only one way can FARC's tactical actions have strategic or even operational significance: if they can be parlayed into political consequence. Strategic, operational, and even tactical techniques for using violent action to effect political gain are a central element of the people's war approach used by FARC. They are recognized as such in FARC doctrine, and they were critical to the FMLN effort in El Salvador that was so important to FARC's doctrinal evolution. A key issue is whether FARC is attempting to use its tactical efforts to exploit rifts in the Colombian political spectrum. Captured documents and information gleaned from prisoner inter rogations demonstrate that FARC is well aware that by inflicting casualties and appearing to be "alive," despite all that the security forces have done, it can provoke political problems of sufficient magnitude to damage or even end Democratic Security and Defense Policy.

It is ironic that the strategic progress of Democratic Security and Defense Policy is unlikely to negate FARC's tactical ability to initiate guerrilla and terror actions completely. But the group's "successes" in these low-level actions really count for little. For instance, there have been many mine casualties among the security forces, but that has little to do with anything save FARC's extensive use of the internationally banned weapons. Mines do not hold towns and villages, and they do not create sympathy for the insurgents; they are indiscriminate defensive weapons. Most COLAR casualties from mines, in fact, have been suffered as the army pushes ever deeper into insurgent base areas and dismantles the FARC counter-state.

Eliminating the strategic rear guard is crucial. There is a common misconception that guerrillas are self-sustaining, obtaining all they need either by generating it or capturing it from the government. In reality, insurgents can rarely if ever obtain crucial components of their war effort, notably arms and ammunition, from within the battlespace and thus must pursue outside acquisition. FARC indeed gets most of its weapons and ammunition from abroad. Even food, as demonstrated by massive caches uncovered in the strategic rearguard throughout 2004 and 2005, is stockpiled and pushed forward to combatants. Eliminating the base areas and their stockpiles therefore eliminates FARC's ability to mass and forces it to engage in terror and guerrilla warfare, which can be much more easily managed by the enhanced capabilities and presence of the state.

Faced with this profound threat to its viability as an insurgent movement, FARC must respond. As a consequence, there should be no doubt that violence in Colombia will continue indefinitely. Yet the state should continue to do precisely what it is already doing: meet the insurgency in a correct and sustainable manner. The Uribe approach is certainly correct in the way it conceptualizes the problem and seeks to respond to it. The approach is also sustainable, because it demands no unacceptable investments of human or material resources or of will. It will face adjustments if the U.S. contribution ends, but it is unlikely this will happen for some time.

What has not registered fully on the Colombian political class is that a correct and sustainable approach is always put in place to play for the breaks. There is no formula for how long the process will take. In the Philippines, OPLAN Lambat Betag (Net Trap) took approximately six years to produce dramatic results; in Thailand, Prime Minister (PM) Order No. 66/23, "The Policy for the Fight to Defeat the Communists," required roughly half that after its implementation.

Still, if the spectacularly successful Peruvian approach against Sendero Luminoso took just somewhere in between the length of these two campaigns, normalcy in Ulster was achieved only through a grueling 25-year effort. And Ulster was but the size of the small American state of Connecticut, with just half its population. Colombia is the size of California, Nevada, Utah, and Idaho, with a population of 42 million. Hence, patience must be as much a part of the equation as a desire to create precisely the correct mix of techniques that will produce demonstrable results.

LESSONS LEARNED

Formal announcements in the first quarter of the new Uribe administration seemed to portend a necessary shift in emphasis in Democratic Security and Defense Policy implementation, from strike to consolidation. Yet the announcements occurred even as a string of distressing events shook public confidence in the administration. Particularly disturbing were several highly publicized episodes of institutional corruption, apparently driven by the need to produce quantifiable results in response to political demands, as well as evidence of political links between prominent backers of Uribe and the outlawed AUC. Nevertheless, the unease and its attendant debate served the useful purpose of highlighting two issues that emerge time and again in the assessment of any counterinsurgency.

LEADERSHIP MATTERS

Uribe has proven to be the right man at the right time, as have figures in other places and times—one thinks of Magsaysay in the Philippines or Templer in Malaya. Four and a half years (2002-2006), Uribe has not enough time to see through a counterinsurgency. Uribe is keenly aware that his success in winning a second term has brought with it the responsibility not merely to do more of the same, but to recalibrate success in such a manner as to deliver "victory." Defining victory in a counterinsurgency, as indicated above, is tricky, but clearly the metrics any political actor uses to measure his standing will be the benchmarks. Overall, Uribe has offered a model of skillful, dynamic leadership.

It is the armed forces that have been the key element because they provide the security upon which all else that has happened depends. Can they continue to function in the manner of the past eight years? Have the myriad reforms been institutionalized? The answer would seem to be affirmative on both counts. It might especially be noted that institutionalization is as much a function of individuals as are structure and procedures. Colombia's military reformers have been followed by others who, in their career particulars, look much like Mora and Ospina.

Despite the optimistic assessment above, we should not underestimate the extent of the challenge facing the military, mainly COLAR, as a result of its expansion and increased operational tempo. COLAR was previously a draftee force of "in and out" enlisted ranks led by a professional officer corps. It now is one-third volunteer. These individuals expect to make the military a career. A host of issues, from family welfare to promotion requirements to NCO rank, must be codified and then allowed to mature.

Adding to the challenge is the continuous nature of the small-unit operations conducted to keep FARC on the run. Everything from block-leave procedures to family counseling (e.g., to cope with a rising level of turmoil within families in a force that historically has had relatively few disciplinary problems) has had to be instituted. Topping all this is the ever-present threat of corruption in an environment saturated with the easy money of the narcotics trade.

In the field, the strategic initiative has seen some tactical setbacks. This was predictable. The insurgents, after all, also have a learning curve. As FARC has been forced to break up into small units, the security forces have done likewise. This has created opportunities for FARC to surprise isolated or tactically sloppy government units with rapid, medium-sized concentrations that then disperse. The technique is not new, but recent actions have seen FARC grappling for a middle ground between large and small concentrations, so that it can attack platoon or squad-size positions without exposing itself too much. Such measures, though seeking tactical initiative, are strategically and operationally defensive–and an indication of just how successful the government has been. Before the military reforms kicked in, in the Samper/early Pastrana years, FARC fielded large columns that would attack even reinforced companies.

Beginning in February 2005, FARC units, responding to instructions from the organization's secretariat, began an effort to inflict maximum casualties. Their intent, obviously, was to exploit the pressure for "no bad news" placed upon the military by the political structure. They sought to spook at least a proportion of the Colombian "chattering classes" into viewing the normal give-and-take of tactical action as a sign of larger strategic defect. Although they could have a strategic impact by manipulating perception and spurring on the debate about "sustainability," in reality, FARC's small, hard-to-prevent tactical successes have meant nothing to the strategic situation.

The current favorable strategic situation, some have argued, could be undone in a flash by follow-on personalities. Is this likely? No, for all of the reasons discussed above. In particular, both the reforms and the demands of internal war have accelerated change in military (particularly COLAR) leadership. Warfighters who would be as comfortable in the U.S. system as their own have begun to dominate promotion boards, with "service in the field" as the salient factor in selection. This is a critical element, since the military is the shield for all else that occurs in counter insurgency.

As combat-tested officers have begun to dominate the services, the question emerges as to what sort of soldiers they are. In terms of the institution they have made, the results disprove the constant drumbeat about lax standards and abuses that outsiders, especially international human rights organizations, often make. To the contrary, the military, under its reform-minded leadership, consistently emerged in Colombian polls as one of the most respected institutions in the country, with favorable numbers reaching near the 80th percentile.

In sum, the reforms have endeavored to demand more from officers professionally, particularly as regards the mechanics and theory of warfighting. This has resulted in greater knowledge at the strategic and operational levels of war as well as increased tactical expertise.

Put together, military popularity and effectiveness undoubtedly contributed to President Uribe's own consistently high rating with the public. It remains to be seen how recent scandals will affect his position, but the damage is unlikely to be long-lived or deep.

For his part, Uribe has dealt with the military in an increasingly sophisticated and collegial manner. He especially grew to respect the professional judgment of Carlos Ospina, when Ospina was CG Joint Command. This allowed Ospina to exercise a degree of influence and to be heeded when he counseled caution at appropriate times. It remains to be seen, in the post-Ospina command environment, if Uribe will be so dominant as to upset the civil military balance necessary for the armed political campaign that is counterinsurgency.

THE STRATEGIC APPROACH IS CRITICAL

The strategic approach, with its operational (lines of action and campaigns) implementation, must be the foremost concern of leadership in a counterinsurgency. To this end, Uribe was fortunate to have officers of the caliber of Mora and Ospina. If Mora saw COLAR through its early transformation, Ospina not only finished the job, but implemented the central operations of Plan Patriota He had to do this even as resources remained constrained and demands rose for greater emphasis upon other national priorities.

It is not enough, say critics, to regain control of the population; areas seized and held must be consolidated. The military is keenly aware of the point at issue—and has U.S.-supported programs designed to address this dimension of the conflict. The real questions revolve around resource allocation and timing. Here, Uribe has stood his ground, remaining true to the spirit of his strategy: security is the necessary basis for all that follows. Now, in his second term, he has indicated that he intends to exploit counterinsurgency gains and put additional emphasis upon consolidation.

It is precisely the substantial progress made in restoring a semblance of "normal life" that has allowed internal debate over other issues to surface, to include discussion of trends in civil–military relations. The latter is often overlooked in judging the effectiveness of military leaders, but here, too, Colombia has been well served. Ospina, in particular, sought to implement a very "American" vision of the military's subordinate relationship to civil authority.

However, as with the emphasis on being a combat veteran as the key determinant for promotion, the reinforcement of civilian authority as the final word does not sit well with some military elements. It is Uribe's understanding that healthy civil–military relations depend upon an invisible line not being crossed—by either side—that has tempered military discontent and made operations function as smoothly as they have under various defense ministers. The military has maintained firmly its right to determine operational and tactical particulars, and Uribe seems to have acquiesced.

That COLAR continues to transition from its German heritage (transmitted historically through Chilean vectors) to an American model has been stated directly in command briefings to officers. (The air force has long looked to America for inspiration, the navy to the British.) Yet this has not led to an uncritical adoption of either U.S. forms or procedures. American difficulties in Iraq, stemming at least in part from the intervention of civilian leadership in military operational efforts, have been a poignant reminder that a balance must be struck between obedience to civilian authority and to institutional independence.

In Colombia, what this balance should be has been left deliberately indeterminate.

| ***American difficulties in Iraq, stemming at least in part from the intervention of civilian leadership in military operational efforts, have been a poignant reminder that a balance must be struck between obedience to civilian authority and to institutional independence.*** |

CHALLENGES TO COME

In the larger sense, Uribe's national policy has always stood on three legs—not merely security but also fiscal health and social development. Fiscal health is necessary for all else to proceed and has given no grounds for complaint. Social development remains at the heart of all illegal actors' ability to recruit manpower. It, too, has been addressed by progress in the other two sides of the triangle. That one would wish for greater emphasis or speed is a judgment call that imprudently ignores demonstrable progress.

Although the Democratic Security and Defense Policy approach might not require major adjustments, there are strategic areas that bear close monitoring, especially by Washington in this, a critical theater of the battle against global insurgency:

THE BATTLE IS NOT OVER

U.S. support, both materiel and personnel, will play an important role for the foreseeable future. It must be maintained. Unfortunately, a tendency has emerged in U.S. circles that seeks to interpret realities on the ground in terms that speak to the artificial deadlines created by funding legislation. This is extraordinarily dangerous, particularly the notion that the war is won and it is time to talk of winding down U.S. aid and converting Colombian forces to other uses (such as United Nations peacekeeping).

THE US. GOVERNMENT NEEDS TO GRASP THE TRUE NATURE OF COLOMBIA'S STRUGGLE

In some U.S. political and media circles, the conflict is still labeled counternarcotics, or counterterrorism, or counterinsurgency, or something else. It is all of these things and must be approached in a unified manner. This is precisely what the Colombians have been fighting to achieve and they have made dramatic strides, although these have come at considerable political and personal cost for key players such as President Uribe, former Minister Uribe, and former CG Joint Command Ospina.

Image 4.8 U.S. Army Special Forces Soldiers with Special Operations Command South march toward a parade field December 7, 2016 in Florencia, Colombia. Source: Photo by Osvaldo Equite/Defense Visual Information Distribution Service

THE DRIVE TOWARD UNITY OF EFFORT MUST EXTEND TO THE U.S. SIDE.

Greater effort is necessary to raise the level of awareness in Washington that what happens in Colombia underpins our Latin American position. This is not a new domino theory so much as a recognition that, in the present strategic environment, Latin America is the forgotten theater, U.S. Southern Command the forgotten command, and Colombia our forgotten but closest, most reliable ally. At a time when the forces of the radical left are again on the march throughout the hemisphere, to include advocating a severely restricted fight against drugs, Colombia's interests coincide with those of the United States. More than that, Colombia remains a stable democratic state committed to reform and the market economy. Its contrast with an increasingly unstable and strategically dangerous Venezuela could not be greater.

ENHANCING THE RELATIONSHIP BETWEEN U.S. AND COLOMBIAN FORCES AND THE TWO COUNTRIES' STRATEGIC CULTURES

Military cooperation could be enhanced in myriad ways, in particular by augmenting training programs so that they more accurately reflect the close relations between Washington and Bogota. Simultaneously, both governments should encourage closer relations between U.S. and Colombian centers of strategic thought, risk assessment, and regional analysis. Colombia has a level of expertise and analytical capability surpassing any in Latin America, but its talents have been underutilized.

| Colombia has a level of expertise and analytical capability surpassing any in Latin America, but its talents have been underutilized. |

They could make a greater contribution to Democratic Security and Defense Policy, as well as the larger war against terrorism.

There are other areas one could highlight, such as the desire for even greater force strengths and mobility assets. Yet these must be carefully balanced against available resources and the system's ability to absorb any more inputs. Burnishing what the Uribe administration has already done should pay greater gains than seeking to load any more requirements onto the system.

What bears repeating is the point to which this analysis has returned often: the present effort is both correct and sustainable; it is the right strategic posture required for progress and popular security. Hence, continued care must be exercised to ensure that Democratic Security and Defense Policy remains a multifaceted approach—a strengthening of the state's governance, finances, and democratic capacity enabled by the ever more powerful and capable shield provided by the security forces. By themselves, these facets are not the solution—that lies in the use of legitimacy to mobilize response against those using political violence for illegitimate ends—but they will certainly enable it.

About the Author

Dr. Thomas A. Marks is Professor of Terrorism, Insurgency, and Counterinsurgency at the School for National Security Executive Education (SNSEE), National Defense University, where he is also Chair of the Irregular Warfare Department. He holds a B.S. from the United States Military Academy and a PhD from the University of Hawaii. A former government officer, Dr. Marks is the author of numerous articles and monographs on the Colombian conflict. His latest book is Maoist People's War in Post-Vietnam Asia (Bangkok: White Lotus, forthcoming). He has lectured and published widely on terrorism and insurgency.

NOTES

1. Thomas A. Marks, "A Model Counterinsurgency: Uribe's Colombia (2002-2006) vs FARC," *Military Review*, (March-April 2007): 49–64.

2. At one point Colombia was third in U.S. foreign aid, behind only Israel and Egypt.

3. A sixth division was organized during the Uribe administration from what previ ously had been the Joint Task Force (which had been positioned in the extreme south). The COLAR order of battle thus became I Division (2, 4, 11, 17 Brigades); II Div (5, 14, 16, 18 Brigades); III Division (3, 8 Brigades); IV Division (7. 9 Brigades); V Division (1, 6, 13 Brigades); and VI Division (12, 26, 27 Brigades). Later. in July 2005, a seventh division was created when the very large I Division area was split. The new VII Division (based in Medellin) had assigned to it 17. 11 (both from I Div) and 14 Brigade (from 11 Div). The former Caribbean-bounded I Div heartland became a joint command. Additionally, the national reaction force, or FUDRA (*Fuerza de Despliegue Rapido*), which matured during the Pastrana administration, is a light division equivalent, with 3 mobile brigades and 1 Special Forces brigade (of 4 SF battalions). An independent task force (Omega) of virtual division strength operates in the south.

4. Literally, "Commanding General of the Military Forces," which accurately defines the authority and responsibility inherent to the position. I have rendered it as CG Joint Command to facilitate my analysis.

5. See Democratic Security and Defense Policy, 23–30.

6. See Democratic Security and Defense Policy, 42.

7. Recent official documents have dropped "nacional" from their translations of *Ministerio de Defensa Nacional.*

8. Initially, the local-forces were called So/dados Campesinos (Peasant Soldiers), a name the troops themselves disliked-Colombia, despite its substantial agricultural sector, is classified as approximately three-quarters urban, and the units were universally located in rural towns. Hence, So/dados de mi Pueblo ("Home Guards" would be the most useful rendering) came to be used simultaneously.

9. *Masa dispersa*, or "dispersed mass" is a slang rendering of the technique. It is not a formal term.

10. Major General Mario Montoya was promoted to lieutenant general (the highest rank in the Colombian military system) in early December 2006.

11. Quotes for emphasis.

PART II:

APPLICATION OF RESILIENCE AND RESISTANCE

CHAPTER 5: MILITARY SCIENCE

I. Assessing Resistance for the Purpose of Informing International Policy

By Robert S. Burrell and John Collison

ABSTRACT: This section details an applied methodology to operationalize an irregular approach to conflict and competition, in particular, external support to intrastate resilience or resistance. It furthers the application of two foundational learning concepts: (a) the Resilience and Resistance Model and (b) the Resistance Continuum. Using the Resistance Continuum, analysts can categorize the general nature of resistance movements across a spectrum from non-violent protest through belligerency. This essay offers several ways to identify and then assess resistance organizations and prescribes methods to make recommendations concerning potential external support in another state's intrastate conflict consisting of three primary options: to (a) support current governance, (b) support opposition to governance or occupation, or (c) prepare the environment for a future policy supporting resiliency or resistance. Portions of this essay were previously published in *Expeditions with Marine Corps University Press.*[1]

INTRODUCTION

In the 21st Century, irregular conflicts have caused the U.S. a great deal of angst.[2] As recently demonstrated in the Hamas attack and Israel's response in November 2023, irregular threats and asymmetric attacks continue to prevail globally.[3] Fortuitously, the U.S. Congress in 2021 demanded that the Department of Defense develop education to prepare for future irregular struggles.[4] In response, and to address the academic gap in preparing for irregular forms of conflict, in 2024 *Small Wars & Insurgencies* journal published "A Guide for Measuring Resiliency and Resistance," which analyzes nation–states in more advanced human-centric terms to assess current levels of governmental and societal resiliency to subversion and coercion, as well as internal and/or external aggression.[5]

Measuring state resiliency and potential for resistance comprises only *phase one* of a comprehensive approach to addressing irregular threats. *Phase two* includes identification of resistance movements and categorizing their nature according to the typology recommended in the *Resistance Continuum. Phase three* describes resistance movements in terms of leadership, cause, environment, organization, and actions. And finally, *phase four* examines potential external support options to partners in intrastate conflict. This essay expounds on the state-centric resilience and resistance analysis model articulated in *Small Wars & Insurgencies* journal (i.e., phase one). As an extension, it introduces methodologies to facilitate deeper understanding of resistance movements themselves (phases two and three). This improved methodology provides a more informed approach to assist decision/policymakers to consider irregular approaches to competition, deterrence, and war (phase four).[6] Throughout the proposed methodology, interdisciplinary methods remain vital, as military science alone cannot prepare practitioners for future conflict.

IDENTIFYING RESISTANCE MOVEMENTS

This chapter outlines methods to leverage current scholarship and research carried out by some of the top universities and nonprofit organizations that study resistance to help practitioners identify existing resistance movements and trends of success or failure within nation-states. The organizations acknowledged include Swarthmore College in Pennsylvania, Harvard University, Carnegie Endowment for International Peace in Washington D.C., a nonprofit organization called the Armed Conflict Location & Event Data Project, and Uppsala University in Sweden. The data generated by all these organizations remains publicly available on the web.

Once an analyst identifies resistance movements, its general nature can be categorized along this typology of as either (a) nonviolent legal, (b) nonviolent illegal, (c) rebellion, (d) insurgency, or (e) belligerency. Fortunately, there are many ways to identify resistance movements in any country or region. As a starting point, several recommended sources are offered below.

- A quick search of a nation-state in the *Global Nonviolent Action Database*, originally created by researchers at Swarthmore College, can provide excellent results on the activities of nonviolent organizations.[7] It is updated regularly. https://nvdatabase.swarthmore.edu
- Similarly, Harvard University also maintains a database called *Nonviolent and Violent Campaigns and Outcomes* (NAVCO), but the data currently appears generally limited to 1900–2013.[8] Compiled by dozens of separate researchers, it contains several datasets to evaluate. These show durations of conflicts, the number of participants, the relative percentage of participants to the population, and the success or failure rates. https://dataverse.harvard.edu/dataverse/navco
- Harvard also maintains a site called *Mass Mobilization Protest Data*.[9] This downloadable dataset illustrates protests internationally with dates and numbers of people mobilized and the purposes for the protest. However, it does not directly identify the organizations participating (but these could be quickly surmised with a subsequent search about the event on the web). https://dataverse.harvard.edu/dataverse/MMdata
- Carnegie Endowment for International Peace maintains a *Global Protest Tracker* with location, dates, size, and duration of mass protests around the world—no datasets for download required and entirely web-based.[10] It then expands in more detail about particular events with (a) the trigger, (b) motivation, (c) key participants, and (d) outcomes. https://carnegieendowment.org/publications/interactive/protest-tracker

- The non-profit funded *Armed Conflict Location & Event Data Project* (ACLED) offers data on violence in every country in terms of (a) battles, (b) riots, (c) explosions, or (d) violence against civilians.[11] More detailed information requires downloading the datasets. https://acleddata.com
- Finally, Sweden maintains the *Uppsala Conflict Data Program* (UCDP), which illustrates organized violence around the world, particularly intrastate conflict.[12] It contains an interactive map in which a researcher can click on a state and offers up a summary of the conflict, the history, and number of deaths. At least 25 battlefield-related deaths, taking place between the recognized government and an armed group, are required for categorization. This site is continually updated by Uppsala's Department of Conflict and Peace Research. https://ucdp.uu.se/exploratory

These recommended sites are not all-inclusive for identifying resistance movements within a state or region, but they provide a good start for research.[13] Once identified, the student, scholar, or practitioner can bin these movements along the Resistance Continuum, using the descriptions prescribed previously, as a means of categorizing the general nature of activities preferred by each organization. These organizations should be binned on the Resistance Continuum.

ASSESSING RESISTANCE MOVEMENTS

After identifying resistance movements along the Resistance Continuum, an assessment can summarize the potential of each. Several studies and theories attempt to define ways of deconstructing and assessing political movements, insurgencies, or resistance organizations in general.[14] This essay incorporates the typology of resistance introduced by Jonathon Cosgrove and Erin Hahn in *Conceptual Typology of Resistance* (this excellent study remains worthy of a fuller examination than offered here).[15] Essentially, when assessing a resistance movement, we recommend operationalizing the research of Cosgrove and Hahn, who argue that a resistance has five attributes: (1) actors, (2) causes, (3) environment, (4) organization, and (5) actions (ACEOA).[16] Cosgrove and Hahn define these characteristics in the following terms.[17]

| when assessing a resistance movement, we recommend operationalizing the research of Cosgrove and Hahn, who argue that a resistance has five attributes: (1) actors, (2) causes, (3) environment, (4) organization, and (5) actions |

Actors: The individual and potential participants in an organized resistance, as well as external contributors and either competing or cooperating resistance groups.

Causes: The collectively expressed rationales for resistance and the individual motivations for participation.

Environment: The preexisting and emerging conditions within the political, social, physical, or interpersonal contexts that enable or constrain the mobilization of resistance, directly or indirectly.

Organization: "The internal characteristics of a movement: its membership, policies, structures, and culture."[18]

Actions: The means by which actors carry out resistance as they engage in behaviors and activities in opposition to a resisted structure; can encompass both the specific tactics used by a resistance movement and the broader characteristics or repertoires for action (i.e., strategy).

1. *Actors.* The actor category consists of (a) leaders, (b) participants, the (c) population, (d) other resistance movements, and (e) external support.[19] Subsequent to more detailed description of actor components, see Table 5.1.
   a. Leaders can be categorized as either agitators, prophets, reformers, statesmen, or administrators. Each can be evaluated as to their potential in comparison with others of their type (for instance, Martin Luther King Jr. might be considered a reformer and a nearly perfect archetype in this sense).
   b. Participants are either full-time members (e.g., in the armed component or underground) or part-time supporters (e.g., in the auxiliary).
   c. The population can be evaluated generally as having one of three tendencies: those who support the government, those who support change through resistance, and those who prefer to remain uncommitted. The Army War College's Study of Internal Conflict has demonstrated that support from 15 percent or more of the population can prove decisive.[20]
   d. Other resistance movements that share desired change and can collaborate with one another can prove to have more potential than when evaluated separately.
   e. External support, either material or nonmaterial, can prove extremely important for a successful resistance. As described previously, external supporters can consist of multiple organizations, not simply governments. Table 5.1 outlines the basic tenets for actor analysis.

TABLE 5.1. ASSESSING POTENTIAL OF RESISTANCE ACTORS

ACTORS	COMPONENTS	QUALIFIERS
The individual and potential participants in an organized resistance, as well as external contributors and either competing or cooperating resistance groups.	Type of Leader	Agitator, prophet, reformer, statesman or administrator
		Categorize the type of leader or leaders hold sway over this movement. Assess their potential by comparing them with historical examples of the same. Are they recognized nationally or internationally?
	Participants	Full time members and part-time supporters
		Evaluate the loyalty, enthusiasm, and popularity of the cadre in the inner circle as well as other supporters with the organization? What is the potential that their commitment might have exponential effects on the outcomes of the movement.
	Population	Who supports the government? Who support change through resistance? Who prefer to remain uncommitted.
		What percentage of the population actively or passively supports resistance? Remember, 15 percent or more support of resistance by the population could prove decisive.
	Other Resistance Organizations	The sum of resistance organizations can add up to a more powerful and united whole.
		Are they any other resistance movements which, when organized together, could more effectively contribute to a shared desired change? Conversely, are there other movements with incompatible goals? If so, assess the advantages or disadvantages this support could have on the resistance efforts.
	External Support	Which types of organizations outside the geographic boundaries of the country support this resistance movement?
		Assess the totality of external support organizations who favor the resistance and could provide substantial material or non-material support. Could this support prove decisive?

Source: Author

2. *Causes.* Resistance supports a cause that represents the collectively expressed rationales for opposition to authority, as well as the individual motivations for participating in such a group. Rationale for resistance consists of either a desire for sweeping change in authority or society, or specified changes for individual groups or communities.[21] The scholar or practitioner should identify the organizations' stated public narrative, which normally delegitimizes the current authority and legitimizes its own claims.

   The second key component to analyzing the cause consists of the motivations of the participants. A group of scholars at Artis International, particularly Scott Atran, have conducted several studies on the difference between "devoted actors" and those motivated by self-interest have in conflict. A cause that fuses participants' cultural defining values, spiritual formidability, and trust in the group and/or leader can inspire actors to endure long periods of discomfort as well as personal sacrifice.[22] Consequently, the rationale of particular causes can inspire a greater will to fight. Table 5.2 outlines the basic tenets for causation analysis.

TABLE 5.2. ASSESSING THE CAUSE OF RESISTANCE

CAUSE	COMPONENTS	QUALIFIERS
The collectively expressed rationales for resistance and the individual motivations for participation.	Rational	Rational for resistance consists of either a desire for sweeping change in authority or society or specified changes for individual groups or communities.
		Identify and restate the prevailing rational of the resistance, including ends, ways, and means utilized to attain success. Utilize this rational when analyzing its power of motivating the core values of a population below.
	Motivations	What motivations are utilized by the resistance to recruit and maintain its supporters through sacrifice and difficulty?
		Does the stated rational of the movement identify with sacred values in the population? If so, what percentage of the population identifies with those values? Remember, 15 percent or more could prove decisive.

Source: Author

3. *Environment.* One analogy of the environment's relationships with resilience and resistance is that the environment represents the chessboard on which the king, rook, bishop, queen, knight, and pawn compete. Assessing the environment's influence or constraints on both resilience and resistance activities requires an interdisciplinary approach but should include at a minimum an evaluation of: (1) environmental, (2)

governmental, (3) socio-political, (4) technological, and (5) relationship factors.

The environment consists of geographic limitations, like maritime boundaries, mountainous terrain, or urban terrain but also space and cyberspace. One could describe the environment in terms of domains (i.e., land, maritime, air, space, and the information environment) like those articulated in military doctrine.[23] Governance represents the current rule of law, or lack thereof, as a system of control of the nation–state. For instance, in western nations, the accepted rule of law can facilitate nonviolent action, wherein opposition to an authoritarian regime may require more secrecy or violence to implement. There are five prevalent forms of government: monarchy, democracy, oligarchy, authoritarianism, and totalitarianism.[24]

In addition to governance, every society has socio-economic factors that determine the accepted norms; challenging these norms or accepting them will affect the ways in which resistance activities take place. Further, a society's access to technology provides various means to communicate with or influence them but also provides platforms for nonviolent or violent action.[25] As one example, communicating with a resistance in North Korea may require primarily face-to-face interaction, where Joshua Wong's resistance in Hong Kong relied on web-based platforms.[26] Lastly, resistance takes place in human terrain (in addition to the domains listed previously), and "preexisting and emerging relationships among individuals, organizations," and social groups prove fundamental to how resiliency and resistance interact with one another.[27] Table 5.3 outlines the basic tenets for environmental analysis.

4. *Organization.* One study completed by Johns Hopkins University generally categorizes resistance organizations into two bins: (1) mass organization and (2) elite organizations. Each has advantages and disadvantages. Mass organizations have few bars to entry for recruiting, taking advantage of size to compete with authority. These can prove excellent archetypes for non-violent protest, like social movements or unions, or as a belligerent organization in a civil war. However, a mass organization is difficult to train and control, easier for authorities to infiltrate, and members can prove undisciplined. In contrast, elite organizations take advantage of extensive vetting, selective recruiting, superior training, and a high degree of motivation. These types of movements are normally secretive, operating with undergrounds or, when overt, maintaining covert or clandestine activities. An elite organization can influence mass organizations and even hijack or control their behaviors. Elite organizations designed to blossom into a mass organization given the right circumstances are called *elite-fronts*; an example would include traditional communist parties.[28]

TABLE 5.3. THE ENVIRONMENT IN WHICH RESISTANCE TAKES PLACE

ENVIRONMENT	SUBCOMPONENTS	QUALIFIERS
The preexisting and emerging conditions within the political, social, physical, or interpersonal contexts that enable or constrain the mobilization of resistance, directly or indirectly.	Domains	Land, maritime, air, and space domains and the information environment
		Explain how the land, maritime, air, and space domains, as well as the information environment, provide opportunities or constraints for this particular movement.
	Governance	What type of governance exists in the state?
		Assess the rule of law in the state. *Is it best categorized as a monarchy, democracy, oligarchy, authoritarianism, and totalitarianism? What are the secondary effects of that governance structure on resistance?*
	Socio-political Aspects	Full-time members and part-time supporters.
		Evaluate the loyalty, enthusiasm, and popularity of the cadre in the inner circle as well as other casual supporters within the organization. What is the potential of their commitment, which could affect the outcomes of the movement?
	Technology	Technological capabilities of the society
		Describe communication platforms dominant in the society as well as access to smartphones and the World Wide Web. Will any platforms enable clandestine means of communication? What is the capacity of authorities to monitor and detect resistance? What commercial off-the- shelf platforms might support resistance activities (e.g., radios, medical supplies, or UAVs)?
	Relationships	Preexisting or emerging relationships
		Describe any ongoing relationships between members of the resistance or the organization itself with other influencers inside the state or internationally. What potential could these relationships have on the success of the movement?

Source: Author

Resistance movements can be sub-organized in a myriad of ways. In most military doctrines, these can include: (a) an underground, (b) an armed component, (c) an auxiliary, and (d) a public component.[29] However, nonviolent resistance organizations might forego the need for an underground and an armed component—opting for an overt organization without violent means. Hence, resistance movements may contain all or some of the four components listed. Each component can be described in terms of who comprises its membership, what types of policies guide the members, the structure of each component (like a cellular or hierarchal organization), and the prevailing culture, values, and motivations. Table 5.4 outlines the basic tenets for organizational analysis.

TABLE 5.4. ORGANIZATIONAL STRUCTURE OF RESISTANCE

ORGANIZATION	SUBCOMPONENTS	QUALIFIERS
The internal characteristics of a movement: its membership, policies, structures, and culture.	Organization type	Mass organization or elite organization
		Does this organization require large numbers to oppose authority, or does it require secrecy to maintain a small elite group? Describe how and why the organization operates and the advantages and disadvantages of that choice.
	Resistance components	Underground, armed component, auxiliary, and public component
		How many of the above components does this resistance movement have? Describe each of the above components in terms of membership, policies, structures, and culture.

Source: Author

5. *Actions.* Much of the discussion on ways, or actions, of varying types of resistance movements has been discussed previously in this section. These ways help to define the movement along the Resistance Continuum. The methods of action, typical to a particular resistance, should be restated in the formal assessment (e.g., nonviolent protest, assassination). Additionally, methods of fundraising and equipping should be analyzed. Procurement can consist of the legal market, the black market, battlefield recovery, theft, taxes (or fundraising), manufacturing raids, and external partners outside of the state.[30] One should determine if the movement utilizes self-procurement for most needs, or if it relies primarily on external sponsorship. Table 5.5 outlines the basic tenets for actions or ways analysis.

TABLE 5.5. ACTIONS CARRIED OUT BY RESISTANCE[31]		
ACTIONS	**SUBCOMPONENTS**	**QUALIFIERS**
The means by which actors carry out resistance as they engage in behaviors and activities in opposition to a resisted structure; it can encompass both the specific tactics used by a resistance movement and the broader characteristics or repertoires for action (i.e., strategy).	Ways of resistance	The Resistance Continuum consists of (1) nonviolent legal, (2) nonviolent illegal, (3) rebellion, (4) insurgency, and (5) belligerency
		Qualify the ways utilized for resistance in terms of non-violent protest, illegal protest, rebellious lethal activities, insurgency, or belligerency. Provide examples.
	Funding and procurement	Which methods does the resistance utilize to sustain its activities?
		Of these methods, which are utilized by the resistance: legal market, the black market, battlefield recovery, theft, taxes (or fundraising), manufacturing raids, and external partners?

Source: Author

In summation, we recommend assessing resistance organization with the typology of five central components: (1) actors, (2) causes, (3) environment, (4) organization, and (5) actions (ACEOA).[32] An assessment of a resistance might be as short as five paragraphs, each devoted to one of the subcomponents listed. As a given, such an assessment proves subjective. Quantifying aspects of this approach remains under further research and analysis.

PROVIDING A COMPREHENSIVE EXAMINATION

The first three phases in a comprehensive analysis include: (phase one) measure the resiliency and resistance potential at the state level (as discussed previously in chapter one);[33] (phase two) identify prevalent or influential resistance organizations within the state utilizing the methods prescribed, and then categorize these organizations along the resistance continuum to classify their general nature; and (phase three) assess one or more of those resistance movements by taking a deeper look at its leadership, motivation, operating environment, organization, and activities (ACEOA). The final phase (phase four) remains—subjectively assess the information gathered to make recommendations concerning potential external support in another state's intrastate conflict, which consists of three primary options:

to (a) support a governing authority's resilience, (b) support resistance to current governance or occupation, or (c) prepare the environment for a future policy supporting resiliency or resistance (at a minimum, the comprehensive analysis consists of the 12 steps listed in the Table 5.6).

TABLE 5.6. THE 12-STEP RESILIENCE AND RESISTANCE ANALYSIS PROCESS

PHASE	STEPS
One	1. Measure the state's resiliency.
	2. Identify the potential for a state-sponsored resistance strategy (if applicable).
	3. Measure the potential for external support to resiliency.
	4. Measure the potential for resistance to current authority.
	5. Measure the potential for external support to resistance.
Two	6. Identify the prevalent resistance groups within the state and placing them on the Resistance Continuum.
Three	7. Assess one or more resistance groups in terms of leadership,
	8. cause,
	9. environment,
	10. organization,
	11. and actions.
Four	12. Make a recommendation concerning potential external support to resiliency or resistance, which normally proposes one of three options: (a) support resilience, (b) support resistance, or (c) prepare the environment for a future policy supporting resiliency or resistance.

Source: Author

PHASE FOUR: STRATEGIC OPTIONS IN SUPPORT OF RESILIENCE OR RESISTANCE

To provide adequate analysis and then a recommendation in support of foreign policy options for an external sponsor, we recommend employing the entire 12-step process to complete a holistic overview of a nation–state in terms of resilience and resistance potential (steps 1-5), as well as a deeper understanding of each major resistance organization (steps 6-11), and, finally, recommendations for action (step 12). The typical suggestion for action proposes one of three options: (a) support resilience, (b) support resistance, or (c) prepare the environment for a future policy supporting resiliency or resistance.

| The typical suggestion for action proposes one of three options: (a) support resilience, (b) support resistance, or (c) prepare the environment for a future policy supporting resiliency or resistance. |

A full proposal should also consider the underlying factors of resistance movements that have given space for adversaries to operate; how to counter the frames and narratives of the adversary (either an external actor, state authority, or the resistance); consideration of timings for various aspects of the response; measures of effectiveness; and risk assessment and mitigation.

a. **Support resilience**. Supporting the resiliency of a governing power could include numerous supporting packages that address the sources of instability within a state or even external threats to sovereignty. In the short term, this might include countering the objectives and the means of each of the major resistance movements or their external sponsors, while long-term stability requires addressing the root grievances of the population—essentially subverting resistance. Several U.S. Government publications help outline these approaches, including the *Stabilization Assistance Review,*[34] Joint Publication 3-22 *Foreign Internal Defense,*[35] Joint Publication 3-07 *Stability,*[36] and Joint Publication 3-20 *Security Cooperation,*[37] to name the most prominent. The *Resistance Operating Concept* articulates one type of irregular strategy.[38]
b. **Support resistance**. Proposals to support a particular resistance movement, or multiple movements, should begin with an interagency feasibility assessment. A resistance should have compatible goals with those of the sponsor and behave within acceptable norms of behavior. Via nonviolent resistance, a solid strategy might include the principles of Gene Sharp as articulated in Helvey's *On Strategic Nonviolent Conflict: Thinking About the Fundamentals.*[39] If the external sponsor desires the inclusion of military support to an indigenous insurgency or to the armed component of an occupied state, ATP 3-05.1 *Unconventional Warfare at the Combined Joint Special Operations Task Force Level* provides the best guide.[40]

c. **Prepare the environment for a future policy supporting resiliency or resistance.** Not supporting resilience of current governance or resistance to endogenous governing structures should be the result of a conscious choice following a full evaluation of the options. In many cases, the direct involvement of an external sponsor in intrastate conflict might be regarded with a low likelihood of achieving the desired results. Simultaneously, finding an appropriate partner for resilience or resistance might not be possible. Additionally, overt support has foreign policy considerations and covert support has risks to the same.

CONCLUSION

This section argues for a human-centric approach to conflict, versus a conventional military analysis, and indicates multiple options for either countering or supporting resistance movements to align with foreign policy objectives. This approach makes obvious that external support to resilience or resistance across diplomatic, information, military, and economic lines of effort offer substantial opportunities in shaping international relations, sometimes with irregular methods. By utilizing the resilience and resistance methodology, analysts can provide assessments to military leaders and policymakers that better recognize partners and adversaries and provide options to support a partner's resilience or to support resistance(s) to inspire changes in adversary's behaviors.

Practitioners can operationalize the Resilience and Resistance Model and the Resistance Continuum to identify the most appropriate partner and subsequent complementary support packages in intrastate conflicts. Each resistance movement is best countered or supported by unique methods. For instance, providing lethal force to a resilience or resistance partner is not always the best method of fomenting desired change. In contrast, the type of external support offered can also deliberately change the methods utilized by a resistance and the nature of a conflict. So external partners can deliberately or incidentally change the nature of the intrastate conflict through the means delivered.

Utilizing the methodologies proposed in Part One and Part Two of this book, one can complete a comprehensive analysis of resiliency and resistance factors within a nation-state for the purposes of influencing international foreign policies adopted by external actors.

| Utilizing the methodologies proposed in Part One and Part Two of this book, one can complete a comprehensive analysis of resiliency and resistance factors within a nation–state for the purposes of influencing international foreign policies adopted by external actors. |

As an example of this approach, see the practical application with a real-world case study in Appendix A. The Iran case study demonstrates the utility of this methodology in understanding intrastate conflict and the possibilities offered to external sponsors of change.

About the Authors

Dr. Robert S. Burrell is a resilience and resistance interdisciplinary scholar using data-driven and human-centric methodologies to analyze intrastate conflict ranging from nonviolent protest through belligerency. He is a Senior Research Fellow at the Global and National Security Institute of the University of South Florida. From 2020–2024, he taught irregular warfare at Joint Special Operations University and was the editor in chief of special operations doctrine from 2011–2014.

John Collison is a contractor with Huntington Ingalls Industries and works in J59-Concept Development and Integration (CD&I) at U.S. Special Operations Command.

NOTES

1. Robert S. Burrell and John Collison, "Assessing Resistance for the Purpose of Informing International Policy," *Expeditions with Marine Corps University Press*, 9 Jan. 2024.

2. David Ucko and Thomas A. Marks, "Redefining Irregular Warfare: Legitimacy, Coercion and Power," *Modern Warfare Institute*, 18 Oct 2022. Found at https://mwi.usma.edu/redefining-irregular-warfare-legitimacy-coercion-and-power/

3. Josef Federman and Isaam Adwan, "Israel Strikes and Seals Off Gaza After Incursion by Hamas, Which Vows to Execute Hostages," Associated Press, 9 Oct. 2023. Found online at https://apnews.com/article/israel-palestinians-gaza-hamas-airstrikes-hostages-4377e096f62bf535bebcdff38cf16049, accessed on 10 Oct. 2023.

4. House Resolution 5130 - *Consortium to Study Irregular Warfare Act of 2021*, 31 Aug 2021.

5. Robert S. Burrell and John Collison, "A Guide for Measuring Resiliency and Resistance," *Small Wars & Insurgencies* 35, no 1 (January 2024):147–172, https://doi.org/10.1080/09592318.2023.2292942.

6. For a description of irregular competition, see Robert Burrell, "How to Integrate Competition and Irregular Warfare," *Modern War Institute*, 5 August 2021. Available at https://mwi.usma.edu/how-to-integrate-competition-and-irregular-warfare/

7. *Global Nonviolent Action Database*, found at https://nvdatabase.swarthmore.edu/, accessed on 13 Sep. 2023.

8. Harvard Dataverse, *Nonviolent and Violent Campaigns and Outcomes*, found at https://dataverse.harvard.edu/dataverse/navco, accessed on 13 Sep. 2023.

9. Harvard Dataverse, *Mass Mobilization Protest Data*, found at https://dataverse.harvard.edu/dataverse/MMdata, accessed on 13 Sep. 2023.

10. Carnegie Endowment for International Peace, *Global Protest Tracker*, found at https://carnegieendowment.org/publications/interactive/protest-tracker, accessed on 13 Sep. 2023.

11. *Armed Conflict Location & Event Data Project*, found at https://acleddata.com/, accessed on 13 Sep. 2023.

12. Uppsala University, *Uppsala Conflict Data Program*, found at https://ucdp.uu.se/exploratory, accessed on 13 Sep. 2023.

13. As one example, *Global Database of Events, Language and Tone* (GDELT) offers both raw data as well as analysis on events around the world since 2013 and "monitors the world's broadcast, print, and web news from nearly every corner of every country in over 100 languages and identifies the people, locations, organizations, themes, sources, emotions, counts, quotes, images and events driving" social change. With so much data, this website can become overwhelming, but it provides good resources for compiling a more comprehensive narrative.

14. See David Collier, Jody LaPorte, and Jason Seawright, "Putting Typologies to Work: Concept Formation, Measurement, and Analytic Rigor," *Political Research Quarterly*. 2012;(1).

15. Jonathon Cosgrove and Erin Hahn, *Conceptual Typology of Resistance*, (Fort Bragg, NC: U.S. Army Special Operations Command, circa 2018).

16. Cosgrove and Hahn, *Conceptual Typology*, 4-6.

17. Cosgrove and Hahn, *Conceptual Typology*, 6.

18. Cosgrove and Hahn utilize the definition in Jeremy M. Weinstein, *Inside Rebellion: The Politics of Insurgent Violence* (Cambridge: Cambridge University Press, 2006), 19.

19. Cosgrove and Hahn, *Conceptual Typology*, 7-16.

20. For more information, see the U.S. Army War College website at https://ssi.armywarcollege.edu/soic-study-methodology/. Also see Chris Mason, "COIN Doctrine is Wrong," *Parameters* 51, no.2, 2021: 19-34.19-34.

21. Cosgrove and Hahn, *Conceptual Typology*, 17-24.

22. See Scott Atran, "The Devoted Actor: Unconditional Commitment and Intractable Conflict across Cultures," *Current Anthropology*, Vol. 57, 13 Jun. 2016, 192-203.

23. See Joint Publication 1, *Joint Warfighting*, (Washington D.C.: Department of Defense, 27 Aug. 2023).

24. For a brief explanation of these, see *Sociology: Understanding and Changing the Social World*, (University of Minnesota, 2016), found https://pressbooks.howardcc.edu/soci101/chapter/14-2-types-of-political-systems/, accessed 14 Sep. 2023.

25. Cosgrove and Hahn, *Conceptual Typology*, 25-34.

26. See Dan Garrett, "Superheroes in Hong Kong's Political Resistance: Icons, Images, and Opposition," *Political Science & Politics* 47 (1), 2014: 112–19.

27. Cosgrove and Hahn, *Conceptual Typology*, 29.

28. Paul J. Tompkins Jr., and Robert R. Leonhard, *Undergrounds in Insurgent, Revolutionary, and Resistance Warfare*, (North Carolina: U.S. Army Special Operations Command, 2012), 10-12.

29. Army Technical Publication 3-05.1 *Unconventional Warfare at the Combined Joint Special Operations Task Force Level.*

30. Tompkins and Leonhard, *Undergrounds in Insurgent, Revolutionary, and Resistance Warfare.*

31. Table created by authors.

32. Cosgrove and Hahn, *Conceptual Typology*, 4-6.

33. Cosgrove and Hahn, *Conceptual Typology*, 4-6.

34. *Stabilization Assistance Review: A Framework for Maximizing the Effectiveness of U.S. Government Efforts to Stabilize Conflict-Affected Areas*, (Washington D.C.: Department of State, USAID, and Department of Defense, 2018).

35. Joint Publication 3-22: *Foreign Internal Defense*, (Washington D.C.: Department of Defense, 2 Feb. 2021).

36. Joint Publication 3-07: *Stability*, (Washington D.C.: Department of Defense, 11 Feb. 2022).

37. Joint Publication 3-20: *Security Cooperation*, (Washington D.C.: Department of Defense, 9 Sep. 2022).

38. Otto C. Fiala, *Resistance Operating Concept*, (MacDill Air Force Base, FL: Joint Special Operations University Press, 2020).

39. Helvey.

40. Army Technical Publication 3-05.1: *Unconventional Warfare at the Combined Joint Special Operations Task Force Level*, (Fort Bragg, N.C.: U.S. Army Special Operations Command, Apr. 2021).

II. Classical and Modern Deterrence Theories

By Aaron Baty

ABSTRACT: This section explores the evolution of deterrence strategies, examining historical contexts and definitional shifts. The focus is on the post-World War II era, specifically the development of atomic and nuclear deterrence, the Domino Theory in Vietnam, and the Peace Dividend in South America during the 1980s. The paper progresses to the post-9/11 era, emphasizing the shift toward integrated deterrence and its core requirements, including communication. It then navigates through the post-9/11 landscape, emphasizing the shift toward integrated deterrence, analyzing key documents such as the 2006 DoD Joint Operating Concept and the 2022 National Security Strategy. The frameworks of deterrence are explored, encompassing both agreed-upon and non-agreed-upon norms, escalation ladders, cost imposition, and denial. Additionally, the section delves into the related concept of compellence, detailing frameworks, spectrums of conflict, and military operating concepts. Practical applications within the DoD, including the integration of deterrence and compellence in planning and execution, are also examined. The adversaries' use of deterrence, illustrated by examples from China, Russia, North Korea, and Iran, is analyzed. Finally, challenges, criticisms, strengths, and weaknesses are discussed.

EVOLUTION OF DETERRENCE

HISTORICAL FOUNDATIONS AND MODERN ADAPTATIONS

The evolution of deterrence is deeply rooted in the annals of history, with its origins intertwined in the intricate fabric of international relations and strategic thinking. Prior to the modern conceptualization of deterrence, historical instances suggest that early civilizations employed rudimentary forms of deterrence to dissuade potential adversaries. The classical Greek city-states, for instance, utilized military prowess and alliances as a deterrent against potential invaders, demonstrating an early understanding of the strategic importance of dissuasion. However, it was not until the[1] post-Renaissance period that the term "deterrence" gained prominence, evolving from its Latin root *deterreō*, meaning to frighten away or discourage.[2]

As the world approached the 20th century, the definitional evolution of deterrence gained momentum, particularly in the context of rising military capabilities and geopolitical tensions. The pre-World War I era saw the emergence of large standing armies and the formulation of military doctrines aimed at deterring potential adversaries through sheer military might. The devastating consequences of World War I, however, prompted a reevaluation of deterrence strategies. The concept began to incorporate elements of diplomacy, economic influence, and the psychological impact of warfare. This shift set the stage for the interwar period, a time marked by the articulation of more comprehensive deterrence theories that acknowledged the multidimensional nature of security and the need for a nuanced approach to dissuasion.

POST WORLD WAR I AND THE LEAGUE OF NATIONS

The League of Nations, established in 1920 following the conclusion of World War I, was the first international organization aimed at maintaining peace and preventing future conflicts through collective security and diplomacy. Conceived under the Treaty of Versailles, the League sought to create a forum for nations to resolve disputes peacefully and promote a system of collective security. The League's establishment represented a significant departure from traditional power politics and reflected a global commitment to preventing devastating conflicts.[3]

Despite its noble aspirations, the League of Nations faced inherent challenges that contributed to its ultimate failure. One critical factor was the absence of major powers (such as the U.S., which chose not to join), that diminished the League's overall effectiveness. Additionally, the League lacked a standing military force and faced difficulties in enforcing its decisions, relying heavily on the voluntary cooperation of member states. The League's inability to address emerging geopolitical tensions and the aggressive expansionist policies of certain nations, notably Germany and Japan, weakened its credibility and effectiveness as a guarantor of global norms.[4]

The League of Nations' failure to prevent the outbreak of World War II highlighted the limitations of international organizations in establishing and maintaining global norms. The League's inability to deter aggression, along with its structural shortcomings, underscored the need for a more robust and enforceable system of collective security. The U.S. did not have any intention on entering World War Two and viewed the conflict as a European matter. This stance changed on 7 December 1941 when the Japanese attacked Pearl Harbor.

The Japanese attack on Pearl Harbor is often interpreted as a surprise offensive that drew the U.S. into World War II. However, some scholars argue that this attack was, in fact, an attempt at deterrence aimed at coercing the U.S. to reconsider its policies and actions in the Pacific. Japan was facing a series of challenges, including economic sanctions imposed by the U.S. and its allies, and its military expansion in China and Southeast Asia had strained relations with the United States. In this context, some argue that the Japanese government believed that a bold and aggressive move, such as the attack on Pearl Harbor, could intimidate the U.S. into accepting Japan's territorial and strategic ambitions in Asia, ultimately avoiding a full-scale conflict.

One piece of evidence supporting this interpretation is the fact that Japan did not formally declare war on the U.S. until after the attack, which suggests a desire to avoid a direct confrontation. Additionally, Japan may have hoped that by crippling the U.S. Pacific Fleet, it could buy time to consolidate its territorial gains in Asia without immediate U.S. intervention. While this interpretation is debated among historians, it underscores the complexity of the motivations behind the attack on Pearl Harbor and the role of deterrence in international relations during World War II.[5]

The lessons learned from the League of Nations paved the way for the establishment of the United Nations in 1945, which sought to address the deficiencies of its predecessor and create a more effective framework for international cooperation, conflict resolution, and the establishment of global norms.[6]

THE UNITED NATIONS

The creation of the United Nations (UN) in the aftermath of World War II marked a concerted effort by the international community to establish an organization that could prevent future global conflicts and promote cooperation among nations.

| The creation of the United Nations (UN) in the aftermath of World War II marked a concerted effort by the international community to establish an organization that could prevent future global conflicts and promote cooperation among nations. |

The idea for the UN emerged during the wartime conferences among the Allied powers, including the Tehran Conference in 1943 and the Yalta Conference in 1945. The foundational discussions recognized the need for a more effective and inclusive international organization than its predecessor, the League of Nations, which failed to prevent World War II.[7]

Formally established on 24 October 1945 with the signing of the UN Charter in San Francisco, the United Nations aimed to foster international cooperation, maintain peace and security, promote human rights, and provide a platform for dialogue among nations. The UN Charter outlined the principles of collective security, wherein member states pledged to refrain from using force against one another and committed to collective action in response to aggression. The creation of the UN also reflected a collective acknowledgment of the necessity to address global challenges through a multilateral framework, emphasizing diplomacy, dialogue, and the rule of law.[8]

The structure of the UN included the General Assembly, where all member states were represented, and the Security Council, which held primary responsibility for maintaining international peace and security. The creation of the UN represented a commitment to preventing future conflicts and fostering collaboration on issues ranging from humanitarian assistance to economic development, making it a cornerstone of the post-war international order.[9] The establishment of UN also brought forth an era of international norms.

Global norms refer to widely accepted standards, principles, and rules that govern behavior and interactions among nations and international actors in various domains, including politics, economics, security, and human rights. These norms serve as a framework for guiding international relations and cooperation, facilitating the peaceful coexistence of states, and addressing common challenges on a global scale. Global norms are often enshrined in international agreements, treaties, and institutions, shaping the expectations and responsibilities of countries in their interactions with one another.

For example, the concept of state sovereignty, a fundamental global norm, is enshrined in the UN Charter, which recognizes the territorial integrity and political independence of member states. Another prominent example is the norm of human rights protection, as articulated in the Universal Declaration of Human Rights and subsequent international treaties. These norms establish a shared understanding of the rights and responsibilities of states and individuals, promoting a more just and equitable world order. Global norms evolve over time in response to changing circumstances and societal values, reflecting the collective aspirations of the international community for a more peaceful, stable, and inclusive world.

MUTUALLY ASSURED DESTRUCTION AND DOMINO THEORY

Following the development and use of atomic weapons during World War II, the emergence of the Cold War prompted a reevaluation of traditional deterrence theories. The U.S. and the Soviet Union found themselves in a precarious balance of power, with the threat of mutually assured destruction (MAD) serving as the foundation for nuclear deterrence. MAD was a concept in which each side knew the other would destroy them if they initiated a nuclear attack. This period saw the establishment of a delicate equilibrium, where the possession of nuclear weapons became a deterrent against direct aggression.[10] The U.S. began to look outward from the Soviet Union itself to deter possible Soviet aggression abroad.

Massive retaliation was a strategic doctrine associated with the concept of MAD during the Cold War. This doctrine, also known as the Eisenhower Doctrine, emphasized the U.S.

commitment to responding to any aggression by the Soviet Union or its allies with an overwhelming and disproportionate nuclear response. It aimed to deter the Soviet Union from engaging in any form of aggression by making the potential consequences of such actions unacceptable.

As President Dwight D. Eisenhower stated in a speech in 1953, "Local defenses must be reinforced by the further deterrent of massive retaliatory power." This approach relied on the idea that the threat of a devastating nuclear response would discourage the Soviets from initiating any conflict, thereby maintaining precarious peace. Massive retaliation was a central element of U.S. Cold War strategy during the 1950s and early 1960s, reflecting the belief in the effectiveness of deterrence through the specter of nuclear annihilation.[11]

In 1954, Eisenhower introduced the domino theory. This theory posited that the spread of communism in one country would trigger a chain reaction, much like a row of falling dominos, leading neighboring countries to also fall under communist influence or control. See Figure 5.1. The theory suggested that if one nation in a region fell to communism, it would create a "domino effect," where nearby countries would be more susceptible to communist ideology or intervention, leading to the further expansion of communism and potentially threatening U.S. interests and global stability.[12] The domino theory would be a prominent framework throughout the Cold War, one which directly impacted US foreign policy.

Figure 5.1: The domino theory. Source: Beao/commons.wikimeda.org

FLEXIBLE RESPONSE

In 1961, President John F. Kennedy adopted "flexible response" as a deterrence framework aimed to avoid massive retaliation under the MAD framework. Kennedy's decision to adopt flexible response as a strategy in the Cold War marked a significant shift from the previous administration's reliance on the doctrine of massive retaliation, which had been championed by Eisenhower.

Flexible response was a strategic doctrine adopted by the U.S. during the Cold War to provide a more versatile and graduated approach to conflict than the earlier doctrine of massive retaliation. This strategy aimed to counter the spread of communism and address conflicts with varying levels of military force. In Vietnam, the U.S. applied the principles of flexible response as it sought to deter the global expansion of communism while also compelling North Vietnam to abandon its support for communism. The Vietnam War, which began in the 1950s and escalated throughout the 1960s, was a manifestation of this strategy. The U.S. deployed a combination of conventional military forces and counterinsurgency tactics to achieve its goals in Vietnam, reflecting its commitment to deterring communism globally while attempting to compel a non-communist regime in South Vietnam.[13]

In Europe, flexible response was also a key component of U.S. Cold War policy. It aimed to deter Soviet aggression while providing a range of options for responding to various levels of conflict. This included conventional military forces and a commitment to defend NATO allies, making it clear that aggression in Europe would not go unopposed. This approach helped maintain stability in Europe during the Cold War and prevented the outbreak of a large-scale conflict. It showcased the adaptability of flexible response in addressing regional security challenges within the broader context of global deterrence against communism.[14]

President John F. Kennedy Signing Proclamation 3504, Naval Quarantine of Cuba. Source: U.S. Government/public domain

Moreover, Kennedy's decision to adopt flexible response was also influenced by the desire to address the concerns of America's allies, particularly in Western Europe. Many European leaders were uncomfortable with the idea of massive retaliation, as it implied the potential use of nuclear weapons on their soil, which they found unsettling. By adopting flexible response, Kennedy aimed to reassure these allies and strengthen their confidence in the U.S. as a reliable partner in the defense against the Soviet threat.

Kennedy's decision to embrace flexible response was driven by the need for a more adaptable and less escalatory approach to the Cold War, as well as the desire to foster better relations with American allies in Europe.[15] Flexible response would be another prominent framework throughout the Cold War and beyond. The U.S. found itself in a greater position of responsibility than it had previously and needed to incorporate the reservations and desires from their overseas partners and allies.

INDIRECT CONFRONTATION

Indirect confrontation, often referred to as "proxy warfare" or "low-intensity conflict," is a strategy wherein one state or actor seeks to advance its interests or exert influence over another state or region by supporting local proxies or insurgent groups rather than engaging in direct military confrontation.

| Indirect confrontation, often referred to as "proxy warfare" or "low-intensity conflict," is a strategy wherein one state or actor seeks to advance its interests or exert influence over another state or region by supporting local proxies or insurgent groups rather than engaging in direct military confrontation. |

After the Vietnam War, the U.S. continued to compete with the Soviets without direct confrontation. This period witnessed a shift toward integrated deterrence, where military, economic, and diplomatic instruments were strategically combined to influence the behavior of potential adversaries. The U.S. aimed to capitalize on the opportunity to bleed the Soviet Union economically and militarily by providing support to Afghan resistance fighters, who became known as the *Mujahideen*. The belief was that by arming and funding the Mujahideen, the U.S.

could create a protracted and costly conflict for the Soviet Union, akin to the U.S. experience in Vietnam, and thereby weaken Moscow's global position.[16] The U.S. aimed to weaken Soviet influence and military involvement in these areas by indirectly aiding resistance movements, ultimately contributing to the Soviet Union's decline and withdrawal from Afghanistan in 1989.[17] However, the unintended consequences of this support, including the rise of extremist groups and the destabilization of Afghanistan, would have long-term repercussions that the U.S. and the world would grapple with for decades to come.[18]

In South America, the U.S. employed a strategy of using proxies as part of its efforts to counter the spread of communism in the region. One notable example of this strategy was the support provided to anti-communist regimes and paramilitary groups in countries like Nicaragua, El Salvador, and Guatemala. The U.S. aimed to prevent the rise of leftist governments that it perceived as aligned with the Soviet Union or Cuba. In Nicaragua, for instance, the U.S. supported the Contras, an anti-Sandinista rebel group, both financially and logistically, with the intention of destabilizing the Sandinista government. This proxy warfare approach was a central element of U.S. foreign policy in the region during the 1980s, as it sought to protect American interests and prevent the spread of communism.[19] The use of proxies in South America during the Cold War was met with significant controversy and had profound and lasting consequences for the region. While it achieved some short-term strategic goals for the U.S., it also fueled conflicts, human rights abuses, and political instability in several countries. The repercussions of these proxy wars continue to be felt in the social and political landscape of South America to this day, with legacies of violence and political polarization remaining in countries like El Salvador and Guatemala.

The post-Cold War era witnessed significant global instability as well as a growing threat from non-state actors targeting American interests. Events such as the bombings of the U.S. embassies in Nairobi, Kenya, and Dar es Salaam, Tanzania, in 1998; the attempted attack on the World Trade Center in 1993; and the attack on the USS *Cole* in 2000 highlighted the evolving nature of security challenges. These incidents demonstrated that traditional nuclear deterrence strategies, which primarily focused on deterring state actors, were not sufficient to prevent attacks by non-state actors and violent extremist organizations.

The realization that nuclear deterrence was ill-suited to address the threat posed by non-state actors ultimately culminated in the devastating terrorist attacks on September 11, 2001 (9/11). The 9/11 attacks exposed the vulnerability of the U.S. and the inadequacy of traditional deterrence strategies in deterring terrorist organizations like Al-Qaeda.

| The 9/11 attacks exposed the vulnerability of the U.S. and the inadequacy of traditional deterrence strategies in deterring terrorist organizations like Al-Qaeda. |

In response to this new reality, the U.S. began to shift its approach to security and deterrence. The concept of integrated deterrence emerged, emphasizing a comprehensive and flexible approach to deterrence that could address threats from both state and non-state actors. This approach aimed to combine military, diplomatic, economic, and informational tools to shape the behavior of adversaries and potential aggressors, regardless of whether they were state or non-state actors. Integrated deterrence recognized that traditional military might alone was insufficient in an era where threats could come from a wide range of actors with varying motivations and capabilities.[20]

GLOBAL WAR ON TERROR

The Global War on Terror (GWOT) following the 9/11 attacks brought a new set of challenges and necessitated a more comprehensive and adaptable approach to deterrence. The traditional Cold War model was no longer sufficient to address the complexities of asymmetric warfare and terrorism. Deterrence evolved to integrate not only military and diplomatic strategies but also intelligence, law enforcement, and homeland security measures. The interconnected and multifaceted nature of the contemporary security environment demanded a more holistic and integrated approach to deterrence.[21]

Al-Qaeda Attacks New York City on 11 September 2001. Source: U.S. Government/public domain

With the added threats of non-state actors, the concept of integrated deterrence became central to national security strategies. The DoD increasingly emphasized the need for a unified and synergistic approach across all elements of national power—DIME (diplomatic, informational, military, and economic). The integration of various capabilities and the coordination of efforts across different domains became imperative in addressing the dynamic and evolving nature of threats. Integrated deterrence, thus, represents a change in basic assumptions from the simplistic deterrence models of the Cold War to a more nuanced and adaptable strategy that acknowledges the complexity of the contemporary security landscape.[22]

| Integrated deterrence, thus, represents a change in basic assumptions from the simplistic deterrence models of the Cold War to a more nuanced and adaptable strategy that acknowledges the complexity of the contemporary security landscape. |

Although non-state actors and terrorist groups are not new, the environment they operate in, is. Integrated Deterrence becomes ever more challenging with the introduction of social media and global connectivity.

THEORY OF DETERRENCE

WHAT IS DETERRENCE?

In its simplest forms, deterrence is the practice of discouraging or restraining someone— in world politics, usually a nation-state—from taking unwanted actions, such as an armed attack. It involves an effort to stop or prevent an action, as opposed to the closely related but distinct concept of "compellence," which is an effort to force an actor to do something. While the concept seems simple enough, the theories on how to implement them are quite different.

CLASSIC DETERRENCE THEORY

Classical deterrence theory, a cornerstone of international relations, has its roots in the historical context of the Cold War. Emerging as a response to the escalating tensions between the U.S. and the Soviet Union, it sought to understand and rationalize the strategies behind nuclear deterrence. The theory's historical origins can be traced back to the post-World War II era when nuclear weapons became a dominant factor in international politics. Classical deterrence theory is characterized by several fundamental principles and assumptions that underpin its framework. These assumptions include the belief that rational actors weigh the costs and benefits of their actions, that military power is a central tool in deterring adversaries, and that the threat of retaliation serves as a deterrent to aggression. Jervis's work dissects these core concepts, providing a nuanced understanding of how they contribute to the theory's foundational principles and shape the strategies of deterrence.[23]

At the heart of classical deterrence theory lies the pivotal role of military power in deterring potential adversaries.

| At the heart of classical deterrence theory lies the pivotal role of military power in deterring potential adversaries. |

Military capabilities, especially the possession of nuclear weapons, can serve as a powerful deterrent against potential aggressors. The dynamics of brinkmanship highlights how rational actors can manipulate the risk of escalation to deter adversaries effectively. While classical deterrence theory has been influential in shaping international relations and security strategies, it is not without its critiques and limitations. These criticisms challenge some of the theory's assumptions, such as the belief in rationality and the focus on military capabilities. These critiques emphasize the complexity of international relations and the need to consider diverse actors and situations in the global arena. These observations serve as a reminder that, while classical deterrence has been a powerful tool, its applicability must be carefully considered in an evolving international landscape.

Classical deterrence theory, rooted in the historical context of the Cold War, has played a significant role in shaping international relations and security strategies. Its foundational concepts and assumptions, as outlined by scholars like Jervis and Schelling, have provided valuable insights into the strategic calculus of deterrence. Nevertheless, the theory is not without its critiques, as highlighted by those who question its assumptions. As the international landscape evolves, it is essential to critically assess and adapt deterrence strategies, considering the changing dynamics of global politics and security challenges.

MODERN DETERRENCE THEORY

The evolution of Modern Deterrence Theory reflects the need for a multidimensional approach to address the complex security challenges of the 21st century. Unlike its classical, military-centric predecessor, modern deterrence incorporates a broader range of factors, including technological advancements and the interconnected global environment. Modern deterrence must adapt to the technological changes and the ubiquitous information environment, which enables states to achieve strategic objectives without triggering unacceptable counteractions. This requires a deterrence model that goes beyond the traditional cost-benefit analysis of potential aggressors.[24]

Michael Mazarr's report by the RAND Corporation emphasizes that effective deterrence strategies must integrate various elements such as economic interdependencies, diplomatic initiatives, and information warfare. This multidimensional approach reflects the interconnectedness of global affairs and the necessity to address different types of threats simultaneously. The report highlights that economic sanctions, diplomatic measures, and information operations are integral parts of a robust deterrence framework, complementing traditional military strategies.[25]

In addition to the global interdependences, modern theory recognizes the significance of psychological and cyber elements in shaping contemporary deterrence strategies. Martin Libicki's 2009 work explores the role of cyber capabilities in deterrence. This analysis highlights how the interconnected nature of the modern world makes cyber vulnerabilities a critical aspect of deterrence, necessitating a proactive approach to mitigate threats in the cyber domain. Additionally, psychological deterrence acknowledges the importance of perception and cognitive factors in deterring adversaries, emphasizing the need to shape an adversary's beliefs about the costs and risks of aggression.[26]

As exemplified by the 2018 National Defense Strategy of the United States of America, Modern deterrence theory also finds practical application in today's global context. This strategy, outlined in the "Summary of the 2018 National Defense Strategy," recognizes the evolving security environment and emphasizes the importance of a multidimensional deterrence approach. It reflects the U.S. commitment to adapting deterrence strategies to address contemporary challenges, including great power competition and emerging threats in the cyber and information domains.

In sum, modern deterrence theory has emerged as a response to the evolving security landscape of the 21st century. Its development reflects a shift toward a more comprehensive and multidimensional approach to deterrence. Mazarr's analysis underscores the importance of factors beyond military power, while Libicki's highlights the critical role of cyber and psychological elements. The application of modern deterrence theory highlights the need for flexible and adaptable deterrence strategies in the modern era.

KEY DIFFERENCES

Classical deterrence theory, born during the Cold War, primarily revolves around the concept of deterrence through military power. It relies heavily on the threat of massive retaliation to dissuade adversaries from initiating aggression. In contrast, modern deterrence theory encompasses a broader spectrum of instruments beyond military might, recognizing the roles of economic, diplomatic, and informational tools in deterring adversaries.

| Classical deterrence theory, born during the Cold War, primarily revolves around the concept of deterrence through military power. It relies heavily on the threat of massive retaliation to dissuade adversaries from initiating aggression. In contrast, modern deterrence theory encompasses a broader spectrum of instruments beyond military might, recognizing the roles of economic, diplomatic, and informational tools in deterring adversaries. |

While classical deterrence theory assumes rationality in decision-making and focuses on states as primary actors, modern deterrence theory acknowledges the complex interplay of state and non-state actors as well as the importance of perception and psychological factors.

Assessing the strengths and weaknesses of classical and modern deterrence theories is essential to understanding their practical utility. Classical deterrence theory's strength lies in its simplicity and historical success in preventing major conflicts during the Cold War. However, it is often critiqued for its rigidity, assuming rationality in adversaries, and its potential for catastrophic consequences in the event of failure. Modern deterrence theory, with its multidimensional approach, offers versatility and adaptability to diverse threats. It acknowledges the limitations of classical deterrence and seeks to address contemporary challenges. However, it is critiqued for its complexity and the difficulties in coordinating various elements in a unified strategy.

The contrast between classical and modern deterrence theories reveals the evolving nature of deterrence in international relations. While classical deterrence theory paved the way for strategic thinking during the Cold War, modern deterrence theory recognizes the need for a more nuanced and adaptable approach in the 21st century. Both theories have their strengths and weaknesses, and their applicability depends on the specific context and nature of threats faced by nations in the modern era. Understanding these differences is crucial for policymakers as they navigate the complexities of deterrence in an ever-changing global landscape.

MODERN INTEGRATED DETERRENCE THEORY

Modern integrated deterrence theory has emerged as a critical framework in the 21st century's evolving security landscape. In an era marked by complex threats, the relevance of the theory transcends traditional deterrence models. At its core, it emphasizes the importance of integrating DIME elements to achieve effective deterrence. Central to the application of this theory is the establishment and adherence to internationally recognized norms and laws, which serve as the bedrock of global order and the foundation of deterrence strategies.

Two key U.S. documents, the 2006 Joint Operating Construct (JOC) and the 2022 National Defense Strategy (NDS), reflect the evolution of American strategic thought and its alignment with these principles. This section seeks to comprehensively examine this theory through the lens of the U.S., shedding light on its core principles, the evolution of deterrence theories, and the critical role of adaptability, agility, and cooperation in contemporary deterrence strategies.

MODERN INTEGRATED DETERRENCE THEORY

Modern integrated deterrence theory represents a change in basic assumptions in deterrence thinking, acknowledging the multifaceted nature of 21st-century threats. At its core it emphasizes the necessity of integrating diplomatic, informational, military, and economic instruments of power to deter adversaries effectively. Unlike traditional deterrence theories that relied heavily on military might, modern integrated deterrence theory recognizes that threats often manifest in diverse domains, demanding a comprehensive approach to deterrence.[27]

> ***| Unlike traditional deterrence theories that relied heavily on military might, modern integrated deterrence theory recognizes that threats often manifest in diverse domains, demanding a comprehensive approach to deterrence. |***

The evolution of deterrence theories can be traced from the Cold War era to the contemporary security landscape. During the Cold War, deterrence was primarily associated with the threat of nuclear annihilation, characterized by the doctrine of MAD. However, the post-Cold War era brought new challenges, including transnational terrorism, cyber warfare, and hybrid threats. This theory reflects the dynamic nature of security threats and acknowledges the limitations of traditional deterrence models.[28]

In an era marked by rapid technological advancements and shifting geopolitical dynamics, the ability to adapt deterrence strategies is paramount. Adversaries constantly evolve, demanding flexibility in response. Agility entails the capacity to tailor deterrence measures to specific threats, recognizing that a one-size-fits-all approach is inadequate. Moreover, cooperation with allies and international partners is indispensable in achieving deterrence objectives. Collective action amplifies the effectiveness of deterrence and reinforces the stability of the international order.[29]

Modern integrated deterrence theory represents a fundamental shift in deterrence thinking, adapting to the complexities of the 21st century security landscape. It is characterized by a comprehensive approach that integrates multiple instruments of power and emphasizes adaptability, agility, and cooperation. Later, we will delve deeper into the practical application through the analysis of key U.S. strategic documents and the critical role of internationally recognized norms and laws in global deterrence strategies.

Modern integrated deterrence theory has not only reshaped how the U.S. approaches deterrence but has also influenced its military strategy significantly. The 2006 JOC played a crucial role in paving the way for the development and implementation in U.S. military strategy. This construct introduced the concept of "multi-domain operations" and emphasized the need for a more integrated approach to warfare. It recognized that modern conflicts often transcend traditional domains, encompassing not only physical battlefields but also cyberspace and the information domain.[30]

The JOC's contribution lies in its recognition of the complex nature of contemporary threats and the need for a comprehensive approach to deterrence. It laid the groundwork for the integration of DIME elements in U.S. strategy. The JOC emphasized the importance of synchronized and coherent actions across all domains to achieve desired strategic outcomes. This approach reflects the core principle of Modern Integrated Deterrence, which calls for the simultaneous and integrated use of all instruments of national power.[31]

In addition, the 2022 NDS underscores the core principles guiding U.S. strategy. This strategic blueprint places a significant emphasis on deterrence to avert conflicts and safeguard American interests. It acknowledges the evolving nature of threats and the imperative for the U.S. to tailor its deterrence strategies accordingly. The NDS recognizes the critical importance of maintaining advantageous power balances and thwarting adversary objectives through deterrence rather than aggression.[32]

The NDS is in alignment with contemporary thinking on deterrence by highlighting the necessity for a comprehensive and integrated approach. It underscores that deterrence encompasses a spectrum of tools; is not limited to military capabilities; and encompasses diplomatic, economic, and informational means. The strategy emphasizes the significance of cultivating robust alliances and partnerships to strengthen deterrence endeavors, aligning with the cooperative nature of international relations. Furthermore, it underscores the crucial need for adaptability and agility in the face of evolving threats, a principle that resonates with both contemporary deterrence theory and the JOC.[33]

Secretary of Defense Lloyd Austin Unveils the 2022 National Defense Strategy. Source: U.S. Government/public domain

THE IMPORTANCE OF INTERNATIONALLY RECOGNIZED NORMS AND LAWS

The significance of internationally recognized norms and laws in maintaining global order and deterrence cannot be overstated. These norms and laws provide a framework for behavior among nations and help prevent conflicts by establishing boundaries and expectations. Adherence to internationally recognized norms and laws is a key component of effective deterrence, as it contributes to the stability of the international system.

| Adherence to internationally recognized norms and laws is a key component of effective deterrence, as it contributes to the stability of the international system. |

International organizations and agreements play a vital role in shaping and upholding these norms and laws. Organizations like the UN and agreements such as the Geneva Conventions provide a forum for nations to codify and enforce rules of conduct in various domains, including armed conflict, human rights, and the environment. These international bodies serve as mechanisms for conflict resolution and promote peaceful coexistence among states.

Adherence to internationally recognized norms and laws contributes to deterrence by creating predictability and reducing the likelihood of miscalculation. When nations abide by established rules, potential adversaries can better assess the consequences of their actions, making the cost of aggression more transparent. The threat of international condemnation and sanctions further reinforces the deterrence effect. Norms and laws serve as a deterrent by discouraging behavior that violates established international standards.[34]

The integration of modern integrated deterrence theory into U.S. strategy has reshaped how the nation approaches deterrence by emphasizing a comprehensive, multi-domain approach. The 2006 JOC and the 2022 NDS reflect this shift, highlighting the importance of coordinated actions across diplomatic, informational, military, and economic domains. Furthermore, the adherence to internationally recognized norms and laws is paramount for maintaining global order and enhancing deterrence efforts. International organizations and agreements play a critical role in shaping these norms and providing mechanisms for conflict prevention and resolution.

The realm of cyberspace is engendering a shift in societal norms around privacy and security. The interconnectedness facilitated by the internet and digital technologies has blurred the boundaries between public and private spheres, prompting a reassessment of what constitutes personal information and the extent to which individuals are willing to share it. Cybersecurity concerns are challenging traditional notions of trust, both in interpersonal relationships and in the institutions that manage and store vast amounts of personal data. As societies grapple with the implications of cyber threats and the vulnerabilities inherent in a digitally interconnected world, norms surrounding individual responsibility for digital hygiene, data protection, and the ethical use of technology are undergoing a significant metamorphosis. The intersection of AI (artificial intelligence) and cyber technologies is thus fostering a dynamic reevaluation of societal norms, demanding adaptability and ethical considerations in the face of rapidly advancing technological landscapes. As AI capabilities rapidly advance, professionals across cybersecurity, information governance, and eDiscovery must continuously adapt policies and procedures to govern AI systems and data responsibly. Critical issues around synthetic media, autonomy, and accountability will shape legal investigations and discovery processes.[35]

FRAMEWORKS

Deterrence frameworks represent fundamental strategic constructs employed by nations to discourage potential adversaries from pursuing actions contrary to their interests. These frameworks encompass a plethora of theories and principles that guide the formulation and execution of deterrence strategies.[36] Nevertheless, a persistent challenge within deterrence frameworks is the inherent complexity of assessing whether the intended deterrence measures are achieving the desired impact on the adversary. The essence of deterrence lies in altering the adversary's decision-making calculus, rendering it inherently intricate to gauge the success or failure of such actions.[37]

One central facet of deterrence frameworks revolves around preserving the status quo, aiming to dissuade adversaries from disturbing the prevailing balance of power or challenging established norms.[38] However, evaluating the effectiveness of such deterrence actions poses difficulties due to the absence of observable events. Success in maintaining the status quo typically translates into the absence of overt aggression or provocative actions by the adversary, making it challenging to distinguish whether the deterrence strategy itself or external variables contribute to the apparent stability.[39] This complexity is compounded by the dynamic nature of international relations, where a myriad of factors may influence an adversary's conduct beyond the purview of specific deterrence measures employed.

Deterrence frameworks grounded in agreed-upon norms, such as international treaties or alliances, introduce an additional layer of intricacy when evaluating the impact of deterrence actions.[40] Isolating the influence of deterrence from the broader context of diplomatic engagements and multilateral agreements becomes challenging. Moreover, the components integral to deterrence frameworks – the escalation ladder, cost imposition, and denial – all present measurement challenges.[41] The efficacy of these strategies is contingent upon the adversary's perceptions, making it demanding to assess the psychological impact or the extent to which an adversary is dissuaded from pursuing undesirable actions.

The challenge of assessing the effectiveness of deterrence actions within established frameworks underscores the intricate nature of strategic interactions in international relations. The intangible and psychological dimensions of dissuasion render it inherently problematic to measure the consequences of deterrence efforts, particularly when dealing with diverse adversaries on the global stage.

| The intangible and psychological dimensions of dissuasion render it inherently problematic to measure the consequences of deterrence efforts, particularly when dealing with diverse adversaries on the global stage. |

While deterrence remains a pivotal element of statecraft, policymakers must grapple with the nuanced dynamics of influence and coercion in a world where the success of deterrence actions often manifests in the absence of observable events.

COMMUNICATIONS THEORY

Communications theory plays a pivotal role in international relations, serving as a fundamental tool for states to convey their messages, intentions, and policies among themselves and to their partners and allies. This theory encompasses various aspects of effective communication, including the transmission of information, the interpretation of signals, and the management of diplomatic relations. By applying communications theory, states can enhance their diplomacy, prevent misunderstandings, and strengthen their strategic partnerships. Regardless of what type of integrated deterrence model you ascribe to, communications theory will be a vital tool.

SIGNAL TRANSMISSION AND CREDIBILITY

One critical element of communications theory is the concept of signal transmission. In the international arena, states employ various means to transmit messages, such as diplomatic notes, official statements, and even public speeches. The clarity and credibility of these signals are essential, as they can influence the perceptions and actions of other states. Waltz argues that communication is a necessary condition for the operation of the system highlighting the significance of well-structured and comprehensible messages in international diplomacy.[42]

INTERPRETATION AND PERCEPTION

Moreover, communication theory emphasizes the importance of interpretation and perception. States must not only send messages but also ensure that their intended meanings are correctly understood by recipients. Misinterpretation can lead to misunderstandings, miscalculations, or even conflicts. Therefore, diplomatic efforts often involve not only conveying messages but also managing the way these messages are perceived and interpreted by different actors, as highlighted by theories on perception and cognitive psychology.[43]

COALITION BUILDING AND ALLIANCE MANAGEMENT

In addition to direct state-to-state communication, the application of communications theory extends to coalition building and alliance management. States use strategic messaging to strengthen their bonds with partners and allies, fostering cooperation and coordination. Effective communication can build trust, align interests, and solidify strategic partnerships, as seen in the realm of military alliances like NATO and ASEAN (Association of Southeast Asian Nations). Communications theory is a vital tool in international relations, enabling states to convey their messages, manage perceptions, and strengthen alliances. Effective communication, characterized by clear signals and accurate interpretation, is crucial in preventing conflicts and fostering cooperation among states and their partners. By applying the principles of communications theory, states can navigate the complex landscape of international diplomacy and enhance their strategic interactions in an increasingly interconnected world.

INTERNATIONAL RELATIONS THEORY

International relations (IR) theory is a multidisciplinary field that seeks to understand the complex interactions between states and non-state actors in the global arena. This section will discuss three foundational theories for IR.

The United Nations, Geneva, Switzerland. Source: WikiCommons.

The first foundational theory in IR is realism, which posits that states are primarily motivated by self-interest and the pursuit of power. Realist thinkers like Hans Morgenthau argue that international politics is characterized by competition and conflict among states, as they vie for power and security in an anarchic system where there is no central authority to enforce order.[44] Realism's emphasis on state-centricity and the role of power in shaping international relations continues to be influential in analyzing global politics.

Another prominent IR theory is *liberalism*, which takes a more optimistic view of international cooperation and the potential for peaceful resolution of conflicts. Liberal thinkers such as Immanuel Kant argue that the spread of democracy, economic interdependence, and international institutions can promote cooperation and reduce the likelihood of war.[45] The liberal perspective has been instrumental in explaining the role of international organizations like the United Nations (UN) in facilitating diplomacy and conflict resolution.

A third major IR theory is *constructivism*, which emphasizes the role of ideas, norms, and identity in shaping international behavior. Constructivist scholars like Alexander Wendt argue that state actions are not solely driven by material interests but are influenced by socially constructed perceptions and beliefs.[46] This theory highlights the importance of ideational factors in understanding state behavior, including the impact of international norms, culture, and identity, on foreign policy decisions. These three theories—realism, liberalism, and constructivism—represent just a few of the diverse perspectives within the field of international relations, each offering unique insights into the complexities of global politics.

GLOBAL NORMS

Internationally recognized norms and laws serve as the bedrock of global order and are integral to the effectiveness of deterrence strategies. These norms provide a framework that fosters predictability in international relations, allowing states to anticipate the actions and reactions of one another. When states adhere to established norms and laws, it reduces the likelihood of miscalculation and misunderstandings, thus enhancing stability in the international system.

| When states adhere to established norms and laws, it reduces the likelihood of miscalculation and misunderstandings, thus enhancing stability in the international system. |

For example, the principles outlined in the UN Charter, which include the prohibition of the use of force except in self-defense or with Security Council authorization, are fundamental in guiding state behavior and preventing unprovoked aggression.[47]

Moreover, internationally recognized norms and laws make the cost of aggression transparent. When states transgress these norms, it becomes evident to the international community, enabling collective condemnation and potential punitive measures. The threat of diplomatic isolation, economic sanctions, or even military intervention can deter states from pursuing aggressive actions. The international community often rallies around these established norms to signal its commitment to upholding them and to discourage violations. This not only bolsters deterrence but also acts as a safeguard for global peace and security.[48]

International organizations and agreements, such as the UN and the Geneva Conventions, play pivotal roles in shaping and enforcing internationally recognized norms and laws. The UN serves as a forum for states to collectively define and promote norms, while the Geneva Conventions establish humanitarian rules during armed conflicts. These institutions provide mechanisms for dispute resolution, negotiations, and the creation of international laws that strengthen the framework of global order. Adherence to these norms and laws is not just a matter of legal obligation but also serves as a fundamental tool for deterrence, as states are less likely to engage in actions that would lead to their international isolation or legal liability.[49] Internationally recognized norms and laws underpin the foundation of global order and contribute significantly to deterrence by promoting stability and discouraging unlawful behavior.

GAPS WITHIN GLOBAL NORMS

| "We must establish new norms for deterrence to be effective." – Ash Carter |

China, Russia, Iran, and North Korea have indeed exploited gaps in internationally recognized norms to advance their strategic objectives. Russia's interference in the 2016 U.S. presidential election serves as a prominent example of this phenomenon. The Russian government, through various means, conducted a covert campaign aimed at undermining the U.S. electoral process and sowing discord within American society. This interference exploited the absence of specific international norms governing cyber interference in the domestic political affairs of sovereign states, making it challenging for the U.S. and its allies to respond effectively.[50]

Similarly, China has been active in cyber espionage activities, stealing intellectual property and sensitive information from foreign governments and corporations. These actions have raised concerns about the protection of intellectual property and state secrets, but they also highlight the absence of universally agreed-upon norms and legal frameworks governing cyber espionage. China's utilization of cyber tools allows it to pursue its economic and strategic interests while circumventing established norms that might apply to traditional forms of espionage.[51]

Iran and North Korea have also exploited gaps in non-agreed upon norms to pursue their strategic objectives, often in defiance of international sanctions and deterrence mechanisms. Iran's nuclear program, for instance, has been a subject of international concern and sanctions. However, Iran has employed strategies such as cyberattacks and proxy warfare to circumvent these measures and advance its nuclear ambitions. North Korea's persistent nuclear and missile testing, despite international sanctions, demonstrates its ability to leverage these gaps in non-agreed upon norms to maintain its strategic leverage.[52] These actions underscore the evolving nature of deterrence in an era where non-traditional means can be used to pursue strategic objectives beyond the scope of internationally recognized norms.

THE U.S. RESPONSE AND WAY FORWARD

In response to the evolving use of non-agreed upon norms and tactics by China, Russia, Iran, and North Korea, the U.S. must adopt a multifaceted approach to strengthen its deterrence strategies.

| In response to the evolving use of non-agreed upon norms and tactics by China, Russia, Iran, and North Korea, the U.S. must adopt a multifaceted approach to strengthen its deterrence strategies. |

First and foremost, bolstering cybersecurity is paramount. The U.S. government and private sector should collaborate closely to enhance the resilience of critical infrastructure and protect against cyber threats. Investing in advanced cybersecurity technologies, conducting regular assessments, and promoting information sharing can significantly improve the nation's ability to deter and respond to cyberattacks.[53]

Additionally, countering disinformation campaigns is crucial in the modern information age. Russia's interference in the 2016 U.S. presidential election demonstrated the power of disinformation in sowing discord and undermining democratic processes. The U.S. should invest in robust counter-disinformation efforts, including public awareness campaigns, fact-checking initiatives, and cooperation with social media platforms to identify and combat false narratives.[54]

Furthermore, enhancing intelligence capabilities is essential to effectively deter and respond to threats posed by these actors. The U.S. should prioritize intelligence gathering and analysis to gain insights into the strategies and intentions of China, Russia, Iran, and North Korea. By understanding their actions and motives, the U.S. can develop more effective deterrence strategies and response plans.

Lastly, international cooperation and alliances play a critical role in upholding global norms and deterring hostile actors. The U.S. should work closely with its allies and partners to build a united front against actors that seek to exploit gaps in international norms. Strengthening alliances in regions of strategic importance can help deter aggressive behavior and promote adherence to international standards.[55]

Responding to the evolving use of non-agreed upon norms and tactics by China, Russia, Iran, and North Korea requires a comprehensive approach. Bolstering cybersecurity, countering disinformation campaigns, enhancing intelligence capabilities, and fostering international cooperation are all essential components of a robust deterrence strategy. By adopting these measures, the U.S. can better protect its interests, uphold global norms, and deter hostile actors in an ever-evolving security landscape.

About the Author

Aaron Baty is a retired Special Forces officer who supported operations in Afghanistan, Iraq, and Africa. He is currently pursuing his Doctorate of Business Administration degree at the University of South Florida. He lives in Tampa, FL, with his wife Ursula and three children.

NOTES

1. Paul K. Huth, "Conceptualizing Deterrence," in *Extended Deterrence and the Prevention of War* (Yale University Press, 1988), 15–27, https://doi.org/10.2307/j.ctt211qwdk.6.

2. Hans J. Morgenthau, *Politics Among Nations: The Struggle for Power and Peace*, 4th Ed. (Alfred A. Knopf, 1967).

3. (Britannica, The Editors of Encyclopedia. "Treaty of Versailles." *Encyclopedia Britannica*, 17 Nov. 2023)

4. Morgenthau, *Politics Among Nations.*

5. Eri Hotta, *Japan 1941: Countdown to Infamy* (Knopf, 2013).

6. Morgenthau, *Politics Among Nations.*

7. Patrick M. Morgan, *Deterrence Now* in *Cambridge Studies in International Relations* (Cambridge University Press, 2003).

8. Morgan, *Deterrence Now.*

9. Morgan, *Deterrence Now.*

10. Morgan, *Deterrence Now.*

11. Dwight D. Eisenhower, "The Chance for Peace," speech presented at the American Society of Newspaper Editors, Washington, D.C., April 16, 1953.

12. Robert Jervis, *Perception and Misperception in International Politics* (Princeton University Press, 1976).

13. John F. Kennedy, "National Security Action Memorandum No. 34: South Vietnam," memorandum, March 16, 1961.

14. U.S. Department of State, "Flexible Response: The U.S. and the Berlin Crisis of 1961-1962" (Washington, D.C.: U.S. Department of State, Publication Office).

15. Morgenthau, *Politics Among Nations.*

16. Morgan, *Deterrence Now.*

17. P. W. Carter and R. E. Wilson, "The Soviet Withdrawal from Afghanistan: A Turning Point in the Cold War," *Cold War History* 3, no. 3 (2003): 1-27.

18. Morgan, *Deterrence Now.*

19. Greg Grandin, *Empire's Workshop: Latin America, the United States, and the Rise of the New Imperialism* (New York: Macmillan, 2006).

20. 9/11 Commission Report; U.S. Department of Defense publications on integrated deterrence.

21. Morgan, *Deterrence Now.*

22. Morgan, *Deterrence Now.*

23. Thomas C. Schelling, *The Strategy of Conflict* (Cambridge, MA: Harvard University Press, 1960)

24. Jeffrey W. Pickler, "21st Century Warfare Requires 21st Century Deterrence," *Security Insights*, George C. Marshall European Center for Security Studies, 2022, found at https://www.marshallcenter.org/sites/default/files/files/2022-10/Security%20 Insights_72_Pickler_Modern_Deterrence.pdf, accessed in 31 July 2024.

25. Michael J. Mazarr, "Understanding Deterrence," *Perspective*, Rand Corporation, 2018.

26. Martin C. Libicki, *Cyberdeterrence and Cyber War*, (Rand Corporation, 2009).

27. Morgan, *Deterrence Now.*

28. Richard J. Harknett and Emily Goldman, "Getting to the Left of Hackers: Evaluating a Joint Deterrence Response Strategy," *Journal of Cybersecurity* 4, no. 1 (2018).

29. Morgan, *Deterrence Now.*

30. Joint Operating Concept 2006 (Washington, D.C.: Department of Defense, 2006).

31. Joint Operating Concept 2006.

32. U.S. Department of Defense, "2022 National Defense Strategy" (Washington, D.C.: U.S. Department of Defense, 2022).

33. 2022 National Defense Strategy.

34. Anne Orford, "Feminism, Imperialism and the Mission of International Law," *Nordic Journal of International Law* 71, no. 2 (2002): 275-296, https://doi.org/10.1163/157181002761931387.

35. ComplexDiscovery Staff. "AI in 2023: A Year of Multimodal Innovations and Societal Reflection."

36. Patrick M. Morgan, "The Past and Future of Deterrence Theory," in *Cross-Domain Deterrence: Strategy in an Era of Complexity.* Jon R. Lindsay and Erik Gartzke, eds. (Oxford University Press, 2019).

37. Lawrence Freedman, *Deterrence* (Themes for the 21st Century Series; reprint ed.; Wiley, 2004).

38. Richard K. Betts, "Is Strategy an Illusion?" *International Security* 32, no. 4 (2007): 5-50.

39. Jacek Kugler and Douglas Lemke, "The Impact of Political and Economic Instability on Third-Party Intervention in Militarized Interstate Disputes," *Journal of Conflict Resolution* 40, no. 1 (1996): 144-167.

40. Paul Huth and Bruce Russett, "Deterrence Failure and Crisis Escalation," *International Studies Quarterly* 32, no. 1 (1988): 29–45, doi:10.2307/2600411

41. Thomas C. Schelling, *Arms and Influence* (Yale University Press, 1966), https://doi.org/10.2307/j.ctt5vm52s

42. Kenneth N. Waltz, *Theory of International Politics* (Addison-Wesley, 1979).

43. Robert Jervis, *Perception and Misperception in International Politics* (Princeton University Press, 1976).

44. Morgenthau, *Politics Among Nations.*

45. Immanuel Kant, "Perpetual Peace: A Philosophical Sketch," 1795, retrieved from American University website, http://fs2.american.edu/dfagel/www/Classpercent20Readings/Kant/Immanuelpercent20Kant,percent20_Perpetualpercent20Peace.pdf

46. Alexander Wendt, "Anarchy Is What States Make of It: The Social Construction of Power Politics," *International Organization* 46, no. 2 (1992): 391-425.

47. United Nations, "Charter of the United Nations," 1945, retrieved from United Nations website, https://www.un.org/en/charter-united-nations

48. Samuel Charap and Michael O'Hanlon, "Preventing War in Ukraine: The Case for a U.N. Blue Helmet Mission" (Brookings Institution, 2017).

49. Jonathan I. Charney, "Universal International Law," *Texas Law Review* 66 (1987): 785-842.

50. R. S. Mueller, "The Structure of Cyber Conflict," in *The Oxford Handbook of Cyber Security*, ed. Robert A. Colle, Adam M. Bobrow, and Sushil Jajodia (Oxford University Press, 2019).

51. Thomas Rid and Ben Buchanan, "Attributing Cyber-Attacks," *Journal of Strategic Studies* 38, nos. 1-2 (2015): 4-37.

52. David Von Hippel, "The Long History of Nuclear Coercion," *Survival* 58, no. 4 (2016): 85-106.

53. Thomas Rid, "Cyber War Will Not Take Place," *Journal of Strategic Studies* 35, no. 1 (2013): 5-32.

54. Samantha Bradshaw and Philip Howard, "The Global Disinformation Order: 2019 Global Inventory of Organized Social Media Manipulation" (2019).

55. Elisabeth Braw, *The Last Defense: Meet America's New Nuclear Deterrent* (Brookings Institution Press, 2019).

III. Integrated Deterrence as a National Strategy

By Aaron Baty

ABSTRACT: Integrated deterrence as a national strategy is essential for safeguarding U.S. national security interests in an increasingly complex global landscape. This section explores the concept of integrated deterrence within the framework of strategic documents such as the National Security Strategy and the National Defense Strategy. Integrated deterrence emphasizes the coordinated use of diplomatic, economic, military, and informational instruments to deter adversaries and protect American interests. The Department of Defense plays a pivotal role in operationalizing integrated deterrence, leveraging its resources to develop credible military capabilities and establish cooperative security partnerships with allies. However, challenges exist in fully implementing integrated deterrence across all elements of national power, including interagency coordination and resource disparities among government agencies. The section also discusses the Joint Phasing Construct (JPC), which guides military operations through distinct phases, and the importance of recognizing spectrums of conflict to address diverse security challenges effectively. Furthermore, the role of allies and partners in enhancing collective defense capabilities is highlighted, alongside an analysis of how competitors like China, Russia, Iran, and North Korea operate to advance their strategic interests through tactics such as incremental gains, lawfare, hybrid warfare, and saber rattling. The conclusion underscores the need for the U.S. to adapt its strategy to counter evolving security threats and enhance interagency coordination to address trans-regional challenges effectively.

STRATEGIC DOCUMENTS

The National Security Strategy (NSS) is a critical policy document in the U.S. that outlines the overarching national security priorities and strategic direction of the U.S. Government. It serves as a blueprint for guiding integrated deterrence strategies across all elements of national power, including diplomatic, economic, military, and informational tools. The NSS provides a coherent framework for addressing complex global challenges and threats, ensuring that U.S. actions align with its national interests and values.

One of the primary functions of the NSS is to articulate the strategic approach to safeguarding U.S. national security. It outlines the key principles and priorities, such as promoting American prosperity, advancing peace through strength, and upholding American values.

| One of the primary functions of the NSS is to articulate the strategic approach to safeguarding U.S. national security. It outlines the key principles and priorities, such as promoting American prosperity, advancing peace through strength, and upholding American values. |

Integrated deterrence, as a concept, is embedded in this framework, emphasizing the need to employ a combination of diplomatic, economic, military, and informational instruments to deter adversaries and protect American interests.

The Department of Defense (DoD) plays a central role in operationalizing the integrated deterrence concept outlined in the NSS. It is well-suited for this role due to its significant budget, military capabilities, and organizational structure. The DoD can leverage its resources to develop and maintain credible military capabilities, engage in joint exercises and operations, and establish cooperative security partnerships with allies and partners to enhance deterrence. The National Defense Strategy (NDS) is a complementary document that further refines the defense-related aspects of integrated deterrence, providing guidance on the allocation of military resources and the prioritization of capabilities to address emerging threats.[1]

However, there are challenges and limitations in fully implementing integrated deterrence across all elements of national power. Other interagency and intergovernmental organizations, such as the Department of State, Department of Homeland Security, and various intelligence agencies, may not have the same level of budgetary and manpower resources as the DoD. Additionally, these agencies operate under different hierarchies and may have distinct priorities, which can create coordination challenges. Achieving a truly integrated deterrence approach requires effective interagency coordination, clear lines of communication, and a shared understanding of strategic objectives.[2]

The NSS serves as the foundation for integrated deterrence strategies that leverage all elements of national power to protect U.S. national security interests. While the DoD is well-equipped to operationalize the concept, effective interagency coordination and collaboration are crucial to bridge the gaps and shortcomings that may arise due to differences in budget, manpower, and organizational structure among various government agencies. Strategic documents like the NDS and Presidential Policy Memorandums play a vital role in ensuring a unified and coherent approach to integrated deterrence.

JOINT PHASING CONSTRUCT

The U.S military JPC is a strategic framework that offers a structured delineation of the various phases inherent to military operations, serving as a valuable guide for planning and execution.[3] This construct plays a pivotal role in assisting military commanders and planners in comprehending the sequential progression of operations, spanning from initial shaping activities to the resolution of conflicts.

In the early phases outlined within the JPC, emphasis is placed on shaping activities, which encompass a range of diplomatic, economic, and informational efforts aimed at preventing conflict and creating the necessary conditions for regional stability.[4] During this period, the military's role may involve contributing to deterrence through demonstrations of capability and resolve, thereby deterring potential adversaries from pursuing actions contrary to U.S. interests. The overarching objective of these shaping activities is to curb the escalation of tensions and uphold a status quo that aligns with U.S. strategic goals.

As the situation evolves, the JPC progresses through phases that may involve a shift from deterrence to compellence.[5] Compellence activities are strategically designed to influence the behavior of adversaries, compelling them to alter their actions or cease activities deemed detrimental to U.S. interests. The military's involvement in compellence efforts may encompass targeted strikes, the application of coercive measures, or the use of force to achieve specific objectives. This transition signifies a transition from mere deterrence to active efforts aimed at influencing and modifying the adversary's actions.

The culmination of military operations, as articulated within the JPC, manifests in the resolution phase, signifying the conclusion of hostilities and the establishment of new norms or conditions aligned with U.S. interests.[6] Successful compellence initiatives may lead to alterations in the adversary's behavior, prompting a reevaluation of strategic goals and the establishment of a new equilibrium. The resolution phase underscores the broader aim of shaping the post-conflict environment to ensure enduring stability and security.

The JPC's capacity to guide military operations through distinct phases underscores its utility in addressing the dynamic nature of contemporary conflicts and the necessity for a flexible and adaptive approach.[7] By recognizing the transitions from deterrence to compellence and, eventually, to resolution, the JPC equips military planners with a comprehensive framework for addressing evolving threats and contributing to the establishment of new norms that align with U.S. strategic interests. This strategic adaptability is indispensable for navigating the intricacies of modern conflict scenarios and achieving lasting stability following military operations.

SPECTRUMS OF CONFLICT

Spectrums of conflict encompass a wide range of activities and intensities that characterize conflicts, ranging from low-intensity competition to high-intensity warfare. "The Joint Force, and the national security community as a whole, must be ready and able to respond to numerous challenges across the full spectrum of conflict including complex operations during peacetime and war."[8] These spectrums acknowledge the multifaceted nature of conflicts, their diverse manifestations, and the evolving nature of conflict dynamics, necessitating nuanced strategies and responses. This concept is instrumental in grasping the intricacy of contemporary security challenges and in tailoring specific approaches to address each conflict scenario effectively.

At one end of the spectrum, diplomatic negotiations, economic rivalries, and information warfare are prevalent, where adversaries aim to advance their interests without resorting to overt military actions.[9] This lower-intensity spectrum encompasses various forms of statecraft, influence operations, and coercion strategies, stopping short of outright armed conflict. As conflicts progress along this spectrum, the intensity increases, introducing measures such as cyber operations, proxy warfare, and irregular tactic.[10] These mid-intensity conflicts often blur the lines between conventional and unconventional warfare, presenting unique challenges for defense and security planning.

The higher end of the spectrum encompasses traditional military engagements, including conventional warfare characterized by the overt use of force to achieve decisive outcomes.[11] The spectrums of conflict framework underscore the interrelatedness and fluidity of these various levels of conflict, emphasizing the necessity for a comprehensive and adaptable approach to address evolving security threats. Recognizing the spectrums of conflict is imperative for military planners, policymakers, and strategists as it informs the development of integrated and flexible strategies capable of navigating the complexities inherent in the modern security landscape.

COMBATANT COMMANDS

The United States Global Combatant Command Structure represents a fundamental component within the U.S. Department of Defense's organizational framework, with a primary mandate of planning and executing military operations and activities across diverse regions worldwide. This framework is distinctly regionally oriented, subdividing the global landscape into multiple geographic combatant commands, each entrusted with a designated geographical sphere of responsibility. These commands serve as linchpins in providing strategic guidance and operational direction to U.S. military forces within their respective regions, thereby ensuring the alignment of military endeavors with the specific security challenges and intricacies unique to each geographical area.[12]

The establishment of the regional combatant commands as a foundational element of the U.S. military's organizational structure was formally codified through the enactment of the Goldwater-Nichols Department of Defense Reorganization Act of 1986. This pivotal legislative reform ushered in a significant transformation aimed at augmenting the efficiency and effectiveness of the U.S. armed forces. The legislation prescribed the establishment of geographic combatant commands, systematically enshrining the practice of conducting U.S. military operations with a meticulous grasp of regional contexts and requirements. Under the purview of this act, each combatant command, exemplified by entities like U.S. Central Command (CENTCOM) and U.S. Indo-Pacific Command (formerly U.S. Pacific Command or PACOM), was allocated distinct territorial responsibilities. Commanders of these regions were vested with enhanced authority encompassing both planning and execution within their designated spheres.[13]

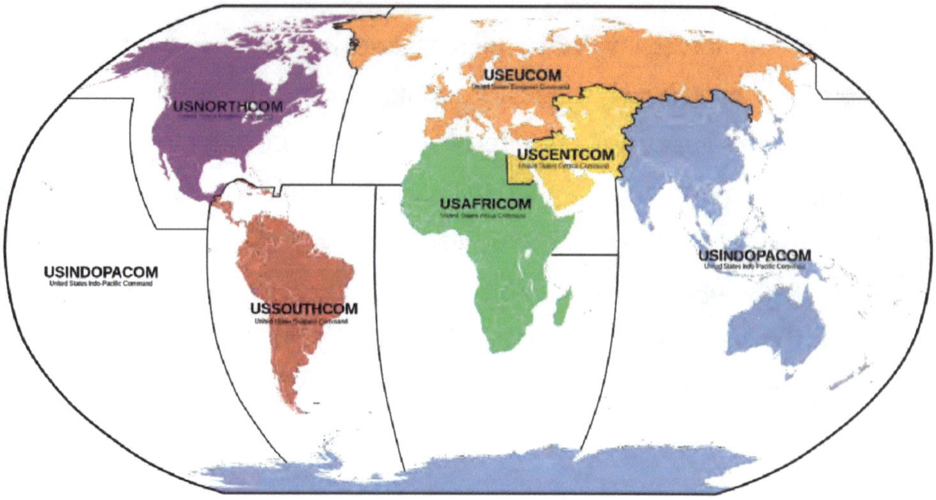

Figure 5.2. U.S. geographic combatant commands. Source: Wikipedia[14]

The enduring regional emphasis within the framework of the combatant command structure has exhibited its pragmatic value in effectively addressing the ever-evolving spectrum of security challenges faced by the United States. This approach promotes nuanced comprehension of regional dynamics, nurturing collaborative relationships with regional partners and fostering the capacity for responses tailored to the unique threats and opportunities found within each geographical domain. The regional combatant commands occupy an indispensable role in the U.S. military's capability to project power, preserve stability, and safeguard American interests globally, all while adhering to the foundational guidance provided by the Goldwater-Nichols Department of Defense Reorganization Act of 1986.[15]

ALLIES AND PARTNERS

U.S. reliance on allies and partners is indeed critical to its Integrated Deterrence Strategy, serving as a cornerstone of its efforts to address global security challenges effectively. Collaborative partnerships with like-minded nations offer numerous advantages. First and foremost, allies and partners enhance the U.S.' collective defense capabilities, pooling military, intelligence, and technological resources to create a stronger deterrent posture. For example, the North Atlantic Treaty Organization (NATO), has played a vital role in ensuring the security of Europe by fostering a robust collective defense mechanism among its member states.

Furthermore, alliances and partnerships contribute to a broader network of intelligence sharing and information exchange, bolstering U.S. situational awareness and early warning capabilities. Collaborative efforts also facilitate burden-sharing, allowing the U.S. to allocate resources more efficiently and effectively. In addition, the diplomatic and political support of allies and partners can be invaluable in building consensus on global security issues, enhancing the legitimacy of international actions, and minimizing the risk of conflicts escalating to dangerous levels.

However, there are also weaknesses and challenges associated with U.S. reliance on allies and partners in its Integrated Deterrence Strategy. One notable challenge is the potential divergence of interests and priorities among different nations within a coalition. Coordination and decision-making can become complex, particularly when addressing multifaceted threats. Additionally, the contributions of allies may vary in terms of capacity and commitment, potentially leading to imbalances within an alliance. Furthermore, the changing geopolitical landscape and evolving threat environment may require continuous adjustments to alliance structures and strategies, which can be time-consuming and challenging to implement. Balancing the interests and expectations of different allies while maintaining a unified front can be a delicate diplomatic endeavor.[16]

While the U.S. benefits significantly from its reliance on allies and partners in its Integrated Deterrence Strategy, there are both advantages and disadvantages to this approach. Collaborative partnerships strengthen collective defense capabilities, intelligence sharing, and political support, but they also present challenges related to coordination, resource allocation, and diverging interests. Striking the right balance and adapting to evolving global dynamics are essential for the continued effectiveness of this strategy.

HOW OUR COMPETITORS OPERATE

CHINA

China employs two prominent strategies, salami slicing and lawfare. *Salami slicing*, a strategic and incremental approach utilized by adversaries to achieve their goals while deliberately remaining below the threshold of overt violence or attribution,

> ***Salami slicing*, a strategic and incremental approach utilized by adversaries to achieve their goals while deliberately remaining below the threshold of overt violence or attribution,**

is a concept that has gained prominence in both geopolitical and cybersecurity contexts, particularly in regard to the behavior of the Chinese Communist Party.[17]

In geopolitics, salami slicing entails making calculated territorial encroachments or policy changes that, when observed individually, may not provoke a robust international response. However, over time, these gradual steps accumulate, leading to a significant shift in the geopolitical landscape.[18] This approach is often associated with expansionist states aiming to alter borders or geopolitical influence without resorting to full-scale warfare.

In the realm of cybersecurity, salami slicing refers to subtle cyber aggression, where adversaries carry out small-scale, incremental cyber operations to achieve their objectives while avoiding a forceful response.[19] For instance, state-sponsored hackers might engage in a series of low-level cyber intrusions or data thefts over an extended period, gradually accumulating strategic advantages.

To address salami slicing, defenders need to recognize and respond to these tactics comprehensively. This involves enhancing situational awareness, improving intelligence gathering, and developing response strategies that consider the cumulative impact of incremental actions. Diplomatic and economic measures can also be employed to deter adversaries from pursuing such tactics, emphasizing the consequences of their actions even when individual steps may seem minor.[20]

Lawfare, a term derived from the combination of "law" and "warfare," constitutes a strategic approach used by adversaries to attain political, military, or economic objectives by exploiting legal mechanisms and institutions.[21]

> ***Lawfare*, a term derived from the combination of "law" and "warfare," constitutes a strategic approach used by adversaries to attain political, military, or economic objectives by exploiting legal mechanisms and institutions.***

This concept expands beyond traditional military conflict, employing legal tools and processes to manipulate or weaken the adversary's position. Lawfare encompasses a broad spectrum of activities, such as leveraging legal challenges, manipulating international legal norms, and utilizing legal procedures to realize strategic goals.[22]

In the realm of lawfare, one dimension involves the manipulation of international legal norms and institutions to discredit adversaries or promote specific narratives. Adversaries may selectively use international forums to propagate their viewpoints, framing disputes or conflicts in a manner that portrays them favorably while undermining the credibility of their opponents. Through the exploitation of legal mechanisms and institutions, adversaries aim to influence global perceptions, garner international sympathy, and shape the response of the international community to their advantage.[23]

Furthermore, lawfare extends its reach into domestic legal systems, where adversaries utilize legal processes to achieve strategic aims. This may entail filing lawsuits, pursuing legal actions, or employing legal tactics to disrupt or divert the attention of their adversaries. By burdening their adversaries with legal challenges, adversaries may aim to divert resources, erode public support, or create political and economic instability.[24]

Notably, lawfare is not exclusive to state actors; non-state entities and transnational organizations also engage in lawfare strategies. For instance, advocacy groups may employ legal avenues to advance their agendas, while terrorist organizations may manipulate legal frameworks to legitimize their actions or attain political objectives. The involvement of non-state actors in lawfare introduces complexities into the strategic landscape, blurring the boundaries between legal and military responses.[25]

A significant aspect of lawfare is its reliance on the rule of law and legal frameworks, emphasizing the critical role of both international and domestic legal systems in contemporary conflicts.[26] The effectiveness of lawfare often hinges on the perception of legitimacy, underscoring the importance of nations and entities comprehending and countering these strategies through legal, diplomatic, and informational means.

Addressing lawfare necessitates a multifaceted response that integrates legal expertise with strategic thinking. Nations and entities grappling with lawfare challenges must fortify their legal defenses, enhance resilience against legal manipulation, and proactively shape narratives within legal frameworks.[27] Furthermore, international collaboration and coordination are imperative to prevent the misuse of legal mechanisms for strategic gains and to ensure that legal processes uphold principles of justice and fairness.

In summary, lawfare constitutes a strategic approach used by adversaries to attain objectives by exploiting legal mechanisms and frameworks. Whether in the international arena or domestic legal systems, adversaries leverage lawfare to shape perceptions, disrupt opponents, and achieve strategic objectives. Recognizing and effectively countering lawfare demands a comprehensive understanding of legal complexities, proactive defense strategies, and international cooperation to safeguard the integrity of legal processes.[28]

RUSSIA

Russia's hybrid warfare strategy, often characterized using "little green men," represents a departure from established norms in the realm of international conflict, and it operates in a deliberately vague and ambiguous manner. This approach blurs the lines between conventional military actions and non-military activities, making it challenging for adversaries to formulate effective responses. Hybrid warfare tactics combine a range of elements, including conventional military forces, irregular militias, cyberattacks, and information warfare, to achieve strategic objectives while avoiding clear attribution.[30]

Russia's hybrid warfare strategy, often characterized using "little green men," represents a departure from established norms in the realm of international conflict, and it operates in a deliberately vague and ambiguous manner. This approach blurs the lines between conventional military actions and non-military activities, making it challenging for adversaries to formulate effective responses. Hybrid warfare tactics combine a range of elements, including conventional military forces, irregular militias, cyberattacks, and information warfare, to achieve strategic objectives while avoiding clear attribution.[30]

| Combine a range of elements, including conventional military forces, irregular militias, cyberattacks, and information warfare, to achieve strategic objectives while avoiding clear attribution. |

Armed men without insignia (Russia's "little green men") at Simferopol airport in 2014. Source: Wikimedia

The Ukraine crisis of 2014 provides a vivid illustration of how Russia employs hybrid warfare tactics. In Crimea and eastern Ukraine, Russia deployed its little green men, uniformed soldiers without official insignia, along with support for separatist groups. This strategy created plausible deniability, as Russia officially denied involvement while effectively controlling the situation on the ground.[32] The ambiguity surrounding the identity and affiliation of these forces made it difficult for Ukraine and the international community to respond decisively.

Furthermore, Russia's hybrid warfare tactics extended into the information domain. They engaged in disinformation campaigns and conducted cyberattacks to sow confusion and disrupt Ukrainian institutions. Using social media, propaganda, and cyber espionage, Russia played a significant role in shaping public perceptions and undermining Ukraine's stability. The Ukrainian case underscores the efficacy of Russia's hybrid warfare tactics in achieving strategic objectives without resorting to overt, conventional warfare. This highlights the need for a comprehensive and adaptable response to hybrid threats, as traditional military measures may prove inadequate in countering the multifaceted challenges posed by such tactics. Consequently, nations worldwide have had to reassess their security strategies and develop a deeper understanding of the complexities associated with hybrid warfare in today's global security landscape.

Russia's hybrid warfare strategy challenges established norms of conflict and relies on deliberate ambiguity. It combines conventional and irregular military forces, cyberattacks, and information warfare to achieve strategic goals while avoiding clear attribution. The Ukrainian case serves as a significant example, emphasizing the necessity for adaptive responses and a comprehensive understanding of hybrid warfare's multifaceted nature in contemporary geopolitics.

NORTH KOREA

North Korea has long employed a strategy of "saber rattling" as a deterrence mechanism to advance its interests, secure its regime, and deter potential adversaries. Saber rattling refers to the practice of issuing threats, making bellicose statements, and displaying military capabilities to project strength and convey a willingness to use force.

> ***Saber rattling* refers to the practice of issuing threats, making bellicose statements, and displaying military capabilities to project strength and convey a willingness to use force.**

This approach is a central element of North Korea's foreign policy and security strategy, with several key components.

First, North Korea's saber rattling often involves provocative actions, such as missile tests and nuclear weapons development. These actions are intended to signal to the international community, particularly its neighbors and the U.S., that it possesses a credible threat of force. By demonstrating advancements in its military capabilities, North Korea aims to deter any perceived threats to its regime by raising the stakes of potential conflict.[33]

Second, North Korea's use of belligerent rhetoric and threats serves to heighten tensions and create a sense of urgency in negotiations. By employing aggressive language and issuing threats, North Korea seeks to extract concessions, gain international attention, and negotiate from a position of perceived strength.[34]

Third, the regime in Pyongyang utilizes saber rattling as a means of rallying domestic support. By portraying external threats and highlighting the need for a strong military, the North Korean leadership reinforces its control over the populace and maintains unity in the face of perceived external enemies.

Finally, North Korea's saber rattling is intertwined with its pursuit of a nuclear deterrent. The regime views nuclear weapons as a critical means of deterring potential adversaries from military intervention. The development and testing of nuclear weapons, accompanied by provocative rhetoric, are intended to establish North Korea as a nuclear-armed state with the capability to inflict significant damage in the event of an attack.[35] North Korea's utilization of saber rattling as a deterrence mechanism involves a combination of provocative actions, belligerent rhetoric, and nuclear development. This strategy aims to deter potential adversaries, extract concessions in negotiations, maintain domestic support, and establish a credible nuclear deterrent. The complex interplay of these elements underscores the multifaceted nature of North Korea's security strategy and its impact on regional stability.

IRAN

Iran employs a strategic approach known as the Iranian Threat Network (ITN) to exert influence over global powers while carefully avoiding the threshold of violence and escalation.

| Iranian Threat Network (ITN) to exert influence over global powers while carefully avoiding the threshold of violence and escalation. |

The ITN represents a complex web of state and non-state actors, including proxies, militias, and political allies, all coordinated to serve Iran's regional and global interests. This approach enables Iran to pursue its strategic objectives while maintaining a degree of plausible deniability, making it challenging for adversaries to respond effectively.[36]

One key element of Iran's ITN strategy is its support for various regional militias and proxy groups across the Middle East. These militias, such as Hezbollah in Lebanon and various Shiite militias in Iraq, receive training, funding, and weaponry from Iran. By backing these proxies, Iran extends its influence beyond its borders and gains leverage in regional conflicts. These groups can carry out destabilizing actions or respond to Iranian adversaries without Iran officially taking responsibility, allowing the Iranian government to remain below the threshold of direct confrontation.[37]

Moreover, Iran utilizes cyber operations as a tool within its ITN to achieve strategic objectives without resorting to overt violence. Iranian state-sponsored hacking groups have been implicated in numerous cyberattacks and espionage activities. These operations can disrupt critical infrastructure, target international adversaries, and collect intelligence. Such activities allow Iran to assert influence in the cyber realm while maintaining ambiguity about its involvement.[38]

In the realm of regional conflicts, Iran leverages its ITN to support groups and factions aligned with its interests, such as the Houthi rebels in Yemen. By providing military assistance, financial support, and political backing to these groups, Iran can exert influence in ongoing conflicts while avoiding direct engagement. This approach allows Iran to project power and extend its reach without crossing the threshold into full-scale warfare.[39]

Furthermore, Iran skillfully employs diplomatic means within its ITN strategy to engage with global powers and influence international negotiations, particularly concerning its nuclear program. By participating in diplomatic talks, Iran can shape narratives, exploit divisions among international actors, and buy time to advance its nuclear capabilities. This diplomatic maneuvering enables Iran to pursue its objectives while staying within the bounds of negotiations and avoiding military escalation.[40]

The ITN is a multifaceted strategy that enables the country to exert influence over global powers while remaining below the threshold of violence and escalation. Through support for regional proxies, cyber operations, involvement in conflicts, diplomatic engagement, and other tactics, Iran effectively pursues its strategic interests while maintaining plausible deniability. This approach poses challenges for international actors seeking to counter Iran's influence and underscores the complexity of modern geopolitical competition.

CONCLUSION

The United States' Integrated Deterrence Strategy is a multifaceted approach aimed at deterring aggression and protecting American interests in an increasingly complex and interconnected world. While this strategy has notable strengths, it also faces challenges, particularly in countering the efforts of competitors such as China, Russia, North Korea, Iran, and non-state actors that exploit ill-defined norms to expand their influence globally. One of the strengths of the Integrated Deterrence Strategy is its adaptability and ability to employ a range of tools beyond military force, including diplomatic, economic, and informational measures. This provides the U.S. with a comprehensive tool kit to respond to different forms of aggression or coercion. However, competitors like China and Russia have proven adept at operating in the seams of ill-defined norms, exploiting legal ambiguities and employing hybrid tactics that blur the lines between military and non-military actions. These competitors may challenge the U.S. in ways that traditional deterrence mechanisms may not fully address.[41]

Another strength of the U.S. strategy is the presence of geographic combatant commands, which are well-organized to address security challenges within their respective regions. These commands have the staffing, expertise, and hierarchies necessary to understand and respond to regional threats effectively. However, a limitation arises when dealing with trans-regional problem sets. Many security challenges today transcend geographic boundaries, requiring a more coordinated and cohesive approach. The traditional structure of the Geographic Combatant Commands may hinder their ability to address these complex, trans-regional issues effectively.[42]

To address these challenges, the U.S. must enhance its ability to adapt to rapidly changing security environments and counter competitors' strategies effectively. This may involve strengthening interagency and intergovernmental coordination to address trans-regional threats and improving the U.S. capacity to define and respond to these complex problem sets. Additionally, the U.S. should continue to invest in diplomatic and economic tools to shape the behavior of competitors and non-state actors that operate in the seams of international norms.

About the Author

Aaron Baty is a retired Special Forces officer who supported operations in Afghanistan, Iraq, and Africa. He is currently pursuing his Doctorate of Business Administration degree at the University of South Florida. He lives in Tampa, FL, with his wife Ursula and three children.

NOTES

1. Department of Defense, "Summary of the 2018 National Defense Strategy of the United States of America" (2018).

2. Executive Office of the President, "National Security Presidential Memorandum 1" (2017).

3. Joint Chiefs of Staff, *Joint Publication 3: Joint Operations*, (Department of Defense, 2020).

4. *Joint Operations.*

5. *Joint Operations.*

6. *Joint Operations.*

7. *Joint Operations*, 3.

8. Frank G. Hoffman, "Gray Zone and Hybrid Challenges," *PRISM* 7, no. 4 (November 2018).

9. U.S. Army, *Multi-Domain Operations: A Concept for Success* (2019), https://www.tradoc.army.mil/wp-content/uploads/2018/12/MDO_WhitePaper.pdf.

10. U.S. Army, *Multi-Domain Operations.*

11. U.S. Army, *Multi-Domain Operations.*

12. Joint Chiefs of Staff, *Joint Publication 5: Joint Planning*, (Department of Defense, 2020).

13. *Goldwater-Nichols Department of Defense Reorganization Act of 1986*, Pub. L. No. 99-433, § 1, 100 Stat. 992 (1986)

14. Image by IronMike6 and from Wikipedia, found at https://en.wikipedia.org/wiki/File:Unified_Combatant_Commands_map.PNG, accessed on 22 July 2024.

15. *Joint Planning.*

16. M. S. Lund and V. A. Franke, "NATO and the Future of Deterrence," *Joint Force Quarterly* 92 (2019): 56-63

17. J. S. Duffield, "Salami slicing: A new Chinese way of war." *Australian Journal of Defence and Strategic Studies* 12, no. 2 (2020): 21-38.

18. Oriana Skylar Mastro, "Beijing's Salami Slicing: It's Getting the Desired Results," *Foreign Affairs*, 2018, accessed February 8, 2024, https://www.foreignaffairs.com/articles/asia/2018-08-06/beijings-salami-slicing

19. Kenneth Geers, "Sun Tzu and Cyber War,"*Naval Criminal Investigative Service (NCIS) and Cooperative Cyber Defence Centre of Excellence* (CCD COE). Accessed February 8, 2024. https://ccdcoe.org/uploads/2018/10/Geers2011_SunTzuandCyberWar.pdf.

20. "Sun Tzu and Cyber War."

21. Oona Hathaway and Scott J. Shapiro, "Outcasting: Enforcement in Domestic and International Law," *Yale Law Journal* 121 (2012): 252

22. Hathaway and Shapiro, "Outcasting," 252.

23. Eric A. Posner and John C. Yoo, "A Theory of International Adjudication," *University of Chicago Law & Economics*, Olin Working Paper No. 206; *UC Berkeley Public Law Research Paper* No. 146.

24. Charles J. Dunlap, Jr., *Lawfare: A Decisive Element of 21st-Century Conflicts*, (Washington, DC: National Defense University, Institute for National Strategic Studies, 2009)

25. Dunlap, "Lawfare."

26. Eric A. Posner and John C. Yoo, "A Theory of International Adjudication," *University of Chicago Law & Economics*, Olin Working Paper No. 206; UC Berkeley Public Law Research Paper No. 146.

27. Hathaway and Shapiro, "Outcasting."

28. Hathaway and Shapiro, "Outcasting,"

29. Mark Galeotti, "Hybrid, Ambiguous, and Non-linear? How New Is Russia's 'New Way of War'?," *Small Wars & Insurgencies* 27, no. 2 (2016): 282-301.

30. Niklas Nilsson, "Between Russia's 'Hybrid' Strategy and Western Ambiguity: Assessing Georgia's Vulnerabilities," *The Journal of Slavic Military Studies* 34, no. 1 (2021): 50-68.

31. Photo by Elizabeth Arrott and Voice of America and thereby copyright free, taken on 28 February 2014, found on Wikimedia at https://commons.wikimedia.org/wiki/File:VOA-Crimea-Simferopol-airport.jpg, accessed on 22 July 2024.

32. Mark Galeotti, "Hybrid, Ambiguous, and Non-linear? How New Is Russia's 'New Way of War'?," *Small Wars & Insurgencies* 27, no. 2 (2016): 282-301

33. Seong-Ryoul Cho, "North Korea's Security Dilemma and Strategic Options," *The Journal of East Asian Affairs* 23, no. 2 (Fall/Winter 2009): 69-102. https://www.jstor.org/stable/pdf/23258186.pdf.

34. Victor D. Cha, "Powerplay: Origins of the U.S. Alliance System in Asia," *International Security* 34, no. 3 (2009): 158–96. http://www.jstor.org/stable/40389236.

35. Edward Kwon, "North Korea's Nuclear Strategy and the Direction of WMD Development," *The Journal of East Asian Affairs* 33, no. 1 (2020): 27–64. http://www.jstor.org/stable/45441011.

36. Kenneth Katzman, "Iran's Foreign and Defense Policies," *Congressional Research Service Report* R44017, January 11, 2021

37. Daniel L. Byman, "Proxy Power: Understanding Iran's Use of Terrorism," *Brookings Institution*, July 26, 2006.

38. Keren L G Snider, Ryan Shandler, Shay Zandani, and Daphna Canetti, "Cyberattacks, cyber threats, and attitudes toward cybersecurity policies," *Journal of Cybersecurity* 7, no. 1 (2021): tyab019, https://doi.org/10.1093/cybsec/tyab019.

39. Ariane M. Tabatabai, Jeffrey Martini, and Becca Wasser, "The Iran Threat Network (ITN): Four Models of Iran's Nonstate Client Partnerships," *RAND Corporation*, 2021.

40. Ray Takeyh and Charles A. Kupchan, "Negotiating with Iran: Racing for the Exit," *Foreign Affairs* (2021).

41. Department of Defense, "Summary of the 2018 National Defense Strategy."

42. Abel Morrobel. "Review of Commanding Excellence: Inspiring Purpose, Passion, and Ingenuity through Leadership that Matters, by Gary Morton." *Journal of Applied Christian Leadership* 12, no. 2 (2018): 111-113.

43. Department of Defense, "Summary of the 2018 National Defense Strategy."

| 353 |

IV. Allied Security Strategies: NATO Special Operations as a Case Study

By Gabriele Pierini

ABSTRACT: This section examines the responses of U.S. allies and partners managing internal and external threats to their security and to the stability of the Euro-Atlantic region. The Cold War version of total defense was resurrected in spirit but revised in practice for the modern era. U.S. allies and partners either witnessed or withstood the Russian invasion of Georgia in 2008 and Ukraine in 2014 and 2022. Whether under partial occupation or facing daily hybrid measures, these countries have led the way in revising their national defense structures, plans, and preparations. This paper highlights and examines total defense strategies planned and implemented by coalitions. In particular, it assesses the concept of Comprehensive Defence, a whole-of-society approach to national defense developed by NATO Allied Special Operations Forces Command, and how SOF may support this effort.

| "Having a strong military is fundamental to our security, but our military cannot be strong if our societies are weak."
– Jens Stoltenberg, 7 October 2020 |

INTRODUCTION

The quote by NATO Secretary General Jens Stoltenberg at the Global Security Forum took place in Bratislava, Slovakia, four years ago. Stoltenberg further stated, "having a strong military is fundamental to our security, but our military cannot be strong if our societies are weak, so our first line of defense must be strong societies able to prevent, endure, adapt and bounce back from whatever happens." In other words, military efforts to defend NATO territory and populations need to be complemented by robust civil preparedness. This is necessary to reduce potential vulnerabilities and the risk of attack in peacetime competition, crisis, and conflict.

This principle is enshrined in Article 3 of the North Atlantic Treaty, signed on 4 April 1949, which forms the legal basis of NATO and is implemented by the Alliance: "In order more effectively to achieve the objectives of this Treaty, the Parties, separately and jointly, by means of continuous and effective self-help and mutual aid, will maintain and develop their individual and collective capacity to resist armed attack."[1] Resilience is both a national responsibility and a collective commitment; each ally must be strong and adaptable enough to handle the full range of crises envisaged by the Alliance. Simultaneously, the individual commitment of every ally to maintain and strengthen its resilience reduces the vulnerability of NATO as a whole.

The scope of this chapter is twofold: on one side, it educates the reader on NATO's main considerations for resilience and resistance, with a particular focus on the role of SOF. On the other side, it sets the stage, by including open questions, to expand the conversation about state resilience.

The first section explores NATO's official definitions of resilience and resistance. The two are always tricky topics within the Alliance—particularly the latter—and the goal is that the chapter will spark some interesting discussions. The second and third sections present the history of the Allied Special Operations Command (SOFCOM)—NATO's component command for SOF—and the concept of *comprehensive defense*; what it does entail and when and why it was created. The fourth section defines NATO SOF missions and tasks. This is followed by some food-for-thought information on what SOF in general, and NATO SOF specifically, can do in support of resilience and resistance under the umbrella of comprehensive defense. The conclusion offers some open questions for further thought.

DEFINING RESILIENCE AND RESISTANCE IN NATO

Definitions will help frame this discussion. The following definitions are from NATO Term, which is the official terminology database of the Alliance, approved by NATO Standardization Office. Twenty years ago, NATO defined resilience as "the ability of a functional unit to continue to perform a required function in the presence of faults and errors."[2] Now, NATO defines the term as "the ability of an entity to continue to perform specified functions during and after an attack or an incident."[3] What differences do we notice?

A specific rationale to explain the "why" of a definition is not provided by the NATO Standardization Office. Therefore, the real difference is that in the first definition—which was taken from the NATO Information Technology Glossary and approved by the Consultation, Command, and Control Board (recently renamed NATO Digital Staff), a senior civilian body that reports to and advises the North Atlantic Council on policy matters—resilience is dealt in generic terms and focuses on digital aspects. The second and more recent definition for resilience specifies that the ability to continue to perform some functions must follow "an attack or an incident," and is thus more related to matters of security and defense. This second definition is in the NATO Glossary of Terms and Definitions and approved by the Military Committee (MC), the senior military authority in NATO.

What about NATO's definition of *resistance*? It reads, "actions of an organized, often ideologically motivated, group or movement that seeks to affect or prevent political change within a country or a region through subversion, and that are focused on persuading or coercing the population through irregular activities."[4]

Some words in the text look very SOF-oriented, for example, "organized [...] group or movement" and "irregular activities." It's unclear whether SOFCOM agrees or disagrees with these definitions. Reading through the latter, the definition neither mentions SOF nor refers to a national component that is legally authorized to resist an invasion. Resistance is arguably one of the most sensitive topics to be discussed within NATO. In fact, the NATO official definition of resistance was agreed upon for the first time only two years ago, in early 2022, one week after Russia invaded Ukraine on 24 February. We can speculate whether this is a coincidence, but there is no doubt it is a sign that something is changing in the strategic environment.

HISTORY OF ALLIED SPECIAL OPERATIONS FORCES COMMAND

During the last 20 years, NATO SOF have been conducting military assistance (MA), special reconnaissance (SR), and direct actions (DA)—the three principle tasks assigned to SOF by NATO doctrine[5] —mainly as counterterrorism activities in Afghanistan. The involvement of the host nation's agencies, such as the ministry of defense, interior, and intelligence services, was fundamental. Some like to define this as a "whole-of-government" approach.

First in Bosnia in the early 1990s and then in Afghanistan in the early 2000s, the allies realized that there was little interoperability in terms of tactics, techniques, and procedures (TTPs); means of communications; and weapons among the SOF community. As a result, at the NATO Summit that took place in Riga, Latvia, in 2006, the Euro-Atlantic leaders decided to create an organization that could link NATO SOF together under the so-called NATO SOF Transformation Initiative.

The initiative consisted of three pillars:

1. The creation of a NATO SOF Coordination Centre (NSCC) to be located in Mons, Belgium, within the premises of NATO Supreme Headquarters Allied Powers Europe (SHAPE), and act, as a Memorandum of Understanding organization, in order to provide strategic advice to the Supreme Allied Commander Europe (SACEUR); synchronize SOF development across NATO; and support the International Security Force Assistance (ISAF) mission in Afghanistan [NSCC was transformed into NATO Special Operations Headquarters (NSHQ) in 2010, with the additional role of acting as a backup SOF operational command and control element; and then into Allied Special Operations Command (SOFCOM) in 2023, as NATO component command for SOF].
2. The creation of an Office for Special Operations (OSO) within SHAPE, to be directed – in a dual-hatting form – by the commander of NSCC as SOF Advisor (SOFAD) to SACEUR.
3. The creation of a NATO Special Operations School (NSOS), to be located at Chièvres Air Base in Belgium, and act as the official learning institution for NATO SOF.

In 2023, NSOS became an accredited university and was renamed NATO Special Operations University.

Recent changes in the strategic environment include power competition fought through military means but also diplomatic, information, and economic means, which oblige SOF to relook its employment—from expeditionary and counterterrorism-oriented (such as in the case of Afghanistan) toward a regional approach that is more functional to deal with near-peer threats such as Russia. Additionally, there is now space and necessity for what can be called a "whole-of-society" approach to national defense. The involvement of host nation's agencies is not enough: the threat now requires civil preparedness in support of the military instrument of power. This is why, three years ago, SOFCOM (NSHQ, at the time) generated the concept of comprehensive defense.

COMPREHENSIVE DEFENSE

Comprehensive defense is an umbrella concept that covers both resilience and resistance. Lt. Gen. Eric Wendt, who served as the 5th commander of NSHQ from 2019 to 2021, tasked a cross-functional action team to develop the first version of the *Comprehensive Defence Handbook*, which serves as the basic reference for this concept and its related activities. This publication is not an official NATO policy or doctrine but has the intent to practically assist NATO allies and partners in the development of national programs designed to enable all members of society to contribution to deterrence and defense. The *Handbook* was published in December 2020 and is currently under revision.

Comprehensive defense is "an official government strategy, which encompasses a whole-of-society approach to protecting the nation against potential threats."[6] These threats may be natural or man-made; the model does not change. However, by "potential threats," we usually refer to any action taken by an individual or organization, external or internal to the society, the effects of which threaten a nation's safety, security, sovereignty, or its people's right to self-determination. The *Handbook* takes into consideration four types of threats:

Types of Threats:

1. **Weaponized information**
(such as misinformation and disinformation)
2. **Cyber attacks**
3. **Terrorism**
4. **Armed incursion**

For the sake of length and scope, this section will not provide an in-depth focus on all the aspects of comprehensive defense but will limit itself to an overview of the main principles, pillars, phases, and layers.

The public sector (governments) of most Western nations, which on average comprises only 2 percent of the population, is not able to deal with the above-mentioned threats alone, as these threats are hybrid in nature and target different sectors of the society. To do so, the public requires the involvement of the rest of the population—the 98 percent, namely the private and civic sectors. This is why we are no longer talking about "whole-of-government" but "whole-of-society" approaches to national defense. The role of the public sector is to educate, inform, and enable the population so that citizens are aware, capable, and—if necessary—committed to support national efforts in countering malicious activities. A powerful, national comprehensive defense is built on the share of trust, cohesion, and motivation between the government and its citizens. The *Handbook* provides a checklist of activities that promote these efforts.

The framework presented in the *Handbook* is based on six pillars that represent essential elements of today's societies. They differ slightly from NATO's seven baseline requirements—officially approved by the allies in 2016 at the Summit that took place in Warsaw, Poland—but they can practically help the reader in understanding the functions and actions needed to prevent, respond to, and recover from crises. See Figure 5.3.

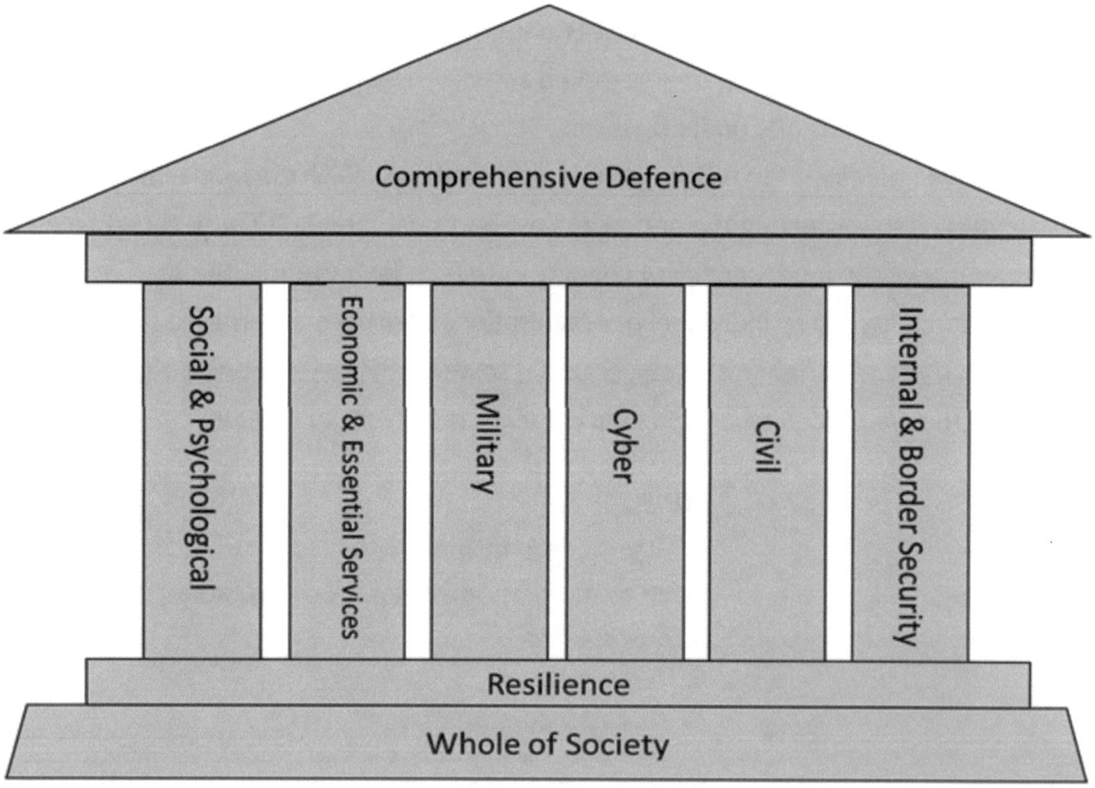

Figure 5.3: Comprehensive defense pillars. Source: NSHQ Comprehensive Defense Handbook.

The resilience of individuals, and of society as a whole, across the physical, virtual, and cognitive dimensions, is the bedrock of the comprehensive defense model.

1. **Social and psychological defense**. A pillar based on the will and ability of citizens and residents to face, cope with, and overcome crises. It is built on national identity, culture and heritage through educational and informational campaigns.
2. **Economic and essential services**. This pillar encompasses government's ability to build and sustain a strong economy and a durable critical infrastructure that can sustain national shocks and provide essential services during shortages or failures.
3. **Military defense**. The civic sector and the military mutually reinforce each other: The civic sector might support the military by providing enabling infrastructure and services and by adapting its production for dual-use purposes; the military might be employed for civil scopes, such as in the cases of natural disaster relief or protection from chemical, bacteriological, radiological, nuclear, and explosive (CBRNE) materials, just to name a few. For more information, see the Handbook.
4. **Cyber defense** focuses on the prevention, detection, and response to threats or attacks to national digital infrastructure and its ability to process critical information.
5. **Civil defense** focuses on protecting civilians from malicious attacks and natural disasters with a focus on emergency operations. This often includes the participation of voluntary paramilitary organizations or civil defense corps designed to assist the government during civil emergencies.
6. **Internal and border security**. This includes public safety and law enforcement activities as well as domestic intelligence and border management.

Comprehensive defense is an integrated layered defense system and is usually planned and executed according to three phases:

1. **Preparation**. The government defines roles and responsibilities, puts structures in place, and establishes and maintains the conditions to allow for resilience. This serves as a strong deterrent to malicious acts that may be contemplated by an adversary.
   Response. This phase may begin before or after a threat has presented itself. The scope is to deal, in the best way possible, with the malicious activities mentioned above.
   It may range from consequence management to information campaigns to armed or unarmed resistance. This could involve emergency response in time of disaster, or military response in the case of armed aggression, or even a combination during a crises or conflict.
2. **Recovery**. The goal is to absorb the shock and move on. Re-establish the conditions that existed prior to crisis and/or conflict, with the implementation of lessons identified, lessons learned, and best practices.

The **four main layers of comprehensive defense** are:

1. **National resilience**. The first line of defense, as per Art. 3 of the North Atlantic Treaty. Resilience is developed by all the actors in accordance with NATO's seven baseline requirements:
    a. Assured continuity of government;
    b. Resilient energy supplies;
    c. Ability to deal effectively with the uncontrolled movement of people;
    d. Resilient food and water resources;
    e. Ability to deal with mass casualties and disruptive health crises;
    f. Resilient civil communications systems; and
    g. Resilient transport systems.
2. **Traditional defense forces**. This includes armed forces and law enforcement agencies.
3. **Irregular defense**. This includes irregular reservist components, such as Home or National Guards.
4. **National resistance**. This is made up of a government in exile, a shadow government, and what this framework calls the "asymmetric defense component" (ADC; depicted in Figure 5.4). The ADC is made of the following elements:
    a. Underground, which represents the leadership that provides command and control to the resistance movement;
    b. Adapted force. The combat component that conducts military operations and sabotages activities; and
    c. Auxiliary, which provides support to national resistance through the development and management of safe houses and medical facilities, non-standard logistics and communication, and intelligence collection and recruitment.

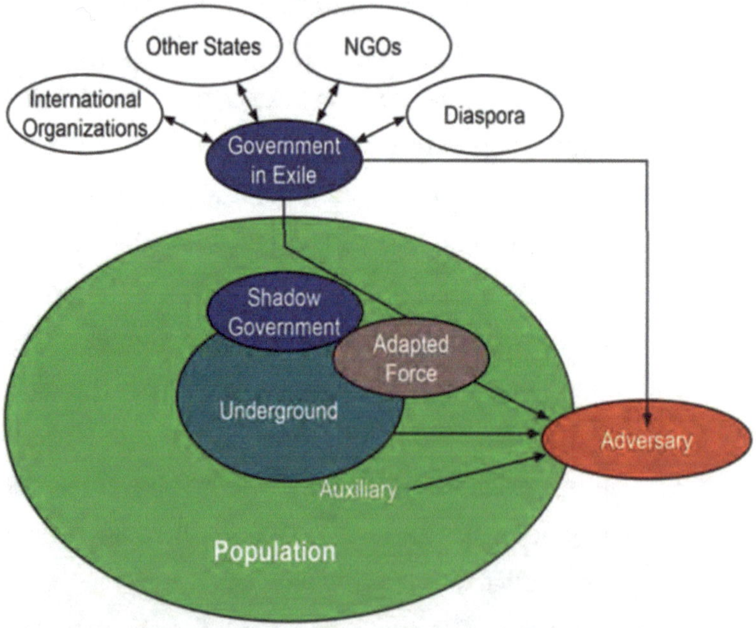

Figure 5.4: Asymmetric defense component structure. Source: NSHQ Comprehensive Defence Handbook.

Much of this will look familiar as doctrine related to SOF. For further information, readers should look for these elements—principles, pillars, phases, and layers—in the *Handbook*.

MISSIONS AND TASKS OF NATO SOF

If the average citizen is asked to think about special operations, they'd probably think about super-fit operators jumping from 24,000 feet and conducting raids, ambushes, and sabotages behind enemy lines. They would be actually quite right: as seen earlier, DA is part of what SOF do. However, there is a general misconception about what special operations are today.

As per NATO doctrine, "special operations are military activities conducted by specially designated, organized, selected, trained, and equipped forces that use unconventional techniques and methods of employment." Kicking in doors and capturing a high-value target is an example of an elite operation. Figure 5.5 presents a simple way to understand the difference between elite and the totality of special operations.

Clearly, the spectrum is broad and creative. Figure 5.5 also postulates how the idea of risk affects decision-making by policymakers, especially if the risk calculus of standard operations in orthodox ways is too high to tolerate. Since special operations are usually conducted in uncertain, hostile, or politically sensitive environments, SOF require innovative tools, tactics, techniques, and procedures.

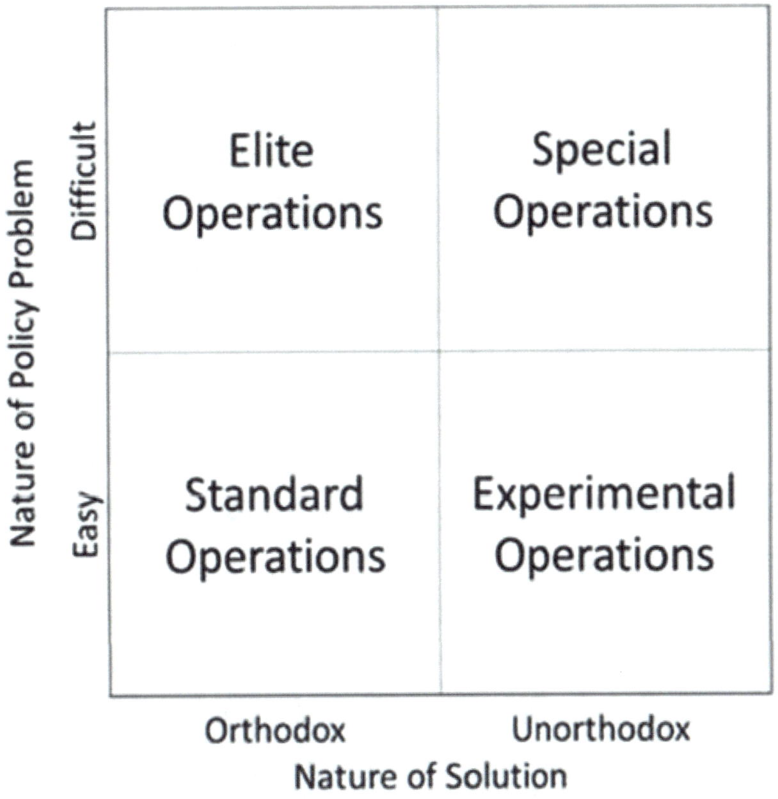

Figure 5.5. Nature of the policy problem vs. nature of the solution options for SOF. Source: U.S. Navy photo.

| Since special operations are usually conducted in uncertain, hostile, or politically sensitive environments, SOF require innovative tools, tactics, techniques, and procedures. |

Nations therefore need a combination of elite as well as special or unorthodox capabilities. Many NATO allies have used this equation to great success in the past. Examples include the German seizure of Fort Eben Emael in Belgium, through the use of gliders, that gave the Germans access to Western Europe on 10 May 1940, and the raid on Alexandria in Egypt, where the Italian frogmen employed innovative manned torpedoes—nicknamed *maiali* (Italian for "pigs")—to disable two Royal Navy battleships on 19 December 1941. These DA events changed the course of their nations' campaigns during World War II. However, beyond sabotage and raids, SOF also found success in delivering influence activities and building trustworthy networks. Two examples are the British "hearts-and-minds" campaign during the Dhofar Rebellion in Oman in the 1960s and 1970s, and the early employment of SOF in Afghanistan following the September 11 attacks in 2001.

THE ROLE OF SOF IN COMPREHENSIVE DEFENSE

The following provide some examples of what may constitute the role of special operations in support of resilience and resistance.

Two key principles of comprehensive defense apply for SOF:

1. Early employment in peacetime competition and crisis is essential for SOF success in crisis or conflict.
2. SOF may be employed primarily where other military or civilian assets are unable to singularly create desired effects.

More in depth, during peacetime competition and crisis, NATO SOF and SOF of NATO nations may conduct activities below the threshold of armed conflict and prepare the environment. The aim of these activities is to ensure interoperability, prepare the operational environment, and boost national resilience.

Through the NATO SOF network, these activities may include, but are not limited to:

1. Connect actors and activities to achieve and maintain situational awareness and understanding of the strategic environment—this is valid across the whole spectrum and supports the scope of persistent campaigning.
2. Provide education and training to build allies and partners' capabilities and to help them develop or refine national policies and plans, mainly through the execution of courses, seminars, and exercises.
3. Facilitate whole-of-government and whole-of society approaches to counter hybrid threats through interagency liaison and coordination and encourage the establishment of frameworks and procedures, such as public-private partnerships, to leverage the support of non-military instruments of power, should their support be needed.

This approach serves a twofold objective: on one side, it ensures that different ministries, agencies, and other national entities become more knowledgeable on the concept of comprehensive defense, and, on the other side, it fosters the development of useful connections and networks should national crisis management options need to be activated.

During conflict, SOF may execute special operations with the aim to enable and contribute to the campaign plan of the Joint Force command. Activities may include, but are not limited to:

1. Create a picture of the operational environment and support the targeting cycle through deep and denied-area sensing, backed by non-military instruments of power, if needed and available.
2. Support the sustainment of deep-area resistance with unconventional TTPs related to logistics and communication.
3. Perform unconventional and non-conventional assisted recoveries (UAR and NAR) to support joint personnel recovery (JPR) efforts in non-permissive environments.

Deep-area preparation activities conducted during peacetime competition and crisis may set the conditions for SOF to leverage resistance networks, not only to provide information on systems and targets but also to offer an alternative capability to strike deep and support the degradation—if necessary—of anti-access/area denial (A2/AD) systems.

These activities are exposed to several challenges and require preconditions for success. Of particular note, one might consider the following problems:

1. How to balance the trade-off between operational and communication security—to protect the resistance network(s)—and strategic communication and messaging—to influence public opinion and international support.
2. How to develop and sustain a command-and-control structure that leverages unity of effort in a non-permissive environment.
3. How to define the myriad of legal challenges that arise while planning and executing military activities that involve civilians.

CONCLUSION

As NATO faces the most complex security environment since the end of the Cold War, allies and partners review their approaches to resilience—and, potentially, resistance—by integrating military and non-military instruments of power in crisis response and defense. Following the outbreak of war in Ukraine, comprehensive defense is now seen as the necessary paradigm by many NATO allies and partners to prepare, respond, and recover from malicious activities. Some are developing or updating laws and lessons learned from resistance models that date back to the Cold War.

While currently NATO does not have an officially approved concept for resistance, Allied Command Transformation (NATO's strategic command responsible for leading the transformation of the Alliance's forces and capabilities) has been tasked by NATO Military Committee to deliver an initial concept for resilience under the title of "layered resilience." This concept is currently under development, with the intent for it to be approved by the Military Commitee in 2026. Due to its complexity, and in respect of the consensus decision-making process of the Alliance, the development of this concept is far from being swift.

By virtue of their nature, SOF need to be ready to tackle emerging challenges and, in the NATO context, allied and partner SOF must evolve at the speed required by their nations. As shown, there is potential to employ SOF in support of comprehensive defense. Particularly, SOFCOM's convening power and subject matter expertise are strengths that allies and partners can exploit to reach a common understanding on resilience and resistance, develop interoperability, and share best practices.

| SOFCOM's convening power and subject matter expertise are strengths that allies and partners can exploit to reach a common understanding on resilience and resistance, develop interoperability, and share best practices. |

The following questions can help further the discussion about allied strategies in support of resilience and resistance:

1. Is external support to resilience and resistance the optimal approach to counter adversaries to NATO?
2. How can NATO SOF support state-sponsored resistance strategies and vice versa?
3. How do allies include non-military instruments of power into NATO operations? Is SOF the right actor to facilitate their integration? If yes, why?

About the Author

Gabriele Pierini is an independent scholar. He holds two master's degrees, in International Relations and in Diplomatic Studies. The views expressed are those of the author and do not reflect the official position of the Allied Special Operations Command or NATO.

RECOMMENDED READINGS

AAP-06, "NATO Glossary of Terms and Definitions," February 2018.

AAP-39, "NATO Handbook of Land Operations Terminology," December 2015.

ADatP-02, "NATO Information Technology Glossary," March 2005.

AJP-3.5, "Allied Joint Doctrine for Special Operations," August 2019.

Jonathan Schroden, "Why Special Operations? A Risk-Based Theory," CNA Occasional Paper, September 2020.

NATO, "The North Atlantic Treaty," April 1949.

NATO Press Releases, "Riga Summit Declaration," November 2006.

MC 437/2, "Special Operations Policy," April 2011.

SOFCOM, "Comprehensive Defence Handbook," Vol. I and II, December 2020.

William H. McRaven, "The Theory of Special Operations," Naval Postgraduate School, AD-A269484, June 1993.

NOTES

1. North Atlantic Treaty, Article 3, 4 April 1949.

2. ADatP-02, Consultation, Command, and Control Board approved 1 March 2005.

3. AAP-06, Military Committee approved 16 February 2022.

4. AAP-06 and AAP-39, Military Committee approved on 2 March 2022.

5. AJP-3.5 dated AUG 2019.

6. NSHQ Comprehensive Defence Handbook, Vol. I, December 2020, page 15.

7. MC 437/2, Special Operations Policy, 21 APR 2011

CHAPTER 6: INTERNATIONAL RELATIONS

I. China's Evolving Way of War: From Deception to Deterrence

By Christopher Marsh

ABSTRACT: This chapter examines the propensity of the People's Republic of China (PRC) to operate in gray zone conflict, political warfare, and/or hybrid warfare. It builds upon the lessons of previous chapters, like full spectrum of conflict design and deterrence strategies. It presents a historiography of varying opinions concerning the Chinese Communist Party's (CCP) strategies and tactics and how they have evolved over time, including its use of intelligence and cyber operations.

INTRODUCTION

In his seminal work on the American way of war, Russell Weigley argued that, throughout its history, America had developed its own unique approach to war, one that relied upon cultural and historical constraints, with a decisive change from a strategy of attrition to one of annihilation occurring during the U.S. Civil War.[1] While recently coming under criticism for the blurring of the annihilation-attrition dichotomy and for failing to allow for evolution during the course of a conflict,[2] Weigley's research has largely stood the test of time and offered scholars of war an approach to understanding the historical-cultural nuances of war that seem to appear in the approach to war taken by various nations.[3]

It is no surprise, therefore, that scholars would question whether the world's longest continuous civilization, China—which has seen more than its fair share of war, both domestically and along its borders—has its own unique approach to war. Some scholars argue that there is a Chinese way of war. Unfortunately, most of these studies are premised upon a flawed understanding of China and its history and how this history has impacted contemporary thinking on the part of the country's leadership, senior policymakers, and military strategists. Additionally, Weigley identified a major shift in the American way of war halfway through its 200 years of history at the time of his writing. How many shifts would a civilization like China's have seen throughout its more than five millennia of history?

In attempting to lay out the case for a Chinese way of war, Geoff Babb simply summarizes China's encounter with the West, the 100 years of humiliation, and the rise of the contemporary PRC, arguing that taken together this equates to a unique "way of war."[4] Timothy Thomas (a longtime analyst at Fort Leavenworth's Foreign Military Studies Office) uses the construct of a Chinese way of war as well, but bases his analysis upon an exhaustive study of recent Chinese military and strategic documents and journals, including the very influential *China Military Science*, concluding that the Chinese way of war has changed "dramatically from what it was two decades ago."[5] So, if there is a Chinese way of war today, it is not the same today as it was 20 years ago, prior to China's rise, nor does it perhaps even resemble that of its ancient history (as most cogently argued by China historian Peter Lorge[6]).

| If there is a Chinese way of war today, it is not the same today as it was 20 years ago, prior to China's rise, nor does it perhaps even resemble that of its ancient history. |

Putting aside the Chinese way of war discussion for the moment, another popular theme in the study of China's approach to war is that the PRC has some secret plan to displace the U.S. as the global hegemonic power. For Michael Pillsbury, the goal is to do so by 2049—the centennial anniversary of the CCP's victory in the Chinese civil war. While not necessarily disputing such a hypothesis, the real questions relate to evidence in support of such a thesis and when such a plan was put into place. According to Rush Doshi in *The Long Game*, the first phase of this plan was put into place shortly after what he calls the "traumatic trifecta"—the Tiananmen Square incident of 1989, the collapse of the Soviet Union, and the First Gulf War (1990–1991). One problem with this theory is that these events all shook China to its core and led the country to focus on domestic stability, economic growth, and fighting the three "evils" of terrorism, separatism (or "splittism"), and religious extremism. So, while China's leaders were focusing on holding together the world's most populous country, Doshi argues that they were secretly planning to overtake the U.S. as global hegemon. If China has been planning such a move, such a date seems too early, and the timing seems off historically.

The 18th National Congress of the Communist Party of China in 2012. Source: Voice of America/Public Domain

The limitation with almost all such studies is that they seek to derive either a Chinese way of war or Chinese strategy from theory; this is because China has not used its military in combat since 1979, making analysis of the Chinese way of war quite historical, most likely grossly outdated, or based upon theoretical assessments prone to manipulation and misleading positions.

DECEPTION, SWAGGERING, AND COGNITIVE WARFARE

Three ways in which this may be the case are deception, swaggering, and cognitive warfare. Deception is easy enough to understand: Chinese scholars write articles in leading journals, from the PRC's own *China Military Science* to the Council on Foreign Relations' *Foreign Affairs*, in which they make either sweeping pronouncements aimed as much at mobilizing the masses as in articulating esoteric military concepts to more quasi-official positions from *Zhongnanhai* (the compound that houses the offices of the Chinese Communist Party's ruling elite). Not all these publications are aimed at deception, but for a country with a very long history of deception, from Sun Tzu to Deng Xiaoping (whose 24-character strategy included "hide [your] capability, bide [your] time)," one can never be sure of the veracity of what one is reading or hearing.[9]

Swaggering is a bit more difficult of a concept to grasp and is much less intuitive. In international relations theory, swaggering is one of Robert Art's four ways of utilizing military force and focuses on displays and demonstrations of force to "enhance the national pride of a people or satisfy the personal ambitions of its ruler,"[10] as if only authoritarian states engaged in swaggering. Art, however, concedes that the "instrumental role of swaggering cannot be totally discounted because of the fundamental relation between force and foreign policy that obtains in an anarchic environment."[11] Thus, swaggering is something China is prone to engage in as it develops new military capabilities, such as anti-ship ballistic missiles and aircraft carriers.

Finally, there is cognitive warfare, which is itself in many ways a continuation of deception and swaggering. Two recent examples of this regard the publication of an article in the Chinese journal *Naval and Merchant Ships*, which in May 2020 released an 11-minute video, the supposed result of a "high-level wargame" about how the mainland would attack Taiwan and overtake the island in a matter of hours (incidentally, much of the video is footage of U.S. military equipment and personnel conducting exercises and actual military operations).[12] The second and more recent example relates to a Sino–U.S. conflict (purportedly over the defense of Taiwan or a military engagement in the South China Sea) and concluded that 24 Chinese hypersonic missiles could take out the USS *Gerald R. Ford* carrier strike group "with certainty."[13] According to a Taiwanese scholar at a prominent military think tank in Taipei, such "wargames" and their publicity are examples of cognitive warfare, and Taiwan must develop resilience against and strategies to counter such efforts.[14]

Taken together, deception, swaggering, and cognitive warfare are indeed central components of Chinese strategy today. Whether we are talking about a Chinese way of war or a strategy to displace the U.S. as global hegemon, both must rest on concrete actions and theoretical approaches to several areas, including the gray zone, deterrence, espionage, and cyberspace operations. Discussion of these topics help frame our understanding of the important dimensions of Chinese warfare.

THE GRAY ZONE: CONTINUUM OR TRIP WIRE?

Much of the research coming out of think tanks, academia, and even the military make the fatal mistake of "mirror-imaging" China with the use of such terms as the gray zone, hybrid warfare, political warfare, and even "competition below the threshold of armed conflict." These are Western conceptual constructs, and while Chinese scholars and policymakers may be familiar with them and even use the terms, it can be a dangerous business to assume that these terms describe how the Chinese themselves see the world.

Let's take the gray zone, for a start. According to most scholarship on the concept, the gray zone is "an environment in the greater global competitive space that is short of war but where tensions may be extremely high—and war may even be imminent" but not yet triggered.[15] This is an area of competition that is not quite peace but does not trip the UN Charter's 2(4) threshold of an armed conflict. This is not the same as political warfare, which diplomat George Kennan defined over half a century ago as "the employment of all the means at a nation's command, short of war, to achieve its national objectives."[16] One could say that gray zone tactics are a form of political warfare, but that political warfare is more expansive than the gray zone.

Whereas we in the U.S. regard China's gray zone activities as occurring in peacetime (or at least left of crisis and conflict), China regards its gray zone operations (or more accurately, its "non-war military activities"—the term used in China's strategic documents) as occurring as *part of* conflict, while still short of war. China also appears to have different levels of war, from armed conflict (low on the conflict continuum) to local war (in the mid-range on the conflict continuum) to large war (high on the conflict continuum). So, while the Sino-Russian border conflict of 1969 was below the armed conflict threshold, being merely a border skirmish, the Sino-Indian conflict of 1962 was just over it, with over 1,000 battlefield deaths.[17] This begs the question, where would an invasion of Taiwan be along the Chinese conflict continuum? Certainly such an attack would count as at least a local war, but, depending upon the international response, perhaps it would escalate into something greater. Ukraine, for example, meets the Chinese definition of a local war currently, even though the West is supplying the Ukrainian armed forces with everything from Javelins to F-16s.

Finally, there seems to be some evolution of the concept of "non-war military activities," which was first introduced under Hu Jintao around 2008. Now, under Xi Jinping, the specific wording has developed into the "peacetime employment of military force." According to two scholars from Maxwell Air Force Base's China Aerospace Studies Institute, Roderick Lee and Marcus Clay, these concepts were developed after close study of our own "military operations other than war," particularly as laid out in FM 100-5; JP 3-07; and JP 3-0. Lee and Clay's argument is that warfare needs to be understood as an escalatory concept, something not inherent—they argue—in the gray zone concept.[18] However, many scholars and analysts see the gray zone in terms of an escalatory ladder. Moreover, their escalatory framework includes a "gap" between non-war use of military activities and armed conflict; is that not just another way of saying the gray zone?

Vladimir Putin and Xi Jinping sign a joint statement on developing comprehensive partnerships and strategic cooperation in 2019. Source: Wikicommons

Whatever the value of the concept of a gray zone, there exists a conflict continuum (spectrum, in their words) and there exists a trip wire for the U.S. to respond and redlines ("bottom lines," actually) for China to act. Thinking in terms of competition below the level of armed conflict—with the assumption that our adversaries are abiding by some unwritten rule that they will not trigger an outright war—is a dangerous conceptual game with real-world ramifications.

> ***| Thinking in terms of competition below the level of armed conflict—with the assumption that our adversaries are abiding by some unwritten rule that they will not trigger an outright war—is a dangerous conceptual game with real-world ramifications. |***

Take the full-scale Russian invasion of Ukraine on February 24, 2022, as an example. Russia was operating in the gray zone all the way up until it wasn't. The Russians used *maskirovka* (military deception) and active measures to try and convince the West that they were not about to launch a full-scale war against Ukraine's very existence—and it worked until it was too late to react.[19] Taiwan is learning valuable lessons from this experience, with Ukrainian researchers working with the government, military, and think tanks in Taiwan to help draw lessons from the Ukrainian conflict for Taiwanese resilience, defense, and resistance operations.[20]

DON'T SAY GRAY: POLITICAL WARFARE AND HYBRID WARFARE

Just as the terms gray zone and political warfare are related but distinct, another term that gets thrown into the mix quite often is hybrid warfare. Quite simply, hybrid warfare is warfare by many means simultaneously. While the gray zone is a conceptual and possibly physical realm where malign activities take place, hybrid warfare is a method of war. Indeed, hybrid warfare can occur in the gray zone but also outside it, in the realm of large-scale combat operations (LSCO). In short, hybrid warfare is a form of war that includes all means at a state's disposal—from covert operations to nuclear war.[21]

This distinction is important, because the term political warfare is currently ascendant; indeed, a very influential study was recently released by the Center for Strategic and International Studies (CSIS) and a team of researchers led by Seth Jones. This study, Competing without Fighting: China's Strategy of Political Warfare, is a remarkable research effort by a very talented team of scholars. Unfortunately, its underlying assumptions are open to serious debate. The title says a lot—"competing without fighting." Is this truly China's grand strategy? As anyone who follows Chinese strategy knows, this is the major question to which there is no simple answer. Moreover, China's leaders, policymakers, and military leadership hate the term competition and view it as a Western nicety at best and outright deceptive stratagem at worst.

Additionally, the CSIS study subsumes all Chinese activity under the concept of political warfare. While there is a nice and handy chart in Figure 2.1 on page 10 of the report, which lists and defines everything from "daring to struggle" to "military operations other than war," the underlying assumption is that China is engaged in a competition in which it does not seek to cross the threshold of war, that cognitive tripwire between the gray zone and LSCO. Upon what evidence is such a position based? Just because China has been engaging in gray zone tactics for the past two decades does not mean they will continue to do so indefinitely. Russia, again, provides an excellent example of what can happen when such an assumption is made.

Jones et al. conduct a very comprehensive review of the literature—both in English and Chinese—but they do not always engage with the arguments that disagree with their assessments. The most important counterargument, which the team references multiple times but never actually engages, is the work of the two scholars from Maxwell Air Force Base's China Aerospace Studies Institute mentioned previously, Lee and Clay, who argue that there is no gray zone in Chinese military thinking, only a spectrum of conflict. This is a significant weakness of the CSIS study, and one wonders why they are citing the work when its argument runs counter to their own—and they do not address it.

Secondly, Jones et al. assume just because the PRC has not yet gone over that elusive conflict threshold that they do not intend to in the future. Again, this is the question that keeps students of Chinese strategy up at night. Is China building capacity and biding its time, or will they stay under this illusory threshold (and if so, why?). Is it some norm that has taken root in the international system, as constructivists might have us believe,[22] or is it fear of a massive retaliatory response, as realists would perhaps argue?[23] When it comes to domestic unrest, the CCP apparently has no qualms about calling out the PLA—why would we assume they would be more restrained in the international environment?

| When it comes to domestic unrest, the CCP apparently has no qualms about calling out the PLA—why would we assume they would be more restrained in the international environment? |

The PRC's perspective is one in which they will go to war to protect their interests; there is no ongoing competition preventing them from resorting to military force, especially regarding their "renegade" province of Taiwan. To think otherwise is to mirror-image our own concepts onto China, the fatal mistake mentioned earlier.

DETERRENCE AND COMPELLENCE

According to Dean Cheng, the Chinese term *zhanlue weishe* (strategic deterrence) includes both the concepts of deterrence (i.e., preventing a state from doing something it otherwise would have done) and compellence (i.e., getting a state to do something it otherwise would not have done, or in the words of the famous Prussian, "an act of force to compel our enemy to do our will"[24]). Both are part of the term Cheng refers to as *dissuasion and coercion*,[25] but these are just two other terms that describe the same phenomena of deterrence and compellence. Whatever the terms used, China's use of *weishe* is focused on achieving certain political goals (i.e., compellence), not just deterring certain actions.

In the U.S., most of the time when one talks about deterrence the discussion is centered on nuclear deterrence—trying to deter an adversary from resorting to a nuclear option in a conflict, or perhaps even engaging in conventional war with a nuclear power, because that state has the power to defend itself with its nuclear arsenal. In China, the term is much broader and ranges across the military spectrum of engagement. Dennis Blasko argues quite convincingly that it is not just about nuclear deterrence, as in the United States. According to his analysis, it is part of China's "strategically defensive" defense posture of "active defense," and seeks to defend China's sovereignty, territorial integrity, and national interests from a variety of threats including "terrorists, separatists, and extremists" and to limit conflict escalation in all domains of modern combat.[26] Additionally, the Chinese view is one of both to deter and win wars – so if deterrence fails, one can move right into the warfighting phase (again, this runs counter to the "below the threshold of armed conflict" cognitive tripwire idea).

The Dongfeng-31 nuclear missile launcher, a three-stage solid fuel rocket on display at the Chinese People's Revolutionary Military Museum in 2017. Source: Tyg728/Wikicommons

The U.S. pays close attention to China's increased and indeed increasing military capabilities. It rarely, however, pauses to consider the deterrence effect of these same developments, as Blasko points out.[27] Continuing, he argues that while China has historically pursued a limited nuclear deterrent, that may be changing as part of the PLA's growth and larger modernization program launched under Xi. This includes new intercontinental ballistic missiles (ICBM), submarine-launched ballistic missles (SLBM), and even a new strategic bomber.
Again, according to Cheng,[28] the U.S. must maintain both a modern nuclear force and nuclear enterprise to deter the PRC from threatening either the U.S. or its key allies in Asia.

ESPIONAGE

China is of course the land of Sun Tzu, according to whom the acme of skill is to subdue one's enemy without waging war. He writes, "To win one hundred victories in one hundred battles is not the acme of skill. To subdue the enemy without fighting is the acme of skill."

| "To win one hundred victories in one hundred battles is not the acme of skill. To subdue the enemy without fighting is the acme of skill." - Sun Tzu |

That bit of text is often quoted with little attention to the fact that the *Art of War* devotes an entire section to espionage, something not lost on the PRC as a whole. From state-owned enterprises (SOEs) to universities, not to mention the CCP and units of the PLA, espionage is a booming business for China. Indeed, China has been employing spies both domestically and abroad for millennia, even long before Sun Tzu. During this time, they have not only perfected certain age-old techniques but have also expanded into all realms to include cyber, satellites, and of course balloons!

Let's first address human intelligence (HUMINT) collection methods. Here we have Confucius Institutes and individual scholars in the West along with students and others. While some are professional intelligence collectors, most simply believe they are being patriotic citizens and supporting their nation. Others are induced by financial gain to betray their host nation. For others, it's a component of their career, and they work in some of these same domains.

Nicholas Eftimiades offers the most comprehensive analysis of all-source intelligence collection efforts. Analyzing nearly 600 cases of espionage over the past 20 years, he has determined that the Chinese are taking a "whole-of-society" approach to intelligence collection, and that its efforts have greatly expanded over the past 20 years. Moreover, nearly half of

China's worldwide collection efforts target space and military technologies. Additionally, more than 90 percent of collection is done by ethnic Chinese, with over 80 percent males. Finally, more than half of espionage efforts and covert action campaigns identified are targeted against Taiwan.

CYBERSPACE OPERATIONS

In addition to HUMINT collection, one of the main tools employed for espionage is cyber. Cyber applications are a central mode by which information is applied as a tool of warfare. China conceives of "informationized warfare,"[29] with the space and cyber domains described as becoming the "commanding heights" of strategic competition.[30] Make no mistake: despite claims to the contrary, China sees itself currently engaged in information warfare against the United States.

| Make no mistake: despite claims to the contrary, China sees itself currently engaged in information warfare against the United States. |

This is not a war *per se*, but it is understood as intelligence preparation of the battlefield in the very least. In fact, strategic competition is playing out principally in the cyber realm. The military application of information as an instrument of war—in isolation and in conjunction with other tools—is a central component of China's modern approach to warfare.

Armed forces of the People's Republic of China, July 2000. Source: defenselink.mil

China sees information operations as central to its conception of competition in cyberspace. In fact, the CCP—and by extension its defender, the PLA—views information operations via space, cyber, and electronic warfare as the "tip of the spear" in any future conflict to shape the narrative and obtain information superiority, thereby paralyzing a more powerful enemy.

China further expanded its comprehension of cyberspace with the creation of the Strategic Support Force (SSF) in 2015. Some analysts view the SSF as an enhanced Chinese counterpart to U.S. Cyber Command (USCYBERCOM). The SSF not only focuses on the traditional D4M operations (that is, deny, degrade, disrupt, destroy, and manipulate), as does USCYBERCOM, but has also added space, electronic, and psychological warfare to the mix. Housing these different but complementary cyber-enabled capabilities within the same command is expected to create synergies that these capabilities could not achieve on their own. Moreover, having a suite of functions under the same command during peacetime will give the CCP and the PLA the ability to seamlessly transition to an integrated campaign during wartime.

While China has not openly published a cyberspace strategy, scholars and practitioners are in widespread agreement as to the CCP's aims. It seeks to control the flow of information to and within China to ensure domestic stability (and halt the efforts of "splittists" who seek the disintegration of the PRC) and preserve economic growth through commercial espionage. By controlling dissent and driving economic growth, the CCP is attempting to ensure that it maintains its grip on power.

In regard to maintaining economic growth, China's operations against economic targets and the commercial sector are viewed by former USCYBERCOM commander General (Ret.) Keith Alexander as "the greatest transfer of wealth in history."[31] While the CCP continues to claim that cyber economic espionage is not the work of the government but rather criminal elements within China, the cybersecurity group FireEye has been able to identify with a high degree of certainty that there are at least 10 advanced persistent threats (APTs) operated by the CCP, nine of which focus on industrial espionage.

In addition to these government-supported APTs, China also has a very large patriotic hacker community that it can mobilize when needed. Due to the extensive cyberspace dragnet that the CCP has put in place, the government is aware of the activities of these hackers and can stop them when it so desires. However, the CCP has also employed this hacker network as an extension of the state while at the same time retaining plausible deniability since these patriotic hackers are not formally part of the state apparatus.

The CCP's firm hold on power is a function of both its ability to maintain economic growth and its control over the flow of information in China. This explains why both economic cyberspace espionage and information operations have seen increased investment by the Chinese military in recent years.

Moreover, there is a coupling of the state and private enterprise as seen in China's

Cybersecurity Law and National Intelligence Law, the latter of which necessitates that "any organization and citizen shall, in accordance with the law, support, provide assistance, and cooperate in national intelligence work, and guard the secrecy of any national intelligence work that they are aware of."

CHINA'S EVOLVING WAYS OF WARFARE

As mentioned in the introduction, Timothy Thomas has done the most to articulate a Chinese way of war as it has evolved over the past two decades, based upon in-depth analysis of Chinese military and strategic documents and journals, leading to his conclusion that the Chinese way of war has changed "dramatically from what it was" only 20 years ago.[32] During this time, China has engaged with modern concepts of warfare and even implemented doctrinal changes to its way of war. While Thomas' study is extremely valuable and important, the fact remains that the PLA is untested in even their ability to shoot, move, and communicate in combat—not to mention conduct multi-domain joint operations in an LSCO environment. Since we cannot evaluate China's military performance to determine their capabilities and assess (and counter) their military operations, we are largely left—like Thomas and others (including M. Taylor Fravel,[33] Peter Mattis,[34] and Rush Doshi)—to study Chinese military concepts and doctrine. But when it comes to espionage and offensive cyberspace operations (OCO), not to mention some dimensions of cognitive warfare, there are concrete actions that we can study and assess. We must be careful, however, not to equate all of this as a Chinese way of war. Though such things certainly comprise components of it, any true assessment of a contemporary Chinese way of war will hopefully not be conducted, because if it does it means that diplomacy and deterrence have failed and that conflict with the world's largest military has occurred. This is something we must try our best to avoid, while still defending American interests at home and abroad.

About the Author

Dr. Christopher Marsh is the Chair of the Joint Special Operations Master of Arts (JSOMA) program at Fort Bragg, a division of the College of International Security Affairs of the National Defense University. Marsh is an expert on Russian approaches to special operations, irregular warfare, military strategy, and foreign policy. Prior to joining NDU, Dr. Marsh was the Director of the Department of Strategic Research & Analysis at the Joint Special Operations University, US Special Operations Command (USSOCOM). He previously held positions at the US Army School of Advanced Military Studies (SAMS), US Air Force Special Operations School, and Baylor University.

NOTES

1. Russell Weigley, *The American Way of War: A History of United States Military Strategy and Policy* (University Press, 1973).

2. Brian McAlister Linn, "The American Way of War: An Overview," *Historically Speaking* 11, no. 5 (2010): 22-23.

3. James Jay Carafano, "Wending Through the Way of War," *Historically Speaking* 11, no. 5 (2010): 25-26.

4. Geoff Babb, "China's Military History and Way of War: A Backgrounder," *Military Review* (March 2023): 1-6.

5. Timothy Thomas, "The Chinese Way of War: How Has it Changed?" (MITRE, June 2020).

6. Peter Lorge, "The Many Ways of Chinese Warfare," *Historically Speaking* 11, no. 5 (2010): 24-25.

7. Michael Pillsbury, *The Hundred-Year Marathon: China's Secret Strategy to Replace America as the Global Superpower* (2016).

8. Rush Doshi, *The Long Game: China's Grand Strategy to Displace American Order* (Oxford University Press, 2023).

9. Ralph Sawyer, *The Tao of Deception: Unorthodox Warfare in Historic and Contemporary China* (Basic Books, 2007).

10. Robert Art, "To What Ends Military Power?" *International Security* 4 (Spring 1980): 9.

11. Art, "To What Ends Military Power,' 9.

12. Kristin Huang, "How Would Mainland China Attack Taiwan? A Video Outlines One Scenario," *South China Morning Post* (May 20, 2020). https://www.scmp.com/news/china/military/article/3085302/how-would-mainland-china-attack-taiwan-video-outlines-one

13. Stephen Chen, "Chinese Scientists War-Game Hypersonic Strike on US Carrier Group in South China Sea," *South China Morning Post* (May 23, 2023). https://www.scmp.com/news/china/science/article/3221495/chinese-scientists-war-game-hypersonic-strike-us-carrier-group-south-china-sea

14. Author interview, name and institution withheld. Taipei, Taiwan: November 16, 2023.

15. Christopher Marsh, "The Gray Zone and Hybrid Conflict: A Conceptual Introduction," In Mitt Regan and Aurel Sari, eds. *Hybrid Threats and Gray Zone Conflict: The Challenge to Liberal Democracies* (Oxford University Press, 2024): 31-32.

16. George Kennan, "Policy Planning Staff Memorandum" (Washington, DC: May 4, 1948). https://history.state.gov/historicaldocuments/frus1945-50Intel/d269

17. Allen Whiting, *The Chinese Calculus of Deterrence: India and Indochina* (University of Michigan Press, 1975).

18. Roderick Lee and Marcus Clay, "Don't Call it a Gray Zone: China's Use-of-Force Spectrum," *War on the Rocks* (May 9, 2022). https://warontherocks.com/2022/05/dont-call-it-a-gray-zone-chinas-use-of-force-spectrum/

19. Christopher Marsh, "*Maskirovka* and the Gray Zone: The Utility of the Gray Zone Concept after Russia's Invasion of Ukraine," in James Kiras and Martijn Kizten, eds., *Into the Void: Special Operations Forces After the War on Terror* (Oxford University Press, 2024).

20. Author interview, names and institutions withheld to protect the operation. Taipei, Taiwan: November 14, 2023.

21. Marsh, "The Gray Zone and Hybrid Conflict," 41-42.

22. For constructivist perspectives, see the seminal work of Nicholas Onuf, *World of our Making: Rules and Rule in Social Theory and International Relations* (University of South Carolina Press, 1989).

23. For realist perspectives, see Kenneth Waltz, *Theory of International Politics* (Addison-Wesley, 1979) and John Mearsheimer, *The Tragedy of Great Power Politics* (W.W. Norton, 2001).

24. Carl von Clausewitz, *On War*, edited and translated by Michael Howard and Peter Paret (Princeton University Press, 1989).

25. Dean Cheng, "Chinese Views on Deterrence," *Joint Force Quarterly* 60, 1st quarter (2017). See also Dean Cheng, "An Overview of Chinese Thinking about Deterrence," in NL ARMS Netherlands *Annual Review of Military Studies: Deterrence in the 21st Century – Insights from Theory and Practice*, Frans Osinga and Tim Sweijs, eds. (T.M.C. Asser Press, 2021): 177-200.

26. Dennis Blasko, "'Peace Through Strength': Deterrence in Chinese Military Doctrine," *War on the Rocks* (March 15, 2017).

27. Blasko, *Peace Through Strength.*

28. Cheng, "Chinese Views on Deterrence."

29. B.A. Friedman, "Finding the Right Model: The Joint Force, the People's Liberation Army, and Information Warfare," *Journal of Indo-Pacific Affairs* 6, no. 3 (March–April 2023): 1–17.

30. U.S. Department of Defense, *Military and Security Developments Involving the People's Republic of China 2023: Annual Report to Congress* (Department of Defense, 2023), 36.

31. Keith Alexander, *Gen. Alexander: Greatest Transfer of Wealth in History* (American Enterprise Institute, 2012).

32. Timothy Thomas, "The Chinese Way of War: How Has it Changed?" (MITRE, June 2020).

33. M. Taylor Fravel, *Active Defense: China's Military Strategy Since 1949* (Princeton University Press, 2019).

34. Peter Mattis, *Analyzing the Chinese Military: A Review Essay and Resource Guide on the People's Liberation Army* (CreateSpace, 2015); see also Peter Mattis and Matthew Brazil, *Chinese Communist Espionage: An Intelligence Primer* (Naval Institute Press, 2019).

II. Russia's "Special" Way of War: Special Operations Forces, *Spetsnaz*, and Irregular Warfare

By Christopher Marsh

ABSTRACT: While in the U.S., the process of defining irregular warfare is an ongoing and contentious affair, Russia does not group its mission sets—such as counterterrorism, counterinsurgency, and unconventional warfare—together in this way. This section examines the development of Russian special military units over time, the evolution of these units, and the creation of their Special Operations Forces Command (KSSO). To shed light on the similarities and differences between Russian and American approaches to irregular warfare, the section includes four case studies: two counterterrorism operations and two unconventional warfare missions. While some similarities were identified in terms of unit mission and core activities, similar does not mean "the same," particularly when it comes to approaches between the West and Russia to conducting special operations and irregular warfare.[1]

INTRODUCTION

In the U.S. and the West more broadly, defining irregular warfare (IW) is an ongoing and contentious affair.[2] Long considered "a violent struggle among state and non-state actors for legitimacy and influence over the relevant populations," the Joint Staff J-7 (Directorate of Joint Force Development) at the Pentagon recently released an updated official definition of irregular warfare. According to them, IW should now be considered across the Department of Defense as "a form of warfare where states and non-state actors campaign to assure or coerce states or other groups through indirect, non-attributable, or asymmetric activities."[3]

In Western Europe, other definitions prevail. A leading effort was made in this regard by scholar Martijn Kitzen, who calls it "a violent struggle involving non-state actors (including violent armed groups acting as state proxies) and states with the purpose of establishing power, control, and legitimacy over relevant populations."[4] This definition, which resembles the traditional U.S. definition of IW, can be broken down even more succinctly—as was done by James Kiras, who earlier identified terrorism and insurgency as being at the crux of irregular warfare.

If terrorism and insurgency are the main modes of irregular warfare, then actions taken to counter such warfare, in the form of mission sets, can be considered irregular warfare as well. Thus, counterinsurgency and counterterrorism are at the center of the U.S. approach to IW.

| If terrorism and insurgency are the main modes of irregular warfare, then actions taken to counter such warfare, in the form of mission sets, can be considered irregular warfare as well. Thus, counterinsurgency and counterterrorism are at the center of the U.S. approach to IW. |

Foreign internal defense (which seeks to aid and support a friendly nation in its fight against terrorism and insurgency), unconventional warfare (which seeks to support a resistance movement in its fight against an occupying force or oppressive regime), and stability operations (efforts which seek to bring peace and security to a region or country, particularly in a post-conflict environment) round out the list of operations typically seen as comprising irregular warfare, that is, before the addition of other—mostly information-related activities—to the irregular warfare toolkit.[6]

Sandor Fabian and Gabrielle Kennedy concluded in their study that most of the institutions they surveyed linked special operations with irregular warfare.[7] As they phrased it, this "linkage is bolstered by the idea that [Special Operations Forces (SOF)] are considered as either the primary tool of irregular warfare or a major component in the practice."[8] As U.S. Army Special Forces officer Lt. Col. Stephan Bolton puts it, "in many instances, SOF tend to be the best instrument to conduct the military aspects of an IW campaign."[9] In the U.S. in particular, and the West more generally, IW campaigns require whole-of-government efforts, and various actors across the interagency play important roles. When it comes to the military aspects of IW, however, these efforts are led by SOF.

In Russia, things are a little different. The mission sets are largely the same as the traditional mission sets of U.S. SOF-led activities to counterinsurgency, terrorism, and instability, with the addition of a strong emphasis on intelligence activities and covert and clandestine operations. This is because Russia is no stranger to terrorism, insurgency, and separatism, fighting all three in combination throughout its history. Beginning with the period of the Soviet Union, the Kremlin became involved in all mission sets identified above, from Chile and Cuba to Angola and Vietnam. And this is without mentioning the domestic struggles with each, from terrorist cells such as the 19th century's *Narodnaya Volya* (the People's Will) to Chechen terrorists and insurgents following the collapse of the Soviet Union.

Despite its long history in fighting irregular warfare, inside Russia there is really no debate on what is or is not irregular warfare—in fact, the term itself is not used. From writings on

| Despite its long history in fighting irregular warfare, inside Russia there is really no debate on what is or is not irregular warfare—in fact, the term itself is not used. |

special operations theory[10] to Gerasimov's insights into non-linear and asymmetrical warfare,[11] the debates seem to center on how Russia can counter its adversaries rather than on fleshing out concepts that describe modes of war with which they are intricately familiar. Moreover, there is no discussion over winning "hearts and minds" or gaining legitimacy. Rather, legitimacy is something that is constructed post-facto to preserve national interests. At other times, parsing out the type of conflict does not seem to matter much to the powers that be in the Kremlin.

As for the term itself, Russians do not use the word *neregulyarnaya voina* (irregular warfare) or *neregulyarnye voiny* (irregular wars) themselves, with the term simply understood as a U.S. "doctrinal construct." In fact, they have a rather accurate understanding of our conceptual framework for IW as laid out in the unclassified annex to the 2020 National Defense Strategy, though they argue that "events in which America is directly involved are presented in a false and distorted light."[12]

Russia views IW as a Western construct and never mentions its own operations in an IW framework. Nevertheless, the country has also been faced with and engaged in activities and situations those in the U.S. would label IW, such as insurgency, terrorism, and instability, and thus have engaged in counterinsurgency and counterterrorism along with foreign internal defense and unconventional warfare. Russia's history with stability operations as understood by the U.S. is more storied and will not be examined in detail here. The remainder of this section will examine the forces that Russia employs to conduct counterinsurgency, counterterrorism, foreign internal defense, and unconventional warfare—that is, primarily its SOF, *spetsnaz*, and conventional forces (including mechanized troops and artillery). Historically, spetsnaz referred to special operations units in the Soviet Union's military intelligence service. However, the focus here is on Russia's modern SOF and spetsnaz – units that conduct some of the country's most secret and sensitive operations.

| Russia views IW as a Western construct and never mentions its own operations in an IW framework. |

Again, as Bolton explains, "in many instances, SOF tend to be the best instrument to conduct the military aspects of an IW campaign."[13] This is largely true for the Russian armed forces as well, particularly when it comes to counterterrorism. But in Russia there are also the spetsnaz units—some of which are roughly equivalent to U.S. and Western tier 1 units—which also conduct counterterrorism and counterinsurgency. If we want to understand Russian approaches to IW, we must first begin with a solid foundational understanding of Russia's "special" way of war.

THE ROOTS OF RUSSIAN SOF

The roots of Russia's SOF rest with the Soviet spetsnaz units. The term comes from *voiska spetsial'nogo naznacheniya*, or special designation troops. Though most spetsnaz units are certainly elite and distinct from conventional forces, they are not the same thing as SOF, despite the frequent confusion by commentators. Nor are they simply the equivalent to U.S. Army Special Forces ("Green Berets"), despite the fact that until recently they have nearly been synonymous terms. To translate spetsnaz as Special Forces is a gross—and misleading—oversimplification and will only lead to confusion since the latter are only one part of the U.S. Army SOF community (the others being psychological operations, civil affairs, rangers, and 160th Special Operations Aviation Regiment). Each other U.S. service has its own special operators (i.e., Air Commandos, Navy SEALs, MARSOC Raiders). To make things simple, in Russia SOF are spetsnaz, but not all spetsnaz are SOF.

There is even greater diversity among Russian spetsnaz units than there is within the U.S. SOF community, because unlike in the U.S. and most NATO countries, where special operations units are limited to departments or ministries of defense, this is not the case in Russia. There, the ministry of interior, Federal Security Service (FSB, the offspring of the KGB), and other divisions have their own spetsnaz units, especially for intelligence collection and diplomatic security. As Lester Grau and Chuck Bartles clearly explain it, "The word 'special' [in spetsnaz] is used in a very broad way that can indicate that the unit has a very narrow area of specialization, such as signals intelligence, engineering, reconnaissance, etc.; or the unit is experimental or temporary in nature; or the unit conducts tasks of special importance such as sensitive political or clandestine operations. This broad usage of the term means that 'spetsnaz' cannot be thought of as equating to the Western concept of Special Operations Forces (SOF)."[14]

This is certainly true, and I would argue that this opinion is even held by the Russian Ministry of Defense, which is why they are now developing their own SOF units, sily spetsial'nykh operatsii, or SSO. But this is a relatively new phenomenon and one that is only beginning to be understood, as discussed below.

SOVIET SPETSNAZ

Before Joseph Stalin had him executed in 1938 during the Great Purge, Russian military theorist Mikhail Svechnikov proposed the creation of a small, highly trained military unit that could work behind the scenes to obstruct enemy operations. This idea was adopted by military officer Ilya Starinov, who became known as the "grandfather of the Spetsnaz."[15] The earliest actual spetsnaz units were formed in 1950, though it is worth mentioning the role of special-type units during the Russian Revolution (1917–1921). As John Dziak observes, while there is no "unbroken link" between such special units and later spetsnaz; the former are often viewed as the conceptual forerunner to the latter.[16] All spetsnaz and SOF trace their heritage back to the *razvedchiki* (reconnaissance scouts) of Russia's Revolutionary War. These soldiers served as behind-the-lines commando elements, providing much-needed intelligence on enemy positions and capabilities, a critical function given the nature of the war the Red Army was fighting.

The most significant of the special units of the Revolutionary period was the ChON, or special purpose brigades (*Chasti osobogo naznachenya*), which by the end of the war numbered some 40,000 soldiers.[17] This was an elite unit formed from broader categories of existing specialized forces.[18] As an ideological movement, the Bolsheviks recruited from among the most loyal and "believing" Communist soldiers to serve in the ChON. Their mission was two-fold. First, they acted as counterintelligence agents at home and in occupied territories, ensuring that those in the ranks were in fact loyal Communists. And secondly, they served as sabotage, assassination, and agitation assets behind enemy lines.[19] These were the first units to be given the label "special designation."[20]

The Russians started experimenting with elite reconnaissance and sabotage units during the Spanish Civil War (1936–1939) and employed such units in the Soviet–Finnish "Winter War" (1939–1940), as well as in Romania, Yugoslavia, and Belarus during the Second World War.[21] But the real roots of Russia's special operators lay with the partisan fighters who fought the German forces that invaded the Soviet Union in World War II. These loosely organized units were tied to the Partisan Directorate, which officially was under the Supreme High Command (*Stavka*). Though they did not have the label spetsnaz, they were resistance fighters of all types (including naval infantry and even sappers) who organized as best as they could to halt the German advance into the Soviet motherland, to support conventional forces when the opportunity arose, and to organize guerilla operations if behind enemy lines. This latter was a crucial role for the partisans during World War II, or, as it is known in Russia, the Great Patriotic War.

The Russians viewed their partisan experience in the Great Patriotic War as a highly effective "unconventional" adjunct to their conventional operations. The Soviets claim that during the war partisans killed, wounded, or took prisoner "hundreds of thousands of German troops, collaborators, and officials of the occupation administration." They also claim to have derailed more than 18,000 trains and destroyed or damaged thousands of trains and tens of thousands of rail cars. These operations supposedly had a devastating effect on German morale and forced the Germans to deflect much-needed resources from frontal operations.[22]

The real era of the spetsnaz, however, was the Cold War. The Soviets developed the first spetsnaz units in 1950—two years prior to the formation of U.S. Army Special Forces.[23] In the Soviet Union's attempt to keep up with U.S. military developments, the Kremlin put a tremendous amount of effort into their spetsnaz and special operations capabilities as they tried to outflank the West at all levels across the political-military spectrum.[24] Methods included covert and clandestine activities of all kinds, with particular emphasis on insurgencies, active measures/deception operations,[25] psychological operations, and even support for terrorist organizations around the world.[26] Spetsnaz squads were trained to be dispatched to reconnoiter in the enemy rear area to locate nuclear delivery systems, enemy forces, headquarters, airfields, signal sites and other important targets.[27] They were also trained for and prepared to conduct raids, ambushes, sabotage, and surgical strike operations against key enemy personnel and infrastructure, and to conduct reconnaissance/intelligence operations at the same time.[28] Moscow's capabilities were augmented by proxy forces where possible, especially when direct Soviet involvement might be imprudent.

Despite this experience, beginning in 1979 Soviet spetsnaz found themselves operating in a very different environment—Afghanistan. Spetsnaz were involved from the very beginning, with Col. Vassily Kolesnik tasked with recruiting a "Moslem battalion" of spetsnaz of Tajik, Turkmen, or Uzbek nationality, with the idea being that its members could pass themselves off as Afghans. The unit had been assembled by June 1979 and began its training with the spetsnaz unit *Al'fa* (more on this unit below), though the "Moslem battalion" reportedly did not know what their mission was going to be. During the December 1979 invasion of Afghanistan, their role was to penetrate the imperial palace and facilitate the infiltration of two other spetsnaz units–*Kaskad* (Cascade) and *Zenit* (Zenith). Together, this combined unit was the lead in conducting "Operation 333," killing President Hafizullah Amin and those loyal to his rule and setting the stage for the Soviet invasion of Afghanistan.[29] See Figure 6.1.

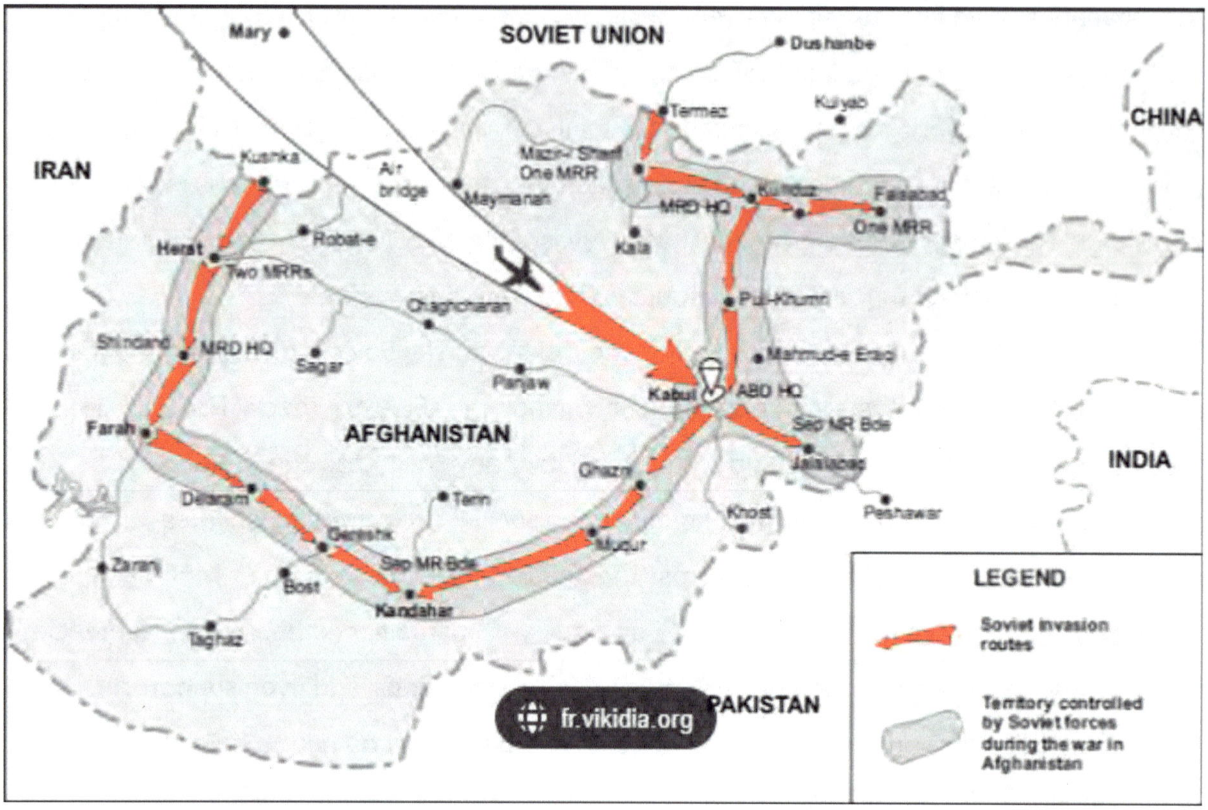

Figure 6.1: Soviet invasion of Afghanistan, 1979. Source: Wikimedia Commons.

As the war in Afghanistan dragged on, spetsnaz units found themselves carrying out a variety of missions, many of which were unlike those for which they had been trained against the West. By October 1980 they were increasingly being brought in to supplement the Limited Contingent of Soviet Forces in Afghanistan (the official name of the Soviet operation in Afghanistan). There, their skills for such things as reconnaissance, ambush, and rapid reaction were put to regular use. Some spetsnaz also became involved in so-called "Stinger hunting missions," once the U.S.-made, shoulder-fired weapon was introduced into the war zone. These "Stinger hunters" sought out Mujahedeen equipped with such weapons and did their best to destroy both the weapon and its operator.[30]

Overall, during their 10-year struggle against the Mujahideen, the Soviets deployed a significant capability based around their highly trained spetsnaz units. A critical examination of their performance by Tony Balasevicius, however, reveals that Soviet spetsnaz were often misemployed and, as a result, were unable to make a significant contribution to the outcome of the war.[31] This misemployment, Balasevicius argues, was based on the Kremlin's desire to focus spetsnaz efforts on propping up their under-prepared conscript army rather than trying to identify how spetsnaz might contribute to a grander operational vision for success. Thus, in the end, they served as key enablers but were not tasked with the proper missions and given the correct resources, including an appropriate amount of time to meet operational and strategic demands.[32]

SPETSNAZ IN THE POST-SOVIET ERA

The two wars in Chechnya were the crucible for Russia's spetsnaz that Afghanistan had been for Soviet predecessors. Often employed simply as elite infantry, they were also tasked with targeting and eliminating high value targets, in this case rebel leaders of the Chechen insurgency that sought to liberate the territory from the Russian Federation. These rebel leaders were not just untrained farmers, however, and included in their own ranks many former Soviet/Russian soldiers and even some former spetsnaz. This obviously greatly problematized the Chechen wars, and Russia's losses were considerable.

At the start of the First Chechen War (1994–1996), spetsnaz found themselves in their usual role of battlefield reconnaissance. But once the illusion of a quick victory disappeared—which did so very quickly—many spetsnaz found themselves being used as shock troops. In fact, Moscow sought capable fighting forces from wherever it could, including naval spetsnaz and even the ministry of the interior's Otryad Mobil'nyy Osobogo Naznacheniya (OMON) counter-riot units. Moscow was greatly relieved when retired Gen. Alexander Lebed was able to negotiate a ceasefire in 1996 in the wake of the presidential elections, after he himself made a strong run for the presidency.[33]

The Second Chechen War (1999–2002) was different from the first in many ways, mostly as what had been a nationalist insurgency had developed into a Muslim religious war for independence.[34] The other difference was that Moscow's invasion of Chechnya in 1999 was well thought out and much more successful than the first. The ground invasion was preceded by an initial air campaign, softening targets and preparing troops for effective operations. From the fall of 1999 through 2009, Moscow directed a sustained campaign that effectively destroyed the Islamic insurgency in Chechnya and reasserted Russian control of the region.[35] The spetsnaz performed their usual roles of deep reconnaissance, interdiction, intelligence gathering, and acting as a rapid reaction unit. In contrast to the First Chechen War, this time spetsnaz were used in their trained roles for the most part and performed well. They were also supporting and supported heavy units, such as artillery and armor, that blasted the capital city of Grozny into rubble.

One of the most notorious of spetsnaz units, not just during the Chechen wars but throughout its existence, is that of the Glavnoje Razvedyvatel'noje Upravlenije (GRU), the Soviet

Union's primary intelligence directorate. Most direct action and reconnaissance spetsnaz units have historically been under the command and control of the GRU, which reports directly to the General Staff.[36] During the Cold War, their mission sets focused on deep reconnaissance and countering weapons of mass destruction, particularly battlefield tactical nuclear weapons, and especially in the event of a war in which Soviet territory was immediately overrun. These units would then be stay-behind forces, conducting sabotage operations and diversion activities and countering tactical nuclear weapons.

Not all spetsnaz fall under the military (or Ministry of Defense). Other very significant spetsnaz units were (and still are) found in other organs of the state. In 1974 the Soviet Union created one of its most elite tier 1 special operations units, simply known as "Group A" (later becoming known as Al'fa, or Alpha). Al'fa was the premier counterterrorist unit in the Soviet Union under the control of the KGB (Committee on State Security), as opposed to most spetsnaz, which fall under the GRU. Today, Al'fa still exists and is still just as elite (the "best of the best," reports *RIA Novosti*), though it is now under the reorganized and renamed KGB, known as the FSB (Federal Security Service).[37] Al'fa remains the premier counterterrorist (antiterror) organization in Russia, thus playing an important role in Russia's approach to irregular warfare.

Also originally falling under the KGB and just as elite and lethal is *Vympel* (meaning "pennant"). Vympel, which also has a counterterrorist role, is most well-known for its unconventional warfare and covert action skills. Again, the roots of this spetsnaz unit date back to the days of the USSR, though in this case it was to deal with the military coup in Chile in September 1973, when the socialist government of Salvador Allende, one friendly to the Soviet Union, was overthrown. "Group V" was created to go in and rescue the General Secretary of the Communist Party of Chile, Luis Corvalan. The special operation was not needed in the end as Corvalan was released in a prisoner exchange, but the incident made it clear that the Soviet Union needed such a unit that could always be on alert and able to carry out covert operations anywhere in the world.[38] By the early 1990s the unit numbered some 300–500 persons and fell under the Ministry of Internal Affairs (MVD).[39] Highly secretive, Vympel plays not only a counterterrorist role but also an unconventional warfare role, though this mission is never mentioned in connection with the unit by the Russian authorities—who deny conducting such operations.[40]

By the time of the collapse of the Soviet Union, the term spetsnaz, when used in reference to the Soviet Union's elite combat units, usually referred to the GRU's Spetsnaz Brigades and Combat Swimmer units (roughly the GRU's naval reconnaissance force with a sabotage/anti-sabotage capability), the Russian Airborne's 45th Spetsnaz Regiment (later brigade), or select elite anti-terrorism units (such as Al'fa and Vympel).[41]

RUSSIA'S SOF JOIN THE FIGHT

On March 6, 2013, General Valery Gerasimov, Chief of the Russian General Staff, announced the creation of Russia's own Special Operations Forces Command (*Komandovanie sil spetsial'nalnykh operatsii*, or KSSO) and Russian special operations forces (or *sily spetsial'nykh operatsii*, or SSO). "Having studied the practice of the formation, training, and application of special operations by the leading foreign powers," he stated, "the leadership of the Ministry of Defense has also begun to create such forces." He continued: "We have set up a special command, which has already begun to put our plans into practice as part of the Armed Forces training program. We have also developed a set of key documents that outline the development priorities, the training program, and the modalities of using these new forces."[42] Minister of Defense Sergei Shoigu also commented on this momentous occasion in developing a Special Operations Forces Command. He said Russia was following "the general trend [in the world] toward specialization and enhanced mobility."[43] Indeed, in their own observation of this event, one Russian news agency pointed out that Russia was only 26 years behind the U.S. in developing such a command, in addition to being behind other Western countries including Germany, France, and Canada.[44]

BACKGROUND

There are many reasons why Russia was so delayed in developing SOF and a SOF command. The previous Minister of Defense of the Russian Federation, Anatoly Serdyukov, had apparently been supportive of the idea, but developments were kept quiet under his leadership. Once he was ousted on criminal charges and replaced by Sergei Shoigu, however, the latter picked up the ball and ran with it, quickly putting into place all the necessary components. Pieces of the puzzle began to be assembled apparently as far back as 2008–2009, in the wake of the 2008 Russia–Georgia War, during which Russian forces did not perform as well as expected and prompted a major reorganization of the Russian military, which came to be known as the "New Look" military reforms.[45] This major structural reorganization of the Russian Armed Forces was announced in October 2008 and began in early 2009 under Serdyukov. The stated aims of the reforms were to reorganize the structure and the chain of command of the Russian army, to reduce it in size, and build the army around a three-link system (military district–operational command–brigade). Another significant aspect of that reform initiative was the establishment of Russia's Special Operations Forces Command (KSSO) and Russian special operations forces (SSO).[46]

Russia began reforms to its special operations forces between 2008-2012. The first step was the establishment in 2009 of the Directorate of Special Operations (*Upravlenie Spetsial'nykh*

Operatsii) centered on a unit based out of a training center in Solnechnogorsk, near lake Senezh. One of its founding fathers was the then-Chief of the General Staff General of the Army Anatoly Kvashnin. This unit had seen significant combat in Chechnya during the Second Chechen War.

The second step was the establishment of another center in Kubinka-2, also on the outskirts of the Moscow region. This center was directly under the control of the GRU and hence it retained its spetsnaz designation, being named the Center of Special Designation (*Tsentr Spetsial'nogo Naznacheniya*). It came to be known as Kubinka. Shortly thereafter, on 1 April 2012, upon the initiative of Gen. Nikolay Makarov, the Directorate of Special Operations was renamed the Special Operations Forces Command (*Komandovanie sil spetsial'nalnykh operatsii*, or KSSO). Finally on 15 March 2013 Kubinka was linked to the special operations forces.

As early as 2012 Makarov had been talking about forming a KSSO, with plans for up to nine special-purpose brigades and expansion of the existing system of military intelligence special forces (i.e., GRU).[47] Intensive physical plant development at both Kubinka and Senezh then began, including infrastructure for basing and military training. Senezh also houses a sniper training school, and both seem to have diver training facilities, although Kubinka apparently includes a special naval operations directorate that controls several special naval operations departments and squads. There is also a cold weather/mountaineering center at Mount Elbrus named "Terskol," in Kabardino-Balkaria, used by Russian special operators for training.[48]

As for the manning of these units, although Russia has spetsnaz units it could just pull from, they are not just renaming them as SOF. Rather, they are selecting the very best from their regular army, particularly their reconnaissance units, having them first serve with spetsnaz units and later undergoing specialized training. Only then do they get designated as Russian special operators.[49] This emphasis on quality special operators speaks to Gen. Gerasimov's comment above about the high quality of the U.S. and Western SOF they encountered in their studies.

Once selected, officers and non-commissioned officers arrive at Senezh, where they undertake a rigorous entrance examination that tests not only the physical conditioning of the SSO operations but also the personality and, perhaps most importantly, the ability to work within a team. The basic principle of Senezh is not to prepare an individual fighter with great skills and abilities, but rather to build teams that can act as a single organism.[50]

| The basic principle of Senezh is not to prepare an individual fighter with great skills and abilities, but rather to build teams that can act as a single organism. |

The Senezh Center is marked by the fact that it builds a culture of teamwork among its trainees, and this is increasingly a factor in recruitment.

The training of the officer recruit special operators is carried out in the Ryazan Higher Airborne Command School (RVVDKU, department of special and military intelligence and the Department of the use of special forces) and the Novosibirsk Higher Military Command School (NVVKU, Department of Special Intelligence and the chair of the special reconnaissance and airborne training). All SOF recruits learn skydiving, mountaineering, swimming and scuba diving, and storming buildings and homes. Depending on the individual tasks the soldiers are being prepared for, the training is more in depth.[51]

Additionally, it should be added that unlike other spetsnaz units that are seen as elite forces that perform missions (e.g., reconnaissance, direct action) for the sole purpose of furthering the movement and maneuver of conventional forces,[52] the structure of Russia's new SOF units suggests that it is intended to act independently. For that purpose, the command has a dedicated special aviation brigade that directly controls combat aviation assets at Torzhok and a squadron of the Il-72 transporters at the Migalovo airfield near Tver. The command also has supporting elements that provide combat support and combat service support functions.[53] Additionally, Alexsey Nikolsky reports that the Senezh compound has a large helipad that can accommodate three Mi-26 heavy transport helicopters (each capable of carrying 70 soldiers with kit). He therefore concludes that this could indicate that about 200 soldiers are on duty at any given time at SSO unit 92154.[54] He also calculates that between Senezh and Kubinka, there are approximately between 2,000 and 2,500 total special operators.[55]

SOF DOCTRINE, MISSIONS, AND ACTIVITIES

The KSSO is reportedly tasked with standardizing doctrine and capabilities for Russia's premiere SOF units in all military forces, and it is supposed to have the capability to provide command and control for these units in wartime.[56] The Russian Ministry of Defense defines the term "special operation" as follows: "the special operation of troops (forces) is a complex of special actions of troops (forces), coordinated by objectives and tasks, time and place of execution, conducted according to a single concept and plan in order to achieve certain goals. Special actions of troops (forces) are activities carried out by specially designated, organized, trained and equipped forces, which apply methods and ways of fighting not typical for conventional forces (special reconnaissance, sabotage, counter-terrorist, counter-sabotage, counterintelligence, guerrilla, counter-guerrilla and other activities)."[57] Here we see the core activities of Russian SOF, which in many ways equate to the traditional core activities of IW in the U.S. and NATO countries.

Russian special operations forces in 2015. Source: Wikimedia

The greatest distinctions between U.S. SOF IW activities and those of Russian SOF is the inclusion of sabotage and counter-sabotage operations. Most other missions are similar, if only worded differently. For example, due to their experience with partisan fighters in World War II, spetsnaz have retained the concept of guerilla (*partizanskie*) and counter-guerrilla (*antipartizanskie*) operations instead of developing a concept of insurgency and counterinsurgency warfare. Likewise, their understanding of conducting guerilla warfare is very close to the U.S. concept of unconventional warfare (UW), especially when it comes to the training of foreign fighters in the conduct of guerilla warfare for the purpose of overthrowing a hostile government.

Military correspondent Alexander Sladkov, while visiting military exercises in the mountains of the North Caucasus, articulated the goals and objectives of Russia's SOF as follows:

> Special Operations Forces (SSO)—troops intended to achieve political and economic goals in any geographical part of the world of interest to the Russian Federation.... They come in cases when diplomatic methods are no longer active. Distracting forces and the attention of certain countries by external problems, problems creating them inside, rocking the political systems of these countries, destabilizing the situation, including through a 'third hand.' Special Operations Forces create, train, and supervise foreign guerrilla movements, eliminate unwanted leaders without any sanctions on foreign soil, and so on... Russian experts' main task is the protection of our citizens abroad, the release of Russians who have fallen hostage somewhere in distant areas, and protecting the interests of our country.[58]

As Nikolsky summarizes his excellent study of Russian SOF, Moscow's units are "proper combat units themselves and can operate independently. They are ready for rapid deployment across a spectrum of counterterrorism and combat missions, on Russian territory and abroad."[59]

SOF AND SPETSNAZ IN ACTION: CASE STUDIES

Despite the fact that Russia does not label its IW activities the same way as does the U.S.,

SOF and spetsnaz are employed to conduct a full range of operations and activities in order to "assure or coerce states or other groups through indirect, non-attributable, or asymmetric activities."[60] How similar, however, are such activities when conducted by Russia's special operations units? The remainder of this section will focus on four case studies, two counterterrorism cases (both part of a larger counterinsurgency campaign) and two more recent cases of SOF and spetsnaz in unconventional warfare campaigns in Crimea and the Donbas. These case studies highlight the vast differences that exist between Russian and U.S. employment of special operations.

NORD-OST' SIEGE, 2002

The first case study is the tragedy at the Nord-Ost' performance in the Moscow Dubrovka theater. The Moscow theater hostage crisis (also known as the Nord-Ost' siege) was the seizure of a crowded theater by 40 to 50 armed Chechens on 23 October 2002. This siege lasted four days and involved 850 ordinary citizens (including 75 non-Russians from some 14 countries) who were held hostage in Moscow's Dubrovka theater by Chechen insurgents who were fighting for the cause of Chechen independence from Russia and demanded the withdrawal of Russian troops from Chechnya and an end to the war. The unit employed to end the siege by assaulting the theater was Al'fa – Moscow's elite tier 1 spetsnaz anti-terror unit, supported by Vympel and SOBR (Special Rapid Response Unit), a domestic rapid-response unit under the Russian national guard. These special operators pumped an undisclosed chemical agent into the building's ventilation system at the start of the assault and failed to inform anyone of the agent used. Once inside the building, the unit shot and killed almost all the self-proclaimed "suicide squad" terrorists, thus ending the siege. The death toll reached at least 170 people (including the terrorists). The high hostage death toll was partly due to the ineptness of the special operators involved, as almost all 130 hostages who died did so because first responders had no idea what agent had been used and were therefore unable to treat the victims effectively.[61]

THE BELSAN TRAGEDY, 2004

The second case study is the Beslan tragedy. The tragedy at Beslan began on 1 September 2004, when Muslim separatists from Russia's North Caucasus seized a grammar school and held everyone hostage (except for adult males, who were taken to a back room and shot). The siege ended three days later when various military units (primarily the spetsnaz units Al'fa and Vympel) attacked the school. This was purportedly set off by an over-zealous sniper who shot one of the terrorists before the planned assault on the building, leading to utter chaos and the death of 330 hostages.[62] Perhaps most alarming were the failures of joint operations execution in this incident; Beslan spetsnaz units were under the control of one command center, while the remainder of forces were under the control of another.

Hostages killed in Beslan gym on 11 July 2008. Source: Wikimedia

The response to both events, however, was universally led by spetsnaz, who both generated praise and criticism for how these events unfolded.

The Beslan tragedy led to serious debate in the Russian Duma, in military circles, and in society at large.[61] Still, any criticism of the spetsnaz-led operations was seen as criticism of the regime, and the Putin regime would not stand for that. In response, narratives were subsequently drawn up for each event that made Russian military forces heroes, brushing aside any criticism of the operations and/or the way the events unfolded. Certainly, the spetsnaz units that conducted these operations were operating with the clear purpose of eliminating the terrorists who had taken hostage hundreds of people, but apparently that was their main objective above preventing casualties and collateral damage.

Nord-Ost' and Beslan are examples of how spetsnaz operated prior to the "New Look" reforms and the creation of the KSSO. Beslan and the Nord-Ost' theater attack are two prime examples of where spetsnaz units—indeed the joint force—failed, mostly due to breakdowns in command and control, as well as inter-service rivalry[64] and the nonexistence of a functioning Russian interagency framework, all things the creation of the KSSO and SOF were meant to address. Much like the American catastrophe at Desert One during Operation EAGLE CLAW in 1980 (when a failed attempt to rescue American hostages being held in Iran resulted in the death of eight special operators), the Beslan and Nord-Ost' tragedies provided a pretext on how Russia's elite units functioned and also didn't function as a joint force, particularly in joint command and control and in interagency operations (a concept wholly unfamiliar to the Russian context). But unlike the U.S., Russia did not draw lessons from their catastrophes. In fact, these failures were covered up and the events heralded as successes for Russian military units, particularly its spetsnaz.

LITTLE GREEN MEN, 2014

The third and fourth case studies are more recent (post-"New Look" reforms and the creation of the KSSO) and focus on the unconventional warfare (*partizanskaya voina*) campaign in Crimea and the Donbas in Ukraine between 2014–2016. As is now well known, on 27 February 2014, the military occupation of Crimea by "little green"—and polite—men began. In a matter of a few days, Russian forces were able to seize power, block, disarm and even win over significant portions of the Ukrainian military, and then legitimize its presence, all the while conducting information operations and working to integrate the region into the Russian Federation.[65] Their campaign centered around a covert unconventional warfare operation. After identifying sympathetic locals (mostly disenfranchised ethnic Russians), they put together a proxy force composed of a variety of groups—local hooligans, want-to-be political leaders, and even Russians from Russia. Then "unidentified men in black uniforms" seized government buildings, including the Crimean parliament, and installed Sergei Aksyonov as the new prime minister of Crimea. SOF operators seized other strategic infrastructure, including the headquarters of the Ukrainian Navy in Sevastapol, the headquarters of the Tactical Aviation Brigade in Belbek, and the Marine Battalion in Feodosia. Spetsnaz personnel were also involved in several of these operations.[66] Finally, Russian forces seized all military bases and related facilities on the peninsula. Within a few short weeks, an entire territorial objective had been seized and politically integrated into the Russian Federation, almost with no shots fired.

WAR IN DONBAS, 2015-2022

The final case study examines separatist movements that emerged in eastern Ukraine, particularly Donetsk and Luhansk, which started immediately following the seizure of Crimea. The separatist forces of the Donetsk People's Republic (DPR) and Luhansk People's Republic (LPR) began a new offensive on Ukrainian-controlled areas, claiming over 9,000 soldiers by summer 2015. From February 2014 to May 2015, spetsnaz, SOF, conventional forces, and private military companies (PMCs) participated in the fighting in the Donbas against Ukrainian government security forces. Both spetsnaz units and Russian SOF (including the VDV, or Russian Airborne troops) were deployed in the region, along with conventional forces, though it is unclear exactly who was doing what. Given their mission sets, it is highly likely that both spetsnaz and SOF were organizing local insurgent forces, engaging in train and equip missions, and serving as military trainers in general. Additionally, it would be naïve to think that they were not also engaged in direct action missions.

One group that was intricately involved was led by Igor Girkin, who was in Ukraine under the alias Igor Strelkov (a surname derived from the Russian word for "shooter"). Strelkov made no efforts to hide the fact that he was engaged in unconventional warfare, with the goal of triggering an armed uprising and separatist movement that would ultimately allow eastern Ukraine to join Russia.[67] This retired FSB colonel led a group of more than 50 fighters, many of whom had been active in Crimea before showing up in eastern Ukraine. While not all had formidable fighting experience, the majority did, with several members even coming from the elite spetsnaz GRU.[68]

| Strelkov made no efforts to hide the fact that he was engaged in unconventional warfare, with the goal of triggering an armed uprising and separatist movement that would ultimately allow eastern Ukraine to join Russia. |

Juxtaposed to the quick and nearly bloodless seizure of Crimea, which can be seen as a successful special operation from Putin's point of view, the battle for eastern Ukraine is part of a protracted UW campaign that continues in a new guise to this day as part of Putin's "special military operation" in Ukraine.

CONCLUSION

Irregular warfare is not the sole domain of SOF. When it comes to conducting the military aspects of an IW campaign, however, SOF are the force of choice. In Russia it is not much different where, despite not labeling the activities as IW, SOF and spetsnaz are employed to conduct a full range of operations and activities to "assure or coerce states or other groups through indirect, non-attributable, or asymmetric activities."[69] Similar, however, does not mean "the same," particularly when it comes to approaches between the West and Russia when conducting special operations. While this section highlighted some of the similarities in this arena through the examination of four case studies, it also identified significant differences.

For instance, Russian special operators seem to place little value on the lives of hostages in CT missions, viewing them simply as collateral damage. And when it comes to the effectiveness of command and control and joint operations—though improved as a result of the "New Look" reforms and the creation of the KSSO—Russian spetsnaz and SOF still have a way to go. Finally, the U.S. does not conduct UW as a means of acquiring land to expand the territory of the U.S., as Russia had done in Crimea and the Donbas.

Though the case studies here are somewhat historical, there is a clear evolution of Russian IW over the past 20 years, with the formation of the KSSO created right at the halfway mark (2012). While it is impossible to predict the future, the course of Russia's "special military operation" in Ukraine has seen the misemployment of Russian SOF and spetsnaz and indeed airborne forces, all of which have been used as light infantry rather than in their intended roles. This is a trend that is likely to continue, as there seems to be no forcing function or institution capable of compelling commanders on the ground, or Putin back in the Kremlin, to utilize these specialized units for the missions for which they were intended.

| While it is impossible to predict the future, the course of Russia's "special military operation" in Ukraine has seen the misemployment of Russian SOF and spetsnaz and indeed airborne forces, all of which have been used as light infantry rather than in their intended roles. This is a trend that is likely to continue, as there seems to be no forcing function or institution capable of compelling commanders on the ground, or Putin back in the Kremlin, to utilize these specialized units for the missions for which they were intended. |

But if Russia does seek to expand the war beyond the currently-occupied regions of Kherson, Zaporizhzhia, Luhansk, and Donetsk into Kharkhiv and Odesa—and perhaps Transdniestria—SOF and spetsnaz are mostly likely to be employed in their capacity as leaders of unconventional warfare. In the end, this might be the future of the Kremlin's approach to irregular warfare and Russia's "special" way of war.

STUDY QUESTIONS

- Why did it take Russia so long to establish a USSOCOM-like organization?
- Are Russian SOF the same as U.S. SOF? What about Russian spetsnaz? What makes them similar and what different?
- Does Russia use the concept of irregular warfare? Why or why not?
- What are the equivalent missions of counterterrorism and counterinsurgency in Russian doctrine?
- What similarities and differences exist between Russian and U.S. special operations related to irregular warfare?

RECOMMENDED READINGS

Tor Bukhvoll, "Military Innovation under Authoritarian Government – The Case of Russian Special Operations Forces," Journal of Strategic Studies, Vol. 38, No. 5 (2015), 602-625.

Tor Bukhvoll, "Russian Special Operations Forces in Crimea and Donbas," Parameters, Vol. 46, No. 2 (2016), 13-22.

Christopher Marsh, Developments in Russian Special Operations: Russia's Spetsnaz, SOF, and Special Operations Forces Command (Kingston, ON: Canadian Special Operations Forces Command, 2016).

Alexsey Nikolsky, "Russian Special Operations Forces – Eight Years and Three Wars," in Elite Warriors: Special Operations Forces from Around the World, eds. Ruslan Pukhov and Christopher Marsh (Minneapolis: East View, 2017), 20-34.

About the Author

Dr. Christopher Marsh is the Chair of the Joint Special Operations Master of Arts (JSOMA) program at Fort Bragg, a division of the College of International Security Affairs of the National Defense University. Marsh is an expert on Russian approaches to special operations, irregular warfare, military strategy, and foreign policy. Prior to joining NDU, Dr. Marsh was the Director of the Department of Strategic Research & Analysis at the Joint Special Operations University, US Special Operations Command (USSOCOM). He previously held positions at the US Army School of Advanced Military Studies (SAMS), US Air Force Special Operations School, and Baylor University.

NOTES

1. The views expressed here are entirely those of the author and do not necessarily reflect the views, policy or position of the United States Government, Department of Defense, United States Army Special Operations Command, or the National Defense University.

2. Sandor Fabian and Gabrielle Kennedy, *The Conceptualization of Irregular Warfare in Europe* (Irregular Warfare Center, 2023).

3. This is the 2023 definition for irregular warfare in joint doctrine, found in "Defense Primer: What is Irregular Warfare," *Congressional Research Service*, 8 January 2024.

4. Martijn Kitzen, "Operations in Irregular Warfare," *Handbook of Military Sciences*, ed. Anders Sookermany, (Cham: Springer International Publishing, 2020): 1-21, https://doi.org/10.1007/978-3-030-02866- 4_81-1.

5. James D. Kiras, "Irregular Warfare: Terrorism and Insurgency," in David Jordan, James Kiras, David Lonsdale, Ian Speller, Christopher Tuck and C. Dale Walton, eds. *Understanding Modern Warfare* (Cambridge University Press, 2009): 186-207.

6. For an excellent review of the evolution of US IW definitions and activities, see Jared Tracy, "From 'Irregular Warfare' to Irregular Warfare: History of a Term," *Veritas* 19, no. 1 (2023): 29-34.

7. Fabian and Kennedy, 6.

8. Fabian and Kennedy, 6.

9. Interview with the Lt. Col. Stephan Bolton, U.S. Army Special Operations Command, Fort Leavenworth, KS, April 28, 2023.

10. Aleksandr Kolpakidi and Vladimir Sever, *Spetsnaz GRU* (Yauza and Eksmo, 2008). See also, Vladimir Kvachkov, *Spetsnaz Rossii* (Voennaya Literatura, 2004).

11. Valeriy Gerasimov, "Tsennost' nauki v predvidenii," *Voyenno Promyshlennyy Kuryer* (February 26, 2013), http://vpk-news.ru/articles/14632. (2013); See also Charles Bartles, "Russia's Indirect and Asymmetric Methods as a Response to the New Western Way of War," *Special Operations Journal*, 2, no. 1 (2016).

12. Vladimir Ivanov, "SShA delaiut stavku na neregulyarnye voiny," *Nezavisimoe voennoe obozrenie* (Oct. 15, 2020). https://nvo.ng.ru/realty/2020-10-15/8_1113_pentagon.html. Accessed 30 April 2023.

13. Interview with Lt. Col. Stephan Bolton.

14. Lester Grau and Charles Bartles, *The Russian Way of War: Force Structure, Tactics, and Modernization of the Russian Ground Forces* (Foreign Military Studies Office, 2016), 231.

15. Elisabeth Edwards, "Spetsnaz: How Russia's Special Forces Became So Deadly," *War History Online* (July 19, 2022). Available at: https://www.warhistoryonline.com/featured/russian-spetsnaz.html?safari=1. Accessed April 29, 2023.

16. John Dziak, "The Soviet Approach to Special Operations," in *Special Operations in US Strategy*, eds. Frank Barnett, B. Hugh Tovar, and Richard Schultz (National Defense University Press, 1984), 100.

17. Mark Galeotti, *Spetsnaz: Russia's Special Forces* (Osprey, 2015), 7.

18. Dziak, "The Soviet Approach to Special Operations," 98.

19. Adam Ulam, *The Bolsheviks* (New York: Collier, 1968).

20. Dziak, "The Soviet Approach to Special Operations," 98.

21. Grau and Bartles, *The Russian Way of War*, 232.

22. Dziak, "The Soviet Approach to Special Operations," 103.

23. Steven Zaloga, *Inside the Blue Berets: A Combat History of Soviet and Russian Airborne Forces, 1930-1995* (Presidio, 1995), 204-205.

24. John Collins, *Green Berets, SEALs & Spetsnaz: U.S. & Soviet Special Military Operations* (Pergamon-Brassey's, 1987), 1.

25. Active measures and disinformation were actions of political warfare that were conducted by the Soviet security services (Cheka, OGPU, NKVD, KGB) during the Cold War to influence the course of world events, in addition to collecting intelligence and producing "politically correct" assessment of it. Active measures ranged "from media manipulations to special actions involving various degrees of violence." They were used both abroad and domestically, and included disinformation, propaganda, counterfeiting official documents, assassinations, and political repression. See Richard Shultz and Roy Godson, *Dezinformatsia: Active Measures in Soviet Strategy* (Pergamon-Brassey's, 1984). See also Oleg Gordievsky and Christopher Andrew, *KGB: The Inside Story* (Hodder & Stoughton, 1990).

26. Director of Central Intelligence, *Soviet Support for International Terrorism and Revolutionary Violence: Special National Intelligence Estimate* (Langley: Central Intelligence Agency, 1981).

27. Grau and Bartles, *The Russian Way of War*, 231.

28. Collins, *Green Berets, SEALs and Spetsnatz*, 4.

29. Galeotti, *Spetsnaz: Russia's Special Forces*, 21-22.

30. Galeotti, *Spetsnaz: Russia's Special Forces*, 21-22.

31. Tony Balasevicius, *Squandering the Capability: Soviet Special Operations Forces in Afghanistan* (Canadian Defence Academy Press, 2011).

32. Balasevicius, *Squandering the Capability*.

33. Anatol Lieven, *Chechnya: Tombstone of Russian Power* (Yale University Press, 1998).

34. See Christopher Marsh, "The Desecularization of Conflict: The Role of Religion in Russia's Confrontation with Chechnya, 1785-present," *Review of Faith and International Affairs* 14, no. 1 (2016): 66-79. See also Robert Schaefer, *The Insurgency in Chechnya and the North Caucasus: From Gazavat to Jihad* (Praeger, 2011).

35. U.S. Army Special Operations Command, *"Little Green Men": A Primer on Modern Unconventional Warfare, Ukraine, 2013-2014* (US Army Special Operations Command, 2016).

36. Kolpakidi and Sever, *Spetsnaz GRU*.

37. Nikolai Protopopov, "Rebyata iz 'Al'fy': Kak rabotaet elitnyi spetsnaz FSB Rossii," *RIA Novosti* (July 29, 2019). https://ria.ru/20190729/1556911713.html. Accessed April 30, 2023.

38. Vladimir Sedov, "Samoe Sekretnoe nodrazdelenie strany: 40 let nazad v SSSR sozdalu unikal'nyi spetsnaz." *Lenta.ru* (19 August 2021), https://turbo.lenta.ru/articles/2021/08/19/vympel/. Accessed April 30, 2023.

39. Sedov, "Samoe Sekretnoe nodrazdelenie strany."

40. Sedov, "Samoe Sekretnoe nodrazdelenie strany."

41. Grau and Bartles, *The Russian Way of War*, 232.

42. Interfax Online, 6 March 2013. As quoted in Lester Grau and Charles Bartles, *The Russian Way of War: Force Structure, Tactics, and Modernization of the Russian Ground Forces* (Fort Leavenworth, KS: Foreign Military Studies Office, 2016), 234.

43. Dmitry Trenin, "Russia's New Tip of the Spear," *Foreign Policy* (May 8, 2013). Available at: http://www.foreignpolicy.com/articles/2013/05/08/russia_new_special_ops_command_afghanistan, accessed 8 August 2016.

44. "Shoigu sozdaet otvergnutye Serdyukovym sily spetsoperatsii, otstav of SShA na 26 let," *Newsru.com* (March 6, 2013). Available at: http://www.newsru.com/russia/06mar2013/minob.html

45. Grau and Bartles, *The Russian Way of War*. See also R. McDermott, *The Reform of Russia's Conventional Armed Forces: Problems, Challenges, & Policy Implications* (Jamestown Foundation, 2022).

46. For a more detailed analysis of Russia's special operations units, see C. Marsh, *Developments in Russian Special Operations: Russia's Spetsnaz, SOF and Special Operations Forces Command* (Canadian Special Operations Forces Command, 2017), https://publications.gc.ca/site/eng/9.831086/publication.html. Accessed 13 April 2023.

47. Roger McDermott, "Putin's Secret Force Multiplier: Special Operations Forces," *Eurasian Daily Monitor* 13, no. 81 (26 April 2016), https://jamestown.org/program/putins-secret-force-multiplier-special-operations-forces/. Accessed 13 April 2023.

48. Aleksey Mikhailov, "Boitsy Chetvertogo Izmereniya," *Voyenno Promyshlennyy Kuryer.* (20 April 2016). http://www.vpk-news.ru/articles/30319. Accessed 10 October 2017.

49. Alexey Nikolsky, "Little, Green and Polite: The Creation of Russian Special Operations Forces," in *Brothers Armed: Military Aspects of the Crisis in Ukraine*, eds. Colby Howard and Ruslan Pukhov (East View Press, 2015).

50. Mikhailov, "Boitsy Chetvertogo Izmereniya."

51. Mikhailov, "Boitsy Chetvertogo Izmereniya."

52. Grau and Bartles, *The Russian Way of War*, 236.

53. Aleksey Mikhailov, "General Staff asks Shoigu to Create 'Commandos'," *Izvestiya Online*, 27 November 2012. Cited in Grau and Bartles, *The Russian Way of War*, 235.

54. Alexsey Nikolsky, "Russian Special Operations Forces – Eight Years and Three Wars," in Ruslan Pukhov and Christopher Marsh, eds., *Elite Warriors: Special Operations Forces from Around the World* (East View, 2017).

55. Nikolsky, "Russian Special Operations Forces."

56. Aleksey Nikolsky, "Russian 'Spetsnaz' Forces – From Saboteurs to Court Bailiffs," *Moscow Defense Brief* 40, no. 2 (2014). Available at: http://mdb.cast.ru/mdb/1-2014/item4/article2/, accessed 6 August 2016.

57. Ministerstvo Oborony Rossiiskoi Federatsii, *Formy primeneniya Vooruzhyonnykh sil Rossiiskoi Federatsii* (Moscow: Ministerstvo Oborony Rossiiskoi Federatsii, n.d.). Available at: http://encyclopedia.mil.ru/encyclopedia/dictionary/details.htm?id=14014

58. Alexander Sladkov, "Novyi Rossiiskii Spetsnaz dlya zagranitsy," Television Channel "Rossiya 24", program "Vesti Nedeli," (28 April 2013).

59. Nikolsky, "Little, Green and Polite," 126-127.

60. "Defense Primer: What Is Irregular Warfare?"

61. Viktor Stepankov, *Bitva za "Nord-Ost"* (Yaza, 2003).

62. See, for example, Timothy Phillips, *Beslan: The Tragedy of School No. 1* (Granta, 2014).

63. Graham Turbiville, *Russian Special Forces: Issues of Loyalty, Corruption, and the Fight against Terror* (JSOU Press, 2005).

64. Nikolsky, "Little, Green and Polite," 131.

65. U.S. Army Special Operations Command, "*Little Green Men.*"

66. Nikolsky, "Little, Green and Polite,"

67. Anton Lavrov, "Civil War in the East: How the Conflict Unfolded Before Minsk," in Howard and Pukhov (2015), 204-205.

68. Lavrov, "Civil War in the East," 204-205.

69. "Defense Primer: What Is Irregular Warfare?"

III. More than a Terror State: The Democratic People's Republic of Korea is a Misunderstood and Neglected Global Malign Actor that Requires an Innovative New Strategy

By David Maxwell

ABSTRACT: This chapter argues that the West has oversimplified the adversarial problem of north Korea by focusing exclusively on nuclear weapons and terrorism. In fact, this malign actor needs further evaluation in terms of state resiliency and internal resistance to that same authority. The state's oppression and control measures over millions of its people is perhaps unparalleled in modern history. Ironically, Koreans are among the most resilient people on earth, yet those in the north simultaneously demonstrate little resistance capability. The scales of resilience and resistance offer the fundamental construct to consider, and its analysis offers insight into what might contribute to changing the lives of the Korean people in the north and the security situation on the Korean peninsula and Northeast Asia.

INTRODUCTION

The label "terrorist state" has been used by international actors and analysts to describe north Korea's government, officially known as the Democratic People's Republic of Korea (DPRK). This designation raises significant questions and concerns about north Korea's activities, including its nuclear program, human rights abuses and use of terror to control the population and eliminate internal opposition, its external use of political violence in support of its political warfare strategy against the Republic of Korea (ROK) and the ROK/U.S. Alliance, systemic kidnapping of ROK and Japanese citizens, and support for terrorist or violent extremist organizations around the world. However, what psychologically terrorizes the international community most of all is its possession of nuclear weapons and their use for coercion.[1] But this is a symptom of the larger problem: the existence of the Kim family regime and its intent to dominate the Korean peninsula.

The U.S. has placed north Korea on its State Sponsors of Terrorism list, removed it from the list to spur diplomacy, and then relisted it when the regime continued its malign activities. So, by definition, north Korea can be considered a terrorist state. However, it is crucial to approach this assertion with careful consideration of the broader geopolitical context and examine how the use of terrorism supports the regime's strategy and objectives and contributes to threats to the rules-based international order through its activities around the world.

| By definition, north Korea can be considered a terrorist state. |

One of the most egregious uses of "terrorist activities" by the regime is the use of political violence to oppress the Korean people in the north. There should be no doubt that Kim Jong Un's "rule by terror," eliminating his opposition in some of the most violent ways in public executions—from hanging to execution by anti-aircraft weapons—sends the message that there can be no resistance to the regime by the elite, the military leadership of the Korean people in the north. This has given rise to the resilience and resistance paradox.

The Korean people have demonstrated a level of resilience in the face of hardship outsiders can hardly imagine, from humanitarian disasters and starvation due to systematic and deliberate policy designs by the regime to the denial of human rights and crimes against humanity that the

| The Korean people in the north are among the most resilient people in history, yet they also have the least resistance capability in the modern world. |

Figure 6.1: Soviet invasion of Afghanistan, 1979. Source: Wikimedia Commons.

UN said has not been seen at such a level since the Nazi rule in the last century. Yet Korean escapees/refugees are often asked why they do not resist such tyranny. They explain that the regime's massive indoctrination effort for more than 70 years, combined with a system of social control that is worse than any caste system ever devised, along with little- to- no knowledge of the outside world, the Korean people simply do not know what to do. They have no frame of reference for resistance; all their effort goes into survival. This is the paradox:
The Korean people in the north are among the most resilient people in history, yet they also have the least resistance capability in the modern world. This is at once the fundamental problem and offers insight into what might contribute to changing the lives of the Korean people in the north and the security situation on the Korean peninsula and Northeast Asia.

APPRECIATE THE CONTEXT

There can be no doubt that north Korea is a terrorist state. It actions align with "as a minimum, terrorism involves the intimidation or coercion of populations or governments through the threat or perpetration of violence, causing death, serious injury or the taking of hostages."[2]

As noted, the U.S. has twice listed north Korea as a state sponsor of terrorism in accordance with the Title 18 U.S. Code definition of terrorism: "involve violent acts or acts dangerous to human life that are a violation of the criminal laws of the U.S. or of any State, or that would be a criminal violation if committed within the jurisdiction of the U.S. or of any State."[3] However, it is the definition developed by Bruce Hoffman that best describes north Korea's use of terrorism for domestic and international strategic purposes:

> We may therefore now attempt to define terrorism as the deliberate creation and exploitation of fear through violence or the threat of violence in the pursuit of political change. All terrorist acts involve violence or the threat of violence. Terrorism is specifically designed to have far-reaching psychological effects beyond the immediate victim(s) or object of the terrorist attack. It is meant to instill fear within, and thereby intimidate, a wider 'target audience' that might include a rival ethnic or religious group, an entire country, a national government or political party, or public opinion in general. Terrorism is designed to create power where there is none or to consolidate power where there is very little. Through the publicity generated by their violence, terrorists seek to obtain the leverage, influence and power they otherwise lack to effect political change on either a local or an international scale.[4]

Hoffman's definition accurately describes the full spectrum of north Korea terrorist activities for both domestic purposes and international objectives. At the heart of the definition and north Korean activities is the desire to employ psychological warfare through terrorist actions that lead to political effects.

To understand why the north Korean regime leverages terrorism, one must understand: (a) the nature of the regime, (b) its objectives, and (c) its strategy. The regime is built on the foundation of the mythical anti-Japanese partisan warfare led by Kim Il Sung as the great revolutionary and guerrilla leader who liberated Korea from the Japanese colonial occupation in 1945. Kim founded the DPRK with the vision of completing the revolution to dominate the Korean peninsula and rid it of foreign influence. The factual error of the myth, of course, is that Kim Il Sung was actually an officer in the Soviet Army and had no part in liberating Korea.[5] Nevertheless, it provides a foundational narrative upon which the legitimacy of the regime rests. In the following short narrative, Adrian Buzo describes how Kim came to power, how he consolidated it, and what the guiding light of the Kim family is:

> In the course of this struggle against factional opponents, for the first time Kim began to emphasize nationalism as a means of rallying the population to the enormous sacrifices needed for post-war recovery. This was a nationalism that first took shape in the environment of the anti-Japanese guerrilla movement and developed into a creed through the destruction of both the non-Communist nationalist forces and much of the leftist intellectual tradition of the domestic Communists. Kim's nationalism did not draw inspiration from Korean history, nor did it dwell on past cultural achievements, for the serious study of history and traditional culture soon effectively ceased in the DPRK.
> Rather, DPRK nationalism drew inspiration from the Spartan outlook of the former Manchurian guerrillas. It was a harsh nationalism that dwelt on past wrongs and promises of retribution for 'national traitors' and their foreign backers. DPRK nationalism stressed the 'purity' of all things Korean against the 'contamination' of foreign ideas, and inculcated in the population a sense of fear and animosity toward the outside world. Above all, DPRK nationalism stressed that the guerrilla ethos was not only the supreme but also the only legitimate basis on which to reconstitute a reunified Korea.[6]

Combining the implementation of the Juche ideology (the ideas of Marxism–Leninism combined with the individual, the nation state, and national sovereignty) with the horrific social classification system (called Songbun) resulted in complete control of the Korean people by isolating north Korea and essentially creating a countrywide prison camp. As such, north Korea can be called the Guerrilla Dynasty and Gulag State. A key tool of guerrillas has historically been the use of terrorism to achieve political effects. Therefore, the use of terrorism resides in the DNA of the Kim family regime.

People's Army guards march in Panmunjom Joint Security Area, 1998. Source: U.S. Air Force photo.

So what are some examples of north Korea's terrorism activities? Here is a representative sample of their direct actions:

- The attempted assassination of the South Korean President Park Chung Hee at the Blue House in 1968.[8]
- The attempted assassination of President Park Chung Hee in 1974 that resulted in the death of his wife.9
- The attempted assassination of South Korean President Chun Doo Hwan in 1983 in Rangoon, Burma. Although he survived, several high-ranking officials and military officers were killed.[10]
- The bombing of Korean Air Flight 848 in 1987 prior to the 1988 Seoul Olympics to try to keep the international community from traveling to Korea and thus make the games fail.[11]
- The assassination of Kim Jong Nam, Kim Jong Un's half-brother, in 2017 in Malaysia to prevent him from becoming a threat to his rule.[12]

There is also north Korean support to international terrorism. Examples include:

- Support to the Red Army faction (a West German far-left militant group) by providing them sanctuary for decades.[13]
- Support to Islamic terrorist organizations such as Hamas, Hezbollah, and the Houthis in the Middle East.[14]

Lastly, one of the most well publicized acts of modern cyberterrorism is the 2014 Sony hack, in which the regime went after the Sony corporation to retaliate against it for releasing the film, *The Interview*, which it found insulting to Kim Jong Un.[15]

There are numerous other examples of terrorist activities such as the detainment, torture, and 2017 murder of American student Otto Warmbier. Although it might not be classified as terrorism under strict interpretation his parents clearly described the regime as terrorist, and the regime certainly attempted to exploit the events for political gain prior to the U.S. White House holding direct talks with Kim Jong Un.[16]

The regime's support to international terrorism has been well documented in the modern era by Bruce Bechtol and throughout the Cold War by Benjamin Young.[17] In addition, the Center for Strategic and International Studies in its *Beyond Parallel* project maintains a comprehensive database of a wide range of north Korean malign activities, from nuclear tests to terrorism activities.[18]

The root cause as to why north Korea uses terrorism derives from the 1953 Armistice that suspended the Korean War. This Armistice sought to solve a Cold War political dilemma by unnaturally dividing the Korean peninsula and the Korean people. These actions remain very prescient in the wording of the Armistice agreement, which called for "the peaceful settlement of the Korean question" but did not provide an acceptable or durable political arrangement, which resulted in subsequent and persistent conflict in various forms.[19] So how can we better understand this political arrangement?

UNDERSTAND THE PROBLEM

The basic problem remains that as long as the Kim family regime remains in power, there will be major strategic threats to the current international world order. These threats include the use of nuclear weapons, as well as large-scale conventional war, in addition to crimes against humanity. Unchallenged, the regime will continue acts of terrorism to support its strategic aims as well as illicit activities. Criminal actions include nuclear weapons proliferation, external support to actors in conflict areas, and support to violent extremist organizations (via violent means and hard currency).

The time has come to finally acknowledge what has long been known but which officials have been reluctant to state: All problems in Korea stem from the continued existence of the mafia-like crime family cult known as the Kim family regime —a regime with the objective of dominating the Korean Peninsula under the rule of the Guerrilla Dynasty and Gulag State.

| The time has come to finally acknowledge what has long been known but which officials have been reluctant to state: All problems in Korea stem from the continued existence of the mafia-like crime family cult known as the Kim family regime. |

The regime clings to a threefold strategy of (a) political warfare, (b) blackmail diplomacy, and (c) developing advanced warfighting capabilities to create dilemmas for the ROK/U.S. alliance. These threats will likely reside near the roots of future conflict in Northeast Asia.[20]

The regime is using political warfareto subvert South Korea as well as drive a wedge in the ROK/U.S. alliance to weaken and potentially collapse the South Korean government and society. Political warfare creates conditions for north Korea's ultimate campaign plan to remove U.S. troops from the peninsula in order for the Kim family to unify Korea by force. Despite the recent announcements by Kim Jong Un that north Korea no longer seeks peaceful unification and that it now treats South Korea as its major enemy, the fact is that north Korea has always treated the South as its enemy and has never had any intent of peaceful unification.[21]

The regime is also conducting a blackmail diplomacy strategy through the use of increased tensions, threats, and provocations to coerce political and economic concessions.

Finally, north Korea continues to develop advanced warfare capabilities, like nuclear weapons and missiles (to include intercontinental ballistic missiles) from which to strike the U.S. mainland. These developments support Kim's political warfare and blackmail diplomacy but they also can support war with the ROK.

The Kim family regime's strategic vision can be summarized in four priorities.

- *Vital Interest*: Survival of the Kim Family Regime
- *Strategic Aim*: Unification of the Peninsula (via subversion, coercion, extortion, use of force)
- *Key Condition*: Split the ROK/U.S. alliance (drive U.S. forces off the Peninsula)
- *Divide to Conquer*: Divide the alliance to conquer the ROK

The regime's main effort remains political warfare, with the intent of subverting the ROK. The following political warfare description from Paul Smith in a publication from National Defense University illustrates the North's activities:

"Political warfare is the use of political means to compel an opponent to do one's will, based on hostile intent. The term political describes the calculated interaction between a government and a target audience to include another state's government, military, and/or general population. Governments use a variety of techniques to coerce certain actions, thereby gaining relative advantage over an opponent. The techniques include propaganda and psychological operations (PSYOP), which service national and military objectives respectively. Propaganda has many aspects and a hostile and coercive political purpose. Psychological operations are for strategic and tactical military objectives and may be intended for hostile military and civilian populations."[22]

An ideological war has been fought since the establishment of the two Koreas in 1948. North Korea conducts an active subversion campaign using its United Front Department and Cultural Engagement Bureau to sow political divisions in the South.[23] This subversion can be described as the "ruthless subversion of democracy." Until this point, all Korean people have not been able to decide for themselves what values they wish to live under. There are the shared values of the ROK/U.S. alliance such as freedom, individual liberty, liberal democracy, free market economy, rule of law, and human rights. Or there are the "values" of the Kim family regime: Juche/ Kimilsungism/(now Kimjongunism), the Socialist Workers Paradise, Songun, Songbun, Byungjin, and the deliberate denial of human rights to sustain Kim Jong Un in power.

| Until or unless there is a deliberate attack from the North, the conflict between North and South is taking place on the battlefield of human terrain. This is very much an information war. |

There is one other serious problem that must be addressed. The real existential threat to the regime comes from within. Threats come from the elite, the second-tier military leadership (the corps and elite unit commanders), and the people. Kim Jong Un must suppress these threats. He has so far been successful in satisfying the elite and the military commanders by providing them with luxury goods and a sufficient lifestyle funded by the regime's global illicit activities of drug trafficking, counterfeiting money, cigarettes and drugs, proliferation of weapons and training to conflict zones around the world, and overseas slave labor.

Kim Jong-Un meets Russian President Vladimir Putin in Vladivostok in 2019. Source: Russian Federation.

As previously noted, Kim Jong Un keeps the people in check through the administration of the Songbun social classification system. He also keeps the elite, the second-tier leadership, and the general population compliant by creating the perception of the South and the U.S. as mortal enemies. He requires this threat to justify the sacrifices and suffering of the people as he prioritizes the development of nuclear weapons, missiles, and advanced military capabilities that he requires for political warfare and blackmail diplomacy and to unify the peninsula by force.

Although the development of this threat perception has been ongoing since the 1953 Armistice, it has recently been brought to new levels as Kim Jong Un has said he no longer seeks unification and has called the ROK the main enemy.[24] There could be many reasons for this; however, the one policymakers and military strategists should be most concerned with is the growing possibility of internal instability and threats from within the regime or among the general population. The internal instability and possible collapse of the regime could have catastrophic consequences for the region.[25]

The COVID-19 pandemic illustrates Kim's fear of internal instability. While the regime was deathly afraid of a widespread outbreak, what might be called the "COVID paradox" appeared to take place. While COVID defense measure were put in place, the regime saw the opportunity to implant far more draconian population and resource control measures.[26] Kim shut down the borders to both legal trade and smuggling, which had significant impacts on the 400 markets that developed since the "Arduous March" of the famine of 1994-1996.[27] These markets have been the fundamental source of resilience of the Korean people in the North.[28] However, these markets and the growth of the "donju," or moneyed class, have become a threat to the regime's control. Kim used COVID to justify myriad measures, including market hour restrictions and closures, taxes on market activity, halting the use of foreign currency, greater restrictions on internal movement, and, most important, a nationwide crackdown on information. These actions have put great stress on the Korean people. They survived the Arduous March, but now the markets that provided two decades of resilience are at risk because Kim Jong Un assesses them to be a threat to regime control.

Kim Jong Un also perceives a new threat that has arisen during the Yoon administration. The ROK has long stressed peaceful unification in various proposed plans. However, in 2023, it really stressed the need to pursue a free and unified Korea. On 26 April, the Yoon/Biden summit stated the following: "The two presidents are committed to build a better future for all Korean people and support a unified Korea that is free and at peace."[29]

Subsequently, in August 2023, the Camp David Summit issued two statements from President Yoon, President Biden, and Prime Minister Kishida of Japan emphasizing Korean unification. The Camp David Principles and the Spirit of Camp David both expressed this sentiment, as follows: "We express support for the goal of the ROK's Audacious Initiative and support a unified Korean Peninsula that is free and at peace."[30] To Kim Jong Un, these statements make a dire threat to his regime and may have led to his bold statement of calling the ROK the enemy and discarding any pursuit of peaceful unification. The ROK has the moral high ground; Kim Jong Un is on untenable terrain and is therefore going on the offensive to try to undermine the ROK and the ROK/U.S. alliance.

International security problems regarding north Korea can be summarized in these six points:

1. There is always the threat of war as Kim seeks to dominate the peninsula under his rule and will use force to attempt to do so.
2. Kim is executing a political warfare strategy to subvert the ROK and the ROK/U.S. alliance to achieve conditions necessary to dominate the peninsula either by the capitulation of the ROK through subversion or the use of force.
3. Kim is executing a blackmail diplomacy strategy to coerce the ROK, the U.S., and the international community into providing political and economic concessions to contribute to regime survival and development of advanced warfare capabilities.

4. There is an ideological war taking place through information between the North and South, and the people will eventually have to decide which values they desire to live by: liberal democracy and freedom versus Juche and no human rights.
5. The Korean people do not know how to resist the regime due to their seven decades of indoctrination and isolation of the outside world, so they have no frame of reference for resistance.
6. The "Korea question" (i.e., the unnatural division of the peninsula) has not been solved.

These six problems require a new approach, beyond countering terrorism.

DEVELOP A NEW APPROACH

Although denuclearization of the North remains a worthy goal, it must be viewed as aspirational as long as the Kim family regime remains in power. The conventional wisdom has always been that denuclearization must come first and unification will follow; there should be no discussion of human rights out of fear that it would prevent Kim Jong Un from making a denuclearization agreement. Today even a blind man can read the tea leaves and know that Kim Jong Un will not denuclearize even though his policies have been an abject failure. His political warfare and blackmail diplomacy strategies completely failed in 2022 and 2023 because Presidents Yoon and Biden, like their predecessors, refused to make the political and economic concessions he demanded just to come to the negotiating table: namely to remove sanctions. It is time for the U.S and the ROK/U.S. alliance to execute a political warfare strategy that flips the conventional wisdom and seeks unification first and then denuclearization.

| It is time for the U.S and the ROK/U.S. alliance to execute a political warfare strategy that flips the conventional wisdom and seeks unification first and then denuclearization. |

Everyone must understand that the only way to end the nuclear program and the human rights abuses is through unification of the Korean peninsula. The ROK and the U.S. must continue to maintain the highest state of military readiness to deter war and then adopt a human rights upfront approach, develop a comprehensive and sophisticated information and influence activities campaign, and focus all efforts on the pursuit of a free and unified Korea—ultimately, a United Republic of Korea (UROK).

A new "Three Plus One" strategy should be developed. It must consist of a political warfare strategy with a human rights upfront approach, a comprehensive and sophisticated information campaign to give the Korean people in the North the tools and knowledge to create change, and the pursuit of a free and unified Korea to solve the "Korea question." The one overarching requirement is that all three must rest on the strong foundation of the ROK/U.S. alliance and its deterrence and warfighting capabilities.

A human rights upfront approach is a key component of a new strategy. It is not only a moral imperative but also a national security issue. Kim Jong Un must deny the human rights of the Korean people in the North to remain in power. The ROK government has reenergized its human rights focus and, with the recent appointment of Ambassador Julie Turner, the U.S. is catching up. Human rights must be incorporated into all aspects of policy execution and strategy. No longer can the excuse be that to mention human rights will hinder potential denuclearization negotiations. Every time the regime conducts a provocation, part of the response must be that the regime prioritizes its military over the welfare of its people and continues to deny their human rights solely to keep them under control. Human rights must be an integral part of a new strategy.

There is a unique relationship between denuclearization, human rights, and unification. The only way to achieve denuclearization and end the human rights abuses being committed against the Korean people in the North is by achieving unification.

| The only way to achieve denuclearization and end the human rights abuses being committed against the Korean people in the North is by achieving unification. |

Perhaps counterintuitively, it is the focus on human rights that must lead to unification, and only when unification is achieved there can be denuclearization. And the connective tissue among the two is information.

Jung Pak, U.S. Special Representative to north Korea, often asks, "who does Kim Jong Un fear the most: the ROK and U.S. military or the Korean people?" It is the Korean people—especially when they are armed with information and knowledge of their universal human rights.

Therefore, Kim Jong Un creates the perception of external threats to justify the suffering and sacrifice of the Korean people.

Some basic considerations for human rights upfront:

- Kim fears the Korean people more than the ROK/U.S. combined military.
- Must be required for any kind of North–South or North–U.S. engagement.
- No normalization is possible without an end to human rights abuses.

Important work has been conducted on information and influence. Here are four crucial reports that provide insights into how information should be employed in the North:

"An Information Based Strategy to Reduce north Korea's Increasing Threat: Recommendations for ROK & U.S. Policy Makers." Fredrick Vincenzo led a working group that developed concrete ideas and actions to influence regime decision-making and mitigate threats.

"A Policy of Public Diplomacy with north Korea: A Principled and Pragmatic Approach to Promote Human Rights and Pursue Denuclearization." Jieun Baek, one of America's leading thinkers and doers on information issues, led a working group to develop public diplomacy actions to influence the regime and the Korean people.

"Maximum Pressure 2.0: A Plan for north Korea." In section six, Mathew Ha and David Maxwell provide a comprehensive list of information and influence activities.

"North Koreans Want External Information, But Kim Jong-Un Seeks to Limit Access." Robert King surveys the information landscape and describes what is being done and what can be done.

An information campaign must be the foundation of any north Korean strategy. Unfortunately, information warfare against north Korea has not been a top priority for Washington or for Seoul. The ROK/U.S. alliance's successful deterrence of war for 70 years has bred complacency, and this has allowed the North to execute a campaign that keeps the regime in power and creates security dilemmas for the alliance and the region.

| An information campaign must be the foundation of any north Korean strategy. |

There is a lone bright spot in the information space. The Korean Services of Voice of America (VOA) and Radio Free Asia (RFA) do yeoman's work trying to penetrate the North with news and practical information.[35] Congress recognized their important work by allocating $40 million in additional resources to expand their work to provide the Korean people in the North with facts, knowledge, and the truth to help the people gain understanding of Korea (both North and South) and the outside world. An overt information warfare campaign supports deterrence and diplomacy and creates dilemmas and pressure on Kim to change his behavior and decision-making or create a change inside north Korea. VOA and RFA significantly contribute to this effort.

A ROK and U.S. information warfare campaign should seek to support three specific outcomes. First, change Kim's behavior toward his nuclear weapons and ballistic missile programs as well as the north Korean people. Second, provide a rationale for the elite and military leadership to force Kim to change his decision-making. Lastly, serve as a catalyst to encourage the Korean people in the North to effect change on their own.36

Some basic considerations for an influence campaign:

- Influence the three target audiences: regime elite, second-tier leadership, and the Korean people in the North.
- Recognize, understand, expose, and attack regime strategy with superior political warfare and information.
- Employ the four principles of influence:
    - massive quantities of information from news to entertainment.
    - practical information from market activity to organization for collective action.
    - the truth about north Korea and the outside world.
    - understanding of the universal human rights for all people.
- Major theme: Kim's strategy has failed to achieve his objectives.

Here are specific actions to contribute to an information campaign.

- Highlight the cost of Kim's nuclear and missile activities. It is estimated that Kim spent $560 million on missile activities alone in 2022, and the Korean people in the North should know that this money could have been used to feed the people.
- Overarching narrative: Every response to the regime's nuclear and missile activities must include a human rights response. For example, Kim Jong Un's deliberate decision to prioritize nuclear and missile development is solely responsible for the suffering of the Korean people in the North.
- Establish a "Korea Desk" at the Global Engagement Center to coordinate U.S. information activities.
- Establish an alliance organization to plan and conduct combined political warfare with a supporting information warfare campaign.
- Establish a Korean Escapee (Defector) Information Institute (KEII) to harness the expertise of key communicators from the North to shape themes and messages and advise on all aspects of the information warfare campaign.
- Effectively employ the $40 million allocated in the 2023 NDAA to the Korean Services of VOA and RFA to increase broadcasts to a level that optimizes access and coverage.
- Fully implement the information planning requirements in the "Otto Warmbier Countering north Korean Censorship and Surveillance Act of 2022."

All of the above require a political warfare strategy to counter the North, as well as to set the conditions for the pursuit of a free and unified Korea. Here is an outline for such a strategy:

FIVE LINES OF EFFORT FOR POLITICAL WARFARE IN NORTH KOREA

1. *Prevent war.* Establish policies for north Korea's second-tier military leadership (military corps commanders and security service leadership). For example, if they do not attack the ROK and maintain control of weapons of mass destruction, they and their families will have a place in a free and unified Korea. When faced with an order to attack, the second-tier leadership must know they have choices other than going to war.
2. *Pressure Kim Jong Un.* Pressure from within could cause Kim to change or at least moderate his policies. As an example, every time the ROK and U.S. must talk about nuclear weapons and missiles, they need to emphasize the human rights abuses.
3. *Offer options to the Korean people.* They must know that there is life outside of Juche and Songbun that they can reach and enjoy. An information campaign must show the successes of escapees living in free countries.

4. *Support for potential emerging leadership who seek change.* When they do act against the regime, the Korean people in the North must know they will be supported by the international community.

5. *Prepare the Korean people in the North for unification.* A long process of education is required to undo Juche, Songbun, and the Ten Principles of Monolithic Ideology and to teach about such things as land ownership, participatory politics, and the rule of law.

Finally, Robert Joseph led a working group and published the seminal work, "National Strategy for Countering north Korea," on the three-part strategy (a human rights upfront approach, an information campaign, and the pursuit of a free and unified Korea) in a report at the National Institute of Public Policy. It should serve as the foundation for a new "Three Plus One" strategy.[43]

CONCLUSION

Terrorism is a tool used by the Guerrilla Dynasty and Gulag State. It is the regime's use of political violence that helps it to maintain power and influence the international community. To stop the employment of terrorism by north Korea, the "Korea question" must be solved. This requires a new strategy that must include a superior political warfare strategy that has at its core a comprehensive, sophisticated, and holistic information campaign with a human rights upfront approach, in short, the "Three Plus One" strategy.

The ROK and U.S. have an opportunity for a new approach to the Korean security challenge. The alliance's way ahead is an integrated deterrence strategy as part of the broader strategic competition that is taking place in the region. There is a need for a Korean "Plan B" strategy that rests on the foundation of combined ROK/U.S. defensive capabilities and includes political warfare, aggressive diplomacy, sanctions, cyber operations, and information and influence activities. While a goal of denuclearization is desirable and required, ultimately the objective must be to solve the "Korea question" (i.e., the unnatural division of the peninsula) with the understanding that denuclearization of the North and an end to human rights abuses and crimes against humanity will only happen when the Korea question is resolved, leading to a free and unified Korea, otherwise known as UROK.

The only way to end north Korea's nuclear program, military threats, and human rights abuses and crimes against humanity—being committed against the Korean people living in the North by the mafia-like crime family cult known as the Kim family regime—is through achievement of unification and the establishment of a free and unified Korea that is secure and stable, non-nuclear, economically vibrant, and unified under a liberal constitutional form of government based on individual liberty, rule of law, and human rights as determined by the Korean people. A free and unified Korea or in short, a United Republic of Korea (UROK).

About the Author

David Maxwell is a retired U.S. Army Special Forces colonel and has spent more than 30 years in Asia as a practitioner and specializes in Northeast Asian Security Affairs and irregular, unconventional, and political warfare. He is the Vice President of the Center for Asia Pacific Strategy and a Senior Fellow at the Global Peace Foundation (where he focuses on a free and unified Korea). He is a member of the board of directors of the Committee for Human Rights In North Korea and he is the editor of Small Wars Journal.

NOTES

1. National Intelligence Estimate, "Scenarios for Leveraging Nuclear Weapons Through 2030," National Intelligence Council, Office of the Director of National Intelligence, January 2023 (declassified 15 June 2023) https://www.dni.gov/files/ODNI/documents/assessments/NIC-Declassified-NIE-North-Korea-Scenarios-For-Leveraging-Nuclear-Weapons-June2023.pdf

2. For the UN definition, see https://www.ohchr.org/en/terrorism, accessed on 22 July 2024.

3. For the US Code Title 18 definition:, see https://uscode.house.gov/view.xhtml?path=/prelim@title18/part1/chapter113B&edition=prelim#:~:text=As%percent20used%percent20in,or%percent20transcend%percent20national, accessed 22 July 2024.

4. Bruce Hoffman, *Inside Terrorism* (Columbia University Press, September 2017). Also see, Andrea Ho, "Professor Bruce Hoffman on Domestic Terrorism," *Georgetown Journal of International Affairs* (21 April 2021).

5. Sydney A. Seiler, *Kim Il-sung 1941-1948: The Creation of a Legend, the Building of a Regime*, (American University Press, 1994).

6. Adrian Buzo, *Politics and Leadership in North Korea: The Guerilla Dynasty* (Routledge, 2017).

7. Robert Collins, *Marked For Life: North Korea's Social Classification System, The Committee for Human Rights in North Korea* (2012), www.hrnk.org/uploads/pdfs/HRNK_Songbun_Web.pdf. Also see, David Maxwell, "The Nature of The Kim Family Regime: The Guerrilla Dynasty and Gulag State, "*Red Diamond* (US Army Training and Doctrine Command, January-March 2019), 18. Found at https://www.fdd.org/analysis/2020/02/19/the-nature-of-the-kim-family-regime-the-guerrilla-dynasty-and-gulag-state/ and here: https://drive.google.com/file/d/12DLljRWQtQNje4hqrpwfGuEMvlpyk5Gt/view

8. Congressional Research Service, "North Korean Actions, 1950-2007: Controversy and Issues," *EveryCRS Report.com* (20 April 2007), https://www.everycrsreport.com/reports/RL30004.html

9. Congressional Research Service, "North Korean Actions 1950-2007."

10. Congressional Research Service, "North Korean Actions 1950-2007."

11. Congressional Research Service, "North Korean Actions 1950-2007."

12. David Maxwell, "Kim Jong Nam Assassination Showcases North Korea's Special Operations Capabilities," *Foundation for Defense of Democracies* (22 April 2018), https://www.fdd.org/analysis/2018/08/22/kim-jong-nam-assassination-showcases-north-koreas-special-operations-capabilities/

13. Maxwell, "Kim Jong Nam Assassination." Also see Congressional Research Service, "North Korean Actions 1950-2007."

14. Bruce Bechtol, *North Korean Military Proliferation in the Middle East and Africa: Enabling Violence And Instability*, (University of Kentucky Press, 2018).

15. Mathew Ha and David Maxwell, *Kim Jong Un's "All-Purpose Sword" North Korean Cyber-Enabled Economic Warfare*, (Foundation for Defense of Democracies, October 2018), 20-22. Found at https://www.fdd.org/wp-content/uploads/2018/09/REPORT_NorthKorea_CEEW.pdf

16. Rick Gladstone, "Parents of Otto Warmbier, North Korea Captive, Say He Was Tortured," *New York Times*, September 26, 2017. Found at https://www.nytimes.com/2017/09/26/world/asia/warmbier-north-korea-torture.html. See also David Nakamura, "Trump puts North Korea back on state sponsors of terrorism list to escalate pressure over nuclear weapons," Washington Post, November 20, 2017. Found at https://www.washingtonpost.com/news/post-politics/wp/2017/11/20/trump-puts-north-korea-back-on-state-sponsors-of-terrorism-list-to-escalate-pressure-over-nuclear-weapons/

17. Bechtol, *North Korean Military Proliferation in the Middle East and Africa.* Also see, Benjamin Young, *Guns, Guerillas, and the Great Leader: North Korea and the Third World* (Stanford University Press, 2021).

18. *Beyond Parallel* (Center for Strategic and International Studies,), https://beyondparallel.csis.org/

19. "Text of the Korean War Armistice Agreement," *Bureau of Arms Control*, 27 July 1953, https://2001-2009.state.gov/t/ac/rls/or/2004/31006.htm

20. Charles Dunlap, "Guest Post: David Maxwell asks "Is Kim Playing Trump?" *Lawfire*, 15 February 2020, https://sites.duke.edu/lawfire/2020/02/15/guest-post-david-maxwell-asks-is-kim-playing-trump/

21. Chun, In- Bum, "How North Korea Wages Political Warfare at Home and Abroad...and How to Respond," *Orbis, Foreign Policy Research Institute*, Spring 2020, https://drive.google.com/file/d/1OM-InGiZHhuUUrl9xsnbR72MN20teWkz/view. Also see, Heather Chen and Yoonjung Seo, "North Korea says it will no longer seek reunification with South Korea, will launch new spy satellites in 2024," CNN, 1 January 2024.

22. Paul A. Smith., *On Political War* (National Defense University Press, 1989), 3. Found at https://apps.dtic.mil/dtic/tr/fulltext/u2/a233501.pdf

23. "North Korea Military Power, A Growing Regional and Global Threat," *Defense Intelligence Agency*, 2021, 57. Found at https://www.dia.mil/Portals/110/Documents/News/NKMP.pdf

24. Shweta Sharma, "Kim Jong-un declares South Korea his No 1 enemy and orders father's unification arch be destroyed," *The Independent*, January 16. 2024, https://www.independent.co.uk/asia/east-asia/north-korea-enemy-south-korea-peaceful-reunification-b2479355.html

25. David S. Maxwell, *Catastrophic Collapse of North Korea: Implications for the United States Military*, (US Army Command and General Staff College and School of Advanced Military Studies, Fort Leavenworth, 1996), https://apps.dtic.mil/sti/citations/ADA314274

26. David Maxwell and Mathew Ha, "Coronavirus may force ROK/U.S. military alliance to contend with North Korean instability," *Military Times*, March 27, 2020, https://www.militarytimes.com/opinion/commentary/2020/03/27/korea/?utm

27. Unattributed, "How Did the North Korean Famine Happen?" *Wilson Center*, April 30, 2002, https://www.wilsoncenter.org/article/how-did-the-north-korean-famine-happen

28. KBS World, "Official Markets in n>Korea," *KBS Radio*, https://world.kbs.co.kr/service/contents_view.htm?lang=e&menu_cate=northkorea&id=&board_seq=447447

29. David Maxwell, "Sometimes less is more: The most important 26 words of ROK-US summit," *Korea Times*, May 8, 2023, https://www.koreatimes.co.kr/www/opinion/2024/01/137_350494.html See also The White House, "Leaders' Joint Statement in Commemoration of the 70th Anniversary of the Alliance between the United States of America and the Republic of Korea,' " April 26, 2023, https://www.whitehouse.gov/briefing-room/statements-releases/2023/04/26/leaders-joint-statement-in-commemoration-of-the-70th-anniversary-of-the-alliance-between-the-united-states-of-america-and-the-republic-of-korea/

30. The White House, "Camp David Principles," August 18, 2023, https://www.whitehouse.gov/briefing-room/statements-releases/2023/08/18/camp-david-principles/ and The White House, "The Spirit of Camp David: Joint Statement of Japan, the Republic of Korea, and the United States," August 18, 2023 https://www.whitehouse.gov/briefing-room/statements-releases/2023/08/18/the-spirit-of-camp-david-joint-statement-of-japan-the-republic-of-korea-and-the-united-states/.

31. Commander Frederick Vincenzo, "An Information Based Strategy to Reduce North Korea's Increasing Threat - Recommendations for ROK & U.S. Policy Makers," *Center for New American Security*, October 3, 2016, pages 1-15. Found at https://www.cnas.org/publications/reports/an-information-based-strategy-to-reduce-north-koreas-increasing-threat.

32. Jieun Baek, "A Policy of Public Diplomacy with North Korea: A Principled and Pragmatic Approach to Promote Human Rights and Pursue Denuclearization," *Harvard Belfer Center*, August 2021, pages 20-28. Found at, https://www.belfercenter.org/publication/policy-public-diplomacy-north-korea

33. Bradley Bowman and David Maxwell, *Maximum Pressure 2.0 A Plan for North Korea*, (Foundation for Defense of Democracies, December 5, 2020), 8-13 and 46-51. https://www.fdd.org/wp-content/uploads/2019/11/fdd-report-maximum-pressure-2-a-plan-for-north-korea.pdf

34. Robert King, "North Koreans Want External Information, But Kim Jong-Un Seeks to Limit Access," Center for Strategic and International Studies, May 15, 2019, https://www.csis.org/analysis/north-koreans-want-external-information-kim-jong-un-seeks-limit-access

35. Voice of America Broadcasting in Korean, https://www.insidevoa.com/p/6438.html and Radio Free Asia, https://www.rfa.org/about/info/listenercomments/comments_korean-08242007111815.html

36. David Maxwell, "North Korea's Provocations: Time For An Asymmetric Approach," *19Forty Five*, October 11, 2022, https://www.19fortyfive.com/2022/10/north-koreas-provocations-time-for-an-asymmetric-approach/.

37. Michael Lee, Jeong Jin-Woo, Chung Yeong-Gyo , "North Korea spends money it needs for food on missiles," *Joongang Ilbo*, December 12, 2022, https://koreajoongangdaily.joins.com/2022/12/12/national/northKorea/Korea-North-Korea-food-insecurity/20221212190148616.html

38. Bowman and Maxwell, *Maximum Pressure*, 50.

39. Bowman and Maxwell, *Maximum Pressure*, 51.

40. Bowman and Maxwell, *Maximum Pressure*, 51.

41. Ethan Jewell, "US earmarks $40M for radio broadcasts into North Korea in honor of Otto Warmbier," *NK News*, December 26, 2022, https://www.nknews.org/2022/12/us-earmarks-40m-for-radio-broadcasts-into-north-korea-in-honor-of-otto-warmbier/?utm

42. Brian Planalp, "President Biden signs legislation in Otto Warmbier's name," *Fox 19 Now*, December 24, 2022 https://www.fox19.com/2022/12/24/president-biden-passes-legislation-otto-warmbiers-name/

43. Robert Joseph, et al., National Strategy for Countering North Korea," *National Institute for Public Policy*, No. 545, January 23, 2023, https://nipp.org/information_series/robert-joseph-robert-collins-joseph-detrani-nicholas-eberstadt-olivia-enos-david-maxwell-and-greg-scarlatoiu-national-strategy-for-countering-north-korea-no-545-january-23-2023/

IV. Hegemon in the Middle East: Iran's Islamic Revolutionary Guards and the Quds Force

By David R. DiOrio

ABSTRACT: This section introduces the Iranian capacity to employ its Islamic Revolutionary Guard Corps (IRGC) and its expeditionary wing, the Quds Force (QF), to advance its hegemonic aspirations. The Islamic Republic of Iran (IRI) believes it is at war with the U.S.—deemed "the Great Satan"—because of its perceived immoral influence and arrogant hegemonic intentions. Since its founding following the overthrow of the pro-U.S. Shah during the Iranian Revolution in 1979, Iran has used an increasingly lethal and expansive network of Shi'a-affiliated surrogates and proxies to expand its reach and influence. The Iran Threat Network (ITN) is a vast constellation of proxies that pursue mutually aligned political objectives while protecting the theocratic Iranian regime that enables them. The ITN conducts political, military, and social development activities—often applying terrorism tactics—to diminish U.S. (and Western) influence in the Middle East and eliminate the State of Israel. Iran seeks to exert their predominance by creating and supporting Tehran-friendly "shadow" governments in places like Syria, Yemen, Lebanon, Bahrain, and Iraq. Iran has leveraged asymmetric and irregular warfare to create political instability and chaos to deter their adversaries and promote their role as the most righteous and legitimate governance authority. The IRGC enables the ITN by providing direction and guidance, command and control, intelligence, military training and armaments, and social or other funding through the QF. Iran resources the hybrid organization to instigate conflict within the gray zone using advanced instruments of hybrid warfare. The IRGC's access to an arsenal of theater ballistic and cruise missiles, drones, cyber tools, maritime interdiction and harassment vessels, and their imminent nuclear weapon capability make them a unique and formidable challenge.

INTRODUCTION

Since its ascension to power after the Iranian Revolution, the Islamic Republic of Iran (IRI) has been aggressively implementing the religious leaders' vision of exporting the revolution across the Arab world.[1] Since 1979, Iran has incrementally constructed a network of irregular partners and asymmetric warfare capabilities to make up for its conventional inadequacy.[2] Their diversified portfolio of hybrid warfare instruments provides a credible threat of lethal and non-lethal operations arrayed against targets of opportunity at the time and place of their choosing while imparting ambiguity and plausible deniability. The Iranian leadership has employed surrogate terrorism, political agitation, and paramilitary violence as the main instruments of power projection.[3] The IRI is exercising a proxy-based stratagem backed by a meager but formidable conventional military order of battle and preeminent nuclear weapon capability.[4]

At its core, the survival of the State and the exportation of the Shi'a worldview are the dominant motivations.

| At its core, the survival of the State and the exportation of the Shi'a worldview are the dominant motivations. |

The ISI views the West as their foremost enemy, aligned with their Western allies and corrupt regional partners, such as Israel and other misguided Islamic States. Iran seeks to upend the international order to benefit all oppressed peoples, both within and outside the Muslim world.[5]

Both regular and irregular forces combine into an Iranian force structure finely tuned for hybrid warfare to provide internal security and support their hegemonic objectives. Tehran has unleashed an array of military forces under the command and support of the Islamic Revolutionary Guard Corps (IRGC), the Quds Forces (QF), and the paramilitary volunteer militia known as the *Basij* (symbol of IRGC; see Figure 6.2). Iran commissioned and designed the IRGC to maintain domestic security and suppress anti-government dissent. This specialized force protects the country's Islamic Republic system and provides a potential counterforce against the Regular Army, which was considered a potential source of a coup. The IRGC has become a considerable military, political, and economic force in Iran, with its own Army, Navy, and Air Force.[6] The QF is a specialized unit subordinate to the IRGC, akin to special operations forces, responsible for training and supporting proxy militias and internal security forces. The Quds are responsible for extraterritorial and clandestine operations to expand Iranian influence and control. The QF provides intelligence, advisory, and direct support to the region's pro-Iranian governments and non-state proxy actors. The Basij is an Iranian paramilitary volunteer militia subordinate to the IRGC.[7] It has a diverse mission set: internal security, law enforcement with an extra-judicial authority, and suppressing popular dissent and opposition to the government.[8] The Basij may also be involved in social services and disaster relief.[9] The IRGC trains Basij cadres in military tactics and Islamic ideology.[10] The Basij typically mobilize during significant internal conflict and political unrest. The combined hybrid force structure is a normal evolution of Iran's operational concepts: The IRGC supports the regular Iranian Army and Navy to protect their territorial integrity in times of general war, and they receive reinforcements from mobilized reserves (Basij) during major internal strife and work with SOF (i.e., QF) to support proxy forces to advance Iranian hegemony.

Figure 6.2. Symbology of the Islamic Revolutionary Guard Corps. Source: Wiki Commons.[11]

Three predominant principles guide the Iranian proxy strategy—gray zone deterrence and compellence using proxy forces, incremental and asymmetrical escalation, and avoiding decisive engagement.[12] A fourth principle of strategic deterrence with a nuclear missile capability is imminent. Iran has built a deterrence warfighting triad consisting of conventional and irregular naval assets capable of disrupting maritime lines of communication; an arsenal of drones and multi-range missiles capable of precision strikes; and a Shi'a foreign service capable of exerting influence as insurgents, counterinsurgents, or terrorists.[13] These Iranian military force elements join

| Three predominant principles guide the Iranian proxy strategy—gray zone deterrence and compellence using proxy forces, incremental and asymmetrical escalation, and avoiding decisive engagement. |

with foreign proxy militias in various proportions to form a hybrid force structure tailored to pursue gray zone warfare principles along the resistance continuum. The mixed force structure is often called the Iranian Threat Network (ITN)—an alliance of surrogates, proxies, and partners that conduct activities to destabilize, disrupt, or destroy anti-Iranian elements throughout the Middle East.[14] The IRGC and QF have ostensibly mastered this hybrid combined-arms concept despite the challenges of controlling a conglomerate of militias on foreign soil and across linguistic barriers. Iran has positioned itself to use small numbers of specialized forces working with substantial proxy elements on foreign battlefields to produce effects disproportionate to their size.[15]

THE STRATEGIC ENVIRONMENT

The struggle for hegemony in the Middle East includes four main regional competitors: Saudi Arabia, Egypt, Turkey, and Iran (see Figure 6.3).[16] Each contender has unique national interests, foreign policy objectives, partners, client states, and organizations to challenge the others. These competitors may align and seek partnerships to advance mutual interests or become confrontational to further their agenda. The relationships between them ebb and flow depending on the geopolitical situation. Saudi Arabia is a monarchy that espouses the strict Islamic worldview of Wahhabism (puritanical Sunni ideology). Saudi strength emanates from the resourcing and exportation of energy and its partnerships with the West. The Gulf Cooperation Council nations largely support Saudi interests, especially those related to mutual economic benefit and the proliferation of Sunni ideology. With their neighboring Emirates, the Saudis exercise an Islamic autocracy form of government (monarchies) that recognizes territorial sovereignty with authoritarian rule. Egypt embraces Islamic modernism, a hybrid governance system that incorporates the attributes of Western democracy with the strict spiritual guidance of a Sunni worldview. Egyptian strength stems from its strategic location in control of the Suez Canal, which secures favors from all global powers. Turkey's governance system espouses Islamic secularism that emerged from the devastation of World War I. This governance system divorces government from religious ideology, offering freedom to pursue a spiritual choice while exercising a more inclusive or democratic approach to policymaking. Turkey status is elevated because it is a member of NATO, located at the crossroads between East and West. Turkey was the epicenter of the last great caliphate—the Ottoman Empire—and seeks to revive its leadership role within the Muslim world by promoting the virtues of a secular nation-state model backed by the security of the NATO alliance. Politics, commerce, and religion are intertwined with Saudi Arabia, and Turkey—each seeking a balance between espousing their religious worldview and partnering with the global community. Iran has a significantly governance with distinctly different strategic objectives.

Figure 6.3: Map of Iran and other regional competitors. Source: Voice of America.[17]

The Iranian government is a complex blending of theocratic and political elements to advance the Islamic Shi'a worldview as the dominant moral authority within the Islamic world. Iran is a consummate revisionist actor that professes an Islamic jihadist governance methodology to promote the tenets of their Islamic Revolution of 1979. Before 1979, the U.S. supported the pro-Western Shah with military assistance to hedge against Soviet influence in the region. Islamic Shi'a clerics and traditional Muslim leaders believed the Western-backed Shah was infringing on their traditional role as the guardians of the faith and principal moral authority. Theocratic truth-seekers advocated a sociopolitical sect based upon traditional Shi'ite dogma, believing that global liberation movements against colonialist oppressors were a justified obligation. The overthrow of the Shah and the ensuing hostage crisis were the pivots upon which the Tehran regime and American psyche changed from benevolence to malevolence.

In 1979, Iran shifted to a spiritual, social order with an autocratic rule of the clerics that fused politics and religion. From that time forward, it was challenging to find common ground with Iran, and the pursuit of national interests was in confrontation characterized by a competition for influence. U.S. policymakers found cooperation even more difficult after 9/11 when Tehran supported and heralded terrorism imposed directly upon the American people. From the Iranian perspective, the U.S., in concert with Western nations, stands as an intolerable impediment to their pursuit of a righteous governance system and moral jurisprudence inspired by their religious ideology. The Grand Ayatollah presides over an authoritarian Imamate to espouse the Shi'a worldview as the only righteous governance path. The competing dominions find imperfections and weaknesses in the other's approach, ranging from oppression to corruption to heresy. Radical Islamic philosophers contend that all these systems are heretical and that restoring a traditional caliphate exercising Islamic jurisprudence (sharia) is the only way to purify the religion and assume a righteous way of life to gain the blessing of Allah. The prevailing hegemonic approach remains to be determined.

| Sunnis fused religion with politics, where all policymaking is within the realm of government officials—democratic or autocratic. In contrast, Shiites believe in the primacy of the Ayatollah and politics as an extension of ideology as interpreted by the clerics. This divergence in governance systems characterizes the ultimate disparity among Muslims; |

Historical, geopolitical, and ideological factors influence Iran's strategic goals. The Sunni-Shia split during the struggle to determine the righteous successor to the Prophet Muhammad in 680 CE led to a political and religious schism that reverberates today. Sunnis and Shiites developed distinct interpretations of Islamic teachings, governance structures, and jurisprudence that manifested utterly different views of leader legitimacy and the role of the Muslim clerics. As Muhammad professed, both factions seek to unify the Muslim *Ummah* (Community) under their separate interpretations of the proper governance structure. Sunnis fused religion with politics, where all policymaking is within the realm of government officials—democratic or autocratic. In contrast, Shiites believe in the primacy of the Ayatollah and politics as an extension of ideology as interpreted by the clerics. This divergence in governance systems characterizes the ultimate disparity among Muslims; however, in some situations, Sunnis and Shiites find common ground on the path to unification. Radical-leaning elements of the Ummah embrace the expulsion of foreign influence and the subjugation of Israel as collective goals. Undeniably, these strategic objectives are the dominant aspects of Iranian policy and action.[23] However, many Muslims expound a *Quietist* view that separates religious and political affairs by focusing solely on individual spiritual reflection.[24] The predominant ideological approach is ambiguous, and frequent opposition protests in Iran portend a disconnect between Iranian public sentiment and foreign policy.[25]

IRANIAN FOREIGN POLICY

Tehran's principal foreign policy objective is to retain tight control of the population by isolating them from detestable foreign regimes. The Shi'a clerics focus on maneuvering the political course of the Middle East by driving out Israel and the U.S. with their allies. Tehran postulates that it cannot withstand an enduring military engagement with the West, so it pursues conflict in the gray zone to gain an operational advantage by inspiring localized conflict in critical regions to increase the cost of a persistent Western presence in both lives and money. Proxy factions, beholding to Tehran, militarily and politically challenge Iranian adversaries to diminish public support and create a governance void. The IRI objective is to inspire a withdrawal of Western forces to generate political space filled with pro-Iranian factions.[27]

| The IRI objective is to inspire a withdrawal of Western forces to generate political space filled with pro-Iranian factions. |

Tehran's decision-making is intentionally opaque, and they appear to augment their militancy and terrorism strategy with an aggressive diplomatic campaign with their neighbors to weaken Arab ties to the U.S. and gain a favored nation status. Tehran is also strengthening relationships with China, Russia, and Turkey to open trade channels and mitigate the impact of Western economic sanctions.[28] Pursuing these neoteric political ties to these global powers will give Iran a more significant standing in the region if—and when—the U.S. withdraws.[29]

A high level of mutual insecurity characterizes the U.S. and Iran relationship.[30] The U.S. perceives the Iranian Revolution as a significant destabilizing movement threatening Americans and the vital lines of communication throughout the region. Indeed, anti-American protests proclaiming "Death to America" have been steadfast and frequent since the 1979 hostage crisis, setting the stage for decades of hostility and recurrent conflict.[31] The U.S. has consistently sided with Iran's rivals during every regional conflict, including the Tanker War, the Iran–Iraq War, and Operation ENDURING FREEDOM. Americans have died at the behest of the Iranians in Lebanon (Beirut) and Saudi Arabia (Khobar) and in the prolonged Israeli conflicts.[32] The U.S. military presence and operational tempo in the region are determined mainly by IRGC actions and order of battle. Tehran is consequently intimidated by the substantial and continuous U.S. presence.[33]

The Supreme Leader, Grand Ayatollah Khamenei, repeatedly claims that the U.S. intends to overthrow the regime and deems America the "Great Satan." Iran regards the economic sanctions against them as a U.S.-instigated economic war.[34] Iran also blames the U.S. for inspiring radical Sunni Islamic movements such as ISIS by supporting corrupt Arab regimes.[35] Tehran believes that the U.S.-controlled regional political structure exacerbates the plight of oppressed peoples such as Palestinians and minority Shi'a populations.[36] Iranian leaders profess that the U.S. intervention in support of Israel has grossly distorted the region's political and economic balance away from the rightful indigenous peoples. The Iranian narrative encourages liberal intellectuals worldwide to arouse pro-Iranian sympathy and portray Iran as a champion of global self-determination despite its sponsorship of terrorism. Iranian policy has consequently focused on eradicating U.S. intervention and inoculating Iranian domestic affairs from U.S. influence.[37]

THE OPERATIONAL ENVIRONMENT

The clash of hegemony in the Middle East has manifested itself into a prolonged gray zone conflict—activities by a state that are detrimental to an opposing state and may be considered acts of war but are not legally deeds of war.[38] Iran has empowered their IRGC with asymmetric tools backed by a formidable conventional military to avoid a confrontation with the West. It is pursuing a nuclear capability to deter potential escalation if a political redline is approached or crossed. Iran acts belligerently to safeguard the revolution and expand its reach and influence—outside the norms of traditional diplomacy but below the threshold of war.[39] The IRGC exploits an advantage where pockets of resistance and rejection are crippling local authority.[40] IRGC-trained militants—regular and irregular—have proven to be very adept in asserting themselves in fragile and disordered environments.[41] Tehran is too weak militarily and lacks sufficient political or economic clout to compel direct change, so it indirectly enables sympathetic proxy forces to subvert the Arab states with blackmail and insurgency while waging a relentless terrorist campaign against the U.S. and Israel.[42] This substitution-style approach gives Tehran plausible deniability while undermining its regional competitors.[43]

In short, Iran has devised an operational approach to circumvent U.S. deterrence endeavors to gain political advantage and obtain diplomatic leverage, control the strategic narrative to influence international opinion, and create security dilemmas to increase their military freedom of action while managing risk to avoid war.[44] Three governmental lines of effort describe the Iranian operational construct. First, the IRI engages politically to confound unified international opposition.[45] Tehran is pursuing security arrangements with other revisionist actors like Russia and China to counterbalance U.S. strength.[46] In parallel, they engage with the Western antagonists and neighboring Arab states to gain some relief from isolation and sanction—and perhaps fragment—those inclined toward an aggressive counter-Iranian position. Second, Tehran provides direct support to rejectionist forces that are resisting status quo Arab powers.[47] Information operations and advanced cyber activities are applied to recruit proxy partners, counter the Western narrative, and inspire an enduring campaign. The diverse resistance movements are evolving and expanding in a tumultuous political environment, presenting a significant challenge to IRGC military planners and befuddling Western military planners seeking to counterbalance the emerging threat. A third line of effort encompasses employing Iran's hybrid conventional and irregular military capability to offset their conventional military limitations.[48] The IRGC tests and probes these lines of effort to determine the opposition resolve-and-response threshold.[49] Tehran discerningly exercises operational art by analyzing and assessing reciprocity and proportionality in each line of effort to manage risk and achieve desired outcomes.[50]

In the context of hybrid warfare, the QF is Iran's most versatile hybrid warfare tool in gray zone deterrence or compellence. However, determining when to deter versus compel is complex and challenging.[51] The IRGC exercises the deterrence strategy by creating security dilemmas, such as threatening "all-out war," which immediately unmasks underlying divisions within the U.S. political bureaucracy between activists and accommodators and generates paralysis among some policymakers.[52] Iranian deterrence has two main facets: Tehran exercises a conventional deterrent strategy with a formidable ballistic missile and rocket capability and a forward deterrence strategy by enabling proxies to induce risk to foreign forces in the region.[53] The QF maintains, trains, and equips resistance movements to create chaos and weaken anti-Iranian governments to exert political manipulation and freedom of military maneuver. Proxy-based tactics exercised by the QF operating in the gray zone restrict or deter U.S. operations while shielding Iranian vulnerabilities or intentions. Iran reduces risk by controlling proxy forces not under formal Iranian command, but ancillary authority presents significant pitfalls. Adversaries can sway proxy allegiances or leverage internal proxy tensions if they understand the disparate political dynamics and divergent sympathies.[54] Suppose deterrence fails or an immediate threat emerges. In that case, the IRGC may abandon the obscurity and protection of proxy gray zone activities to act overtly and directly despite their propensity for caution. The IRGC designs activities to compel an immediate behavioral or policy change, such as the 2019 rocket attack in the Green Zone of Baghdad and the 2020 missile strike on al-Asad Airbase in response to the killing of IRGC-QF Commander Qasem Soleimani.[55] Violent compellence measures risk escalation and a disproportionate enemy response. Also, overt lethal attacks tend to shift international opinion against the protagonist or infuriate the enemy, often rousing the opposite intended reaction. Compellence effects are usually short-lived with a limited shelf life, but they could present significant security dilemmas for their adversaries.[56]

| Tehran's longstanding and methodical infiltration into regional politics and incremental blending of advanced military capabilities with innovative low-cost asymmetrical capabilities forces the U.S. to expend extraordinary effort and resources in adapting to the operational environment. |

From Iran's perspective, its incremental creation of a regional threat network imposes a significant cost to the U.S. without escalating the confrontation to conventional war.[57] If the U.S. deprives the IRGC of its gray zone operations and asymmetric capabilities, the balance of power would be significantly biased against Tehran.[58] The obscurity and informality of the network present a significant challenge to U.S. policymakers and military planners, especially in situations where the U.S. disagrees with the Allies regarding the composition and scope of particular aspects of the network.[59] A combined international strategy to counter the ITN still needs to be discovered. . Tehran's longstanding and methodical infiltration into regional politics and incremental blending of advanced military capabilities with innovative low-cost asymmetrical capabilities forces the U.S. to expend extraordinary effort and resources in adapting to the operational environment.[60]

The IRGC innovatively exploits the inherent advantages of geography and selectively procures weaponry to threaten—but not yet engage—enemy power projection and global commerce, mainly through maritime chokepoints. Iran cannot currently project its military power beyond its frontiers, but their advanced weaponry can inflict severe damage regionally.[61] This persistent threat, absent a decisive engagement, requires the U.S. to resource and sustain a continuous forward presence. The operational situation, therefore, becomes a test of wills, and Iran is determined to outlast their adversaries.

THE TACTICAL ENVIRONMENT – THE IRANIAN THREAT NETWORK

The Iranian Threat Network is an alliance of surrogates, proxies, and partners that conduct activities to destabilize, disrupt, or destroy anti-Iranian entities throughout the Middle East.[62] Figure 6.4 depicts the significant actors within the ITN, and the following briefly describes each.[63] The IRGC-QF recruits, trains, equips, advises, and often controls the ITN at the behest of Tehran and is integral to the IRI formal decision-making process.[64] The IRGC encourages the ITN to cooperate and is willing to work with entities of different religious ideologies to advance shared political objectives.[65] The organization has been traditionally clandestine, but Iran is increasingly becoming more open in legitimizing these groups publicly.[66] The IRGC is skilled in leveraging common grievances to foster rapport and bring groups into the fold, especially during global social movements like the Arab Spring. Tehran endeavors to appeal to ethnic and religious minorities in the region whose disenfranchisement allows Tehran to recruit fighters.[67] Iran considers the ITN a cost-effective force multiplier to complement other asymmetric capabilities, such as ballistic missiles and drones, to compensate for conventional shortfalls.[68] The IRGC-QF directly or indirectly supports it and operates in several countries and all domains—air, sea, land, and cyber.[69]

The ITN engages in lethal and non-lethal actions in a variety of militancy actions—hostage seizing, embassy raids, terrorist attacks, and harassment of maritime trade; autonomous and proxy rocket, missile, and drone strikes; information operations; weapons tests for propaganda; and a myriad of cyber activities.[70] It is evolving and adapting to the current operating environment by becoming more expeditionary and astute in pursuing hybrid warfare, blending conventional and irregular modes of operation to gain synergies.[71] The ITN aggressively and competently employs asymmetric tools, such as improvised explosive devices, rocket-propelled grenades, man-pads, drones, and cyber and information warfare tools to realize disproportionate effects. At the same time, it imposes significant costs on the U.S. regarding lives, materials, and money without escalating the violence to conventional war.[72] In return, the ITN actors become more credible and effective through professional military training and resourcing.[73] From the ITN perspective, aligning with Tehran makes them a more significant player on the world stage.

Figure 6.4. The Iran Threat Network map. Source: Council on Foreign Relations.[74]

Lebanese Hezbollah (LH) is an Iranian-supported Shi'a political and militant group that emerged to counter the State of Israel.[75] The political wing has significant representation in the Lebanese Parliament to direct policy and provide social services. The formidable militant wing—much larger than the Lebanese Army—is primarily poised on the northern Israeli border. LH receives military assistance from Iran and Syria, including funding, weapons, and military training.[76] It is designated a foreign terrorist organization (FTO) by the U.S., European Union, and Arab League.

Fatemiyoun is a significant Iranian proxy militia that is pro-Assad regime and fighting in Syria, often alongside the Syrian Army. It is composed primarily of Afghan Shi'a militia, and many are refugees—sometimes child soldiers—displaced from Afghanistan during the ongoing civil conflict. They are known for their mobility and capacity to transcend borders. The Hezbollah factions throughout the region closely align with the group. It is directly supported by the IRGC-QF, often fights under the command of Iranian officers, and is offered Iranian citizenship as an incentive.[77]

Zeynabiyoun is a relatively small but effective Shi'a Islamist group of Pakistanis living in Iran. The group was recruited and trained by the IRGC and emerged in force from the Arab Spring movement. It currently operates from Syria and operates under IRGC command. The Zeynabiyoun is principally focused on protecting Shi'a holy shrines but will also join in general engagements.[78]

Patriotic Union of Kurdistan (PUK) is an Iraqi Kurdish radical-leaning political and military organization that advocates for Kurdish self-determination. Despite ideological differences, the Iran-PUK relationship is a unique military alliance to defeat common enemies, such as Saddam Hussein and ISIS. The Peshmerga militant wing is primarily trained and resourced by the IRCG-QF.[79]

Badr is an Iraqi Shi'a political and militant organization in Iraq. Iran helped form Badr to oppose Sunni political dominance and undermine U.S.-Sunni cooperation despite the Sunni minority position. The group received training and support from the IRGC-QF in failed efforts to overthrow Saddam Hussein's regime after the Iran-Iraq war. The group currently has significant representation in the Iraqi National Assembly.

Kata'ib Hezbollah (KH) is an Iraqi paramilitary entity of the Popular Mobilization Forces (PMF), an umbrella Shi'a militia organization targeting ISIS. KH is a splinter group from the Mahdi Army militias formed by cleric Moqtada al-Sadr in 2004. The group actively engaged in an insurgency against the U.S.-backed Iraqi authority but assisted the U.S.-supported campaign against ISIS. The IRGC-QF is actively sustaining KH in support of the Assad regime in Syria. The former KH Commander, Abu Mahdi al-Muhandis, was killed alongside IRGC-QF Commander Soleimani.[81]

Asa'ib Ahl Al Haq (AAH) is another Iraqi Shi'a faction and Mahdi Army splinter group that operates as a unit of the PMF. It is notorious for its human rights violations and war crimes. The AAH has representation in the Iraqi Parliament and continues to receive financial and military support from Iran.[82]

Hamas is a Palestinian Islamic-Sunni fundamentalist political and militant organization located in Gaza and established during the First Intifada uprising in 1987. Iran forged an Iran–Hamas partnership to disrupt the Israeli–Palestinian peace process, although frequent rifts occur concerning Hamas sympathies toward Sunni rebels in Syria against Assad. It is an internationally recognized FTO that receives funds, weapons (notably missiles and rockets), training, and military planning capacity from the IRGC.[83] It is currently engaged in an active military campaign with Israel to undermine the Arab–Israeli peace process. Some analysts speculate that the IRGC-QF was directly involved with the Israel attack of October 7, 2023, at an unprecedented level by training and equipping Hamas operatives, planning the operation, and executing the conflict with advanced command and control.

Palestinian Islamic Jihad (PIJ) is a Palestinian militant group that embraces Muslim Brotherhood dogma and operates in both Gaza and the West Bank of Israel. The group professes the establishment of an Islamic State within Israel and is designated an FTO. The group receives significant logistical and military support from the IRCG-QF. It frequently launches terrorist attacks, including suicide raids against Israeli citizens, prompting a diverse range of Israeli Defense Force (IDF) retaliatory responses.[84]

Houthis are a Yemeni Zaidi Shi'a group that espouses a theological movement to resist the expansion of the Sunni worldview emanating from their neighbors. The Houthis are actively engaged with Sunni-backed (mainly Saudi Arabia and the UAE) political entities vying for power in the Yemeni Civil War. Iran aids the Houthis to create a foothold of pro-Iranian influence in their southern sphere and control the maritime trade through the chokepoint in the Bab el-Mandeb Strait.[85]

| Rand Corporation researchers Tabatabai, Martini, and Wasser have developed an insightful model for describing proxy focus, intentions, and subsequent activities. The analysts describe the client-state relationship between Iran and its ITN in four distinct models that embody the gray zone strategy: Targeters, Deterrers, Stabilizers, and Influencers. Targeters focus on targeting foreign forces garrisoned in the region to undercut the Western presence, perhaps making it too costly to maintain troops forward deployed. Deterrers focus on deterring and harassing Iran's regional rivals to shift the balance of power in their favor. Stabilizers help stabilize Iranian allies by augmenting the regular military with capable militia units into a hybrid force structure under direct or indirect IRCG command. Influencers are Iranian partners and proxies integrated into the country's political processes and security establishments. Depending on the evolving political and operational environment, a group may exhibit the characteristics of different models simultaneously or move between them. Iranian support and control provide a measure of legitimacy and credibility to these groups that aggressively inculcate conflict and pursue pro-Iranian agendas throughout the region. |

The color scheme depicted in Figure 6.4 provides a general synopsis of focus within the ITN.[92] The entities typically concentrate on one approach at a time but may change their direction or adapt to the particular security situation, demonstrating their capacity for flexibility, innovation, and depth. A vital aspect of the discrete models is the extent of Iranian command and control (C2) exercised over a given group. Tehran exerts C2 judiciously to prevent exposure and reveal intentions. The diversity in approaches provides Tehran with flexible military and political tools to implement its operational design while preserving plausible deniability.[93] The ITN is likely to continue as the region's primary warfare instrument of power projection.[94] Iran will concurrently improve and test their asymmetric capabilities. If current trends prevail, the network is the most immediate and likely threat to forward-deployed forces, requiring the West to maintain a sizeable forward presence. A large forward presence poses a significant security dilemma whereby deployment in force may deter conflict, reassure partners, influence political processes, and decrease contingency response times; a large military footprint also exposes more targets to the ITN.

IRAN'S NATIONAL SECURITY INSTRUMENTS

Iran implements their national security strategy with a meager albeit capable conventional order of battle and various asymmetric tools. The main instrument of policy is the irregular armed factions that execute the proxy stratagem. However, there are a plethora of conventional and asymmetric capabilities that enable them or protect them. Iran is also developing an impressive array of cyber tools and is adept at leveraging social media and conducting information operations. Iran supplies proxy forces with advanced ballistic missiles, armed or surveillance drones, rocket-propelled grenades, anti-ship/tank/air missiles, or defensive missile systems.[95] Iran may augment with direct fires from IRGC or regular military elements—sometimes from within or near their territorial boundary—with ballistic or cruise missiles, advanced naval mines, manned or crewless fast boats, diesel submarines (Kilo-class) with capable torpedoes, and a handful of capable fighter aircraft.[96] This evolution of IRGC-QF hybrid warfighting techniques, coupled with Iran's avowed intention to purchase advanced fighter bombers and submarines from Russia or China, indicates that they may seek to field a conventional military capability to rival the Western order of battle in a general war.[97]

The IRGC is developing innovative ways of waging conflict beyond its borders, using the Syrian conflict as a laboratory. Iranian regular combat units drawn from conventional brigades and divisions of the IRGC have been fighting alongside and on the front lines with Syrian and Iraqi proxy militias and Lebanese Hezbollah.[98] The hybrid units appear to be deploying as cadres—commanding with Iranian officers, some non-commissioned officers (NCOs), and a relatively small number of enlisted ranks. They are joining with Syrian, Iraqi, and Hezbollah militia that serve as their foot soldiers.[99] Iranian officials attempt to mask the extent of their involvement in the Syrian and Palestinian conflicts, continually insisting that they are only training, assisting, and advising militia forces.[100]

> ***| Iranian officials attempt to mask the extent of their involvement in the Syrian and Palestinian conflicts, continually insisting that they are only training, assisting, and advising militia forces. |***

Casualty analysis and media reports of the unit affiliations of Iranians reported killed in Syria tell a different story. The proportion of Iranians killed compared to the soldiers shows that IRGC-QF officers are leading militia soldiers in primary offensive operations and not purely advising.[101] The IRGC can send a leadership cadre forward, embed it among the militia, and successfully command those proxy militias in exceedingly hard fighting.[102]

CONCLUSION

The IRGC-QF-Basij combined cadres operating with foreign proxy militia supported by Iranian regular forces can present a formidable challenge. The hybrid force structure employing asymmetric tactics complicates U.S. joint and combined military planning by increasing the operational risk of escalation and asymmetric threat to operating forces. Swarming explosive suicide boats armed with anti-ship cruise missiles, surface-to-ship missiles, or submarines employed in narrow maritime chokepoints, suicide drone strikes, cyberattacks on critical infrastructure, or stand-off ballistic missiles might defy conventional defensive measures. The IRGC may leverage a combination of swarming fast boats, sophisticated integrated air defense networks, simultaneous cruise missiles and drone attacks, and matrixed irregular forces to create dilemmas for opposing forces.[103] The hybrid capabilities and the unorthodox employment doctrine emphasize holding forward operating bases and deployed troops in the region at risk from over-the-horizon strikes.[104] There is general agreement among military analysts that Iran's hybrid force structure would not endure a protracted major conflict. However, it can be effective when deployed to leverage gray zone warfare. Iran is also pursuing future capabilities in the space and cyber domains. The ultimate game-changer would be the maturing of their nuclear capability poised on long-range ballistic or cruise missiles. A credible and secure Iranian nuclear capability would dramatically change the deterrence dynamics and response calculus; therefore, averting that capability is a top priority for the United States.

| A credible and secure Iranian nuclear capability would dramatically change the deterrence dynamics and response calculus; therefore, averting that capability is a top priority for the United States. |

The Islamic Republic of Iran has entwined the U.S. in this gray zone dilemma since the 1979 Iranian Revolution.[105] Tehran's steadfast quest for security and its perceived moral obligation to expand the tenets of the Islamic Shi'a worldview are the drivers that shape their actions. We face the reality that Iran may become a significant regional political and military actor post-sanctions.[106] The Iranian foreign policy centers on perpetuating the Iranian Revolution and resisting foreign interference in that endeavor. Iran will persistently oppose U.S. regional influence and reach to create more freedom of action. We must fully expect Tehran to continue threatening its most potent regional rivals' security, authority, and legitimacy to enhance the prospects of being recognized as the primary arbiter of regional power and governance. Iran will likely pursue new or improved relationships with state and non-state regional partners as a deterrent against foreign interference. Iran will attempt to expand its network of regional relationships to broaden its security network and open commercial lines of communication.[107] With the recent reduction of U.S. presence in the region, Tehran discerns a new opportunity to leverage asymmetric and irregular warfare tactics to circumvent or undermine deterrence efforts.[108] Iran is pursuing a proxy stratagem, aligning with regional militias with a compatible ideological affinity and sharing common enemies, like Israel and Saudi Arabia. These activities bring Iran and the U.S. into perpetual competition and conflict in either practice or purpose.[109] Military leaders, practitioners, and planners must clearly understand the weaknesses and failings of a proxy strategy to counter it successfully. Doing so may provide an opportunity to steer Iran toward a more democratic, peaceful, and inclusive posture within the global community.

About the Author

Dr. David R. DiOrio, Captain, U.S. Navy (retired) is a National Security Professional with a Doctor of Philosophy degree in Public Policy and Administration from Walden University. He served as the Deputy Director at the Joint Forces Staff College of National Defense University and is currently Adjunct Faculty at the Joint Special Operations University.

NOTES

1. Ariane M. Tabatabai, Jeffrey Martini, and Becca Wasser, "The Iran Threat Network (ITN): Four Models of Iran's Nonstate Client Partnerships," (RAND Corporation, 2021), 7.
Found at https://www.rand.org/pubs/research_reports/RR4231.html, accessed on September 7, 2023.

2. Tabatabai et al.," The Iran Threat Network," 4.

3. Nathan P. Freier et al., "Outplayed: Regaining Strategic Initiative in the Gray Zone," A Report Sponsored by the Army Capabilities Integration Center in Coordination with Joint Staff J-39/Strategic Multi-Layer Assessment Branch (US Army War College Press, 2016), 53, https://press.armywarcollege.edu/monographs/925, accessed on October 6, 2023.

4. Freier et al., "Outplayed."

5. Michael Eisenstadt, "Deterring Iran in the Gray Zone: Insights from Four Decades of Conflict," The Washington Institute for Near East Policy/Policy Notes 103 (2021), 2. Found at https://www.washingtoninstitute.org/policy-analysis/deterring-iran-gray-zone-insights-four-decades-conflict, accessed on August 18, 2023.

6. U.S. Congress, Congressional Research Service, *Iran's Foreign Defense Policies*, CRS Report R44017, (Washington, DC, January 11, 2021), 17. Found at https://crsreports.congress.gov/product/pdf/R/R44017, accessed October 2, 2023.

7. Ali Alfoneh, "The Basij Resistance Force," see Overview. United States Institute for Peace: The Iran Primer, October 6, 2010. Found at https://iranprimer.usip.org/resource/basij-resistance-force, accessed September 1, 2023.

8. Afoneh, "The Basij Resistance Force."

9. Afoneh, "The Basij Resistance Force.".

10. Afoneh, "The Basij Resistance Force.".

11. Illustration of the IRGC found in wiki commons and public domain, found at https://commons.wikimedia.org/wiki/File:IRGC_Fin_Flash.svg, accessed on 22 July 2024.

12. Congressional Research Service, *Iran's Foreign Defense Policies*, 3.

13. Congressional Research Service, *Iran's Foreign Defense Policies*, 3.

14. Tabatabai et al., "The Iran Threat Network," 2.

15. Congressional Research Service, *Iran's Foreign Defense Policies*, 13-19.

16. Diane Zorri, Houman Sadri, and David Ellis, *Iranian Proxy Groups in Iraq, Syria, and Yemen: A Principal-Agent Comparative Analysis* (JSOU Press, 2020), 16. Map from CIA World Factbook, found at https://www.cia.gov/the-world-factbook, accessed on 28 Dec. 2023.

17. Photo from *Voice of America* and free of copyright, found at https://www.voanews.com/a/human-rights-groups-condemn-journalists-prison-sentences-in-iran/3305083.html, accessed on 22 July 2024.

18. Freier et al., "Outplayed," 49.

19. Zorri et al., *Iranian Proxy Groups*, 15.

20. Zorri et al., *Iranian Proxy Groups*, 15.

21. Zorri et al., *Iranian Proxy Groups*, 15.

22. Zorri et al., *Iranian Proxy Groups*, 15.

23. Congressional Research Service, *Iran's Foreign Defense Policies*, 1-3.

24. Zorri et al., *Iranian Proxy Groups*, 15.

25. Congressional Research Service, *Iran's Foreign Defense Policies*, 3.

26. Kenneth Pollack, "Iran's Grand Strategy Has Fundamentally Shifted," *Foreign Policy*, August 15, 2023, 1. Found at https://foreignpolicy.com/2023/08/15/irans-grand-strategy-has-fundamentally-shifted/, accessed September 22, 2023.

27. Eisenstadt, "Deterring Iran in the Gray Zone," 18.

28. Congressional Research Service, *Iran's Foreign Defense Policies*, 2.

29. Pollack, "Iran's Grand Strategy Has Fundamentally Shifted," 2.

30. Freier et al., "Outplayed," 52.

31. Eisenstadt, "Deterring Iran in the Gray Zone" 2.

32. Freier et al., "Outplayed" 52.

33. Congressional Research Service, *Iran's Foreign Defense Policies*, 1.

34. Congressional Research Service, *Iran's Foreign Defense Policies*, 1.

35. Congressional Research Service, *Iran's Foreign Defense Policies*, 1.

36. Congressional Research Service, *Iran's Foreign Defense Policies*, 2.

37. Congressional Research Service, *Iran's Foreign Defense Policies*, 2.

38. Zorri et al., *Iranian Proxy Groups*, 1.

39. Freier et al., "Outplayed," 52.

40. Freier et al., "Outplayed," 49.

41. Freier et al., "Outplayed," 49.

42. Pollack, "Iran's Grand Strategy Has Fundamentally Shifted," 1.

43. Eisenstadt, "Deterring Iran in the Gray Zone," 19.

44. Ibid., 3.

45. Freier et al., "Outplayed," 50.

46. Freier et al., "Outplayed," 50.

47. Freier et al., "Outplayed," 50.

48. Freier et al., "Outplayed," 51.

49. Eisenstadt, "Deterring Iran in the Gray Zone," 3.

50. Eisenstadt, "Deterring Iran in the Gray Zone," 3.

51. Eisenstadt, "Deterring Iran in the Gray Zone," 5.

52. Eisenstadt, "Deterring Iran in the Gray Zone," 8.

53. Zorri et al., *Iranian Proxy Groups*, 20.

54. Freier et al., "Outplayed," 49.

55. Tabatabai et al., "The Iran Threat Network," 3.

56. Eisenstadt, "Deterring Iran in the Gray Zone," 6.

57. Tabatabai et al., "The Iran Threat Network," 4, 6.

58. Tabatabai et al., "The Iran Threat Network," 4.

59. Tabatabai et al., "The Iran Threat Network," 3.

60. Freier et al., "Outplayed," 51.

61. Freier et al., "Outplayed," 50.

62. Tabatabai et al., "The Iran Threat Network," 2.

63. Tabatabai et al., "The Iran Threat Network," 10.

64. Tabatabai et al., "The Iran Threat Network," 8.

65. Tabatabai et al., "The Iran Threat Network," 8.

66. Tabatabai et al., "The Iran Threat Network," 5.

67. Tabatabai et al., "The Iran Threat Network," 6.

68. Tabatabai et al., "The Iran Threat Network," 5.

69. Tabatabai et al., "The Iran Threat Network," 3.

70. Eisenstadt, "Deterring Iran in the Gray Zone" 3.

71. Tabatabai et al., "The Iran Threat Network," 3.

72. Tabatabai et al., "The Iran Threat Network," 4.

73. Tabatabai et al., "The Iran Threat Network," 5.

74. Kali Robinson and Will Merrow, "Iran's Regional Armed Network," *Council on Foreign Relations*, 31 January 2024, found at https://www.cfr.org/article/irans-regional-armed-network, accessed on 3 April 2024. Per the Council on Foreign Relations policy, it grants free-of-charge access to content.

75. Congressional Research Service, *Iran's Foreign Defense Policies*, 34.

76. Zorri et al., *Iranian Proxy Groups*, 21.

77. Congressional Research Service, *Iran's Foreign Defense Policies*, 16.

78. Congressional Research Service, *Iran's Foreign Defense Policies*, 15.

79. Congressional Research Service, *Iran's Foreign Defense Policies*, 17.

80. Congressional Research Service, *Iran's Foreign Defense Policies*, 31.

81. Congressional Research Service, *Iran's Foreign Defense Policies*, 31.

82. Congressional Research Service, *Iran's Foreign Defense Policies*, 31.

83. Congressional Research Service, *Iran's Foreign Defense Policies*, 35.

84. Congressional Research Service, *Iran's Foreign Defense Policies*, 35.

85. Congressional Research Service, *Iran's Foreign Defense Policies*, 37–38.

86. Tabatabai et al., "The Iran Threat Network."

87. Tabatabai et al., "The Iran Threat Network," 9.

88. Tabatabai et al., "The Iran Threat Network," 10.

89. Tabatabai et al., "The Iran Threat Network," 11.

90. Tabatabai et al., "The Iran Threat Network," 14.

91. Tabatabai et al., "The Iran Threat Network," 16.

92. Tabatabai et al., "The Iran Threat Network," 10.

93. Tabatabai et al., "The Iran Threat Network," 9.

94. Tabatabai et al., "The Iran Threat Network," 19.

95. Congressional Research Service, *Iran's Foreign Defense Policies*, 9-12.

96. Congressional Research Service, *Iran's Foreign Defense Policies*, 9-12.

97. Frederick Kagan and Paul Bucala, "Iran's Evolving Way of War: How the IRGC Fights in Syria," 2. Found at https://www.aei.org/foreign-and-defense-policy/middle-east/irans-evolving-way-of-war-how-the-irgc-fights-in-syria/, accessed October 2, 2023.

98. Kagan and Bucala, "Iran's Evolving Way of War," 1.

99. Kagan and Bucala, "Iran's Evolving Way of War," 1.

100. Kagan and Bucala, "Iran's Evolving Way of War," 1.

101. Kagan and Bucala, "Iran's Evolving Way of War," 1.

102. Kagan and Bucala, "Iran's Evolving Way of War," 2.

103. Eisenstadt, "Deterring Iran in the Gray Zone," 51.

104. Eisenstadt, "Deterring Iran in the Gray Zone," 51.

105. Freier et al., "Outplayed," 49.

106. Kagan and Bucala, "Iran's Evolving Way of War," 2.

107. Freier et al., "Outplayed," 55.

108. Eisenstadt, "Deterring Iran in the Gray Zone," 2.

109. Freier et al., "Outplayed," 55.

V. Defeated? Violent Extremists Organizations and Non-State Actors

By Namrata Goswami

ABSTRACT: Violent extremist organizations (VEOs) and non-state actors challenge the state's legitimacy and its monopoly over the use of legitimate organized violence. The presence of VEOs has a direct effect on the target population, rendering them insecure and responsive to the demands of these groups. Several factors support VEOs ability to sustain themselves. These include organizational structures, population base, terrain, dynamics of allegiances, motivation, training, financial sources, politics, strategy and tactics, structural conditions, external support and bases. This section elaborates on these factors with illustrative cases of VEOs and offers a counterstrategy to defeating VEOs that involve primacy of the political goals, population support, countering propaganda, understanding the terrain, resolute leadership, intelligence, robust military force structures, rule of law, unity of effort, and operational clarity.

INTRODUCTION

Violent extremist organizations (VEOs) and non-state actors exist as a medium to struggle against the state for territorial control, legitimacy, and influence among their target populations. Consequently, the population is a key factor of support in any effort to counter VEOs and non-state actors. Best factors to include in an assessment of population support are intelligence, unity of effort, the understanding of local politics, countering VEO propaganda, developing the appropriate military force structures, rule of law, unity of effort, and operational clarity. These factors, cumulatively utilized, have the effect of defeating VEOs and non-state actors in the long run.

VEOs and non-state actors, across time and space, adopt very similar strategies despite their unique local settings. This may include their organizational structures, their population base, the terrain, dynamics of allegiances, motivation, training, their bases, and financial sources. While VEOs and non-state actors may differ in terms of their areas of operation, culture, and certain modes of functioning, there are inherent similarities in how they prepare for their violent struggles. This implies that we can discuss VEOs and non-state actors within a conceptual framework with descriptive and explanatory usefulness across different groups and organizations. This section offers such an overarching conceptual framework to explain VEOs and non-state actors. The conceptual framework is illustrated by case studies as examples to support a particular conceptual point. Once that is accomplished, the chapter offers some best practices for state actors engaged in fighting against VEOs and non-state actors. These best practices are illustrated by case studies as well.

VIOLENT EXTREMIST ORGANIZATIONS AND NON-STATE ACTORS

What are VEOs and non-state actors? There is no consensus on the definition of what VEOs are, and there is disagreement at the level of the UN as to what constitutes violent extremism. The United Nations High Commissioner for Human Rights report states that it is not always clear from definitions "whether 'violent extremism' presupposes violent action or inciting violent action, or whether lesser forms of conduct that do not normally trigger criminal law sanctions would also be included".[1] The report concludes that violent extremism is broader than terrorism, and, as per United Nations General Assembly Report A/70/674/para 4, and the United Nations Secretary General, "violent extremism encompasses a wider category of manifestations than terrorism since it includes forms of ideologically motivated violence that falls short of constituting terrorist acts".[2] This means both violent and non-violent non-state actors can be involved in radicalizing societies toward the end goal of either political or religious motivated violence that results in civilian and combatant deaths.

| Both violent and non-violent non-state actors can be involved in radicalizing societies toward the end goal of either political or religious motivated violence that results in civilian and combatant deaths. |

A phenomenon like violent extremism can permeate a diversity of groups to include ethnic, religious, or identity-based groups. These groups utilize and take advantage of societal pains like unemployment, religious discontent, minority fears, and clan and ethnic divisions to sow discontent and radicalization toward intolerance. United Nations Security Council Resolution 1624 (2005) calls upon states to prohibit and counter acts of terrorism. In its preamble, Resolution 1624 stresses "the importance of the role of the media, civil and religious society, the business community and educational institutions" toward promoting tolerance, understanding, and enhancing dialogue. This includes engaging with local societies and non-state groups to form a web of institution-based responses to specifically counter extremist narratives. Addressing structural inequalities is part of the process.[3] In his book, *Inside Rebellion: The Politics of Insurgent Violence*, Jeremy M. Weinstein indicates how structural and antecedent conditions could result in insurgent violence. Certain antecedent conditions include deep societal networks, resources for violent extremism and terrorism, and population support (finances, recruitment, and legitimate backing) for VEOs to launch their activities. Resources are critical as they inform the policies, strategies, and tactics of VEOs and non-state actors engaged in propaganda, recruitment, and resource generation for their radicalization and organizational efforts.[4]

This section addresses the term VEO as those groups that organize and train to utilize violence to achieve certain specific political goals. Several factors support the organization and development of VEOs and non-state actors. These include politics, strategy and tactics, population support, structural conditions, terrain, dynamics of allegiance, organization and leadership, feasibility, external support, and bases. Using the method of process tracing,[5] illustrated by comparative case study method,[6] this section offers a detailed analysis of how VEOs and affiliated non-state actors establish themselves and carry out their violent movements. Such an assessment is useful because without understanding how VEOs strategize and what factors shape their existence, a response by the state will be ill informed and inefficient. The section then elaborates on counter measures to VEOs and non-state actors.

HISTORIOGRAPHY

Several works on VEOs and related non-state actors highlight their structures, organizations, ideologies, motivations, and strategies. For instance, Thomas R. Mockaitis in *Violent Extremists: Understanding the Domestic and International Terrorist Threat*[7] examines the ideological motivations that propel VEOs, ranging from lone wolves to cyber threats, weaving a narrative of networks and the range of activities spanning orders. The *Routledge Handbook on Non-Violent Extremism Groups Perspectives and New Debates* specifies the importance of studying non-violent extremism as a precursor to violent extremism. Case studies in the Handbook includes Buddhist, Hindu, Islamic, far right, far left, environmentalist, and feminist manifestations of such ideological systems. The aim of the Handbook is to offer a broad-based assessment of such non-violent extremism.[8] Diego Gambetta and Steffen Hertog in *Engineers of Jihad: The Curious Connection between Violent Extremism and Education* underscore the disproportionate connection between Islamist radicals and engineering. The book examines the link between socioeconomic conditions, ideology, and mindset that results in recruitment to VEOs.[9] Mark Juergensmeyer in *Terror in the Mind of God The Global Rise of Religious Violence* analyzes the religious roots of terrorist violence related to the major religions of the world.[10] Randy Borum, in "Radicalization into Violent Extremism I: A Review of Social Science Theories,"[11] studies the term *radicalization* by examining different pathways that lead to VEOs. He goes on to examine certain theories that can help define and explain this pathway better. Oluf Gøtzsche-Astrup, et al. assert that social psychological group processes and intergroup dynamics play a key role in radicalization and establishment of VEOs. The authors specify that radicalization can occur through religious ideology but also psychological factors, such as personality, to include traits of aggression, anxiety, social domination, and authoritarianism. In this, social interactions, social influences, social context, and identities play a vital role.[12] Pamela G. Faber, et al. examine the role of women in VEOs, both at an operational and organizational level, and how roles shift over time. The study found that women play "supporting, enabling, and operational roles" in

VEOs based on case studies of the Revolutionary Armed Forces of Colombia–People's Army, Liberation Tigers of Tamil Eelam, Kurdistan Workers' Party, Al-Shabaab, National Socialist Underground and National Action (two white supremacist groups in Europe), Boko Haram, Islamic State of Iraq and Syria (ISIS), Lord's Resistance Army, and Abu Sayyaf Group.[13]

The literature on countering VEOs and related non-state actors is insightful.[14] David Galula's classic manual *Counterinsurgency Warfare Theory and Practice* counts as one of the most extensive expositions on the qualities required by counterinsurgency forces to succeed against an insurgency, which can also be generalized to counter VEOs. Galula focuses on the importance of population support, intelligence, and terrain knowledge. Bruce Hoffman and Jennifer Morrison Taw's *A Strategic Framework for Countering Terrorism and Insurgency*; Frank Kitson's *Low Intensity Operations: Sub-version Insurgency & Peacekeeping*; C E Callwell's *Small Wars A Tactical Textbook for Imperial Soldiers* all offer insights into the strategies of countering VEOs and related non-state groups. Tahir Abbas offers a compelling argument on how it is important to realize that VEOs do not emerge in a vacuum, and too much focus on the ideological roots of VEOs may not be good.[15] The UN Interregional Crime and Justice Research Institute book *Strengthening Efforts to Prevent and Counter Violent Extremism: Good Practices and Lessons learned for a Comprehensive Approach to Rehabilitation and Reintegration of VEOs* draws on 15 years of experience in the implementation of programs aimed at countering VEOs. The book recommends that prisons are a place where countering VEOs and their networks is important, as most recruitment and command operations occur during prison time together, which gets replicated once such extremists are released. Shashi Jaykumar's edited book, *Terrorism, Radicalisation & Countering Violent Extremism Practical Considerations & Concerns*, offers an exhaustive study of how radicalization occurs, and, consequently, how to counter violent extremism. Such efforts include mentoring and deradicalization, community engagement, and a persistent effort to root out the causes of violent extremism like displacement, identity crisis, and poverty. The book discusses how social media can instigate and perpetuate radicalization and the turn to violent extremism.[16] Michael J. Williams' book titled *Preventing and Countering Violent Extremism Designing and Evaluating Evidence-Based Programs* is a study on how to design programs that can inform evidence-based practices to counter VEOs. Ideology, culture, socialization, the use of violence to meet needs, grievance, threats to self-esteem, relative deprivation, the role of extremist institutions, funding, the role of biochemistry, and decisions all play a cumulative role in perpetuating VEOs.[17] Sara Zeiger and Anne Aly's edited volume on *Countering Violent Extremism Developing an Evidence-Base for Policy and Practice* analyses the structural reasons for violent extremism with illustrative case studies from Syria, the Horn of Africa, Central Asia, and right-wing extremism in Canada. This section is followed by illustrative examples on how to combat violent extremism, to include both physical and online violent extremism.[18] Most books and articles recommend the importance of involving the local communities from where violent extremism springs.

There is significant literature on VEOs and countering VEOs, as is evident from the review above. This chapter builds on that literature to offer a clear conceptual framework on VEOs. The chapter also offers counter measure on how to counter VEOs and related non-state groups. Case studies are illustrative of a conceptual point and are not meant to be exhaustive. References are provided on cases for those who aspire to know more, in footnotes, for each case cited.

STRATEGIES AND ORGANIZATIONAL DYNAMICS OF VIOLENT EXTREMIST ORGANIZATIONS

VEOs plan their organizational dynamics well. As per the requirement of their environment, this includes such considerations as population support, resources, and terrain, and these groups make decisions about how to organize for the use of violence to meet political end goals. Political end goals are key to keep in mind when thinking about why VEOs exist, as well as when deciphering their motivations and dynamics of allegiance. There are several key factors that inform how VEOs and related non-state actors strategize for the long-term feasibility and sustainability of their movements. These include, but are not limited to, organizational structures, population base, terrain, dynamics of allegiances, motivation, training, financial sources, politics, strategy and tactics, structural conditions, external support and bases (see Figure 6.5 for Al Qa'eda's operational footprint in 2024). Let's consider each factor conceptually, illustrated by some case studies, as a reference point.

ORGANIZATIONAL STRUCTURES

VEO organizational structures tend to be centralized for purposes of decision-making, while implementation of decisions made by leadership is decentralized for reasons of secrecy and fear of being detected. If implementation of a particular decision to use force requires too much back-and-forth communication between the core leadership and affiliated members or groups on the ground, there is the fear of detection by state forces. While much is made of the difference between insurgencies (centralized decision-making) and terrorism (decentralized decision-making), a study of their organizational structures tells us that both forms of violent extremism are based on centralized decision-making. Usually, this assumption of decentralization is based on a conflation of strategy and tactics—strategy being a connection between means and end goals and tactics involving the everyday organizing principles to implement a strategy. In light of this, how do VEOs organize institutionally to implement their strategy of using violence toward meeting their political goals?

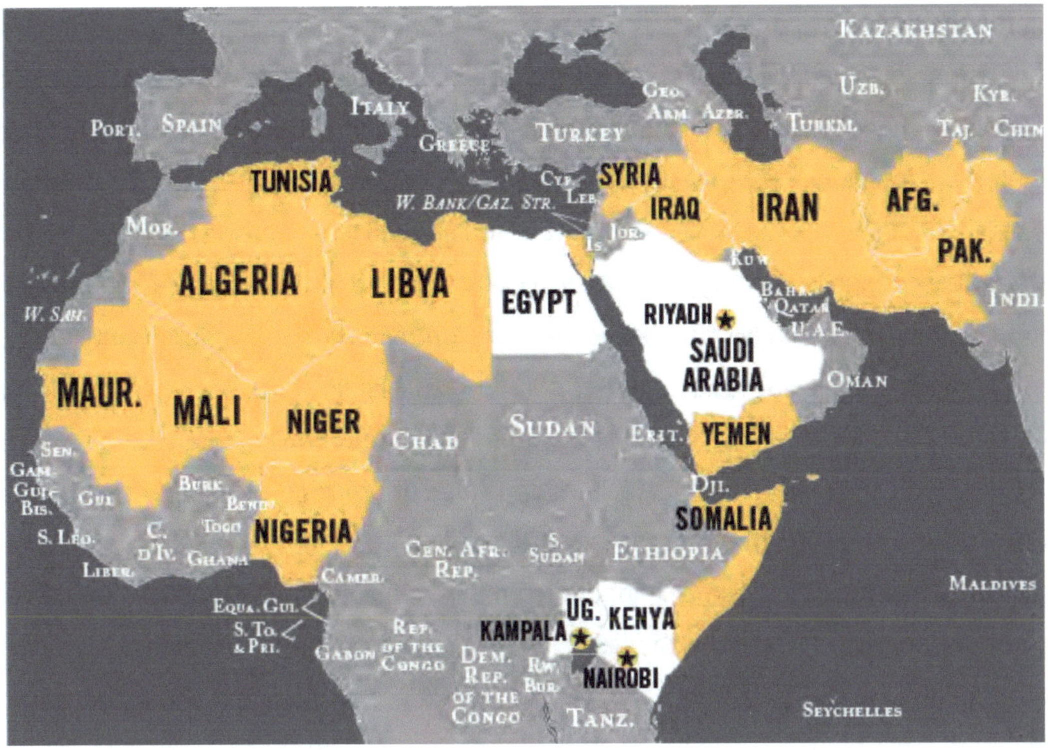

Figure 6.5. Countries in yellow indicate the 2024 presence of al-Qaeda and affiliated groups. Source: National Intelligence Counterterrorism Center/public domain

The first step is to build the core leadership team. The leaders are the ones who get motivated to fight for a particular political cause, and they select the method of fighting as well as who to recruit, how to build the organizational structure, and how to propagate their cause to the target population. Leaders of VEOs are usually connected by ethnicity, religion, preexisting social networks, and a certain proclivity to violence.

| Leaders of VEOs are usually connected by ethnicity, religion, preexisting social networks, and a certain proclivity to violence. |

For instance, al-Qaeda took root during the anti-Soviet jihad in Afghanistan when Osama bin Laden and a few of his close associates organized collectively as al-Qaeda to bring clarity to the global jihad he aspired to wage against the United States. Bin Laden's vision was to reorganize the jihad phenomena into a collective whole to strengthen the various movements. In this, when other local jihadi groups gave allegiance to al-Qaeda, what they got in return was access to the al-Qaeda brand and resources including much-needed weapons and money but also expertise in terms of training for jihad.[19] Training included weapons training, propaganda training, and training in religious texts and politics and terrain adaptation and intelligence gathering. The tactics of using violence, like the attacks on the two U.S. embassies in Africa, turned al-Qaeda into a visible, highly potent transnational terrorist organization, but to the target audience, it represented a successful VEO that was able to execute its strategy. Al-Qaeda did not have the goal of establishing an Islamic state immediately and instead was focused on bringing about Western policy change in the Middle east, especially targeted at thwarting Western presence, to include military troops, in Muslim countries.[20] In contrast, the ISIS leadership structure led by Abu Bakr al-Baghdadi was focused on establishing an Islamic state, with territorial control, in defiance of al-Qaeda leader Abu Musab al-Zawahiri who cautioned against such a political end goal. Al-Baghdadi, along with his core leadership, aimed at utilizing a combatant-style attack on Iraq from Syria and took over Iraqi cities like Mosul. In 2014, after the takeover of Mosul, al-Baghdadi announced the Islamic caliphate from the steps of the al-Nuri Mosque in Mosul. Taking inspiration from al-Zarqawi, the leader of Al-Qaeda in Iraq, al-Baghdadi asserted ruthless centralized control over ISIS strategy and decision-making, announcing a Shura council and breaking away from al-Qaeda and its ideological moorings.[21] The second step is to generate the finances, garner resources, and establish presence by the use of violence and propaganda. The final step is to engage in competitive outbidding of other rival groups.[22] These aspects are elaborated upon in the sections below.

POPULATION BASE

VEOs and non-state groups exist within a pool of supportive and/or sympathetic population that have suffered under the governance of a state viewed as hostile to their aspirations for political, economic, and identity-based aspirations. In fact, the leaders and core recruits of VEOs come from the local population or affiliated ethnic, religious, and identity-based organizations. The affected population where VEOs exist must make survival decisions in a situation where the very existence and visible presence of a VEO signals the absence or weak presence of the legitimate state. In that situation, the strategy of survival kicks in and populations make decisions based on the local power dynamics and the breakdown in the monopoly of organized violence; this should ideally be with the state but has been diffused and challenged in this context. Leaders of VEOs are known to the population, as most of them come from the local population and are deeply intertwined with the pre-existing networks.

For instance, an insurgent group like the National Socialist Council of Nagalim-Isak-Muivah (NSCN[IM]) has been in existence in the conflict-affected Naga inhabited areas of India since its formation in 1988. The leaders are drawn from local tribes, the insurgent army is well trained in guerilla tactics, and the strategy of the NSCN(IM) is to violently fight for better Naga political representation by seceding from the Indian state. Historically, Naga-inhabited areas were independent but were included as part of British India in the latter part of the 19th century. The tensions between the post-colonial Indian state and these tribes continue, and the Naga population has local anxieties of being subjugated by a majority Indian state where their population is a minority. The insurgent group claims to represent such aspirations, and the target population views the insurgents as representing their political anxieties. Moreover, remote Naga areas where insurgent presence is heavy are not well governed, with a sporadic presence of the Indian state, thereby resulting in support to the Naga armed groups as a survival mechanism for the local population.[23]

A similar case can be seen with Boko Haram in northeast Nigeria. Local agency and structural conditions led to Muhammad Ali establishing Boko Haram. Ali, as per eyewitness accounts, was radicalized in Maiduguri, Northeast Nigeria, where fellow Islamic radicals rejected Western education. Indeed, Ali applied to study medicine at the University of Maiduguri and was accepted, but he refused to enroll. Critically, Ali's path to radicalization occurred in Northeast Nigeria, amid local Salafi circles. He was inspired by Taliban and al-Qaeda strategies, but it was the poor implementation of the sharia system in Northern Nigeria in 2000 that led to his deep-seated anxiety and radicalization. The events of 9/11, the invasion of Afghanistan by the U.S., and the subsequent speeches of Taliban and al-Qaeda leaders, especially the writings of Osama Bin Laden and al-Zawahiri, influenced Ali's thinking. Ali translated several of these speeches to local Kanuri and Hausa languages. The Muslim population in Northern Nigeria viewed South Nigeria, dominated by Christians, as discriminatory toward them, and this resulted in grievances (e.g., poverty, inequality, police brutality) and sympathy for Boko Haram's cause.24

TERRAIN

Terrain is vital in waging violent extremism, as groups need to base themselves for an organized fight. Such terrain should be remote, inaccessible to counter forces, and house a supportive or coerced population. This occurs when the VEOs are the only visible presence in a particular area. Forests, mountains, and deserts can offer vital base support. This is even more true when VEOs are located within population groups in a particular country where the use of localized air power by the state based on satellite imagery would be seen as illegitimate and disproportionate. Case in point is India's fight against VEOs. The Naxalites, or left-wing extremist groups, are present in India across several Indian states like Chhattisgarh, Andhra Pradesh, Jharkhand, Bihar, and West Bengal.[25] The use of violence against Indian security personnel is rampant, resulting in the deaths of local police and paramilitary.[26] These areas are also remote, forested, and difficult to reach, as per field work conducted by the author in 2013.

This is the main reason left-wing extremist groups have chosen these areas. However, the use of the Indian Air Force against the Naxalites would be viewed as illegitimate as the VEO exists within local population groups—in other words, Indian citizens. Terrain is also a key factor for Boko Haram, especially its hideout in Sambisa Forest, Borno State during the height of the kidnapping of the school girls,[27] the Taliban's ability to regroup, and for insurgent groups like the Moro Islamic Liberation Front in the Southern Philippines (see Figure 6.6 for location of Boko Haram in 2024).[28]

DYNAMICS OF ALLEGIANCE

Allegiance is a major part of the organizational, leadership, and ideological aspects of VEOs. The dynamics behind giving allegiance is insightful, as several factors play a role in people/groups giving their allegiance to a particular group. Structural and antecedent conditions play a role but so do leadership, ideology, religion, ethnicity, and resources.

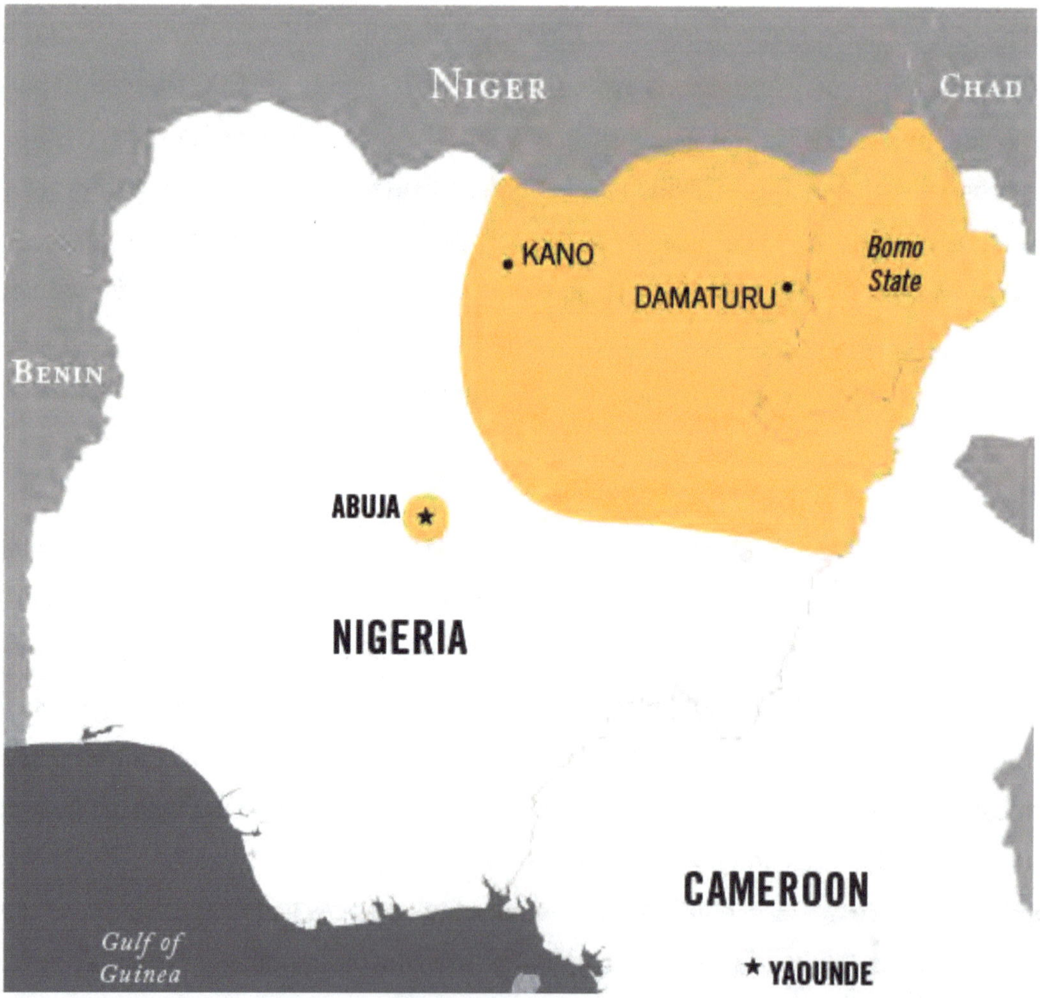

Figure 6.6. Map of Nigeria. Region of Boko Haram operations in 2024 highlighted in orange. Source: Director of National Intelligence/public domain

VEOs seek allegiances with like-minded groups because it extends their presence, gives them greater access to terrain and resources, and improves their ability to build presence and hold on a particular territory and target population.

| VEOs seek allegiances with like-minded groups because it extends their presence, gives them greater access to terrain and resources, and improves their ability to build presence and hold on a particular territory and target population. |

Giving allegiance to a particular VEO by other groups depends on three aspects and is a dynamic factor. First is leadership. Who the leader of a particular VEO is matters, as charismatic and visionary leaders draw in allegiances. One could see this play out with Osama Bin Laden, al-Baghdadi, and guerilla groups like the Viet Minh, where Ho Chi Minh's personality and charisma attracted several local allegiances.[29] Second is ideology. If there is convergence of religious, political, and ethnic ideology, groups will merge to form stronger alliances as is the case with Boko Haram giving its allegiance to ISIS,[30] and/or smaller Naga armed groups giving their allegiance to the NSCN(IM) in India. Finally, structural conditions like poverty, lack of economic development, absence of the state in these ungoverned spaces, and fear of minority persecution by a majority population in a particular state can result in local populations giving their allegiance to a particular local VEO. Moreover, antecedent conditions and social networks based on ethnicity, religion, terrain, and familial relationships can result in allegiance to VEOs. Allegiances can bring about access to better training—for example, with smaller Jihadi groups giving allegiance to al-Qaeda resulting in their access to al-Qaeda training camps in Afghanistan,[31] as well as increased motivation for the VEOs.

FINANCIAL SOURCES, RECRUITMENT, AND ACCESS TO WEAPONRY

The feasibility of running an armed insurgency or a terrorist group—and the resulting violence—requires resources, an ability to recruit, and access to weaponry. VEOs become feasible when these conditions are met: a financial network, a population from which to recruit and/or draw in foreign recruits through propaganda and promise of money and ideology, and access to weapons. In the case of ISIS, access to oil, to include control and sale of oil, brought in about $1 million a day at its peak in 2014. ISIS also utilized tactics like kidnapping, ransom, and extortion to fund its violent activities. At present, ISIS sources of funding depend on areas they operate in but also external donors in the Gulf. Countries like Qatar, Kuwait, and United Arab Emirates (UAE) do not have strict legislations to counter such funding, and money can be channeled to VEOs through Hawala networks (informal banking). Illegal taxation and extortion are also utilized by VEOs. ISIS, for example, earned around $12 million a month in 2014 from such taxation in cities like Mosul. Highways were areas for taxation of vehicles and trade between Iraq and Jordan.[32] In 2015, ISIS had an annual budget of $2 billion.[33] Extortions include both voluntary and involuntary donations. VEOs in Southeast Asia, for instance, fix the rate of taxes to be given from a particular area depending on the stature and income of a particular person.[34] Goods in VEO-dominated areas are taxed as well, and drug trafficking is a major source of VEOs' finances. In Colombia, a leftist insurgent group, the Revolutionary Armed Forces of Colombia, controlled 60 percent of the country's drug trade in 2013.

In addition to finances, a VEO's ability to recruit and get access to weapons is also important. Recruitment can occur from the local population as well as from foreign sources. Local recruitments are conducted based on ethnic and ideological kinship, along with the lure of making money and having security and stability in these remote, ungoverned spaces. ISIS witnessed an enormous flow of foreign fighters, drawn to its declaration of an Islamic caliphate based on sharia law. Though estimates differ, approximately 40,000 foreign fighters from nearly 110 countries have travelled to the Islamic caliphate since it was established in 2014.[35] The draw of religious ideology and the promise of a good life under the caliphate, demonstrated via sleek propaganda videos of life under ISIS and ISIS training videos, attracted those who journeyed to the caliphate. Access to weaponry remains a key component of the feasibility of VEOs. Naga insurgents have access to a black market of weaponry, especially Chinese-originated and manufactured weapons in Southeast Asia;[36] ISIS had access to weaponry in Iraq due to lackluster tracking of weapon transfers by several countries including the U.S., the UK, Russia, and China. Lack of oversight as to where the weapons were going was another way ISIS attained weapons originally meant for the Iraqi army. An Amnesty International Report states, "We've found ISIS use of arms and ammunition from at least 25 different countries—a large proportion were originally sourced by the Iraqi military from the USA, Russia and former Soviet bloc states. The huge scope reflects decades of irresponsible arms transfers to Iraq, coupled with the failure to install oversight mechanisms during the U.S.-led occupation after 2003. Slack controls over military stockpiles and endemic corruption by successive Iraqi governments have added to the problem. Among ISIS's arsenal are portable air defense systems, guided anti-tank missiles and armored fighting vehicles, as well as assault rifles like the Russian AK series and the US M-16 and Bushmaster." [37] Among weapons found with ISIS included the M-16 but also the CQ-556 rifle made in China, a copy of the M-16. ISIS also had access to conventional weapons like mortars, rockets, and grenades by designing their own munitions using machinery in oil fields.[38]

POLITICS, STRATEGY, AND TACTICS

The end goal of VEOs and non-state groups is politics. The political end goal could be policy change (al-Qaeda), regime change (ISIS), territorial change (ISIS, Lasker E-Taiba), status quo maintenance (Taliban), and spoiling and outbidding (ISIS versus al-Qaeda).[39] Usually, the difference in how to get to a political end goal depends on the kind of movement a group wants to adopt (the strategy), for instance, violence or non-violence, and tactics (training, resources, bases) to achieve the end goal. The ISIS end goal is the establishment of an Islamic caliphate; Lasker E Taiba wants territorial change in Kashmir; al-Qaeda aims for policy change and the absence of Western presence in the Middle East; and Naga armed groups want territorial change. The strategy of VEOs is to build the means (institutions, recruits, organizational philosophy and ideology, sources of funding and recruitment, and weapons manufacture) to then inform the tactics (violence; ambushes; attacks; extortions, like taxation and coerced donation; and daily showcasing of presence to the local population), to achieve their political end goals.

ISIS in the Sinai, 2 March 2023. Source: Zezoyo/WikiCommons[40]

EXTERNAL SUPPORT AND BASES

One of the most critical factors for the long-term sustainability of VEOs is external support and bases. At a time when state counterterrorism/insurgency forces are able to clear and hold the territory of a state affected by VEOs, porous borders and external bases help VEOs regroup.

| At a time when state counterterrorism/ insurgency forces are able to clear and hold the territory of a state affected by VEOs, porous borders and external bases help VEOs regroup. |

Without a base in Syria, al-Baghdadi would not have been able to build ISIS into the group it became in 2014 following the death of al-Zarqawi in 2004.[41] Al-Zarqawi was the leader of al-Qaeda in Iraq, but the U.S. surge in troops in 2007 resulted in the decimation of al-Qaeda in Iraq. Between 2007 and 2010, Al-Baghdadi took leadership roles in al-Qaeda in Iraq, which led to the emergence of the Islamic State of Iraq and Syria (ISIS). At that time, Syria was going through a civil unrest. A similar situation replicated in Afghanistan where the entire Taliban core leadership went to Quetta, Pakistan, after the U.S. invasion of Afghanistan post 9/11. There, they regrouped and retrained, and slowly but steadily built back their base both in Pakistan and in Afghanistan.[42] The Naga insurgency in Northeast India had access to bases and weapons in the border regions of Burma, consequently resulting in a long-term manifestation of violence.

COUNTERING VIOLENT EXTREMIST ORGANIZATIONS AND RELATED NON-STATE ACTORS

Once the context that gives rise to VEOs and non-state actors becomes clear, we can think through and formulate what an effective counter VEO strategy would look like. Based on the assessment above, several factors should be part of such a counter strategy. These include the primacy of the political goals, population support, countering propaganda, intelligence, robust military force structures, rule of law, and unity of effort and operational clarity.

| These include the primacy of the political goals, population support, countering propaganda, intelligence, robust military force structures, rule of law, and unity of effort and operational clarity. |

Let us dwell into each of these factors with illustrative examples.

PRIMACY OF POLITICAL GOALS

There is a misguided assumption that countering VEOs is the task of the military, be it within a state or of a transnational character. Based on field work in conflict-affected areas, the author argues that the military response should be guided by a clear political structure visible to the state/states/international organizations that engage in countering violent extremism. This understanding of political goals of clearing an area of VEOs, making the area secure for the civilian institutions of the affected area to function again, and securing the population is vital to ensure effective counter strategies. This strategy would include a variety of institutions and resources, of which the military response is but one. One can gauge this from the cases of Afghanistan, Iraq, Vietnam, Burma, Northeast India, and Libya that a cumulative visible goal based on political understanding enables effective counter VEO strategies.

The Indonesian example of using their political mantra of Pancasila,[43] which includes monotheism, civilized humanity, national unity, democracy, and social justice, has been utilized to counter the ideology of hate and violence that ISIS attempted to spread in this Sunni Muslim majority country, which houses the largest Muslim population in the world. The strategy was effective, to the extent that ISIS ideology and propaganda was challenged by respected Indonesian leaders like former President Abdurrahman Wahid, whose influence runs deep in Indonesian society.[44] Several manuals on countering insurgencies, like the British counterinsurgency in Malaya, demonstrate the importance of primacy of politics framing a response to violent extremism.[45] The counterinsurgency manuals of Indian,[46] British, and U.S. military forces identify the significance of the primacy of politics.47 U.S. Joint Publication 3-24 on Counterinsurgency makes it clear that the core of counterinsurgency and/or violent extremism is "the political strategy, which should articulate how the [Host Nation] will address the root causes (opportunity, motives, and means) that drive the insurgency. The strategy provides a framework around which all other programs and activities are organized."[48] In Iraq, while U.S. and Iraqi forces were superior in terms of training, equipment, and resources, it was the political conundrum of Iraqi politics, further heightened by its first pos- Saddam Hussein Prime Minister, Nouri al-Maliki.[49] Al-Maliki's sectarian politics resulted in a deep-seated Sunni-Shia divide resulting in societal tensions, the rise of violence, and complete lawlessness in major Iraqi cities—the end result being an ISIS onslaught in 2014.[50]

POPULATION SUPPORT

The strategy of VEOs is to undermine the state's legitimacy by promoting disorder and population insecurity, forcing the lack of economic development, and deterring access to basic necessities provided by the state to the population. This is a deliberate strategy so the target population's grievances against the state increase and the state is viewed as corrupt, dysfunctional, and—most importantly—uncaring and unresponsive to the affected population's needs. VEOs then fill that vacuum by creating parallel government structures, visibly demonstrating their presence, taxing the population, providing their version of law and order, and threatening the local population with dire consequences if they do not support the VEOs with support and resources required for the fight. It is within this backdrop that state agencies (civilian, military, or a combination) engaged in countering VEOs operate. The realization that VEOs are mingled within the population, and the distinction between combatant and non-combatant, is blurred in this situation and should result in a nuanced and careful approach to countering such violence without alienating the affected population. There are several ways to gain the support of the population. This must include daily contact, organizing collective grievance sessions—and potential solutions to those grievances, identifying who the VEOs are with effective intelligence, and providing security to the affected population to the extent possible. The state forces must be perceived as working for the affected population, not against them. Shaping the choices and preferences of the local affected population involves deep engagement not just with local leaders but also the *right* ones and addressing directly the rumors and misperceptions that have spread about the countering forces. Remaining quiet or unresponsive to such rumors about countering forces is a disastrous strategy. Knowing the nature of both vocal and silent support to the countering forces is important. For a silent population to speak out against VEOs, state forces have to work to provide a level of structural security that enables such a discourse. For instance, an example from history is the British counterinsurgency in Malaya, especially from the 1950s onward; this was a population-centric operation where the Chinese minority population was reallocated to protected villages. The communist insurgency was organized around the ethnic Chinese minority demanding a separate state outside of Malaya. What the British counterinsurgency forces achieved, with support from the Government of Malaya and the Malayan police, was to ensure that the conditions in the newly established protected villages were better than the earlier villages, which were based on the "hearts and mind" strategy conceptualized by Sir Gerald Templer.[51] This might not always work, especially if the VEOs represent a majority population, but one lesson from the British experience in Malaya was to meet the structural conditions of better living standards that succeeded in quelling the grievances. A counter example is the insecurity felt by the Iraqi and Afghan populations due to the presence of VEOs in their midst and the complete failure of the Iraqi and Afghan state agencies to secure these affected populations. The result was a breakdown of the rule of law and VEOs succeeding in the end, with the case of Afghanistan under the Taliban today as a glaring example.[52]

COUNTERING PROPAGANDA

Propaganda is a vital tool for VEOs as they aim to control the population's choices. VEOs show themselves as representing the population's grievances, and promise a better life, political and economic empowerment, and ideological purity when they take over the reins of the political regime. For VEOs, propaganda is easy as they do not have to achieve any of their policy goals today, but only in the future. For those countering this propaganda, they have to deliver on the promises today. Moreover, promising to achieve stability when that could be difficult in the short term is not good counter-propaganda. Instead, the countering forces should expose the weakness and false promises of the VEOs by real-time examples. ISIS claims of establishing a well-functioning Islamic State, where everyone who gave allegiance would thrive, was countered by states and civil societies that exposed their shallowness. Those dispirited members of ISIS who returned to their countries of origin also played a vital role in exposing the realities of ISIS.[53] VEOs exaggerate state injustices to the target population. Such false narratives must be countered by visible examples. Failure to do so perpetuates the power and hold of VEOs as can be witnessed in the ineffective counter to VEO propaganda in northeast India and Burma.[54] Consequently, VEO narratives of being the legitimate representation of population concerns gains credence.

INTELLIGENCE

Intelligence is key in launching effective counters to VEOs. Intelligence includes knowledge of the terrain, logistics, climate, the people, the nature and durability of VEOs, local culture, social customs, the failures of local state structures, perceptions of the counter VEO forces, and population interests. This kind of specific intelligence is vital. Tactical and strategic intelligence will require specialist intelligence staff, a clear understanding of the operating environment, the external support and influences on VEOs, the role of social media and propaganda, and the source of funding. Specialized intelligence on VEO lines of communication, strengths and weaknesses, safe havens, intelligence networks, financial transfers, and recruitment base must inform the architecture of counterintelligence.

| Specialized intelligence on VEO lines of communication, strengths and weaknesses, safe havens, intelligence networks, financial transfers, and recruitment base must inform the architecture of counterintelligence. |

Intelligence has to be bottoms-up and collected by trained agents on the ground. Geospatial intelligence is important along with HUMINT, but it is absolutely vital that intelligence is verified across the intelligence architecture to avoid inaccuracies. Good intelligence will be effective in informing both the policy and strategy of statecraft at the ground level to counter VEOs, while faulty intelligence will further alienate the population as the wrong individual may be picked up as having links to VEOs where none exist. Innocent people accused of violent extremism have ended up getting radicalized in prisons and/or through interrogations. Local language skills help in intelligence gathering but is not enough. The ability to analyze the data and see patterns is also vital. Countering VEOs is tough, especially since the VEOs are also collecting intelligence and adapting to the local needs.[55]

ROBUST MILITARY FORCE STRUCTURES

VEOs design their operations in a manner aimed at addressing the weakness of non-state armed violence, including lack of resources, lack of advantageous base areas, a dearth of ready recruits, weapons, manufacturing capabilities, and absence of a steady tax base to fund the movement. Consequently, VEOs work to build deception, surprise, and swiftness and have an ability to make do with what they can generate for their violence. In comparison, state forces have access to more resources, bigger manpower, larger deployments, an ability to build robust intelligence networks, and a state's industrial base. However, with the wrong policy and faulty strategy, such advantages can be lost—as was determined from the Nigerian military's response to Boko Haram.[56] The U.S. and Afghan military response to the Taliban and the U.S. experience in Vietnam tell a similar story.[57] The Indian experience with counterinsurgency reveals lessons on how force structure matters.[58] The advantages of a counter-VEO, state-based force structure is that it has access to enough manpower and resources. Force structures, however, must be organized in such a way that the minimum use of force suffices to quell violent extremism of a non-state variety. One of the biggest security concerns of local populations is that states will respond with the massive use of force resulting in civilian deaths. Civilian deaths can be costly for the legitimacy of a particular operation, as seen in international reactions to Israel's war on Hamas. It was Hamas who attacked Israel and Israeli citizens during a concert in October 2023.[59] Yet the Israeli bombing of civilian infrastructure in the Gaza Strip and the resultant death of civilians has made the war costly for Israel.[60] In India, the deployment of the military in areas affected by violent extremism has resulted in deep-seated fears among the local population that they would be targeted. Therefore, special operations forces or military units with very clear rules of engagement—who are trained in fighting VEOs and can take advantage of force multipliers like surrendered VEO members for intelligence—supported by good technological surveillance and a deep cultural understanding of the local context and terrain, and who have ability to be coalesced into smaller units[61] and are adept at improvising, have proven to be more effective.

RULE OF LAW

For a successful and legitimate counter-VEO operation, rule of law is critical. The conduct of forces in conflict-affected areas must meet the highest standards, and lawyers form a critical part of the planning process. The rule of law deters lawless use of force and cautions against disproportionate use of force, which can alienate the population. Terms of employment for use of force must be clearly laid out and must form a critical part of the overall policy directives. Those engaged in countering violent extremism must be briefed on rules of engagement regarding how arrests, searches, warrants, interrogation techniques, and intelligence gathering are to be executed by issuing a standard operating procedure. International laws relating to internal armed conflicts are found in Common Article 3 of the Geneva Conventions of 1949 and Additional Protocol II of 1977.[62] Surveillance, to include aerial and space-based surveillance, must be within a legal framework. In India, special provisions like the Armed Forces (Special Powers) Acts of 1958 (amended in 1972) that protect security personnel deployed in conflict-affected areas have created a popular imagination among the conflict-affected population that the military is lawless.[63] Ironically, the military needs such laws to fight VEOs; the special laws render the population hostile to the Indian military.

UNITY OF EFFORT AND OPERATIONAL CLARITY

Unity of effort and operational clarity is critical for the success of countering VEOs. There has to be a common diagnosis of the issue at hand with operational clarity. Such clarity enables an effective strategy to showcase economic resources, responsive administrative structures, counter propaganda, securing an area, and taking the initiative. Different agencies involved in counter efforts must have good communication and close consultation through a central management division. There must not be any ambiguity regarding the command structure informed by a single strategic narrative that countering VEOs is part of the overall plan, specifically between deployed forces and the host nation. Unity of effort brings together political, social, economic, and military actors all working toward a key goal of eradicating violent extremism. Unity of effort can be brought about by appropriating a single command and coordination center. For instance, the British command and coordination structure in existence in 1948 in Malaya proved inadequate in the beginning of the Malayan Communist insurgency due to the absence of a unity of effort. It was only when the Director of Operations position was established, manned by Army Lieutenant General Sir Harold Briggs, in 1950 that coordination occurred under a newly established Federal War Council.[64]

Operational clarity requires a diagnosis of the VEO; there must be operational flexibility and adaptability depending on how the VEOs function, especially for those countering VEOs on the ground; task allocations have to be clear especially in training and force structures and on which areas to hold and clear. Countering VEOs, finally, means that the aim of the counter forces is to secure the population. Without securing the population, violent extremism will always find a home.

| Without securing the population, violent extremism will always find a home. |

The case of Marawi in Mindanao, Philippines, in 2017 offers insights into how the ISIS-linked Maute Group secured a foothold due to the insecurity felt by the Muslim population vis-à-vis a Christian majority country. The counteroffensive by the Philippine military succeeded but left deep-rooted wounds among the Muslim population (the Maute Group aligned itself with Abu Sayyaf; see Figure 6.7 for the 2024 presence of Abu Sayyaf in the Philippine Islands).

Figure 6.7. The Abu Sayyaf Group's area of operations in 2024 indicated in orange in the Philippines. Source: National Counterterrorism Center/public domain

CONCLUSION

In conclusion, it is pertinent to emphasize that understanding the structural reasons for the presence of VEOs along with their organizational setup, ideology, and resource structure will help in formulating an effective counter-VEO strategy. This strategy must be based on a people-centric approach, especially for those affected by the presence of a VEO, supported by a solid intelligence architecture, terrain knowledge, cultural understanding, well-trained military force structures, and operational clarity. An architecture that is inclusive of policies, institutions, resources, and collaborative efforts between outside forces and the host nation can be effective in dealing with VEOs. However, addressing the root causes that result in such violent movements can go a long way in addressing structural issues and increasing the ability to stamp out such violence before it becomes full-blown violent extremism.

About the Author

Dr. Namrata Goswami is an author and educator specializing in space policy, international relations, and ethnic identity. Currently, Dr. Goswami teaches at the Schriever and West Space Scholar Programs, the School of Advanced International Studies, Johns Hopkins University, Thunderbird School of Global Management, Arizona State University, and at the Joint Special Forces University. She is a guest lecturer at Emory University for seminars on Technology, Society & Governance, and India today. She worked as a Research Fellow at MP-Institute for Defence Studies and Analyses, New Delhi; a Visiting Fellow at Peace Research Institute, Oslo, Norway; La Trobe University, Melbourne, Australia; University of Heidelberg, Germany; Jennings-Randolph Senior Fellow, United States Institute of Peace; and was a Fulbright Senior Fellowship Awardee. She was awarded the Minerva grant by the Office of the U.S. Secretary of Defense to study great power competition in outer space. In April 2019, Dr. Goswami testified before the U.S-China Economic and Security Review Commission on China's space program. Her co-authored book, Scramble for the Skies: The Great Power Competition to Control the Resources of Outer Space was published in 2020 by Lexington Press; Rowman, and Littlefield. Her book on The Naga Ethnic Movement for a Separate Homeland was published in 2020 by Oxford University Press. She has published widely including in The Diplomat, the Economic Times, The Washington Post, Ad Astra, Asia Policy, Live Encounters Magazine, Cairo Review. She was invited in November 2019 to share about her life and her work at a Tedx event held at the Rosa Parks Museum, in Montgomery, Alabama.

NOTES

1. United Nations Office on Drugs and Crime, "Radicalization and Violent Extremism", n.d., https://www.unodc.org/e4j/zh/terrorism/module-2/key-issues/radicalization-violent-extremism.html

2. United Nations Secretary General, "Plan of Action to Prevent Violent Extremism: Report of the Secretary General," 2015, https://digitallibrary.un.org/record/816212?ln=en (accessed on October 25, 2023).

3. United Nations Security Council Counter Terrorism Committee, "Countering Violent Extremism and Terrorist Narratives," n.d., https://www.un.org/securitycouncil/ctc/content/countering-violent-extremism-and-terrorist-narratives

4. Jeremy M. Weinstein, *Inside Rebellion: The Politics of Insurgent Violence* (Cambridge University Press, 2006).

5. For more on methods, see Stephen Van Evera, *Guide to Methods for Students of Political Science* (Cornell University Press, 1997).

6. Alexander L. George and Andrew Bennet, *Case Studies and Theory Development in the Social Sciences* (MIT Press, 2005).

7. (Praeger Security International, 2019).

8. Elisa Orofino and William Allchorn, eds., *Routledge Handbook of Non-Violent Extremism Groups, Perspectives and New Debates* (Routledge, 2023).

9. Diego Gambetta and Steffen Hertog, *Engineers of Jihad: The Curious Connection between Violent Extremism and Education* (Princeton University Press, 2017).

10. Mark Juergensmeyer, *Terror in the Mind of God The Global Rise of Religious Violence* (University of California Press, 2017).

11. Randy Borum, "Radicalization into Violent Extremism I: A Review of Social Science Theories," *Journal of Strategic Security* 4, no. 4 (2012): 7-36.

12. Oluf Gøtzsche-Astrup et.al., "Radicalization and Violent Extremism: Perspectives from Research on Group Processes and Intergroup Relations," *Group Processes & Intergroup Relations* 23, no. 8 (December 2020): 1127-1136.

13. Pamela G. Faber et.al., "Understanding Gender and Violent Extremism," Center for Naval Analysis, March 10, 2021, https://www.cna.org/reports/2021/04/understanding-gender-and-violent

14. There is significant literature on VEOs and countering VEOs, as is evident from the review above. This chapter builds on that literature to offer a clear conceptual framework on VEOs. The chapter also offers counter measure on how to counter VEOs and related non-state groups. Case studies are illustrative of a conceptual point and are not meant to be exhaustive. References are provided on cases for those who aspire to know more, in footnotes, for each case cited.

15. Tahir Abbas, *Countering Violent Extremism The International Deradicalization Agenda* (Bloomsbury, 2021).

16. Shashi Jaykumar, ed., *Terrorism, Radicalisation & Countering Violent Extremism Practical Considerations & Concerns* (Palgrave, 2019).

17. Michael J. Williams, *Preventing and Countering Violent Extremism Designing and Evaluating Evidence-Based Programs* (Routledge, 2020).

18. Sara Zeiger and Anne Aly, eds., *Countering Violent Extremism Developing an Evidence-Base for Policy and Practice* (Perth: Curtin University, 2015).

19. Daniel L. Byman, "Comparing Al Qaeda and ISIS: Different Goals, Different Targets" (Brookings, April 29, 2015), https://www.brookings.edu/articles/comparing-al-qaeda-and-isis-different-goals-different-targets/#:~:text=Alpercent20Qaedapercent20emergedpercent20outpercent20of,builtpercent20topercent20takepercent20jihadpercent20global.

20. Abdel Bari Atwan, *The Secret History of Al Qaeda* (Berkeley, CA: University of California Press, 2008).

21 Namrata Goswami, *ISIS 2.0: South and Southeast Asia Opportunities and Vulnerabilities* (JSOU Press, 2018); Also see Ali E. Abbas, eds., *Improving Homeland Security Decisions* (Cambridge University Press, 2017), chapter 16.

22. Megan Farrell et al., "The Logic of Transnational Outbidding: Pledging Allegiance and the Escalation of Violence," *Journal of Peace Research* 57, no. 3 (2019), https://journals.sagepub.com/doi/abs/10.1177/0022343319880939?journalCode=jpra

23. Namrata Goswami, *The Naga Ethnic Movement for a Separate Homeland: Stories from the Field* (Oxford University Press, 2020).

24. Bulama Bukarti, "Revisiting the Beginning of Boko Haram," *War on the Rocks*, January 24, 2022, https://warontherocks.com/2022/01/revisiting-the-beginning-of-boko-haram/

25. Prakash Singh, *The Naxalite Movement in India* (New Delhi: Rupa Publications, 2016).

26. Rishabh M Pratap and Swati Gupta, "At Least 22 Indian Security Personnel Killed in Clash with Maoist Insurgents," CNN, April 6, 2021, https://www.cnn.com/2021/04/05/india/india-clash-maoist-insurgents-intl-hnk/index.html (Accessed on November 1, 2023).

27. Bodunrin Kayode, "Inside Nigeria's Sambisa Forest, the Boko Haram Hideout where Kidnapped School Girls are Believed to be Held," *The Guardian*, April 29, 2014, https://www.theguardian.com/world/2014/apr/29/nigeria-sambisa-forest-boko-haram-hideout-kidnapped-school-girls-believed-to-be-held (Accessed on November 1, 2023).

28. Shamsuddin L. Taya, *Strategies and Tactics of the Moro Islamic Liberation Front (MILF) in the Southern Philippines* (UUM Press, 2009).

29. Nelly Lahoud, "Bin Laden's Catastrophic Success Al Qaeda Changed the World-But Not in the Way it Expected," Foreign Affairs, August 13, 2021, https://www.foreignaffairs.com/articles/afghanistan/2021-08-13/osama-bin-ladens-911-catastrophic-success (Accessed on November 1, 2023). Peter L. Bergen, *The Osama Bin Laden I Know: An Oral History of Al Qaeda's Leader* (Free Press, 2006). Chester A. Bain, "Calculation and Charisma: The Leadership Style of Ho Chi Minh," *The Virginia Quarterly Review* 49, no. 3 (Summer): 346-356.

30. Adam Chandler, "Boko Haram Pledges Allegiance to the Islamic State," *The Atlantic*, March 9, 2015, https://www.theatlantic.com/international/archive/2015/03/boko-haram-pledges-allegiance-islamic-state/387235/

31. Daniel L. Byman, "Comparing Al Qaeda and ISIS."

32. Markaz, "Cutting off ISIS' Cash Flow," Brookings, October 24, 2014, https://www.brookings.edu/articles/cutting-off-isis-cash-flow/ (Accessed on November 1, 2023).

33. Jose Pagliery, "Inside the $2 Billion ISIS War Machine," CNN, December 11, 2015, https://money.cnn.com/2015/12/06/news/isis-funding/index.html (Accessed on November 30, 2023).

34. "Extortion Highest in Nagaland, Tribal Body Calls it "Loyalty Tax," *The New Indian Express*, September 08, 2022, at https://www.newindianexpress.com/nation/northeast/2022/sep/08/extortion-highest-in-nagaland-tribal-body-calls-it-loyalty-tax-2496181.html (Accessed on November 01, 2023).

35. Richard Barret, "Beyond the Caliphate: Foreign Fighters and the Threat of Returnees," The Soufan Center, October 2017, https://thesoufancenter.org/wp-content/uploads/2017/11/Beyond-the-Caliphate-Foreign-Fighters-and-the-Threat-of-Returnees-TSC-Report-October-2017-v3.pdf

36. Sanjib Kr Baruah, "NSCN on Arms Buying Spree, Building Cadre," The Hindustan Times, January 18, 2011, https://www.hindustantimes.com/delhi/nscn-on-arms-buying-spree-building-cadre/story-2gQE76CuZ95rTSiDzjSEmJ.html (Accessed on November 1, 2023).

37. Amnesty International, "How Islamic State Got its Weapons," January 12, 2018, https://www.amnesty.org.uk/how-isis-islamic-state-isil-got-its-weapons-iraq-syria (Accessed on November 01, 2023).

38. Andrea Dicenzo, "Exclusive: Tracing ISIS' Weapons Supply Chain: Back to the US," *Wired*, December 12, 2017, https://www.wired.com/story/terror-industrial-complex-isis-munitions-supply-chain/ (Accessed on November 01, 2023).

39. Andrew H. Kydd and Barbara F. Walter, "The Strategies of Terrorism," *International Security* 31, no. 1 (Summer 2006): 49-80, at https://www.belfercenter.org/sites/default/files/files/publication/is3101_pp049-080_kydd_walter.pdf (Accessed on November 01, 2023).

40. Photo from Wiki commons by Zezoyo and free copyright infringement, found at https://commons.wikimedia.org/wiki/File:IS_in_Sinai.jpg#/media/File:IS_in_Sinai.jpg

41. "Timeline: The Rise, Spread, and fall of the Islamic State," Wilson Center, October 28, 2019, https://www.wilsoncenter.org/article/timeline-the-rise-spread-and-fall-the-islamic-state (Accessed on November 01, 2023).

42. Bruce Riedel, "Pakistan, Taliban, and the Afghan Quagmire," Brookings, August 24, 2013, https://www.brookings.edu/articles/pakistan-taliban-and-the-afghan-quagmire/ (Accessed on November 01, 2023).

43. "Indonesia's State Ideology of Pancasila," United States Commission on International Religious Freedom, https://www.uscirf.gov/news-room/uscirf-spotlight/indonesias-state-ideology-pancasila (Accessed on November 02, 2023). Uri Friedman, "How Indonesia Beat Back Terrorism-for now," *The Atlantic*, September 25, 2016, https://www.theatlantic.com/international/archive/2016/09/indonesia-isis-islamic-terrorism/500951/ (Accessed on November 02, 2023).

44. Greg Barton, *Abdurrahman Wahid: Muslim Democrat, Indonesian President* (University of Hawaii Press, 2002). Please read Abdurrahman Wahid, *The Illusion of an Islamic State: How an Alliance of Moderates Launched a Successful Jihad Against Radicalization and Terrorism in the World's largest Muslim-Majority Country* (Jakarta: The Wahid Institute, 2011).

45. Robert Thompson, *Defeating Communist Insurgency: Experiences from Malaya and Vietnam* (London: Chatto & Windus, 1966).

46. Integrated Headquarters of Ministry of Defence (Army), "Doctrine for Sub Conventional Operations," December 2006, https://indianstrategicknowledgeonline.com/web/doctrinepercent20subpercent20convpercent20w.pdf (Accessed on November 02, 2023).

47. Alexander Alderson, "The British Manual to COIN and Stabilisation A Retrospective on Developments since 2001," *The RUSI Journal* 157, no. 4 (2012): 62-71, https://www.tandfonline.com/doi/abs/10.1080/03071847.2012.714195 (Accessed on November 02, 2023).

48. Joint Publication 3-24, "Counterinsurgency" April 25, 2018, Validated April 30, 2021, https://www.jcs.mil/Portals/36/Documents/Doctrine/pubs/jp3_24.pdf (Accessed on November 02, 2023).

49. Terrence McCoy, "Maliki Shattered Iraq-and Now he Refuses To Let it Go," *The Washington Post*, August 11, 2014, https://www.washingtonpost.com/news/morning-mix/wp/2014/08/11/maliki-shattered-iraq-and-now-he-refuses-to-let-it-go/ (Accessed on November 02, 2023).

50. Priyanka Boghani, "In Their Own Words: Sunnis on their treatment in Maliki's Iraq," PBS, Frontline, October 28, 2014, https://www.pbs.org/wgbh/frontline/article/in-their-own-words-sunnis-on-their-treatment-in-malikis-iraq/ (Accessed on November 02, 2023).

51. R W Kromer, *The Malayan Emergency in Retrospect: Organization of a Successful Counterinsurgency Effort*, (RAND, 1972).

52. Carter Malkasian, "How the Good War Went Bad," *Foreign Affairs*, March/April 2020, https://www.foreignaffairs.com/articles/afghanistan/2020-02-10/how-good-war-went-bad (Accessed on November 02, 2023). Shivan Fazil and Alaa Tartir, "Iraq in 2023: Challenges and Prospects for Peace and Human Security," Stockholm International Peace Research Institute, March 17, 2023, https://www.sipri.org/commentary/topical-backgrounder/2023/iraq-2023-challenges-and-prospects-peace-and-human-security

53. Mohd Mizan Aslam, "Fake Paradise for Malaysian Foreign Terrorist Fighters," Middle East Institute, February 16, 2021, https://www.mei.edu/publications/fake-paradise-malaysian-foreign-terrorist-fighters

54. Author Fieldwork in Northeast India and Burma, 2007-2014.

55. Moshe Kress and Roberto Szechtman, "Why Defeating Insurgencies is Hard: The Effect of Intelligence in Counterinsurgency Operations-A Best Case Scenario," *Operations Research* 57, no. 3 (May-June 2009): 578-585, https://www.jstor.org/stable/25614776

56. Aryn Baker, "Nigeria's Military Quails When Faced with Boko Haram," *Time*, February 10, 2015, https://time.com/3702849/nigerias-army-boko-haram/ (Accessed on November 02, 2023).

57. Mike Jason, "What We Got Wrong in Afghanistan," *The Atlantic*, August 12, 2021, https://www.theatlantic.com/ideas/archive/2021/08/how-america-failed-afghanistan/619740/; Katie Lange, "Highlighting History: How 'Tet' Began the End of Vietnam," U.S. Department of Defense, February 07, 2023, https://www.defense.gov/News/Feature-Stories/story/Article/3291950/highlighting-history-how-tet-began-the-end-of-vietnam/

58. Namrata Goswami, "Counterinsurgency Best Practices: Applicability to Northeast India," *Small Wars Journal*, December 06, 2012, https://smallwarsjournal.com/jrnl/art/counterinsurgency-best-practices-applicability-to-northeast-india (Accessed on November 30, 2023).

59. Daniel Estrin, "Hamas Attack on Israeli Techno Festival Leaves at Least 260 Dead and Many Missing," NPR, October 10, 2023, https://www.npr.org/2023/10/10/1204950063/hamas-attack-on-israeli-techno-festival-leaves-at-least-260-dead-and-many-missin

60. Gabrielle Tétrault-Farber and Emma Farge, "UN Experts Say Ceasefire Needed as Palestinians at 'Grave Risk of Genocide," Reuters, November 02, 2023, https://www.reuters.com/world/middle-east/un-experts-say-ceasefire-needed-palestinians-grave-risk-genocide-2023-11-02/

61. Mark Moyer, "Making the Most of Special Operational Forces," Center for Strategic and International Studies, July 18, 2017, https://www.csis.org/analysis/making-most-special-operations-forces

62. See ICRC, "International Humanitarian Law…Treaties & Documents" at http://www.icrc.org/ihl.nsf/INTRO/475

63. M P Nathanael, "A Draconian Law that Needs to Disappear," *The Hindu*, August 27, 2022, https://www.thehindu.com/opinion/lead/a-draconian-law-that-needs-to-disappear/article65815551.ece; Ministry of Home, Government of India, "The Armed Forces (Special Powers) Act, 1958, https://www.mha.gov.in/sites/default/files/armed_forces_special_powers_act1958.pdf

64. Goswami, "Counterinsurgency Best Practices," 56.

65. International Crisis Group, "The Philippines: Militancy and the New Bangsamoro," June 27, 2019, https://www.crisisgroup.org/asia/south-east-asia/philippines/301-philippines-militancy-and-new-bangsamoro

| 477 |

CHAPTER 7: ENGINEERING

I. Cyber, Artificial Intelligence, and Space: Impact on the Future of Special Operations

By Namrata Goswami

ABSTRACT: Irregular warfare requires a specialized understanding of context, population, and technology. This kind of warfare aims to discredit, disrupt, sabotage, delegitimize, deter, and destroy an adversary's capabilities while ensuring that a conflict, especially between Great Powers, remains within the threshold of unconventional warfare. Toward achieving this goal, the utilization of cyber, artificial intelligence (AI), and space by countries like China, Russia, Iran, and North Korea and by violent extremist organizations like ISIS are becoming more common. This section highlights the relevance of cyber, AI, and space for special operations. It illustrates how these technologies are being utilized in the operational and strategic domains today. The section concludes by offering two distinct but vital future scenarios on the cumulative effect of cyber, AI, and space and highlights the roles U.S. Special Operations Command and Special Operations Forces might be called upon to accomplish in these situations.

INTRODUCTION

Irregular warfare (IW) is a way of war that involves a specialized understanding of context, location, geography, population, and technology. IW is such that the battlefield keeps evolving, and adaptation, initiative, and technological prowess are key factors in winning that fight.[1] As per the 2020 Irregular Warfare Annex to the U.S. National Defense Strategy, IW is defined as "a struggle among state and non-state actors to influence populations and affect legitimacy. IW favors indirect and asymmetric approaches, though it may employ the full range of military and other capabilities, in order to erode an adversary's power, influence, and will. It includes the specific missions of unconventional warfare (UW), stabilization, foreign internal defense (FID), counterterrorism (CT), and counterinsurgency (COIN). Related activities such as military information support operations [MISO], cyberspace operations, countering threat networks, counter-threat finance, civil-military operations, and security cooperation also shape the information environment and other population-focused arenas of competition and conflict."[2] The terms 'influence population' and 'affect legitimacy' are vital and include a full spectrum of tactics to achieve a strategic end goal of undermining the adversary. Disinformation, economic coercion, guerilla and proxy wars, and covert operations are a timeless part of these kinds of warfare. Countries like China, Russia, and Iran; non-state groups like the Islamic State of Iraq and Syria (ISIS); and al-Qaeda (AQ) are engaging in such kinds of warfare and adapting to a security environment where emerging technologies like cyber, artificial intelligence (AI), space, and drones are starting to play a role in enabling irregular warfare strategies. The Irregular Warfare Annex to the U.S. National Defense Strategy cautions that the U.S. has been cynical and ineffective in ensuring that it is equipped to deal with the reality that IW is part of the great power competition environment the country finds itself in.[3] The Annex recommended that the Department of Defense (DoD) must institutionalize IW as a core competency for which it is prepared and avoid being reactive when adversaries utilize IW to gain strategic advantage. The Annex cautions that the U.S. is underprepared for IW. It is important to repeat that caution here.

> Our adversaries seek to undercut our global influence, degrade our relationships with key allies and partners, and shape the global environment to their advantage without provoking a U.S. conventional response. As we reorient the Department toward great power competition, we do not have the luxury of discarding our well-honed ability to wage irregular war as we have done in the past. For example, the U.S. entered irregular wars in Vietnam, Afghanistan, and Iraq unprepared to conduct the major counterinsurgency (COIN) campaigns and counterterrorism campaigns that followed.[4]

An effective counter to such an adversary strategy is a combination of conventional and special operations that are equipped to deal with such a fluid environment. This would involve a whole gamut of military activities like force movements, partner nations' engagement, intelligence, information operations, and training. These tactical and strategic environments are, more often than not, influenced by emerging technologies like cyber, AI, and space.[5] While it is not just up to the United States Special Operations Command (USSOCOM) to deal with IW, the adaptability and fluidity of Special Operations Forces (SOF) implies that they will be called upon to do so, not only in dealing with groups like ISIS but also to affect the environment of great power competition in favor of the U.S. To do this effectively in the 21st century, understanding the role of emerging technologies concerning special operations is key.

WHAT ARE EMERGING TECHNOLOGIES?

Emerging technologies are defined by five key attributes: they are radical in their novelty, there is a fast growth in their emergence, they are coherent, they have prominent impact, and they result in uncertainty and ambiguity.[6]

> ***| Emerging technologies are defined by five key attributes: they are radical in their novelty, there is a fast growth in their emergence, they are coherent, they have prominent impact, and they result in uncertainty and ambiguity. |***

Danielle Rotolo, Diana Hicks, and Ben. R. Martin's article on "What is Emerging Technology" aims to provide a measurement of what indeed is an emerging technology—for instance, nanotechnology. Measurement in terms of impact is key (e.g., resource allocation, characteristics, regulation, and utilization), specifically as the impact is understated, unknown, or even obscured but becomes a dynamic factor once the technology is universally accepted and utilized.[7] Utilization is vital, especially from the point of view of SOF as it could impact operations and how technologies like AI can affect results. In a 2024 report by PricewaterhouseCoopers (PwC), eight emerging technologies were included as most essential. These include artificial intelligence (AI), internet of things (IoT), blockchain, augmented reality (AR), quantum computing, advanced robotics, virtual reality (VR), and neuromorphic computing.[8]

This chapter argues that besides these eight technologies, space and cyber technologies combined with AI and drones, are transforming how warfare tactics, strategy, and operations are shaping up. So how would cyber, AI, and space technologies play into, and affect the future of SOF and its understanding of warfare, its context, training, tactics, strategy, doctrine, operations, and end goals? This chapter analyzes each technology specifically and then offers two scenarios of the cumulative effects of the technologies working together on USSOCOM and SOF.

Artistic representation of artificial intelligence and machine learning. Source: WikiCommons[9]

CYBER

Cybersecurity is a vital capability to defend U.S. critical infrastructure, especially disrupting malicious activities before they affect the U.S. homeland. Cybersecurity adds to the joint fight and is a vital capability required for USSOCOM and SOF in the 21st century. As the Department of Defense's 2023 Cyber Strategy states, "military capabilities are most effective when used in concert with other instruments of national power, creating a deterrent greater than the sum of its parts. In this way, cyberspace operations represent an indispensable element of U.S. and Allied military strength and form a core component of integrated deterrence."[10] Operations in cyberspace will be conducted for campaigning, limiting and frustrating adversary actions just below the threshold of armed conflict. This can be achieved in partnership with U.S. allies and partners. The DoD Cyber Strategy clearly states that "both the People's Republic of China (PRC) and Russia have embraced malicious cyber activity as a means to counter U.S. conventional military power and degrade the combat capability of the Joint Force. The PRC in particular sees superiority in cyberspace as core to its theories of victory and represents the Department's pacing challenge in cyberspace. Using cyber means, the PRC has engaged in prolonged campaigns of espionage, theft, and compromise against key defense networks and broader U.S. critical infrastructure, especially the Defense Industrial Base (DIB). Globally, malicious cyber activity continues to grow in both volume and severity, impacting the U.S. Homeland and placing Americans at risk."[11] China views cyber as a means to degrade the combat capability of the U.S. Joint Forces as part of its theory of victory.

| China views cyber as a means to degrade the combat capability of the U.S. Joint Forces as part of its theory of victory. |

In times of conflict, China is equipped to launch cyberattacks on the U.S., especially to deter military modernization and sow chaos.[12] In a similar vein, Russia is developing cyber capabilities that attack U.S. institutional structures, including U.S. elections and U.S. critical infrastructures.[13] Other actors developing their cyber capabilities to target the U.S. through espionage, ransomware, compromising cryptocurrencies, and targeting U.S. media, government, academic, and business websites include North Korea and Iran. Violent extremist organizations (VEOs) utilize cyber for purposes of recruitment, geo-spatial mapping, espionage, generating resources, and propaganda.[14] The DoD Cyber Strategy focuses on four critical tasks in the cyber domain: defend the Nation; prepare to fight and win the Nation's wars; protect the cyber domain with allies and partners; and build enduring advantages in cyberspace.[15]

| The DoD Cyber Strategy focuses on four critical tasks in the cyber domain: defend the Nation; prepare to fight and win the Nation's wars; protect the cyber domain with allies and partners; and build enduring advantages in cyberspace. |

Defending U.S. critical infrastructure and the DIB is a key focus of the U.S. 2023 National Cybersecurity Strategy. Malicious cyberattacks routinely target this vital industrial base. While defense companies must implement the Cybersecurity Maturity Model Certification Program, which mandates that companies with DoD defense contracts implement information security standards, such standards need to constantly evolve and adapt as adversaries get better at cyberattacks. The U.S. 2023 National Cybersecurity Strategy cautions that while the digital ecosystem has transformed lives for the better, it has also "been misused to enable transnational repression and digital authoritarianism; steal data and intellectual property; distribute disinformation; disrupt critical infrastructure; proliferate online harassment, exploitation, and abuse; enable criminals and foster violent extremism; and threaten peace and stability."[16] While the IOT builds a comprehensive ecosystem from consumer goods and digitized industrial capabilities to directing and utilizing a constellation of satellites, thereby bringing about efficiency, it can also be compromised, requiring a robust defense mechanism. "The widespread introduction of artificial intelligence systems—which can act in ways unexpected to even their creators—is heightening the complexity and risk associated with many of our most important technological systems."[17] The National Cybersecurity Strategy highlights the Russian 2017 NotPetya cyberattack on Ukraine that quickly spread to 60 countries causing damage of approximately $10 billion.[18] Maersk, the global transportation company, saw 49,000 of its laptops and print capabilities destroyed. Adam Banks, the head of technology at Maersk, stated that "all of our 1,200 applications were inaccessible, and approximately 1,000 were destroyed. Data was preserved on back-ups, but the applications themselves couldn't be restored from those, as they would immediately have been reinfected. Around 3,500 of our 6,200 servers were destroyed – and again, they couldn't be reinstalled."[19] The CIA concluded that the Russian military was behind the NotPetya cyberattacks on Ukraine.[20] In 2020, a federal Grand Jury in Pittsburgh charged six Russian Main Intelligence Directorate (GRU) officers from Unit 74455 as the perpetrators of the malware.[21] Similarly, in 2020, Russia's Intelligence Service (SVR) was behind the malware code attack on SolarWinds's software program, Orion, utilizing that breach to target U.S. government agencies that included the U.S. Departments of Justice, Energy, Defense, and Homeland Security; the U.S. Cybersecurity and Infrastructure Agency

(CISA); and major private companies ranging from Microsoft and Intel to Cisco. The Russian effort was aimed at espionage and destroying, degrading, and altering data to spread misinformation. Speed of compromise is a key element, as a defender might not be able to match the speed of the universal breach. In the case of SolarWinds, the Russian agents succeeded in carrying out a nine-month breach into U.S. computer networks before it was discovered. Moreover, the hackers cleaned the crime scene so well it was difficult to attribute, and while the U.S. has stated that Russian intelligence was behind it, Russia has denied any culpability.[22] The SolarWinds hack was spectacular as it was a complete supply chain hack.

Artistic depiction of cybersecurity. Source: WikiCommons/public domain

WHAT SHOULD SPECIAL OPERATIONS PREPARE FOR?

As per former USSOCOM Commander, Army General Richard Clarke, 60 percent of the focus of special operations is on the domain of digital information, where cyber is vital.[23] In the era of great power competition and a fight against VEOs, cyber can play the role of a force multiplier regarding who has the strategic advantage in terms of information and toward shaping the environment. Therefore, USSOCOM should be developing dedicated cyber units that can work jointly with U.S. Cyber Command to utilize and support information warfare. For China and Russia, informatization is a key focus of their military training and is aimed at utilizing a combination of cyber, AI, and space, to develop force advantage.[24] The idea behind China's notion of information warfare is to manipulate and influence adversary decision-making and behavior. Concepts like big data and brain science play into China's development of psychological warfare. Such information manipulation is viewed by China as offering a leading edge for national power and great power competition.[25] Russia has successfully utilized information warfare and cyber to diffuse and generate different perspectives on the Russian war in Ukraine. Russia has utilized propaganda and created fake social media accounts and manipulated videos and images to support and vindicate the Russian war effort. Such accounts push narratives that Ukraine is run by Nazis and the U.S. supports that, that Ukrainian videos showcasing Russian attacks on Ukrainian civilian areas are fake and doctored by the Ukrainian authorities, and that the Western weapons Ukraine has received are being sold on the dark web.[26] Channels such as Telegram and a pro-Kremlin English-language account, ASB Military News, have posted about Western equipment being resold, and these web accounts have been shared widely. A BBC investigation found out that old, doctored images of weapons were passed off as Ukrainian military selling U.S.-donated weapons on the dark web. The cybercrime threat intelligence company KELA discovered that this dark web marketplace for weapons was polluted by pro-Russian propaganda sources and that false narratives were created by these sources.[27] A similar cyber operation to defuse narratives about Russia's invasion of Ukraine is underway whereby the world is divided as to the legitimacy of the invasion.[28]

A joint Special Forces team moves together out of a U.S. Air Force CV-22 on Osprey Feb. 26, 2018, at Melrose Training Range, New Mexico. Source: U.S. Air Force

VEOs like ISIS and Taliban 2.0 have widely used cyber as part of their information operations and to draw in recruits, and finances as well as train their terrorist cells. During the period 2014 to 2015, ISIS utilized cyber extensively to draw in foreign recruits, generate funding, and radicalize young people in Western countries to then motivate them to attempt to travel to the Islamic caliphate.[29] ISIS utilized X (Twitter at that time), Surespot, and Telegram to share messages about the vitality of the group and why the U.S. and its military were corrupt. These social media messages were shared extensively on the internet.[30] In fact, the content for ISIS was made in a decentralized way, from sources in West Africa, Latin America, Southeast Asia, and the Caucasus. Around 2016, ISIS released 38 new items daily and bombarded the cyber world with documentaries, images, videos, and audio clips in languages from Russian, Hindi, and Arabic to English to Bengali. The group dovetailed its content to target audiences. This content included a narrative of a stable caliphate replete with economic development, social welfare schemes, and military victories against the U.S and allied forces. A 2004 book that inspired ISIS was written by Abu Bakr Naji and titled The Management of Savagery. It advised jihadist groups like ISIS to go to regions beset by anarchy to establish Islamic sharia law-based governance and utilize the internet to spread those sharia-based ideas of governance.[31]

CYBER

In this context of the utilization of cyber across different platforms by adversaries, SOF integration of cyber into their training and operations, especially from the perspective of a local and/or regional context is useful and a force multiplier. There is the "triad" concept, where the U.S. Army is working to integrate its Special Operations Command, its Space and Missile Defense Command, and its Cyber Command more closely. This is an attempt to weave together multiple battlefield concepts and develop a response to adversaries operating in multiple domains, all at once. Lt. Gen Jonathan Braga, commander of U.S. Army Special Operations Command, specified during the Space and Missile Defense Symposium in 2022, "we have examples in the counter-ISIS fight of combining lethal and non-lethal effects for a much larger holistic effect that I would say had a larger impact than anyone predicted prior to us truly synchronizing our efforts for having an effect on the adversary there."[32] Concepts like Project Convergence, the annual sensor-to-shoot test, aspire to include this idea of the triad (special operations, cyber, and space). Such convergence would, however, require specific knowledge about how cyber, AI, and space interact with special operations to build strategic and tactical advantages for USSOCOM. But it could also pose a challenge if not done effectively and well.

SOF can be utilized to support U.S. cyber and other information warfare missions and their implementation in a localized context. This would require training not only in the cyber domain but also on how to operate in a particular local and/or regional context. Development of specialized units like the 915th Cyber Warfare Battalion in 2019 by the U.S. Army is a good step that can be further augmented by developing Special Mission Units dedicated to cyber warfare. Learning about how to hack IoT is a much-required capability for USSOCOM and SOF. Such capabilities can help analyze, collect, and utilize specific information to achieve information dominance in the realm of cyber. As per McKinsey & Company, IoT "describes physical objects embedded with sensors and actuators that communicate with computing systems via wired or wireless networks—allowing the physical world to be digitally monitored or even controlled."[33] The potential value of IoT is estimated to be about $12.5 trillion by 2030, meaning more and more societies will be utilizing these devices. IoT helps in data management, including a cloud infrastructure that can store data and be utilized effectively through cloud computing. Cybersecurity is a critical component of such cloud-based data storage and computing. The interconnectedness of these platforms means that they can be attacked by adversaries—but, if adversaries utilize such a cloud-based system, they can be hacked as well. This is where special operations can play a critical role. An example of such an attempt to hack IoT was the Russian Military Intelligence Service (or GRU) cyber operation unit 26165 targeting the Organization for the Prohibition of Chemical Weapons (OPCW) in the Hague in 2018. GRU agents aimed to hack the Wi-Fi system of the OPCW from proximity. This amounted to an attempt to sabotage the network and manipulate the data.[34] This Russian cyber operation was disrupted by The Netherlands Defense Intelligence and Security Service, and the four Russian intelligence officers involved in the cyberattack were apprehended and escorted out of The Netherlands.[35] The U.S. and the United Kingdom collaborated with The Netherlands for this particular operation and took the unprecedented step of holding a press conference to share the details of this Russian cyberattack on the OPCW. The OPCW was collecting evidence on Syria's use of chemical weapons and subsequently the usage of the Novichok nerve agent against former Russian spy Sergei Skripal, who was living in Salisbury in the UK at that time. The UK asked for technical assistance from the OPCW experts, who confirmed the use of the nerve agent in the attack on Skripal.[36] It was through close coordination between the U.S., the UK, and the Netherlands that the Russian cyber operation against the OPCW was thwarted. However, such close coordination between a host nation and SOF may not always be possible.

| SOF can disrupt, deny, and damage the cyber networks of adversary nations and thwart or destroy adversary communication nodes in the cyber domain. |

Consequently, developing SOF cyber capability to access adversaries and VEOs' cyber networks can help gather vital information and/or thwart cyberattacks. This is a much-required capability, especially in a hostile environment, but which nonetheless poses a threat to the U.S. and its allies. SOF can disrupt, deny, and damage the cyber networks of adversary nations and thwart or destroy adversary communication nodes in the cyber domain.[37]

MISO can play a role in this regard. Given the critical importance of convergence in this domain, a USSOCOM strategy of building institutional mechanisms between its own cyber expertise and U.S. Cyber Command would enable the development of effective countermeasures to deter or destroy adversary cyber operations.

ARTIFICIAL INTELLIGENCE

The great power competition we are witnessing is being structurally operationalized in ways that it remains within the threshold of conventional war. Great powers are utilizing technologies like cyber, AI, and space, either individually or in combination, to create advantages that are invisible to the layman but have devastating effects. In this new era of competition, the end goal is to gather legitimacy and affect population choices. In this context, such irregular warfare concepts like unconventional warfare, FID, stabilization efforts, counter-terror, and counterinsurgency gain credence. SOF will be called upon to carry out military information operations and cyberspace operations, countering network threats and anticipating the challenge of AI.[38] AI-enabled data analysis, large language models, autonomous drones and weapons, precision guidance of munition, and terrain mapping and analysis have made a dent into the operational dimension of war.

| AI-enabled data analysis, large language models, autonomous drones and weapons, precision guidance of munition, and terrain mapping and analysis have made a dent into the operational dimension of war. |

There are concerns that unmanned vehicles equipped with AI—whether on land, air, sea, or space—can make lethal decisions affecting a strategic space. China has conducted simulations of an unmanned orbital platform equipped with AI and carrying hundreds of CubeSats that can both remove space debris and act as a weapon and/or deterrence against any threats to Chinese space systems (CubeSats describes nanosatellites that utilize a standard size and form factor).39 Chinese AI systems will determine when and where to release cubesats to deter enemy satellites.[40] A similar AI-enabled technology called underwater drone or unmanned undersea vehicles (UUVs) was demonstrated by China in the South China Sea (SCS). An underwater drone, enabled by AI, was able to locate, recognize, follow, and attack a simulated enemy submarine—completely based on autonomous AI decision-making and without any interventions by humans.[41] Similar UUVs have made their presence felt in the Indian Ocean, close to India's maritime borders, where the People's Liberation Army Navy (PLAN) utilized underwater drones for research purposes and to gather oceanographic data, a capability that can be utilized for military purposes.[42] Such operations are conducted as gray zone operations defined as "maneuver in the ambiguous no-man's-land between peace and war, reflecting the sort of aggressive, persistent, determined campaigns characteristic of warfare but without the overt use of military force."[43] In fact, the Chinese approach of combining different technologies like cyber, AI, and space; construction of artificial islands in the South China Sea; and use of paramilitary and maritime militias to aggressively follow foreign military vessels in the area all fall under the realm of unconventional warfare.[44] ISIS has utilized drone attacks on the Peshmerga and in 2017 established a division called the unmanned aircraft of the Mujahedeen.[45] Drones can be guided by GPS/Google Earth and equipped with rudimentary AI facial recognition systems to target specific individuals and/or military infrastructure. A swarm of 13 drones equipped with explosive devices were used by Syrian-based, ISIS-affiliated groups to target two Russian military bases (Hmeimim and Russian Naval CSS point) in Syria in 2018.[46] Russia took down most of the drones with its anti-aircraft missiles. In 2019, 25 drones and missiles were utilized in an attack on two Aramco sites, forcing Saudi Arabia to shut down half of its oil production, amounting to 5.7 million barrels of crude oil per day.[47] The Houthis of Yemen claimed responsibility for the attacks. In March 2022, the Houthis again launched separate missile and drone attacks on Aramco's petroleum product distribution center in Jeddah, and refineries in Ras Tanura and Rabigh refineries. Saudis accuse Iran of arming the Houthis with advanced missiles and drones.[48]

During the 2023–2024 Israel–Hamas war, Houthis have launched drones and missiles targeted at commercial shipping in the Red Sea.[49] The strategic concern is that non-state armed groups like ISIS and the Houthis will be able to create autonomous drone swarms enabled by rudimentary AI that can target sites/people/cities/shipping by making decisions on their own based on the sensory data they have been fed.

> ***| The strategic concern is that non-state armed groups like ISIS and the Houthis will be able to create autonomous drone swarms enabled by rudimentary AI that can target sites/people/cities/shipping by making decisions on their own based on the sensory data they have been fed. |***

This will dramatically increase their ability to create mass violence and political instability, which is something terrorists and insurgent groups aim for. The aim behind such terrorist tactics is to challenge the legitimacy of the state and influence the affected population. If a state actor is unable to provide security and protect lives, its legitimacy and credibility is at stake. The biggest concern is AI-enabled drones delivering explosives with precision to the intended target. From a USSOCOM and SOF perspective, this would mean that these terrorist groups can execute rapid coordinated attacks and conduct such unmanned operations from a distance. As per Jacob Ware, writing for War on the Rocks, "killer robots, combining drone technology with more advanced AI, will possess the means and power to autonomously and independently engage humans. The lethal autonomous weapon has been called the 'third revolution in warfare,' following gunpowder and nuclear weapons, and is expected to reinvent conflict, not least terrorist tactics."[50] A hypothetical scenario utilizing swarms of drones was depicted in a YouTube video titled "Slaughterbots,"[51] which targeted individuals posting certain political views on social media as well as U.S. policymakers on Capitol Hill. Given the lethal possibilities of such technologies, terrorist groups will attempt to acquire them. Drones have already been utilized for propaganda imagery by ISIS, by the Taliban to capture images of suicide bombers driving a Humvee into a police base in Helmand in October 2016 and for intelligence, surveillance, and reconnaissance (ISR). Taliban spokesperson Zabihullah Mujahid stated that drone technology "is very useful in discovering the enemy's most important bases."[52] Given the availability of AI and drone technologies, terrorist groups will plan on acquiring them given their low signature (meaning there will be little or no trace of who the perpetrator is when such technologies are used), their cheap costs, their fast reaction times, their processing capabilities, and their lethality.

The U.S. Army's MQ-1C Warrior Unmanned Aerial Vehicle. Source: U.S. Government

The key challenge for USSOCOM is that U.S. adversaries are already using AI to expand their operational and strategic capabilities. China views such capabilities as force multipliers as it utilizes a combination of AI, space, robotics, and quantum communication to create a strategic advantage. Lawfare is part of that mission set, with China taking advantage of legal mechanisms to create strategic vantage points for itself. In fact, within China, given their policy goal of civil-military fusion, civilian tech companies are mandated by law to create technologies that have dual capabilities to include AI.[53] Consequently, information that translates to intelligence is vital in SOF's effective operations. AI plays a key role in that information mapping by using big data analysis and machine learning to derive pattern recognition. The DoD's Project Maven, under the National Geospatial Agency, not only aims to apply AI to ISR platforms and sensors and utilize algorithms on the battlefield but also recognize threats to the supply chain of AI.[54] Such pattern recognition using AI can utilize the automated exploitation of audio, video, and big data to effectively create a clear map of what SOF is up against. This equips USSOCOM to both respond quickly and effectively to adversaries' utilization of similar tools and recognize patterns of how they behave, how they trade, how they move in the world of the dark web, how they communicate, and where they are located, exposing these subversive activities.

This implies that there should be a proactive USSOCOM strategy of responding to threats to avoid being caught in the reactive cycle where the adversaries decide how and where they attack, thereby gaining strategic initiative and surprise. AI can integrate raw data to create deep pattern recognition, assess habits, and offer a fast and efficient ISR capability to SOF.[55] One of the challenges for SOF and USSOCOM would, however, be that adversaries can manipulate the data collected, which could lead to false pattern recognition and faulty scenarios being generated, both in cyberspace and in a local context. Therefore, human intervention is critical to maintaining a reality check. SOF will face another challenge, in that adversaries utilizing facial recognition software can discover SOF operations and weapons. Therefore, smart utilization of AI where deception is part of the training process is key to successful SOF missions. The greatest challenge for USSOCOM is a mindset, adaptability, effective training, and the adoption of key technologies like AI. The innovation in AI is happening now;[56] the training, adoption, and integration into missions has to happen in real-time, not sometime in the future when AI is a mature part of adversary planning, strategy, targeting, and deception.[57]

| The innovation in AI is happening now; the training, adoption, and integration into missions has to happen in real-time, not sometime in the future when AI is a mature part of adversary planning, strategy, targeting, and deception. |

An example from the ground where AI was used to analyze sensor data occurred in Ukraine, where NATO utilized AI and machine learning to determine where the allied forces were located and where the adversary (Russia) was. This helped NATO deliver precise location maps of enemy positions to Ukrainian commanders. AI-enabled pattern recognition helped when deciding which weapons could be most effective against a particular enemy position.[58] According to Mykhailo Fedorov, Ukraine's former Minister of Digital Transformation (2019-2023), it was this kind of data analysis that proved vital in the liberation of the Kherson, Izium, Kharkiv, and Kyiv regions of Ukraine from Russian forces. Russia has, with time, learned lessons from the use of big data like Google Earth, AI, machine learning, and pattern recognition combination that helped Ukrainian forces. Subsequently, Russian forces have adopted a strategy of planting land mines, ensuring a heavy presence of troops with a propensity to take large casualties—a World War I-kind of trench warfare that involves hand-to-hand combat.[59] AI could rapidly create deep fakes, using compromising or incriminating information (*kompromat*),[60] typically for blackmail used strategically to discredit and generate fake news and propaganda. Kompromat is aimed at disparaging not only military forces but also influencing the target population. After all, the main intention of irregular warfare is to inform population choices and create legitimacy for a particular activity. Russia's takeover of Crimea, the invasion of Ukraine, and the Russian influence operations in Ukraine[61] as well as China's narratives on the South China Sea, the U.S., and Taiwan are filled with and supported by such generative AI. Generative AI can be utilized to tailor messages resonating with the target population. Such efforts can penetrate adversary networks, government, and civilian institutions; the media; and social networks. Infiltration by generative AI means domination of a particular domain to create disruptions in society. From the perspective of generative AI, several scenarios can be drawn out for USSOCOM to consider. Generative AI could enable rapid crafting of messaging, rapid translation of messages (and dissemination may be through space-enabled means). ChatGPT (generative pretrained transformer) and Dall-E can be utilized to create content like code, audio, images, simulations, and videos[62] that can drastically change and affect the irregular warfare battlefield.

Optimizing AI for USSOCOM missions is another area that can work as a tremendous force multiplier. AI can help in planning and selecting multiple options through the development of war gaming. For instance, SOF can utilize AI-generated war gaming to understand the impact of killing a terrorist leader of an organization like ISIS or Boko Haram. This is a consequence-driven war game. AI, using social network analysis (SNA), can create a reliable assessment of the impact a particular terrorist leader has on the development, organization, and recruitment of that terrorist organization. AI can be optimized for command-and-control decision-making as a support to human leadership and for real-time analysis of different options in an area where there is disinformation, conflict, absence of state-driven legitimate governance, and the presence of hostile forces.

| AI can be optimized for command-and-control decision-making as a support to human leadership and for real-time analysis of different options in an area where there is disinformation, conflict, absence of state-driven legitimate governance, and the presence of hostile forces. |

In a not-too-distant future, AI can be utilized for automated weapons systems making decisions on which targets to secure, which to attack, and which to deter from launching an attack on friendly forces. This kind of adaptation to new and rapidly emerging technologies is vital. A

SOF culture has to be developed that understands the criticality of these rapidly emerging technologies and their contribution to how warfare is being conducted today.[63]

SPACE

Russia's invasion of Ukraine made it evident how space capabilities can add to the lethality of forces on the ground. According to Kyrylo Budanov, the Head of the Main Ukrainian Intelligence Directorate, SpaceX's Starlink constellation of satellites located in low Earth orbit (LEO) has been utilized by the Ukrainian military in the front line in all aspects: ISR, communications, targeting, movement of weapons and troops, and battlefield operational mapping. Starlink is a private sector-run satellite constellation launched by SpaceX in 2019 providing high-speed, low-latency broadband internet. In fact, so powerful has been the impact of Starlink that there are allegations in Walter Isaacson's biography of SpaceX CEO Elon Musk that Musk ordered SpaceX to turn off Starlink communications to disrupt a sneak attack by Ukraine on Russia's naval fleet near the Crimean coast. This resulted in Ukrainian submarine drones, armed with explosives, to lose communication, resulting in failure of the operations.[64] The argument given by Isaacson is that Musk feared Russia would retaliate with nuclear weapons. Musk responded to this allegation on X by stating that Starlink never provided communications over Crimea. "There was an emergency request from government authorities to activate Starlink all the way to Sevastopol. The obvious intent being to sink most of the Russian fleet at anchor. If I had agreed to their request, then SpaceX would be explicitly complicit in a major act of war and conflict escalation."[65] This brings home the fact that space-based satellite communications play a vital role in the technologies used in the warfare of today.

| This brings home the fact that space-based satellite communications play a vital role in the technologies used in the warfare of today. |

However, such a technology at the hand of a private space company comes with costs, specifically when it is not directed by U.S. official space policy and/or U.S. official response to Russia's invasion of Ukraine. For instance, after providing satellite internet to Ukraine until February 2023, SpaceX made a shift in company policy by stating that Starlink communication was never meant to support the Ukrainian military's offensive operations.[66] This shift in company policy has direct effects for Ukrainian military effectiveness. Consequently, it's better and wiser to have a solid government-to-government agreement in place and/or develop satellite communications and other related capabilities through public–private cooperative mechanisms.[67]

U.S. Space Force extremely high frequency communications satellite. Source: U.S. Air Force illustration

The Ukraine case brought home the vitality of space and the modern battlefield despite the issues raised due to the private sector undertaking of satellite communications. Satellite imagery has played a vital role in augmenting the battlefield. It was space-based ISR that provided clear evidence of the Russian buildup before its 24 February 2022 invasion of Ukraine; imagery pinpointed the kind of weapons and hardware utilized by Russia. Also, early warning radar supported by satellites identified and tracked the launch of ballistic missiles, and satellite imagery and intelligence helped the Ukrainian military sink Russian warships in the Black Sea. Space has played a role in launching weapons supported by space-based precision, navigation, and timing (PNT), specifically the High Mobility Rocket System (HIMARS) rockets guided by GPS with a range of 50 miles. GPS-guided Excalibur artillery shells have been provided by the U.S. to Ukraine.[68] Russia has, however, been successful in deterring and thwarting the GPS-guided HIMARS by utilizing electronic jammers, which resulted in the HIMARS rockets missing their target.

WHAT CAN SPECIAL OPERATIONS LEARN FROM THIS?

USSOCOM Commander, Army General Bryan P. Fenton, stated that a triad between SOF-space-cyber is value addition to SOF operations. Moreover, SOF has to navigate through advanced air defenses using cyber and space, which brings in the point of developing a satellite payload capability of its own. This can help in building ISR, communications, and targeting. Leveraging the satellite-based capabilities that the National Reconnaissance Office (NRO) provides can help build that required intelligence map.[69] The shift in grand strategic thinking for the U.S., from COIN to near-peer competition, is enabling a focus on space as a force multiplier. USSOCOM should partner with other services and develop its own CubeSats and hosted payloads.[70]

| The shift in grand strategic thinking for the U.S., from COIN to near-peer competition, is enabling a focus on space as a force multiplier. USSOCOM should partner with other services and develop its own CubeSats and hosted payloads. |

Some of the specific USSOCOM and SOF space requirements, according to Special Reconnaissance Program Executive Officer David Breede, could include, but are not limited to, "data exfiltration, the remote command and control of sensors, of doing [radio frequency] survey from a soft cube payload...[building] organic small CubeSat capability that will be (designed, built, and owned) by SOCOM. We're seeing how that applies to addressing future capabilities."[71] USSOCOM can work with the Defense Advanced Research Project Agency (DARPA) by hosting payloads in its Blackjack program that aim to develop a LEO-based, global high-speed internet for the DoD, providing a "highly connected, resilient, and persistent coverage".[72] DARPA can accomplish its high-speed internet goal in partnership with the U.S. commercial sector.

Space will play a critical part in the future battlefield, with countries like China and Russia integrating space into their military strategies and warfare doctrines.[73] Moreover, both China and Russia will be enabling non-state actors with space-based support in areas they deem their spheres of influence. Such support is evident with Russia in Crimea and now Ukraine as well as China in the South China Sea island disputes, Taiwan, Tibet, and the East China Sea.[74] This utilization of space to broadcast power will go beyond China and Russia's immediate regions, to an extension globally. This aspect was evident in the Belt and Road Initiative summit held in Beijing in October 2023 where Russian President Vladimir Putin was the guest of honor.[75]

Space technologies like China's BeiDou Navigation System, satellite internet, launch, ISR and Russia's counterspace capabilities will be utilized to support their own special operations in areas of core strategic interests.

In this context, it is useful to conceptualize how space assets can enable SOF.

SPACE-ENABLING SPECIAL OPERATIONS

Space can help execute a wide variety of typical doctrinal missions for SOF. This includes such missions like unconventional warfare and foreign internal defense,[76] counterinsurgency, special reconnaissance, counterterrorism, information operations, counter proliferation of weapons of mass destruction (WMD), and security force assistance and can deny adversaries access to ungoverned spaces and hostile forces the ability to move freely.[77] Space is a force multiplier in all these missions. Military engagements and deterrence will require a range of capabilities to promote and secure specific U.S. interests and develop partner nations and host nation capabilities. This implies that SOF would form a part of the overall U.S. security effort and would bring their special capabilities to an operational theatre. This is especially true in regard to foreign internal defense, which is one of the core special operations tasks of USSOCOM. Space can play a vital role in building this capability—and in supporting a multi-agency strategy to achieve effectiveness in this mission that also requires interoperability, meaning "the ability to act together, coherently, effectively, and efficiently to achieve tactical, operational, and strategic objectives."[78]

Space capabilities within the U.S. have changed with the development of reusable launch capabilities like SpaceX's Falcon 9 and Falcon Heavy, capable of lifting 18.4 tons and 24.75 tons to LEO respectively.[79] Such a revolution in space launch has translated to cheaper launches, which means SOF can send up custom-made payloads and/or hosted payloads on mega constellations within the concept of proliferated LEO constellations.[80] This ensures that satellite payloads and satellites themselves are protected against adversarial counterspace capabilities. Such a strategy of diversification can extend to middle Earth orbit (MEO) or geosynchronous orbit (GEO).[81] The advent of LEO mega-constellations decreases the cost and increases the availability of overhead sensing and global communications. SOF can now receive high bandwidth internet globally through Starlink[82] and will be able to communicate on unmodified cell-phones via Lynk.[83] SOF can procure near-real-time overhead imagery (Planet, Blacksky, Maxxar),[84] RF mapping (Hawkeye360, Spire),[85] synthetic aperture radar (IceEye), hyperspectral, infrared (Northrop Grumman) or, through the NRO, the National Geospatial Intelligence Agency and the U.S. Space Force (USSF), commercial contracts. Thermal imagery that can detect both fires and/or artillery fire can be procured from the National Oceanic and Atmospheric Administration.

The use of mega-constellations to remotely control drones and create battlespace maps is well documented in Ukraine. Palantir, an American technology company, created detailed maps of the Ukrainian battlefield in 2022 by utilizing commercial satellites. This map provided targeting intelligence. Based on such a digital map, the Ukrainian military could select a missile, an armed drone, or an artillery piece to target a Russian position. Drones would confirm the strike, and satellite imagery could be utilized to assess damage. Palantir has a tool called Megaconstellation that helps understand what kind of commercial satellite data is available for a battlespace. To develop a real-time assessment of one Ukrainian military mission for Kherson, 40 commercial satellites that passed over the area in 24 hours were utilized to craft the mission map. If required, 306 commercial satellites pass over an area, and their images can be bought (Images cover a range of 3.3 meters). Communications and broadband connectivity were provided by the Starlink mega constellation.[86] A similar utilization of technology is being detected in Israel's war against Hamas.[87] In fact, Planet satellite imagery traced the movement of Israeli troops and detected impact craters from missile strikes in Gaza City, similar to the Maxar imagery of the 40-mile Russian convoy that entered Ukraine in February 2022.[88] Planet satellite imagery showcased the previous position of Israeli tanks and armored personnel that cut off the city from the rest of the Gaza Strip.[89] A downside to all of this is that space can restrict SOF movement; it will also require quick and resilient adaptation and new operational concepts as adversary satellites can pick up SOF movements. As a rule, overhead space sensing systems make it difficult to concentrate and move forces of all kinds without detection. Adversary proliferation of space-based sensors, communication, and geolocation—especially coupled with AI pattern recognition and anomaly detection—increases the possibility of detection, tracking, and targeting.

A potentially extremely powerful combination is the linking of SOF with space capabilities for unconventional warfare, where space enables a normally weak force to oppose a stronger government force. Space can suppress government transmission systems (e.g., jamming, space-delivered cyberattacks), and substitute with space-delivered (and perhaps AI-generated) messaging via TV, radio, internet, and text. As demonstrated in Ukraine, space can be highly enabling to the lesser power. It can create a secure command and control network. It can enable partisan forces to have PNT, targeting, and command.

This is also likely the case for FID and security force assistance (SFA), where SOF might take the lead to grow expertise in weak states to use space capabilities to reduce ungoverned spaces. Thus, just as the Building Partner Air Capacity Seminar (BPACS) course that helps grow aviation enterprise development (AED),[90] we could imagine SOF doing "space enterprise development"—teaching host nation personnel to make use of (likely commercially available) space capabilities such as overhead sensing to monitor ungoverned areas and communications systems to coordinate government security actions and command forward elements, and provide forward services such as tele-education, telehealth.

A huge attack surface for space systems exists on planet Earth. By CARVER—criticality, accessibility, recoverability, vulnerability, effect, recognizability—[91] (a system that assesses and ranks threats and opportunities), it may be a preference of counter-space planners to use SOF for special reconnaissance or direct action to map, sabotage, introduce computer network attack code, install local jammers, and target critical personnel networks to mitigate risks emanating from adversary systems. Disabling adversary ground stations that support space operations can be another area where SOF can play a vital role. As the U.S. and allied economy extends into space, including private space stations, several authors,[92] have articulated and anticipated a need for a SOF mission set, which might involve, for example, special reconnaissance against adversary facilities, direct action to rescue American citizens or interests (or sabotage adversary facilities), countering any in-space proliferation of WMD (for example, a foreign space station experimenting with bioweapons or a Lunar uranium enrichment facility or misuse of a Lunar catapult), or counterterror against the seizure of U.S. commercial or space tourism spacecraft or facilities.

Based on the analysis above, the section concludes with two hypothetical scenarios of how cyber, AI, and space affect ways of warfare, and their impact on SOF.

SCENARIOS

There are two scenarios that USSOCOM and SOF might have to deal with in the next few years. One could include the sabotage and destruction of VEOs and their networked cyber, AI, and space. The second draws out SOF roles in a Taiwan scenario that falls under great power competition.

SCENARIO I: THE CALIPHATE STRIKES AGAIN

ISIS learns several lessons from Ukraine's resistance to Russia and explores how it can make use of free PNT and commercially available space-based internet, drones, and imagery for its end goals. ISIS utilizes front organizations that utilize virtual private networks and encrypted networks to purchase space-based internet and imagery from a variety of sources. ISIS purchases a large number of "off the shelf" Chinese and Iranian-made drones. ISIS can target U.S. and host nation assets using commercial imagery; they can coordinate and command the drones to target specific adversary infrastructure via space-based internet. The ISIS command center in Syria is hidden within a commercial cloud facility. ISIS has prepositioned its drones in everyday trucks with no easily distinguishing marks. Some of the drones are equipped with rudimentary facial recognition AI, and they are predesignated to target certain key Iraqi officials. The data for the facial recognition is stored in a closed network in a very secure ISIS facility. ISIS's planned attack seeks to decimate the Iraqi senior leadership.

U.S. Special Operations Command Central (SOCCENT) receives intelligence of such an ISIS plan to carry out a coordinated strike against Iraq, aiming to target Iraqi key officials, military leaders, and critical infrastructure. The President of the United States is briefed on this ISIS plan and determines that the attack must be thwarted; he designates USSOCOM as supported command with U.S. Central Command, U.S. Cyber Command, and U.S. Space Command as the supporting commands. SOCCENT is to lead in coordinating efforts.

RESPONSE BY SPECIAL OPERATIONS

USSOCOM would have to do a CARVER analysis and determine what to disable regarding the ISIS plan. A SOF response could include getting to the location of the ISIS command center to disable, sabotage, and/or corrupt their targeting system. SOCCENT must coordinate with U.S. Central Command, U.S. Cyber Command, and U.S. Space Command to synchronize a response strategy. U.S. Central Command can enhance intelligence sharing with Iraqi security forces on the imminent ISIS attack and provide assistance to disrupt and dismantle ISIS's plan. U.S. Cyber Command can target ISIS propaganda, recruitment, and communication networks. U.S. Space Command can track and provide analyses on satellite imagery and cut of GPS signals that ISIS could potentially use for a drone attack. SOF should be on the ground training, equipping, and advising Iraqi forces to thwart this ISIS threat.

SCENARIO II: RETAKING OUR LOST TERRITORY

The second scenario that USSOCOM and SOF will have to prepare for lies in the realm of the great power competition and Taiwan.[93] This will involve two great powers in the system—the U.S. and China. It will unfold in a manner that would challenge U.S. effectiveness in responding to a peer competitor.

In the runup to the 2028 general elections in Taiwan, the anti-China Democratic Progressive Party (DPP) promises Taiwanese voters a referendum on Taiwan's independence within the first month of its win. The DPP has been a strong advocate of an independent Taiwanese national identity and does not see itself as having any historic connections with mainland China. In the 21st National Congress of the Communist Party of China in 2027, President Xi Jinping warns that any movement toward Taiwanese independence will be met with China's use of force against the island resulting in military intervention and forceful reintegration of Taiwan with the mainland. U.S. intelligence briefs the President of the U.S. that China is preparing for an invasion of Taiwan if the DPP wins. Taiwanese election polls indicate that the DPP will win in a landslide in the 2028 elections. The President tasks USSOCOM to prepare Taiwan for resisting a Chinese invasion.

RESPONSE BY SPECIAL OPERATIONS

SOF is asked to accomplish three critical tasks. One is to train and equip Taiwanese forces. This would include not only the transfer of military hardware but also focus on training, the development of military doctrine, and establishing effective operational coordination. Training would involve SOF building a Taiwanese defense force that can jam satellites; effectively use commercial satellite imagery, satellite internet, and navigation; carry out cyberattacks against Chinese ground stations; and use electronic warfare and electromagnetic mapping. Taiwanese forces would benefit from SOF assistance in developing and using munitions guidance systems that can target China's naval vessels and aircraft. This would require training on the use of precision guidance munitions and real-time target tracking. Second, to accomplish this, SOF must carry out joint operational training and exercises with Taiwanese defense forces that can utilize U.S.-supplied counter-space and other military hardware. Third, SOF would advise and equip Taiwanese forces with robust doctrine development skilled at multi-domain operations across cyber, AI, and space, whether on land, sea, air, space, or cyber.

In addition to these three tasks, Taiwanese defense forces will have to develop capabilities to disable Chinese propaganda against Taiwan—meaning they will have to go deep into mainland China with direct kits that can disrupt such propaganda. This will involve access to local networks and disrupting the spread of false Chinese narratives and deep fakes from servers located within China. If there are costs involved with a Chinese attack on Taiwan, disrupting and infecting China's social credit system and social media channels like Weibo can act as a deterrent.

Taiwanese forces must also understand and disrupt Chinese overhead capabilities regarding space. For instance, they should be able to disrupt Chinese satellite communications and target ground stations with cyberattacks that sabotage these systems. Taiwanese forces will have to possess specific intelligence about the kind of training that Chinese military forces have undergone in their quest to reintegrate Taiwan militarily and understand Chinese military capacities regarding the Taiwan Straits. SOF, in particular, must prepare for supporting supply chain logistics to Taiwan given the distance of the U.S. from Taiwan. The Taiwan conflict is different from Ukraine. NATO shares land borders with Ukraine, which has been instrumental in supply chain logistics. Most critical will be intelligence sharing, battlefield mapping, and satellite imagery and analysis. Training and interoperability with Ukrainian forces and SOF have been underway since 2014 after the Russian occupation of Crimea. A similar strategy must be in place between SOF and Taiwan.

CONCLUSION

Cyber, AI, and space, individually, and/or in combination, have challenged conventional military doctrines and required militaries to establish training programs on how to use these technologies effectively. In conflict areas like Ukraine and Taiwan, Russia and China are training their military forces to utilize cyber, AI, and space not only for offensive and defensive operations but also for taking strategic initiative. USSOCOM and SOF have to adapt and build capabilities that anticipate the cumulative effect of these technologies on the areas of their potential presence and operations. While cyber, AI, and space act as enablers for SOF, these technologies can be utilized by competent adversaries to locate, disrupt, and deter SOF operations. Having a clear understanding of both the positive and negative effects of these technologies on special operations and building resilience and adaptability is the logical way forward. ♠

About the Author

Dr. Namrata Goswami is an author and educator specializing in space policy, international relations, and ethnic identity. Currently, Dr. Goswami teaches at the Schriever and West Space Scholar Programs, the School of Advanced International Studies, Johns Hopkins University, Thunderbird School of Global Management, Arizona State University, and at the Joint Special Forces University. She is a guest lecturer at Emory University for seminars on Technology, Society & Governance, and India today. She worked as a Research Fellow at MP-Institute for Defence Studies and Analyses, New Delhi; a Visiting Fellow at Peace Research Institute, Oslo, Norway; La Trobe University, Melbourne, Australia; University of Heidelberg, Germany; Jennings-Randolph Senior Fellow, United States Institute of Peace; and was a Fulbright Senior Fellowship Awardee. She was awarded the Minerva grant by the Office of the U.S. Secretary of Defense to study great power competition in outer space. In April 2019, Dr. Goswami testified before the U.S-China Economic and Security Review Commission on China's space program. Her co-authored book, Scramble for the Skies: The Great Power Competition to Control the Resources of Outer Space was published in 2020 by Lexington Press; Rowman, and Littlefield. Her book on The Naga Ethnic Movement for a Separate Homeland was published in 2020 by Oxford University Press. She has published widely including in The Diplomat, the Economic Times, The Washington Post, Ad Astra, Asia Policy, Live Encounters Magazine, Cairo Review. She was invited in November 2019 to share about her life and her work at a Tedx event held at the Rosa Parks Museum, in Montgomery, Alabama.

NOTES

1. Stephen Biddle, *Non-State Warfare: The Military Methods of Guerillas, Warlords and Militias* (Princeton University Press, 2021).

2. U.S. Department of Defense, "Summary of the Irregular Warfare Annex to the National Defense Strategy, (2020), https://media.defense.gov/2020/Oct/02/2002510472/-1/-1/0/Irregular-Warfare-Annex-to-the-National-Defense-Strategy-Summary.PDF

3. U.S. Department of Defense, "Summary of the Irregular Warfare Annex," 2.

4. U.S. Department of Defense, "Summary of the Irregular Warfare Annex," 4.

5. Warren Chin, "Technology, War and the State: Past, Present and Future," *International Affairs* 95, no. 4 (July 2019): 765–783.

6. Daniele Rotolo, Diana Hicks, and Ben R. Martin, "What is an Emerging Technology," *Research Policy* 44, no.10 (December 2015): 1827–1843, https://www.sciencedirect.com/science/article/abs/pii/S0048733315001031

7. Rotolo et al., "What is Emerging Technology," 1827–1843.

8. PricewaterhouseCoopers, "The New Essential Eight Technologies: What You Need to Know," https://www.pwc.com/us/en/tech-effect/emerging-tech/essential-eight-technologies.html#:~:text=Theypercent20include percent3Apercent20artificialpercent20intelligencepercent20(AI,pandemicpercent20acceleratingpercent20emerging percent20techpercent20adoption

10. U.S. Department of Defense, "Summary 2023 Cyber Strategy of The Department of Defense (September 2023): 2, https://media.defense.gov/2023/Sep/12/2003299076/-1/-1/1/2023_ DoD_Cyber_Strategy_Summary.PDF

11. U.S. Department of Defense, "2023 Cyber Strategy," 2, 4, 5.

12. Jon Bateman, *U.S.-China Technological "Decoupling" A Strategy and Policy Framework*, Carnegie Endowment for International Peace (April 25, 2022): 2, https://carnegieendowment.org/research/2022/04/us-china-technological-decoupling-a-strategy-and-policy-framework?lang=en

13. Colleen Walsh, "Wide Range of Possible Targets for Russian Cyber Strikes, From Infrastructure to Smartphones," *The Harvard Gazette*, February 24, 2022, https://news.harvard.edu/gazette/story/2022/02/harvard-cyber-expert-assesses-russia-threat/

14. U.S. Department of Defense, "2023 Cyber Strategy," 5.

15. U.S. Department of Defense, "2023 Cyber Strategy," 2–3.

16. The White House, "National Cybersecurity Strategy," March 2023, https://www.whitehouse.gov/wp-content/uploads/2023/03/National-Cybersecurity-Strategy-2023.pdf, 1.

17. The White House, "National Cybersecurity Strategy," 2.

18. Mark Simos, "Overview of Petya, a Rapid Cyber Attack," *Microsoft*, February 5, 2018, https://www.microsoft.com/en-us/security/blog/2018/02/05/overview-of-petya-a-rapid-cyberattack/

19. "NotPetya: The Cyberattack that Shook the World," *The Economic Times*, March 4, 2022, https://economictimes.indiatimes.com/tech/newsletters/ettech-unwrapped/notpetya-the-cyberattack-that-shook-the-world/articleshow/89997076.cms?from=mdr; Also see Cybersecurity & Infrastructure Security Agency, "Petya Ransomware," February 15, 2018, https://www.cisa.gov/news-events/alerts/2017/07/01/petya-ransomware

20. Ellen Nakashima, "Russian Military was behind 'NotPetya' Cyberattack in Ukraine, CIA Concludes", The Washington Post, January 12, 2018, https://www.washingtonpost.com/world/national-security/russian-military-was-behind-notpetya-cyberattack-in-ukraine-cia-concludes/2018/01/12/048d8506-f7ca-11e7-b34a-b85626af34ef_story.html

21. Office of Public Affairs, U.S. Department of Justice, "Six Russian GRU Officers Charged in Connection with Worldwide Deployment of Destructive Malware and Other Disruptive Actions in Cyberspace," press release number 20-1, 117, October 19, 2020, https://www.justice.gov/opa/pr/six-russian-gru-officers-charged-connection-worldwide-deployment-destructive-malware-and

22. Dina Temple-Raston, "A 'Worst Nightmare' Cyberattack: The Untold Story of the SolarWinds Hack," NPR, April 16, 2021, https://www.npr.org/2021/04/16/985439655/a-worst-nightmare-cyberattack-the-untold-story-of-the-solarwinds-hack

23. Jared Keller, "SOCOM Chief: Door Keepers are Out, Cyber Operators are In," *Task & Purpose*, May 12, 2020, https://taskandpurpose.com/news/special-operations-forces-cyber-warfare/

24. Nathan Beauchamp-Mustafaga, "Chinese Next-Generation Psychological Warfare," RAND, June 1, 2023, https://www.rand.org/pubs/research_reports/RRA853-1.html

25. Dean Cheng, *Cyber Dragon Inside China's Information Warfare and Cyber Operations*, The Changing Face of War, (Praeger, 2016).

26. Maria Korenyuk, Lucy Swinnen, and Jack Goodman, "Undercover with Russia's Fake Arms Dealers," BBC, (September 24, 2022), https://www.bbc.com/news/world-62983444

27. Korenyuk et al., "Undercover with Russia's Fake Arms Dealers."

28. Shannon Bond, "How Russia is Losing—and Winning—the Information War in Ukraine," NPR, February 28, 2023, https://www.npr.org/2023/02/28/1159712623/how-russia-is-losing-and-winning-the-information-war-in-ukraine; Vera Bergengruen, "Inside the Kremlin's Year of Ukraine Propaganda," *Time*, February 22, 2023, https://time.com/6257372/russia-ukraine-war-disinformation/

29. Mia Bloom, "Telegram and Online Addiction to Terrorist Propaganda," Minerva Research Initiative, May 29, 2019, https://minerva.defense.gov/Owl-In-the-Olive-Tree/Owl_View/Article/1859857/telegram-and-online-addiction-to-terrorist-propaganda/

30. Brendan I Koerner, "Why ISIS is Winning the Social Media War," *Wired*, March 2016, https://www.wired.com/2016/03/isis-winning-social-media-war-heres-beat/

31. Koerner, "Why ISIS is Winning the Social Media War."

32. Andrew Eversden, "Army Cyber, Space and Special Operations Command Integrating under New 'Triad' Concept", Breaking Defense, August 09, 2022, https://breakingdefense.com/2022/08/army-cyber-space-and-special-operations-commands-integrating-under-new-triad-concept/

33. McKinsey & Company, "What is the Internet of Things (IoT)?," August 17, 2022, https://www.mckinsey.com/featured-insights/mckinsey-explainers/what-is-the-internet-of-things, accessed 10 December 2023; Sean Costigan and Gustav Lindstrom, "Policy and the Internet of Things," *Connections* 15, no. 2 (Spring 2016): 9–18, https://www.jstor.org/stable/26326436.

34. Remarks Minister of Defense, The Hague, October 4, 2018, https://www.justice.gov/opa/page/file/1098576/download

35. Jon Henley, "Visual Guide: How Dutch Intelligence Thwarted a Russian Hacking Operation," *The Guardian*, October 4, 2018, https://www.theguardian.com/world/2018/oct/04/visual-guide-how-dutch-intelligence-thwarted-a-russian-hacking-operation

36. Kenneth D. Ward, "Russia, Syria, and the Global Chemical Weapons Crisis," *Arms Control Association*, September 2021, https://www.armscontrol.org/act/2021-09/features/syria-russia-global-chemical-weapons-crisis

37. For more on this, see Josh Golding, "Byte, With, and Through: How Special Operations and Cyber Command Can Support Each Other," *War on the Rocks*, November 11, 2022, https://warontherocks.com/2022/11/byte-with-and-through-how-special-operations-and-cyber-command-can-support-each-other/

38. Department of Defense Directive, "Irregular Warfare," May 12, 2017, https://www.esd.whs.mil/Portals/54/Documents/DD/issuances/ DoDd/300007p.pdf?ver=2019-02-04-124731-573

39. For more on CubeSats, see "What are SmallSats and CubeSats?" *National Aeronautics and Space Administration*, found at https://www.nasa.gov/what-are-smallsats-and-cubesats/#:~:text=CubeSats%20are%20a%20class%20of,%2C%206%2C%20and%20even%2012U, accessed on 25 July 2024.

40. Zhang Tong, "Scientists Test AI for an Orbital Carrier to Defend China's Space Assets," *South China Morning Post*, July 13, 2022, https://www.scmp.com/news/china/science/article/3185176/scientists-test-ai-orbital-carrier-defend-chinas-space-assets

41. Stephen Chen, "China Reveals Secret Programme of Unmanned Drone Submarines dating Back to 1990s," *South China Morning Post*, July 8, 2021, https://www.scmp.com/news/china/military/article/3140220/china-reveals-secret-programme-unmanned-drone-submarines-dating

42. Rathindra Kuruwita, "What Swirls Beneath Research Activity of Chinese Ships in the Indian Ocean?," *The Diplomat*, November 9, 2023, https://thediplomat.com/2023/11/what-swirls-beneath-research-activity-of-chinese-ships-in-the-indian-ocean/

43. Michael J. Mazarr, "Mastering the Gray Zone: Understanding a Changing Era of Conflict," Advancing Strategic Thought Series, United States Army War College Press, 2014, https://apps.dtic.mil/sti/pdfs/AD1000186.pdf

44. Prakash Panneerselvam, "Unmanned Systems in China's Maritime 'Gray Zone Operations,'" *The Diplomat*, January 23, 2023, https://thediplomat.com/2023/01/unmanned-systems-in-chinas-maritime-gray-zone-operations/

45. Emil Archambault and Yannick Veilleux-Lepage, "Drone Imagery in Islamic State Propaganda: Flying like a State," *International Affairs* 96, no. 4 (July 1, 2020): 955-973, https://academic.oup.com/ia/article/96/4/955/5813533

46. David Reid, "A Swarm of Armed Drones Attacked a Russian Military Base in Syria," CNBC, January 11, 2018, https://www.cnbc.com/2018/01/11/swarm-of-armed-diy-drones-attacks-russian-military-base-in-syria.html

47. Congressional Research Service, "Attacks on Saudi Oil Facilities: Effects and Responses," October 1, 2019, https://crsreports.congress.gov/product/pdf/IN/IN11173

48. Aziz El Yaakoubi and Maha El Dahan, "Saudi Aramco Petroleum Storage Site Hit by Houthi Attack, Fire Erupts," *Reuters*, March 26, 2022, https://www.reuters.com/world/middle-east/saudi-air-defences-destroy-houthi-drones-state-tv-2022-03-25/

49. "Yemen's Houthis Launch their Largest Red Sea Aerial Attack, but No Damage is Reported", *Associated Press*, January 10, 2024, https://www.npr.org/2024/01/10/1223906769/yemens-houthis-red-sea-attack-missiles-drones-us-navy-un-resolution

50. Jacob Ware, "Terrorist Groups, Artificial Intelligence, and Killer Drones", *War on the Rocks*, September 24, 2019, https://warontherocks.com/2019/09/terrorist-groups-artificial-intelligence-and-killer-drones/

51. "Slaughterbots", You Tube, https://www.youtube.com/watch?v=O-2tpwW0kmU

52. Austin Bodetti and Hamidullah Barakzai, "Here Comes the Taliban Drones," *The Diplomat*, November 02, 2016, https://thediplomat.com/2016/11/here-come-the-taliban-drones/

53. Namrata Goswami, "Great Power Competition and/or Cooperation in Space: The State of Play," in *Challenges and Opportunities at the Dawn of the New Space Age*, Perry World House, University of Pennsylvania, Spring 2022, https://global.upenn.edu/sites/default/files/perry-world-house/space-goswamithoughtpiece.pdf

54. Jaspreet Gill, "NGA Making 'Significant Advances' Months into AI-Focused Project Maven Takeover," BreakingDefense, May 24, 2023, https://breakingdefense.com/2023/05/nga-making-significant-advances-months-into-ai-focused-project-maven-takeover/

55. Daniel Egel, et.al., "AI and Irregular Warfare: An Evolution, Not a Revolution," *War on the Rocks*, October 31, 2019, https://warontherocks.com/2019/10/ai-and-irregular-warfare-an-evolution-not-a-revolution/

56. Nick Bostrom, *Superintelligence Paths, Dangers, Strategies* (Oxford University Press, 2016).

57. Mathew Moellering, "Hiding in the Noise: Preparing the Irregular Warfare Community for the Age of AI," Modern War Institute at West Point, September 26, 2022, https://mwi.usma.edu/hiding-in-the-noise-preparing-the-irregular-warfare-community-for-the-age-of-ai/

58. David Ignatius, "How the Algorithm Tipped the Balance in Ukraine," *The Washington Post*, December 20, 2022, https://www.almendron.com/tribuna/how-the-algorithm-tipped-the-balance-in-ukraine/; The Atlantic Council, "Ukraine's Digital Resilience: A Conversation with Deputy Prime Minister of Ukraine Mykhailo Fedorov," YouTube, https://www.youtube.com/watch?v=VI75e0QU6uE

59. Yulia Latynina, "Ukraine is Waking up to Reality," *The Hill*, November 12, 2023, https://thehill.com/opinion/international/4305675-ukraine-is-waking-up-to-reality/

60. James P. Choy, "Kompromat: A Theory of Blackmail as a System of Governance," *Journal of Development Economics* 147 (September 22, 2020), https://www.sciencedirect.com/science/article/abs/pii/S0304387820301103

61. "Fighting in the 'Grey Zone': Lessons from Russian Influence Operations in Ukraine," Senate Armed Services Committee, Subcommittee on Emerging Threats and Capabilities, Testimony by Michael Carpenter, March 29, 2017, https://www.armed-services.senate.gov/imo/media/doc/Carpenter_03-29-17.pdf

62. "What is Generative AI?", McKinsey & Company, January 19, 2023, https://www.mckinsey.com/featured-insights/mckinsey-explainers/what-is-generative-ai

63. Michelle A. Flournoy, "AI is Already at War", *Foreign Affairs*, October 24, 2023, https://www.foreignaffairs.com/united-states/ai-already-war-flournoy

64. Sean Lyngaas, "'How Am I in this War?': New Musk Biography Offers Fresh Details about the Billionaire's Ukraine Dilemma," CNN, September 11, 2023, https://www.cnn.com/2023/09/07/politics/elon-musk-biography-walter-isaacson-ukraine-starlink/index.html#:~:text=Elonpercent20Muskpercent20secretlypercent20orderedpercent20his,thepercent20eccentricpercent20billionairepercent20titledpercent20percentE2percent80percent9CElon; Also see Walter Isaacson, *Elon Musk* (Simon & Schuster, 2023).

65. Elon Musk (@elonmusk), Twitter, September 7, 2023, https://twitter.com/elonmusk/status/1699917639043404146?lang=en

66. Joey Roulette, "SpaceX Curbed Ukraine's Use of Starlink Internet for Drones-Company President," Reuters, February 9, 2023, https://www.reuters.com/business/aerospace-defense/spacex-curbed-ukraines-use-starlink-internet-drones-company-president-2023-02-09/

67. Amritha Jayanti, "Starlink and the Russia-Ukraine War: A Case of Commercial Technology and Public Purpose?," Belfer Center for Science and International Affairs, Harvard Kennedy School of Government, March 9, 2023, https://www.belfercenter.org/publication/starlink-and-russia-ukraine-war-case-commercial-technology-and-public-purpose

68. Jonathan Beale, "Space, the Unseen Frontier in the war in Ukraine," BBC, October 5, 2022, https://www.bbc.com/news/technology-63109532

69. Stew Magnuson, "Just In: Special Operators Look to Space to Boost Capabilities," *National Defense*, May 11, 2023, https://www.nationaldefensemagazine.org/articles/2023/5/11/just-in-special-operators-look-to-space-to-boost-capabilities

70. Nathan Strout, "Special Operations Command is Diving into Space," C4ISRNET, June 6, 2020, https://www.c4isrnet.com/battlefield-tech/space/2020/06/06/special-operations-command-is-diving-into-space/

71. Strout, "Special Operations Command is Diving into Space."

72. Stephen Forbes, Blackjack," https://www.darpa.mil/program/blackjack

73. The State Council, The People's Republic of China, "Full Text: China's National Defense in the New Era," July 24, 2019, https://english.www.gov.cn/archive/whitepaper/201907/24/content_WS5d3941ddc6d08408f502283d.html; Juliana Suess, "Space and the Future of War According to Russia," RUSI, November 26, 2021, https://www.rusi.org/explore-our-research/publications/commentary/space-and-future-war-according-russia

74. Mark Stokes et al., "China's Space and Counterspace Capabilities and Activities," The U.S.-China Economic and Security Review Commission, March 30, 2020, https://www.uscc.gov/sites/default/files/2020-05/China_Space_and_Counterspace_Activities.pdf

75. Huaxia, "Xi, Putin Hold Talks in Beijing," Xinhua, October 18, 2023, https://english.news.cn/20231018/493db611b7a044ea8bd3e4411b47dfcd/c.html

76. Joint Publication 3-22, "Foreign Internal Defense," August 17, 2018, Validated February 02, 2021, https://www.jcs.mil/Portals/36/Documents/Doctrine/pubs/jp3_22.pdf?ver=2018-10-10-112450-103

77. Air Force Doctrine Publication 3-2 "Irregular Warfare," U.S. Air Force, August 10, 2020.

78. "Irregular Warfare," I-2 to I-3.

79. SpaceX, "Falcon 9 and Falcon Heavy," https://www.spacex.com/vehicles/falcon-9/

80. Frank Wolfe, "Saltzman: Proliferated LEO Shows Mettle in Ukraine," *Defense Daily*, October 18, 2023, https://www.defensedaily.com/saltzman-proliferated-leo-shows-mettle-in-ukraine/space/

81. Lisa Sodders, "LEO, MEO or GEO? Diversifying Orbits is not a One Size Fits All Mission," Space Operations Command, July 18, 2023, https://www.spoc.spaceforce.mil/News/Article-Display/Article/3462529/leo-meo-or-geo-diversifying-orbits-is-not-a-one-size-fits-all-mission-part-1-of

82. Starlink, https://www.starlink.com/

83. Lynk, https://lynk.world/

84. Planet, https://www.planet.com/; Blacksky, https://www.blacksky.com/; Maxxar, https://www.maxar.com/

85. HawkEye, https://www.he360.com/

86. David Ignatius, "How the Algorithm Tipped the Balance in Ukraine," *The Washington Post*, December 19, 2022, https://www.washingtonpost.com/opinions/2022/12/19/palantir-algorithm-data-ukraine-war/

87. Dov Lieber, "Israel Dominates Battlefield with Technological Advantage over Hamas," *The Wall Street Journal*, November 14, 2023, https://www.wsj.com/world/middle-east/israel-dominates-battlefield-with-technological-advantage-over-hamas-e697b470

88. Nicole Werbeck, "Satellite Images Show 40-Mile-Long Russian Military Convoy Nearing Kyiv," NPR, February 28, 2022, https://www.npr.org/sections/pictureshow/2022/02/28/1083650286/satellite-images-show-40-mile-long-russian-military-convoy-nearing-kyiv

89. Jon Gambrell, "Satellite Photos Analyzed by AP Show an Axis of Israeli Push Earlier this Week into the Gaza Strip,"AP, November 9, 2023, https://apnews.com/article/israel-palestinians-hamas-gaza-satellite-images-ae2e8593342c2eed508c3c5d99838fd9

90. Army Staff Sgt. Brandon Nelson, "2021 Building Partner Aviation Capacity Seminar in Krakow, Poland", 352ND Special Operations Wing", September 10, 2021, https://www.352sow.af.mil/News/Article-Display/Article/2770576/2021-building-partner-aviation-capacity-seminar-in-krakow-poland/; Chris Watcher, "Air-Mindedness: The Core of Successful Air Enterprise Development" *Small Wars Journal*, January 10, 2012, https://smallwarsjournal.com/jrnl/art/air-mindedness-the-core-of-successful-air-enterprise-development

91. Luke Bencie and Sami Araboghli, "A 6-Part Tool in Ranking and Assessing Risks," *Harvard Business Review*, September 21, 2018, https://hbr.org/2018/09/a-6-part-tool-for-ranking-and-assessing-risks

92. P. W. Singer and August Cole, *Ghostfleet: A Novel of the Next World War* (William Morrow, 2016); John M. Collins, *Military Space Forces: The Next 50 Years* (Brassey's Inc, 1989).

93. Gordon Lubold, "U.S. Troops have Been Deployed in Taiwan for at Least a Year," *The Wall Street Journal*, October 7, 2021, https://www.wsj.com/articles/u-s-troops-have-been-deployed-in-taiwan-for-at-least-a-year-11633614043; A B Abrams, "Building a US Special Forces 'Stealth Network' on Taiwan", The Diplomat, May 03, 2023, https://thediplomat.com/2023/05/building-a-us-special-forces-stealth-network-on-taiwan/

APPENDICES

APPENDIX A: A 2024 GUIDE TO ASSESSING RESILIENCY AND RESISTANCE IN IRAN

By Robert S. Burrell and David R. DiOrio

ABSTRACT: This section applies a data-driven and human-centric methodology to examine resiliency and resistance aspects in the Islamic Republic of Iran. It applies a four-phase process to analyze state resiliency, assess resistance to governance, identify various resistance movements, and provide possible options for an external actor to influence Iran's stability or to support resistance to the Islamic Republic.[1] It was previously published by the Global and National Security Institute.[2] In analyzing Iran, the data finds the Islamic Republic is a fragile regime governing a state with multiple and deep societal divisions. Resistance in Iran has great potential but external support to the Islamic State also has potential. Support for a reform movement akin to the Green Path of Hope could prove decisive.

Protesters chant slogans during a demonstration over the death of Mahsa Amini who was detained by the morality police, in Tehran, Iran, on Sept. 21, 2022. Source: Voice of America

INTRODUCTION

The sudden death of Iranian President Ebrahim Raisi in a helicopter crash on 19 May 2024 may provide an opportunity to usher in a new destiny for the Iranian people. Many considered the hard-liner to be the Supreme Leader Khamenei's enforcer in consolidating the clerics' power through the executions of dissidents and the jailing of political prisoners. Raisi mobilized America's rivals by pursuing a military alliance with Russia and economic ties with China to weaken the grip of Western political and commercial dominance in the region. The former president was the mastermind of a proxy-based militarization campaign to make a Western presence in the Middle East so costly that the U.S. and its allies would withdraw. Despite the recent escalation of hostilities, the U.S. remains steadfast in maintaining a forward presence to strengthen regional partnerships and protect vital trade routes.

The Islamic Republic of Iran has been a destabilizing force in the Middle East since its ascension to power after the Iranian Revolution in 1979. Iranian leadership has provoked violent conflict and destructive activities to assert its hegemonic aspirations.[3] Iran's government is a complex blending of theocratic and political elements that pursues expanding Islamification in conformity with "Khomeinism," a radicalized ideology to reassert Shi'ism as the dominant Islamic moral authority.[4] Tehran views the U.S. and Israel as their main threats and focuses their foreign policy on eliminating their regional influence.[5]

| Tehran views the U.S. and Israel as their main threats and focuses their foreign policy on eliminating their regional influence. |

With a relatively small regular military, the regime relies on specialized forces to lead a network of proxies that engage in surrogate terrorism, political agitation, and paramilitary violence as the main instruments of power projection.[6]

The Islamic Republic's June 2024 presidential elections are a clear sign of the underlying resentment harbored by a dissatisfied Iranian electorate. The Republic's Guardian Council (a conservative 12-member oversight board staunchly dedicated to the Supreme Leader) hand-picked the candidates. Of the six selected, five were hard-line protégés of the Supreme Leader, with only one moderate capable of policy reforms. The Supreme Leader urged the people to vote in a symbolic gesture to strengthen his mandate, but, frustrated with the lack of choice and previous electoral fraud, voter turnout was the lowest ever recorded since the republic's formation (40 percent).[7] Iranians striving for meaningful governmental change supported the moderate Masoud Pezeshkian, while others desiring the status quo supported the conservative Saeed Jalili. The first phase of the election narrowed the field to these two candidates who will compete in a run-off election.

Tehran's low governance performance and public skepticism pose a significant dilemma for the Islamic Republic. Strategic decision-makers should assess the resilience of the Islamic Republic of Iran by examining its governance performance, perceived legitimacy, and resistance potential. Pezeshkian is open to a renewed dialogue with the West to improve relations, ease sanctions, and generate a new nuclear deal. Jalili considers Western nations untrustworthy and looks to China and Russia for partnerships and assistance. Considering the potential sea change in Iranian politics, now poses an opportune moment to review Western foreign policies. Should the West promote: (1) a more resilient Islamic Republic if Pezeshkian is elected, (2) support resistance to the Islamic Republic in favor of governmental reforms irrespective of the election results, or (3) actively shape the strategic environment and defer to a future opportunity? A comprehensive assessment of the resilience metrics and exploring resistance strategies may lead U.S. policymakers to a more effective approach.

Determining the resilience of, and resistance to, Iran's governance system warrants a comprehensive analysis supported by a fact-based methodology (resiliency deals with the Islamic Republic's ability to overcome adversity imposed from internal or external subversion, coercion and/or aggression; resistance deals with Iranian society's, the population's, or a subgroup's opposition to malign indigenous power structures or to external occupation). Such an analysis and assessment can better inform DOD activities, force posture, and interagency collaboration to achieve U.S. national objectives, not just in the case of war but in competition. It leverages analytical data from top universities, financial institutions, governmental agencies, and nongovernmental organizations to inform a four-phase process. Phase one measures state resiliency, resistance to the same, and assesses the likely success of external support to either resilience or resistance. Phase two identifies prevalent or influential resistance organizations within the state, and then categorizes these organizations across a continuum to classify their general nature. Phase three assesses one of those resistance movements, the Green Path of Hope, by taking a deeper look at leadership, motivation, operating environment, organization, and activities. Phase four analytically assesses the information gathered to make recommendations concerning potential external support in another state's intrastate conflict.

PHASE ONE: MEASURING IRAN'S RESILIENCE AND RESISTANCE

In this assessment, we present percentages based on internationally recognized and publicly available statistics. At its heart, Iran's significance as a case study for measuring resiliency and resistance lies in its potential for: (a) possible uses of external support to increase the Islamic Republic's resiliency and (b) possibilities for subversion and destabilization of the Islamic Republic by a competitor.[9]

MEASURING IRAN'S RESILIENCE

Historical geopolitical, ideological, and demographic factors govern Iran's resilience and resistance posture. The Islamic Republic originated following Ayatollah Ruhollah Khomeini's overthrow of the monarchy during the Iranian Revolution of 1979—a reaction to perceived immoral westernization and secularization efforts. Iran shifted to a spiritual, social order with an autocratic rule of the clerics that fused politics and religion.[10] Subsequently, conservative Shi'a clerics established a theocratic republic. The ultimate political authority remains in the form of a religious scholar referred to as the Supreme Leader, who remains accountable to a popularly elected 88-member body of clerics.[11] Thus, the Iranian government is a complex blending of theocratic and political elements to protect the regime, advance the Islamic Shi'a worldview, and exert regional influence to supplant Western influence.[12] The people of Iran and the Muslim Shi'a community at large have mixed views on the regime's strategic goals and professed legitimacy.[13] The clerics profess that the Islamic Republic is the only righteous governance path within the Islamic world. Theocratic truth-seekers advocated a sociopolitical sect based upon traditional Shi'a jurisprudence, believing that global liberation movements against colonialist oppressors were a justified obligation.[14]

While the Islamic State attempts to maintain the image of a homogeneous and monolithic culture to validate its ambitions for religious, political, and military hegemony in the Middle East, it remains quite fractious in terms of demographics. See Figure A.1. The United States Institute for Peace estimates that 61 percent of the population is Persian, 16 percent Azeri, 10 percent Kurd, and other minorities making up 13 percent.[15] Languages are even more diverse. A census of the country in 1991 reported an estimated 46.2 percent speaking Persian (or Farsi), 20.6 percent Azerbaijani, 10 percent Kurdish, 8.9 percent Luri, 7.2 percent Gilaki and Mazandarani, and 7 percent others.[16] In terms of religion, however, the population remains fairly unified with 90–95 percent Shi'a, 5–10 percent Sunni, and 1 percent others.[17]

Figure A.1. Map of Iran. Source: CIA World Factbook

Governance indicators demonstrate that the Islamic State is ineffective and fragile, causing instability that a resistance movement may exploit. The following percentiles rank Iran in comparison to other countries in the world, with 0 percent as the lowest and 100 percent as the highest. Iran ranks poorly in most respects: 8.7 percent in government accountability; 8.49 percent in political stability; 18.4 percent in government effectiveness; 4.25 percent in regulation efficiency; 17.45 percent in rule of law; and 14.15 percent in control of corruption. These indicators have worsened from 2017 to 2022 (implying that the resiliency of the Islamic Republic is worsening).[18] On a similar note, the Fund for Peace's state fragility index ranks Iran in the bottom third of stable nations at 40 of 179 (or 22.35 percent in comparison with others) and just in-between that of Angola and Bangladesh.[19] Figure A.2 illustrates Iran's governance factors in comparison with Egypt, Turkey, and Saudi Arabia.

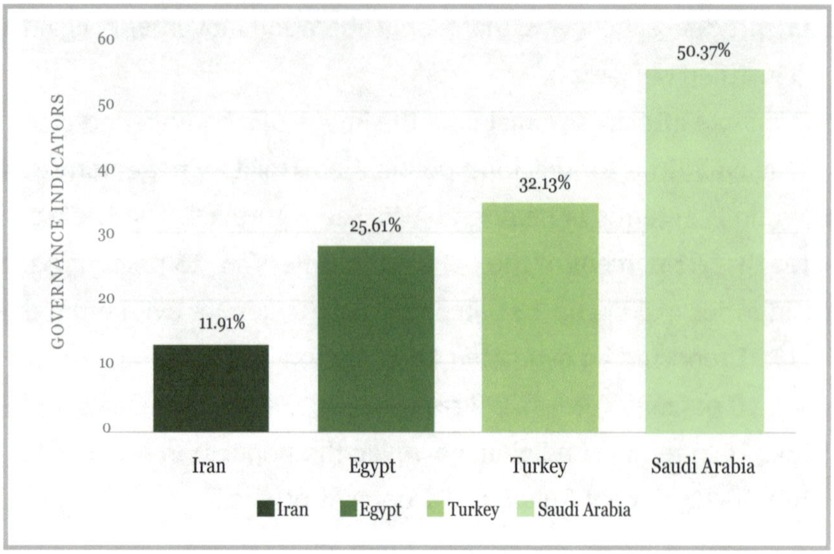

Figure A.2. Comparison of governance in the Middle East. Source: Authors
Note: All states measured utilizing the same indicators. For more information on the method, see Burrell and Collison, "A Guide for Measuring Resiliency and Resistance."

The governance data exposes the weaknesses and vulnerabilities of the Iranian government, but evaluating the national will to fight for and retain current systems of authority and governance proves more subjective. Popular support for Tehran is marred by human rights abuses, severe restrictions on personal freedoms, a corrupt judiciary, and violent suppression of dissenting views.[20] These abuses foster resentment within the population and significantly degrade national morale and confidence in Iranian leadership.

To measure national morale, we lean on the work of Delbert C. Miller.[21] Table A.1 utilizes Miller's analysis methods and his five categories (as closely as possible in line with available polling) to determine the national morale of the Islamic State. (1) We consider Shi'a Persian as the ingroup of Iran, in which when polled, 85 percent of the Iranian population appears to support its cultural, religious, and linguistic dominance.[22] (2) In the same study, most respondents wanted a change in governance, with 65 percent of the population supporting some type of resistance (either nonviolent protest or violent opposition) to change the current form of governance.[23] (3) In the third category concerning the competency of national leaders, an opinion poll in Iran in 2023 demonstrated that 80 percent of respondents prefer a democracy over the Islamic Republic.[24] (4) In terms of Iranian confidence in current resources to defend the interests of Persians, only 21 percent have limited trust in the Islamic Republic of Iranian Army and only 23 percent have limited trust for the Islamic Revolutionary Guard, implying that trust in these national organizations remains relatively high, averaging 78 percent.[25] Lastly, (5) nearly 73 percent of those polled believe "religion must be separated from the state," which directly opposes the current constitution upheld by the Islamic Republic and specifically identified as the national goal therein.[26] This data implies that only 27 percent of the population supports the theocratic form of governance. Table A.1 outlines the five factors used to assess national morale.

TABLE A.1. BASIC FACTORS OF NATIONAL MORALE IN THE ISLAMIC REPUBLIC OF IRAN		
FIVE FACTORS	**RATING**	**PERCENT**
1. Belief in the superiority of the social structure in the ingroup	high	85 percent
2. Degree and manner by which personal goals are identified with national goals	low	35 percent
3. Judgments of the competence of national leaders	low	20 percent
4. Belief that resources are available to hurl back any threats to the ingroup	high	78 percent
5. Confidence in the permanence of the national goal	low	27 percent
TOTAL	average	49 percent

Source: Authors

Tallying the six factors of governance (accountability, stability, effectiveness, regulation controls, rule of law, and controlling corruption) and adding two other aspects of national morale and state fragility equally (eight metrics in total), the resiliency of the Islamic Republic of Iran is estimated at 17.85 percent.

MEASURING THE POTENTIAL FOR EXTERNAL SUPPORT TO IRAN'S RESILIENCY

To measure the potential of external support to the Islamic Republic, we subjectively examine: (1) its relations with nearby regional states, (2) its broader international relations, and (3) its relations with nonstate groups. The Middle East is locked in a historical struggle for influence and legitimacy between four formidable powers—Egypt, Saudi Arabia, Iran, and Turkey.[27] Tehran's grand strategy appears to currently focus on their proxy militancy campaign against the West while nurturing diplomatic ties to ease sectarian tensions with Sunni neighbors. However, a fractured hegemonic competition in the Middle East among Iran, Turkey, and an Arab coalition is likely to endure.[28] Iran has a formal alliance with Syria. Syria is the key to commercial and military access across the fertile crescent and has potential of providing direct and/or indirect support to enhance the Islamic Republic's resilience. However, Syria's current government has serious issues of internal unrest and is unlikely to offer much external support to the Islamic Republic. A continued U.S. presence in the region backed by a NATO alliance, particularly Turkey, is expected to challenge Iranian hegemonic intentions. Iran's goals for an Islamic religious unification could solicit regional sponsors, but it remains equally likely that it will stand alone. Consequently, Iran's bilateral relations in the Islamic world make the chances of receiving external support for the resiliency of the Islamic Republic at 50 percent.

The Islamic Republic's resiliency is also directly linked to international geopolitics. A direct and protracted military confrontation with the U.S. and its allies remains the greatest threat to the regime's resilience. Tehran's proxy strategy is primarily focused on eliminating Western influence by driving out Israel and the United States, with their allies.[29] The stratagem is to inspire a withdrawal of Western forces to generate political space for pro-Iranian factions.[30] Simultaneously, Tehran is strengthening relationships with China and Russia to open trade channels, share military resources, and enhance collective defense measures to openly challenge the U.S. military presence and mitigate the impact of Western economic sanctions.[31] China has a friendly relationship with Iran and signed a cooperation agreement in 2021.[32] Additionally, Russia has military cooperation with Iran and strategic interests in the region.[33] Pursuing these neoteric political ties to these global powers may provide Iran with more substantial resilience against Western influence.

Iranian resilience is linked to global commerce in addition to its growing ties with Russia and China. The Islamic Revolutionary Guard Corps (IRGC) and the Quds Forces (QF)—with their proxy allies—have postured to either protect or hinder key global lines of communication. Many trade partners desire a stable Iran to access the nation's energy resources, leverage its important geostrategic position, and advance the Shi'a religious movement with the subsequent windfall. Iran belongs to several organizations, including the Organization of the Petroleum Exporting Countries (OPEC). OPEC may have an interest in supporting the Islamic Republic in times of need, but the ultimate Iranian strategic goal is to control energy access in the Persian Gulf rather than share it with other OPEC members. This strategic objective is shared with China, which views Iran as a key strategic partner in their Belt and Road Initiative. Yemen offers crucial access to maritime lines of communication via the Red Sea and provides a Shi'a bulwark on the theater's Saudi-Wahabbist southern flank. In summation, Iran's relationships with powerful external states, like China and Russia, as well as its geostrategic position and economic resources, make it a strong candidate for securing external support for resiliency, assessed as 75 percent.[34]

Several non-state groups have vested interests in the Islamic Republic, with the potential to support its resiliency. Some have described these pro-Iranian proxies as the Iranian threat network (ITN), an alliance of surrogates and partners that conduct activities to destabilize, disrupt, or destroy anti-Iranian entities throughout the Middle East.[35] The IRGC-QF recruits, trains, equips, advises, and often controls the ITN at the behest of Tehran and is integral to the Islamic Republic's formal decision-making process.[36] The organization has been traditionally covert, but Iran is increasingly becoming more open in acknowledging these groups publicly.[37] The IRGC is skilled in leveraging common grievances to foster rapport and bring groups into the fold, especially during global social movements like the Arab Spring. Tehran endeavors to appeal to ethnic and religious minorities in the region whose disenfranchisement allows Tehran to recruit fighters.[38] Iran considers the ITN a cost-effective force multiplier to complement other asymmetric capabilities, such as ballistic missiles and drones, to compensate for conventional shortfalls.[39] The IRGC-QF directly or indirectly supports the ITN and operates in several countries and all domains—air, sea, land, and cyber.[40] The ITN includes a dozen major organizations: Al-Astar Brigades in Bahrain; Kataib Hezbollah, Badr Organization, and Asaib Ahl al-Haq in Iraq; Hezbollah in Lebanon; Hamas and Palestinian Islamic Jihad in Palestinian territories; Famiyoun Brigade, Zainabiyoun Brigade, Quwat al-Ridha, and Baqir Brigade in Syria; and the Houthi movement in Yemen.[41] While these violent resistance movements currently receive support from Tehran, relationships within the ITN can be volatile and shifts in allegiances or support fluctuate (see Figure A.3). However, this network of established allegiances enhances the potential for non-state support to Iran's resiliency, assessed at 75 percent.

Figure A.3. Notable Iran-backed militias as of October 2023. Source: Council on Foreign Relations

In summary, the geo-political situation in the region is volatile and precarious making external support to resiliency unpredictable. While Iran is generally self-sufficient and without the need for foreign aid to enhance resiliency, Tehran has a network of potential partners that could be leveraged in times of crisis; these include the possibility of regional, international, and nonstate sponsors. We average these three metrics to find the potential of external sponsorship to the Islamic Republic at 66.7 percent.

MEASURING THE POTENTIAL FOR DOMESTIC RESISTANCE IN THE ISLAMIC REPUBLIC

In the past two decades, nonviolent and violent resistance in Iran has proven endemic. Extensive data exists to measure the potential for internal resistance to the authority of the Islamic Republic. This data ranks Iran in comparison with other nations, with 0 percent indicating little to no intrastate resistance to authority and 100 percent indicating the highest. Using this dataset, the potential for resistance to the Islamic State of Iran is quite high: (a) in terms of current governance not adhering to the rule of law, 82.38 percent; (b) in political instability, 91.19 percent; (c) in the perception of the Republic not controlling corruption, 83.05 percent; (d) in a dismal record of political rights, 79.26 percent; (e) in not respecting civil liberties, 88.60 percent, and (f) in its inability to regulate the shadow economy, 22.15 percent.[42]

In addition to this, over the past two decades, nonviolent and violent resistance in Iran has proven endemic. Between 2019 and 2024, over 23,000 acts of violence have occurred in Iran with 11,324 fatalities.[43] Further, 12 major nonviolent protests occurred between the years 2017 and 2023 in Tehran, five of which garnered a violent response from the regime.[44] Between the years 2017 and 2020, 39 acts of violent terrorism occurred in Iran by numerous non-state groups.[45] Additionally, in 2024, the Global Peace Index places Iran as 133 out of 163 (or 81.60 percent unpeaceful in comparison with others).[46]

| The Islamic Republic scores 71.36 percent in resistance potential to current authority, implying it is ripe for regime change. |

As well, Freedom House ranks Iran categorically as "unfree" and one of the worst at 89 percent in comparison with others.[47] Lastly, the CIA routinely assesses nations in terms of food insecurity, but Iran is not on the list, making the threat of food insecurity only a 25 percent factor in potential resistance.[48] Averaging nine of the preceding data figures equally, the Islamic Republic scores 71.36 percent in resistance potential to current authority, implying it is ripe for regime change.

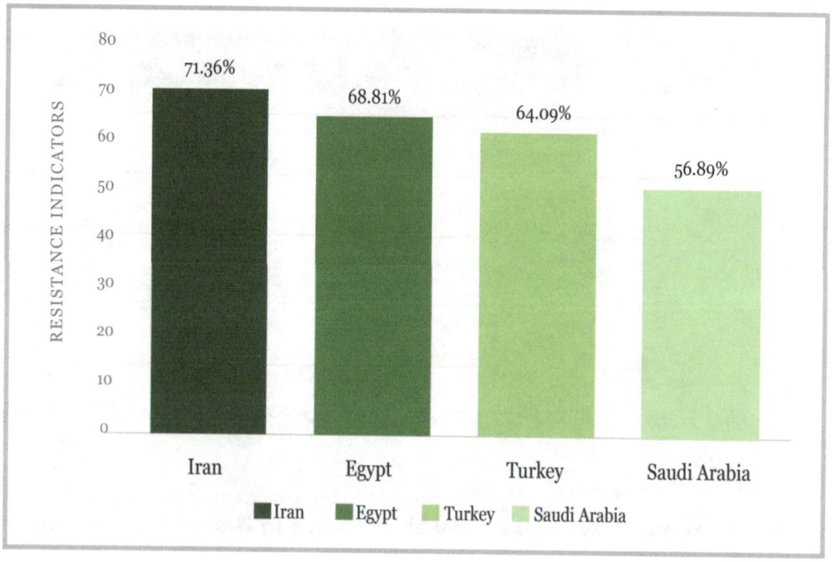

Figure A.4. Comparison of resistance potential in the Middle East. Source: Authors Note: All states measured utilizing the same indicators. For more information on the method, see Burrell and Collison, "A Guide for Measuring Resiliency and Resistance."

Figure A.4 illustrates Iran's resistance potential in comparison with other Middle Eastern states.

MEASURING THE POTENTIAL FOR EXTERNAL SUPPORT TO RESISTANCE IN IRAN

The U.S. severed official relations with the Islamic Republic in 1979 following the attack on the U.S. Embassy in Tehran.[49] It is safe to say that the U.S. could include subversion as a strategy for regime change and that other Western nations may support such a strategy as well. Additionally, the U.S. has referred to Iran as a threat in its *National Security Strategy*.[50]

Accordingly, we rank the Islamic Republic as a possible target of external support to resistance from an adversary as plausible due to adversarial diplomatic relations – making subversion an amenable foreign policy option (100 percent).[51]

The historical case study analysis completed by the Study of Internal Conflict at the Army War College poses four important indicators of possible success or failure of an insurgency in Iran.[52] (1) First, more than 15 percent of the Iranian population does not identify as citizen of the state. In the current environment, a scenario in which 15 percent of the population does not identify as a citizen of the Islamic Republic remains foreseeable, in which case the success of violent opposition could be as high as 96 percent.[53] (2) Second, more than 15 percent of the population does not acknowledge the legitimacy of the regime. It is fair to say more than 15 percent of the population does not believe current governance is legitimate, implying the success of resistance as high as 94 percent.[54] (3) Third, more than 15 percent of the population has meaningful communication with a resistance movement. This is possible and has certainly happened in the past with the Green Path of Hope discussed later. If such communication could be sustained to 15 percent of the population, the success of resistance is 96 percent. (4) Fourth, could a sanctuary exist for an armed component of resistance in a neighboring state? The answer here is likely yes. States that could provide sanctuary include Afghanistan, Pakistan, Iraq, and possibly others. The geography of Iran's isolated mountainous regions also supports insurgency potential. In either case, a sanctuary for armed resistance could equate to the definitive success of an insurgency.

Considering the preceding, we include three data points regarding external support to resistance potential: (1) plausibility of subversion by another nation state—100 percent, (2) 15 percent of the population does not acknowledge the legitimacy of the Islamic Republic—a 96 percent historical success rate of insurgency, and (3) the chance of one or multiple border states providing a resistance sanctuary—75 percent.[55] Averaging these three metrics, we assess that external support to an Iranian-based resistance movement in the Islamic Republic has a possible success rate of 90.33 percent.

| We assess that external support to an Iranian-based resistance movement in the Islamic Republic has a possible success rate of 90.33 percent. |

PHASE ONE: SUMMARY

In summation of phase one, based on the quantitative survey information and qualitative analysis presented, the Tehran regime has poor resiliency to indigenous threats (17.85 percent). Meanwhile, China, Russia and/or non-state actors have an above-average chance to reinforce the resiliency of the Islamic Republic should they desire to do so (66.67 percent). At present, the prospect for internal resistance to the Islamic Republic has unrealized potential (71.36 percent), particularly due to the population's high desire for a change in governance. Due to historical, geographic, and political factors, the overall probability of external support to Iranian resistance is 90.33 percent.[56]

PHASE TWO: IDENTIFYING IRAN'S RESISTANCE MOVEMENTS

Both nonviolent protest and violent revolution in Iran have a mixed record of success in the 20th and 21st centuries.[57] In short, Iranians have executed 12 major resistance campaigns from 1905 to 2023, three of which were successful in instilling governmental change (one violent and two nonviolent movements), a rate of 25 percent.

Currently, Tehran is dealing with several contemporary nonviolent resistance movements. Between 1990 and 2020, over 500,000 people mobilized on over 100 occasions in the Islamic Republic, utilizing non-violent protest.[58] The vast majority of the protests desired political change, social change, and police reforms. In 2017, another two million people mobilized for Akbar Hashemi Rafsanjani's funeral possession, which also witnessed large numbers of protesters for governmental reform.[59] A key leader opposing the current governance is Mir-Hossein Mousavi, a member of the Green Path of Hope (also called the Green Party or Green Movement). Mousavi, the former Prime Minister of Iran (1981–1989), desires the end of clerical rule.[60] The mobilization concerning compulsory hijab laws is another consideration.[61] Following the death of a Kurdish woman, Jina Mahsa Amini, who was arrested over her wearing of the hijab and died in police custody in 2022, the group Women, Life, Freedom (including support from students of University of Tehran) has garnered widespread support and significant protester turnout both in Iran and abroad, winning the Freedom House Award in 2023.[62]

Several resistance organizations contest the Islamic Republic through violence, further alienating the Iranian people from their government and security establishment. Between 1989–2023, one-sided acts of violence have occurred throughout Iran resulting in 636 civilian deaths.[63] Violence between the Islamic Republic and armed groups has resulted in another 1,949 deaths.[64] Armed groups in Iran include: (1) the Islamic State, which Iran opposed in favor of its alliance with Syria; (2) Jaish al-Adl, a Sunni militant group of Baloch ethnicity in Iran's southeastern Sistan and Baluchestan Provinces; (3) Jondullah, another Sunni/Baloch militant group from which Jaish al-Adl derived; (4) Democratic Party of Iranian Kurdistan (KDPI), a Kurdish underground and militant group; (5) Mujahideen e Khalq (MEK), which started as an insurgency but essentially operates in the Iranian diaspora today and advocates change in

Iran through nonviolent methods; (6) Kurdistan Free Life Party (PJAK), a Kurdish militant group; and (7) Kurdistan Workers Party (PKK), another Kurdish militant and politically socialist group.

PHASE TWO: SUMMARY

In summary, several active resistance movements undermine the stability and legitimacy of the Iranian governance system. Figure A.5 organizes the various Iranian resistance organizations across a continuum, including nonviolent groups referring legal forms of protest, nonviolent groups amenable to illegal forms of protest, and the insurgent groups. In the middle of the scale, the threat of rebellion, extended or limited in size and duration, persists in Iran but not currently realized. Additionally, on the far right of the scale, the known insurgent groups have yet to rise to the level of belligerency by exhibiting the functions of an opposing state.[65]

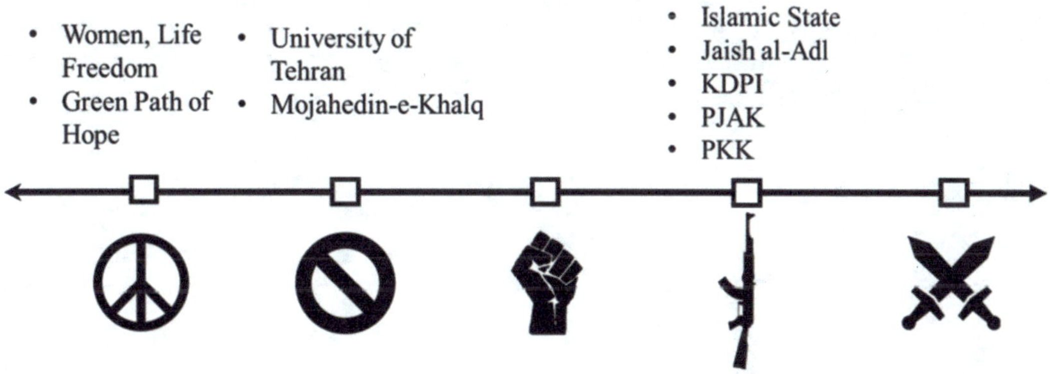

Figure A.5 Diagram of Iran's resistance continuum. Source: Authors

PHASE THREE: ASSESSING IRAN'S RESISTANCE MOVEMENTS

After identifying resistance movements along a resistance continuum, we have chosen to assess the potential of the Green Path of Hope.[66] In assessing the Green Path of Hope, we examine five attributes: (1) actors, (2) causes, (3) environment, (4) organization, and (5) actions.[67]

ACTORS IN THE GREEN PATH OF HOPE.

The Green Path of Hope has a few influential leaders, most notably Mir-Hossein Mousavi and Mehdi Karroubi (a Shi'a cleric, reformist, and founder of the National Trust Party). The Green Path of Hope arose out of opposition to the 2009 election of Mahmoud Ahmedinejad. Opponents believed Ahmedinejad "suppressed civil liberties and political freedoms and politicized numerous social, cultural, and economic issues."[68] A member of the Persian/Shi'a ingroup, Mousavi served as the Prime Minister from 1981–1989 and as the senior advisor to the President from 1997–2005. Viewed as a dire threat to the Islamic Republic, Mousavi and Karroubi have been under house arrest since 2011.

Since the Iranian Revolution, Mousavi and his wife, Zahra Rahnavard, have played active roles in both domestic politics and international diplomacy. Mousavi's ally, Karroubi equally has strong relationships in politics domestically and internationally. Mousavi acts as a reformer who has sought to change the Islamic Republic from within.[69] Following the 2009 Presidential election, he disputed the results. He believes that the ideals of the Iranian Revolution were not achieved and instead resulted in dictatorship. Mousavi has stated that "dictatorships in the name of religion are the worst kind of dictatorships."[70] Due to their advanced ages (both are in their 80s), Mousavi and Karroubi's ability to lead in a revived movement remains questionable. Meanwhile, Rahnavard is a bit younger (in her 70s), charismatic, and committed. Highly educated, Rahnavard is the former Chancellor of Alzahra University and the author of 15 books.[71]

Green Path of Hope Campaign Rally in Tehran, 2009. Source: mongostar/WikiCommons

Determining membership of the Green Path of Hope remains elusive. During the protests in June of 2009, more than 100,000 protestors mobilized (some sources say millions were mobilized).[72] The Islamic Republic violently suppressed the Green Path of Hope in 2009 and 2010, placing 65 journalists in prison and conducting a trial on 250 activists—sentencing some of them to death.[73] Due to inactivity, the commitment of the organizational members remains questionable today. Still, this organization derives significant support from the Persian/Shi'a community, including the Iranian diaspora and prominent exiles.[74]

Mousavi accepted partnerships with reformers and conservatives alike, forming a broad coalition that made change possible. The Green Path of Hope cooperated with other resistance organizations, including the National Trust Party. Activists from the University of Tehran were appealing partners as well, and many students did support the 2009 protests.[75] However, the nature of the Green Path of Hope is nonviolence, which would make a partnership with any of Iran's insurgent groups unlikely. Also, the base of Green Path of Hope is Persian/Shi'a, so cooperation with outgroups like the Kurds or Baloch populations is equally unlikely. The most potential might include cooperation with the Women, Life, Freedom movement.

Thus far, neither international organizations nor foreign nations have offered recognition or support for the Green Path of Hope. However, many Iranians in the diaspora have (and would likely continue) to support the movement. The Iranian diaspora can act as a spokesperson for the movement to the outside world. This includes traditional media, Iranian social media, and artists.[76] In 2009, U.S. President Barack Obama denounced the regime's violent repression of the Green Movement's resistance. Other U.S. politicians wanted more American overt support for the movement.[77] Ultimately, however, external support, if offered, might best be accomplished through low visibility means, as a method of protecting the Green Path of Hope's domestic legitimacy.[78]

It remains difficult to assess how many Iranians would support the Green Path of Hope or a similar movement today. However, a poll of Iranians inside the country taken in 2023 shows that 80 percent prefer democracy to the Islamic Republic.[79] This proves an enticing figure, as anything over 15 percent of the population can prove decisive. However, how many Iranians would accept the potential risks of resistance activities remains undetermined.

CAUSE OF THE GREEN PATH OF HOPE.

Mousavi framed key grievances into a coherent narrative that leveraged shared emotions into a unified resistance effort. Democratic rule in Iran remains the population's prevailing desire. Mousavi ran for the presidential election in 2009 as a reformer and in direct opposition to Islamic conservatives retaining power. When the elections proved fraudulent, he sought mass protest to ensure voting mattered. To produce this, the Green Path of Hope coordinated several nonviolent activities. These included: (1) mass mobilization of protestors, (2) media attention, (3) graffiti on buses, street curbs, and public walls, and (4) stamping currency with antiregime slogans.[80] Following the repression from 2009–2011, the Green Path of Hope's resistance activities have faltered and have yet to reemerge under the same banner. However, democracy remains the predominant desire for most Iranians.

| Democracy remains the predominant desire for most Iranians. |

In the case of the Green Path of Hope, it desires a change in governance from an Islamic theocracy to a democracy.[81] In 2010, Mir-Hossein Mousavi stated three goals for the Green Movement: (1) a return to the original ideas of the Islamic Revolution for justice, independence, and freedom; (2) reexamine the Constitution to ensure protection of human, religious, moral, and national values; (3) and reforms which embody a continuation of the Iranian revolutionary spirit from the past, including the Constitutional Revolution, the Oil Nationalization Movement, and the Islamic Revolution.[82]

THE RESISTANCE ENVIRONMENT IN IRAN

Assessing the environment's influence or constraints on the Green Path of Hope's activities includes an evaluation of: (a) environmental, (b) governmental, (c) socio-political, (d) technological, and (e) relationship factors. In 2009, the leaders of the Green Movement had significant support from the Iranian people. Most of the protestors included women, students, urban-based residents, and those with access to the internet and social media. While large in numbers, the vast majority of these protestors were unorganized and untrained but successfully assembled for six months. The protest momentum could not be sustained, and the Islamic Republic suppressed the movement through force. Subsequently, other groups mobilized 12 significant protests occurring during the years 2017–2020 with 5,000 or more persons mobilized each time.[83] Dissatisfaction with the regime explosively erupted during the Women, Life, Freedom protests in 2022—implying a subsurface of Iranian nonviolent resistance remains.

As mobilization in Tehran akin to the Green Path of Hope will likely continue with nonviolent resistance, the physical geography of Iran has less impact on its potential success, making the space and information domains the most influential. As summarized by Shirazi and Wilkins, "contemporary media industries and technologies in Iran comprise a variety of communication possibilities through which protest participants have potential to engage and mobilize, but also risk surveillance and threat."[84]

The Islamic Republic has several information tools to address the resistance narrative. The Supreme Leader can effectively use the religious pulpit to spread his message. In 2009, Tehran had little ability to limit or monitor internet and social media use. After the crackdown on the Green Path of Hope, it expended many resources to monitor and control Iranian citizens. Iran established the Iranian Cyber Police (FATA) in 2011 to monitor the internet. With the advice and assistance of the Chinese Communist Party, Tehran finished its National Information Network in 2019, which acts as a censorship apparatus. Western social media platforms (like Twitter) were replaced with Iranian ones to further government control. Virtual private networks (VPN) are now outlawed.[85] The government also has access to state-run media sources. It can monitor and block conventional forms of communications and has developed abilities to monitor and censor the internet. Nevertheless, the information environment retains opportunities for resistance to recruit, communicate, and organize, but it requires the use of modern tradecraft to succeed.

Tehran has an extensive and capable security network to suppress activists exposed by social media or other intelligence means. The Supreme Leader controls a national police force of about 260,000 personnel that has aggressively exercised its authority to arrest, put on trial, and execute resistance members. It also has a morality police (or religious police) called Guidance Patrol to carry out extralegal acts. It can call out a paramilitary militia called Basij of 90,000 members to intimidate, bully, or attack opponents. Finally, it has the Islamic Revolutionary Guard Corps of 250,000, well-trained in irregular methods of warfare. While all these capabilities combined appear somewhat intimidating, the loyalties of members within each organization in a domestic struggle, particularly within the ingroup, are not predetermined.[86]

The strength of nonviolent resistance, as demonstrated by the Green Path of Hope, lies in achieving change through popular support, which, ironically, the Islamic Republic's repressive policies strengthen. Misagh Parsa, a professor at Dartmouth College, explored the possibilities of increased democratization in Iran; he assessed that "democratic force, currently demobilized, are certain to resurface and press for democratization as internal social, economic, and political contradictions converge."[87] An organization like the Green Path of Hope may achieve successful reforms when an opportune environment for governmental change presents itself.

ORGANIZATION OF THE GREEN PATH OF HOPE

Resistance can be generally categorized into two bins: (1) mass organization and (2) elite organizations.[88] Each has advantages and disadvantages. Mass organizations have few bars to entry for recruiting and take advantage of size to compete with authority. Mass proves an excellent archetype for non-violent protest, like social movements. However, a mass organization is difficult to train and control, easier for authorities to infiltrate, and members can prove undisciplined. In contrast, elite organizations take advantage of extensive vetting, selective recruiting, superior training, and a high degree of motivation. These types of movements are normally secretive, operating with undergrounds or, when overt, maintaining covert or clandestine activities. An elite organization can influence mass organizations and even hijack or influence their behaviors. Elite organizations designed to blossom into a mass organization given the right circumstances are called elite-fronts.[89]

The Green Path of Hope is a mass organization, with all the advantages and disadvantages of this typology. Mass organizations have more success in democracies than in totalitarian regimes. So, despite its success in 2009, the Green Movement failed to achieve reforms and subsequently has been contained. It might consider incorporating some of the aspects of an elite front model, which better leverages aspects of an underground but can still harness popular mass support when required.

Resistance movements can be sub-organized in a myriad of ways. In special operations doctrine, these can include (a) an underground, (b) an armed component, (c) an auxiliary, and (d) a public component.[90] The Green Path of Hope utilizes two of these, an underground and a public component. The underground organized protestors; it coordinated green attire and used graffiti with ingenious means of spreading the message. The question is how active or effective this underground is today. The public component within Iran consisted of the leadership: Mousavi, Karroubi, and Rahnavard. However, the Iranian diaspora also served as spokespersons abroad for the movement.

The Green Path of Hope generally lacks both an armed component and an auxiliary. As a reformist organization, it has opted for nonviolence, so an armed component might never manifest. However, nonviolent resistance can equally benefit from a healthy auxiliary. Auxiliaries are distinctively different from part-time and overt supporters. Auxiliary examples (covert supporters hiding in plain sight) might include software engineers, communications specialists, journalists, doctors, educators, transportation employees, engineers, judges, industrial workers, policemen, and soldiers. These professionals would seek not to compromise their overt loyalty to the Islamic Republic while secretly supporting resistance activities. This type of subversion can prove quite powerful but requires a well-established and functioning underground for its recruitment, employment, and coordination.

ACTIONS OF THE GREEN PATH OF HOPE

The Green Movement seeks nonviolent but legal forms of resistance, particularly freedom of assembly, which the United Nations has identified as a human right and applies in Iran as it is a member.[91] From 2009–2011, there were 27 mass protests against the regime connected with the Green Path of Hope. These occurred mostly in Tehran but also included Isfahan, Tabriz, Yazd, Shiraz, and Mashhad.[92] Numbers varied but included outbreaks of over 100,000 activists. The Islamic Republic ignored only three of these and reacted to the rest with beatings, shootings, and arrests. The Green Movement created a charter in 2010, which identified its goals with an attempt at a coalition. It also encouraged grassroots social networks of supporters, but these appear more ad hoc and without central organization.

Burning bus during Green Movement protests in Tehran, 13 June 2009. Source: lapost/wikicommons

PHASE THREE: SUMMARY

In summary, the Green Path of Hope demonstrated incredible potential in 2009–2010 and "shook the foundation of the Islamic Republic like no other event in the thirty years since the revolution."[93] Its leader, Mousavi, seeks nonviolent governance reforms, particularly in terms of adherence to electoral freedoms. He formed a coherent message that articulated shared grievances to rally the masses, but Tehran successfully weaponized information tools to counter the narrative. The movement experienced broad support from activists that mobilized in the hundreds of thousands. The base mainly included urban-based youth, students, and women but was comprised of a loose coalition. The primary domains in which it competed included urban areas, space-based media, traditional media, and social media platforms. Movement leaders remain under house arrest, and the Iranian security forces have forced most activists underground. Although currently dormant, the movement has significant potential given the persistent weakness and corruption of Iranian governance, the existence of an organizational framework capable of leveraging the internet, the support of like-minded movements, and the possibility of receiving significant international support.

PHASE FOUR: OPTIONS IN SUPPORT OF RESILIENCE OR RESISTANCE

In phase four, we utilize the data gathered in the previous three phases to better inform foreign policies regarding the Islamic Republic. We explore three possible options: (a) support the resilience of the Islamic State, (b) support domestic resistance to the Islamic State, or (c) choose to support neither, but prepare the environment for adoption of a future policy to support resilience or resistance.

SUPPORTING ISLAMIC STATE'S RESILIENCE

Some U.S. strategic analysts suggest more restraint toward Iranian aggression by reducing military forward presence and minimizing activities in the region. The approach would shift to leverage Saudi, Israeli, or other regional efforts to counter Iranian influence rather than placing U.S. resources and personnel at risk. Part of this strategy to is "offshore balance" military assets to sea or basing in neutral countries.[94] This approach would theoretically reduce tensions and release the U.S. from current foreign entanglements. Disengagement is popular domestically and may appear to be attractive but would likely strengthen the Iranian regime as a result.[95] However, the perceived illegitimacy of the Islamic Republic by its people makes long-term stabilization of Iran through this approach unlikely.

SUPPORTING IRANIAN RESISTANCE

The currently identified violent paramilitary groups actively opposing the Islamic Republic do not appear promising. These organizations are outside the Persian/Shi'a ingroup and do not have the domestic popularity in terms of numbers normally required to implement change. Additionally, violence creates more fragility in a region already in need of stability, making this a risky policy option.

In contrast, supporting a nonviolent resistance comprised of the ingroup, one akin to the Green Path of Hope, has promise and a real chance of success. At least one U.S. senator, Marco Rubio, has advocated for support to Iranian resistance.[96] Current nonviolent resistance movements in Iran have demonstrated potential but thus far exhibited some weakness. The Green Movement failed in three major ways: (1) it followed events versus leading them, (2) the movement's leadership sought reform of current governance while other opposition groups wanted more radical change, and (3) the movement proved unable to appeal to a large coalition of activists, which limited its potential.[97]

Learning from the past, a future Iranian nonviolent movement could organize and operate more effectively. It might establish an underground, designed in a cellular network to decrease the risk of compromise by the regime. It might leverage secrecy to organize, communicate, and compete in resistance activities, sustained over an extended period. It must understand how to operate clandestinely in digital and urban space, perhaps taking note of the Umbrella Movement lessons learned in Hong Kong versus the Chinese Communist Party.[98] It should methodically organize and train in advance, rather than spontaneously and haphazardly.[99] As well, it must better organize, train, and equip for nonviolent action.[100] Additionally, a future movement might attempt to create larger appeal, not only to the energetic efforts of women and youth, but to clergy, small business owners, and industrial workers.[101]

If Western foreign policies desire a popular and democratic system of Iranian governance, a resistance movement seeking these goals must maintain domestic legitimacy. External support for Iranian resistance (either from nation-states, corporations, religious groups, nongovernmental organizations, etc.) might best be offered with low visibility or no visibility, as outside influence may erode the legitimacy of the resistance. The narrative of a domestic, internal movement is essential to its success. However, space and cyberspace are key battlegrounds in which an external supporter might assist with, as well as resistance training, material assistance, or legitimization efforts.

CHOOSE NEITHER BUT PREPARE THE ENVIRONMENT FOR FUTURE POLICY

Preparing the environment for future policy requires building relationships with influential Iranians of many ideological perspectives and understanding the domestic environment of Iran to foresee possible opportunities—in support of either resilience or resistance, depending on the prospects of what the future situation offers. However, by breaking diplomatic relations and imposing sanctions on Iran, many Western nations have restricted their ability to conduct these bond-building activities.[102] Imposing sanctions as the primary means of hampering the Islamic Republic has proven an anemic method of coercion.[103] Damage has been done to the Iranian economy, but the effects tend to hurt the population and do little to stem the military resourcing of the proxy militias.[104] One might argue that sanctions increase Iranian resistance and decrease the Islamic Republic's resilience, but without an accompanying strategy to support change, this has accomplished little in over 50 years.

CONCLUSION

Understanding Iran in terms of its resilience and resistance can provide a starting point for contemplating current or future foreign policy options for the United States, Western nations, and regional partners. This data-centric analysis demonstrates that while resiliency in the Islamic Republic remains low, the ability of external support to the regime's resiliency has above-average potential. Meanwhile, Iran retains high potential for resistance, as well as strong indicators for successful external support to resistance. Of the many resistance movements identified, a nonviolent organization may provide the best chance at implementing changes in governance and increasing regional stability. However, taking advantage of such opportunities will take more than the sudden death of Iranian President Ebrahim Raisi to realize. Unless Western nations prepare the environment, a dedicated and comprehensive effort that could take several years or more, they will remain unprepared to take advantage of these sudden and fleeting opportunities.

About the Authors

Dr. Robert S. Burrell is a resilience and resistance interdisciplinary scholar using data-driven and human-centric methodologies to analyze intrastate conflict ranging from nonviolent protest through belligerency. He is a Senior Research Fellow at the Global and National Security Institute of the University of South Florida. From 2020-2024, he taught irregular warfare at Joint Special Operations University and was the editor–in–chief of special operations doctrine from 2011-2014.

Dr. David R. DiOrio (CAPT Ret.) is a national security professional with a Doctor of Philosophy degree in Public Policy and Administration from Walden University. He served as the Deputy Director at the Joint Forces Staff College of the National Defense University and is currently Adjunct Faculty at the Joint Special Operations University.

NOTES

1. The methodologies utilized in this chapter were previously published by *Small Wars & Insurgencies* and *Expeditions with MCUP*. See Robert S. Burrell and John Collison, "A Guide for Measuring Resiliency and Resistance," *Small Wars & Insurgencies*, December 14, 2023; Robert S. Burrell and John Collison, "Assessing Resistance for the Purpose of Informing International Policy," *Expeditions with MCUP*, January 9, 2024.

2. Robert S. Burrell and David DiOrio, "A 2024 Guide to Assessing Resiliency and Resistance in Iran," Global and National Security Institute, University of South Florida, July 3, 2024, https://www.usf.edu/gnsi/publications/research-articles/2024-guide-resiliency-iran-interactive.pdf.

3. U.S. Department of State, "Outlaw Regime: A Chronicle of Iran's Destructive Activities," Iran Action Group Report, 2020, 11–18, https://www.state.gov/wp-content/uploads/2020/09/Outlaw-Regime-2020-A-Chronicle-of-Irans-Destabilizing-Activity.pdf, accessed May 23, 2024.

4. Ali Parchami, "An Iranian Worldview: The Strategic Culture of the Islamic Republic," *Journal of Advanced Military Studies: Special Issue on Strategic Culture* (2022): 9–15.

5. U.S. Congress, Congressional Research Service, *Iran: Background and U.S. Policy*, CRS Report R47321, Washington, DC, April 22, 2024, https://crsreports.congress.gov/product/pdf/R/R47321, accessed May 30, 2024.

6. Ariane M. Tabatabai, Jeffrey Martini, and Becca Wasser, "The Iran Threat Network (ITN): Four Models of Iran's Nonstate Client Partnerships," RAND Corporation, 2021, 7, https://www.rand.org/pubs/research_reports/RR4231.html, accessed September 7, 2023.

7. Maziar Motamedi, "Iran Heads to Presidential run-off on July 5Amid Record Low Turnout," *Al Jazeera News*, June 29, 2024, https://www.aljazeera.com/news/2024/6/29/iran-poised-for-presidential-run-off-in-tight-race

8. This resistance and resilience methodology was developed by Robert S. Burrell and John Collison. The foundational publications of this approach are found in Burrell and Collison "Assessing Resistance," and Burrell and Collison, "A Guide for Measuring Resiliency and Resistance." For a quick look, see Robert S. Burrell, "A Guide for Measuring Resiliency," *Irregular Warfare Initiative*, January 16, 2024, https://irregularwarfare.org/articles/a-guide-for-measuring-resiliency/.

9. We have decided not to include measuring the Islamic Republic's potential to implement a state-sponsored resistance strategy to foreign occupation, as occupation of Iranian territory remains likely.

10. Diane Zorri, Houman Sadri, and David Ellis, *Iranian Proxy Groups in Iraq, Syria, and Yemen: A Principal-Agent Comparative Analysis* (Joint Special Operations University Press, 2020): 15-16.

11. See the CIA World Factbook at https://www.cia.gov/the-world-factbook/countries/iran/#introduction, accessed on April 2, 2023.

12. Zorri et al., *Iranian Proxy Groups*, 16–18.

13. Seth G. Jones, "Containing Iran," Center for Strategic and International Studies (January 2020): 44–47, https://csis-website-prod.s3.amazonaws.com/s3fs-public/publication/200110_Jones_ContainingIran_WEB_v2.pdf, accessed on May 26, 2024.

14. Jones, "Containing Iran," 44–47.

15. "Iran Minorities 2: Ethnic Diversity," *The Iran Primer*, September 3, 2013, https://iranprimer.usip.org/blog/2013/sep/03/iran-minorities-2-ethnic-diversity, accessed April 2, 2024.

16. "Languages of Iran," Wikipedia, https://en.wikipedia.org/wiki/Languages_of_Iran, accessed April 2, 2024.

17. *Iran 2022 International Religious Freedom Report*, Department of State, 2022, https://www.state.gov/reports/2022-report-on-international-religious-freedom/iran/, accessed April 2, 2022.

18. Data derived from The World Bank, https://www.worldbank.org/en/publication/worldwide-governance-indicators, accessed April 2, 2024.

19. See Fragile States Index, https://fragilestatesindex.org/, accessed April 2, 2024. All following percentiles are indicted as percentages in comparison with other nations.

20. "2023 Country Reports on Human Rights Practices: Iran," U.S. Department of State, 2023, see Executive Summary at https://www.state.gov/reports/2023-country-reports-on-human-rights-practices/iran/, accessed May 25, 2024.

21. Several scholars have attempted to create rubrics for analyzing the will of military units and the nation as a whole. One study by RAND in 2018 shows promise as a method but remains overly oriented on particular military units and not the population itself. Another in the *Journal of Advanced Military Studies* in 2022 expands upon the former study, which focused on military units, and takes a more strategic look at national militaries as a whole. See Ben Connable et al., *Will to Fight: Analyzing, Modeling, and Simulating the Will to Fight of Military Units*, (RAND Corporation 2018), as well as Ben Connable, "Structuring Cultural Analyses: Applying the Holistic Will-to-Fight Models," *Journal of Advanced Military Studies: Special Issue on Strategic Culture* (2022): 153–167. We have chosen to use the work by Delbert C. Miller, "The Measurement of National Morale," *American Sociological Review* 6, no. 4 (August 1, 1941): 487–498.

22. "Iranians' Attitudes toward Political Systems: A 2022 Survey Report," GAMAAN: The Group for Analyzing and Measuring Attitudes in Iran, March 31, 2022, https://gamaan.org/2022/03/31/political-systems-survey-english/, accessed April 3, 2024.

23. We have taken the inverse of this number to determine the populations support for, and personal identification with, the goals of the Islamic Republic. Ibid.

24. "Opinion Survey Reveals Overwhelming Majority Rejecting Iran's Regime," *Iran International*, April 2, 2023, https://www.iranintl.com/en/202302036145, accessed April 3, 2024.

25. See Ali Alfoneh, "Iran: Trust in Family, Distrust in Government," *Iran Media Review*, March 1, 2024, https://agsiw.org/iran-trust-in-family-distrust-in-government/, accessed April 3, 2024. We have taken the inverse of these polls and averaged them to determine the opinion of the ingroup that national resource (the military) can hurl back enemies, equating to 78 percent.

26. Ali Alfoneh, "Official Government Poll: 72.9 percent of Iranians Favor Separation of Religion and State," *Iran Media Review*, March 1, 2024, https://agsiw.org/iran-trust-in-family-distrust-in-government/, accessed February 27, 2024. Also see the Constitution of the Islamic Republic of Iran, https://www.refworld.org/legal/legislation/natlegbod/1979/en/72964, accessed April 5, 2024. Again, we have taken the inverse of this poll, which implies that 27 percent of Iranians agree with the principles of the Iranian Republic as a religious state.

27. Zorri et al., *Iranian Proxy Groups*, 16.

28. Jones, "Containing Iran," 18.

29. Kenneth M. Pollack, "Iran's Grand Strategy Has Fundamentally Shifted," *Foreign Policy*, August 15, 2023, 1, https://foreignpolicy.com/2023/08/15/irans-grand-strategy-has-fundamentally-shifted/, accessed September 22, 2023.

30. Michael Eisenstadt, "Deterring Iran in the Gray Zone: Insights from Four Decades of Conflict," The Washington Institute for Near East Policy/Policy Notes 103, 2021: 18, https://www.washingtoninstitute.org/policy-analysis/deterring-iran-gray-zone-insights-four-decades-conflict, accessed August 18, 2023.

31. Congressional Research Service, "Iran's Foreign Defense Policies," 2.

32. Maziar Motamedi, "Iran Says 25-Year China Agreement Enters Implementation Stage," *Al Jazeera*, January 15, 2022, https://www.aljazeera.com/news/2022/1/15/iran-says-25-year-china-agreement-enters-implementation-stage, accessed April 3, 2024.

33. Emil Avdaliani, "Iran and Russia Enter A New Level of Military Cooperation," Stimson Center, March 6, 2024, https://www.stimson.org/2024/iran-and-russia-enter-a-new-level-of-military-cooperation/, accessed April 3, 2024.

34. Pollack, "Iran's Grand Strategy Has Fundamentally Shifted," 2.

35. Ariane M. Tabatabai, Jeffrey Martini, and Becca Wasser, *The Iran Threat Network (ITN): Four Models of Iran's Nonstate Client Partnerships*, (RAND Corporation, April 14, 2021), 2.

36. Tabatabai et al., *The Iran Threat Network*, 8.

37. Tabatabai et al., 5.

38. Tabatabai et al., 6.

39. Tabatabai et al., 5.

40. Tabatabai et al., 3.

41. Kali Robinson and Will Merrow, "Iran's Regional Armed Network," Council on Foreign Relations, January 31, 2024, https://www.cfr.org/article/irans-regional-armed-network, accessed April 3, 2024.

42. The Global Economy, https://www.theglobaleconomy.com/, accessed June 24, 2024.

43. Armed Conflict Location & Event Data Project, https://acleddata.com, accessed June 10, 2024.

44. Carnegie Endowment for International Peace, https://carnegieendowment.org/features/global-protest-tracker, accessed June 14, 2024.

45. Global Terrorism Database, https://www.start.umd.edu/gtd/, accessed April 3, 2024.

46. Vision of Humanity's Global Peach Index, https://www.visionofhumanity.org/maps/#/, accessed April 3, 2024.

47. Freedom House, https://freedomhouse.org/, accessed April 3, 2024.

48. In subjectively assessing the population's access to food as a measure of resistance, We consider that if the country is on the CIA list, then = 75 percent, as food insecurity exists. If not, then = 25 percent, as the potential for food insecurity remains.

49. U.S. Department of State, https://www.state.gov/u-s-relations-with-iran/, 1 April 2022.

50. President, *National Security Strategy* (Washington, DC: White House, 2022): 11-12, https://www.whitehouse.gov/wp-content/uploads/2022/11/8-November-Combined-PDF-for-Upload.pdf

51. For this metric, we assess the following: (a) if a strategic rival of the U.S., then = 100 percent, (b) if not on good diplomatic terms with the U.S but not a declared rival, then = 50 percent, (c) if on good diplomatic terms with the U.S., then = 25 percent, and (d) if U.S. ally then = 0 percent.

52. Burrell and Collison, "A Guide for Measuring Resiliency and Resistance." For more detailed information on this methodology, see Chapter 2, Section 4 of this book.

53. One scholar made this assessment in 1993: "the formation of the Islamic Republic with ambiguous national and ethnic policies—which has coincided with the fall of the Soviet empire and rising national aspirations among numerous ethnic groups throughout the world—has exacerbated the ethnic tensions and ideological conflicts among the advocates of various modes of national and ethnic identities in Iran." See Ahmad Ashraf, "The Crisis of National and Ethnic Identities in Contemporary Iran." *Iranian Studies* 26, no. 1/2 (1993): 159–164. 163. Also see, Kian Sharifi "Iran's Clerical Ruler, Face 'Legitimacy Crisis' Ahead of Elections," Radio Free Europe, February 22, 2024. https://www.rferl.org/a/iran-s-clerical-rulers-face-legitimacy-crisis-ahead-of-elections/32829440.html, accessed April 3, 2022.

54. Sanam Vakil, "Iran's Crisis of Legitimacy," *Foreign Affairs*, September 28, 2022.

55. We use multiple points of reference to analyze the relationship of neighboring countries to Iran and look for the potential geographic regions for possible sanctuary inside the national boundaries or with a neighboring country. If (a) Iran has strong regional partnerships and resistance sanctuary from a neighbor appears unlikely = 0 percent; if (b) Iran has poor regional international relations but no resistance sanctuary appears possible = 25 percent; if (c) at least one regional partner might potentially offer a resistance sanctuary = 50 percent; and if (d) multiple regional states could possibly offer resistance sanctuary = 75 percent.

56. Outside support would likely need to maintain secrecy, however, to protect the legitimacy of the resistance.

57. Harvard's Nonviolent and Violent Campaigns and Outcomes (NAVCO) lists eleven major campaigns since 1905. (1) The nonviolent Persian Constitutional Revolution of 1905–1906 successfully made changes to the monarchist regime of Mozaffar ad-Din Shah. (2) This reform was followed by a civil war from 1908–1909, in which Constitutionalists forced Mohammad Ali Shah to abdicate in favor of his son. (3) From 1945–1946, the Azerbaijan People's Government, backed by the Soviet Union, unsuccessfully attempted to secede from Iran. (4) From 1977–1979, the generally nonviolent Iranian Revolution occurred, which ousted Shah Reza Pahlavi from power. (5) Starting in 1979, the Democratic Party of Iranian Kurdistan (KDPI) conducted an insurgency with the goal of secession, which officially ended in 1996 and NAVCO assesses as a failure. (6) From 1981 through 2001, Mujahidin-e-Khaliq (MEK) carried out a Marxist insurgency, which steadily lost popularity and failed. (7) In 1999, students from Tehran University, called the Tir 18 Riot for Democracy, attempted to bring about reforms in favor of more freedom, which President Mohammad Khatami suppressed with law enforcement and possibly militias. (8) Since 2005, the Kurdistan Free Life Party (PJAK) has conducted an insurgency for Kurdish independence. Both Iran and Turkey have used military force against PJAK, but the conflict remains unresolved. (9) From 2009–2013, the aforementioned Green Path of Hope sought reforms to the Islamic Republic through nonviolent protest. NAVCO lists these efforts as achieving limited success. (10) From 2017–2018, the Dey protests occurred in the city of Mashad, opposing the Islamic Republic's theocratic system of governance, but failed to bring about change. (11) Another wave of nonviolent protests occurred called Bloody November between 2019–2020, again favoring more democratic rule, failed. See https://dataverse.harvard.edu/dataverse/navco. (12) Not listed by NAVCO due to its recency, the 12th major campaign of resistance in Iran occurred in 2022–2023 following the death of the Kurdish woman Jina Mahsa Amini, which was suppressed without changes to governance. See Carnegie Endowment for International Peace's Global Protest Tracker, which assesses over 10,000 Mahsa Amini protestors (the previously discussed Women, Life, Freedom movement). See Carnegie Endowment for International Peace's Global Protest Tracker with location, dates, size, and duration of mass protests around the world at https://carnegieendowment.org/publications/interactive/protest-tracker.

58. Harvard Dataverse, Mass Mobilization Protest Data, https://dataverse.harvard.edu/dataverse/MMdata, accessed September 13, 2023. Also see the Global Nonviolent Action Database by Swarthmore College https://nvdatabase.swarthmore.edu.

59. Richard Spencer, "As Protest Flare, Iran Bids Farewell to Rafsanjani," *The New York Times*, January 10, 2017.

60. Maryam Sinaiee, "Hundreds of Activists Support Mousavi's Call to End Clerical Rule," *Iran International*, 13 February 2023, https://www.iranintl.com/en/202302132030, accessed April 4, 2024.

61. Thomas Erdbrink and Richard Pérez-Peña, "Iran Arrests 20 Tied to Protests Against Compulsory Islamic Veil, *The New York Times*, February 3, 2018.

62. "'Women, Life, Freedom' Movement Wins 2023 Freedom House Award," *Iran International*, November 5, 2023, https://www.iranintl.com/en/202305102647, accessed April 4, 2024.

63. Uppsala Conflict Data Program, https://ucdp.uu.se/exploratory, accessed June 24, 2024.

64. Uppsala Conflict Data Program.

65. The figure illustrates significant Iranian resistance movements across a resistance continuum including nonviolent legal ones as Women, Life, Freedom and the Green Path of Hope. The epicenter for nonviolent but illegal forms of protest might derive from the University of Tehran. Another group that has changed its methods from violence to nonviolence but remains illegal includes the Marxist Mojahedin-e-Khalq (MEK), which both Iraq and Iran have designated as a terrorist group. The threat of rebellion as a means of violent opposition but limited due to size and duration still endures in Iran; however, no groups currently fit that typology. Insurgent groups include the Islamic State, (6) Jaish al-Adl, (7) KDPI, (8) PJAK, and (9) the Marxist PKK. None of the insurgent groups have risen to the level of belligerency yet.

66. Several studies and theories attempt to define ways of deconstructing and assessing political movements, insurgencies, or resistance organizations in general. See David Collier, Jody LaPorte, and Jason Seawright, "Putting Typologies to Work: Concept Formation, Measurement, and Analytic Rigor," *Political Research Quarterly* 1 (2012).

67. We use the resistance methodology developed by the U.S. Army Special Operations Command and Johns Hopkins University, published by Jonathon Cosgrove and Erin Hahn, *Conceptual Typology of Resistance*, (Fort Bragg, NC: U.S. Army Special Operations Command, circa 2018). For the definitions of each, see Cosgrove and Hahn, 6. The primary sources to complete this assessment include Misagh Parsa, *Democracy in Iran: Why It Failed and How It Might Succeed*, (Harvard University Press, 2016); Marco Rubio, "Why Protests Matter: The Battle between Authoritarianism and Democracy, a War We Must Win," *Journal of International Affairs* 73, no. 2 (July 2020): 251–60; Faegheh Shirazi and Karin Wilkins. "Mapping the Political Discourse of the Iranian Green Movement," *Anthropology of the Contemporary Middle East & Central Eurasia* 4 no.1 (2016): 32–55.

68. Parsa, *Democracy in Iran*, 206

69. Shirazi and Wilkins, "Mapping the Political Discourse," 51.

70. Parsa, *Democracy in Iran*, 233.

71. Hamid Dabashi, "What happened to the Green Movement in Iran?," *Al Jazeera*, June 12, 2013, https://www.aljazeera.com/opinions/2013/6/12/what-happened-to-the-green-movement-in-iran, accessed April 11, 2023.

72. Harvard Dataverse, Mass Mobilization Protest Data, https://dataverse.harvard.edu/dataverse/MMdata, accessed April 9, 2024. For a source stating millions, see Shirazi and Wilkins, "Mapping the Political Discourse."

73. Parsa, *Democracy in Iran*, 233–234.

74. Hamid Dabashi, "What happened to the Green Movement in Iran?," *Al Jazeera*, June 12, 2013, https://www.aljazeera.com/opinions/2013/6/12/what-happened-to-the-green-movement-in-iran, accessed April 11, 2023.

75. See Parsa, *Democracy in Iran*, 215.

76. Shirazi and Wilkins, "Mapping the Political Discourse."

77. Rubio, "Why Protests Matter," 251–252.

78. International support for the Green Path of Hope could erode its domestic legitimacy, particularly in light of U.S. interference in Iran's past domestic politics. Hooman Majd, "Think Again: Iran's Green Movement," *Foreign Policy*, January 6, 2024, https://foreignpolicy.com/2010/01/06/think-again-irans-green-movement/, accessed April 9, 2024.

79. "Opinion Survey Reveals Overwhelming Majority Rejecting Iran's Regime."

80. Shirazi and Wilkins, "Mapping the Political Discourse," 33.

81. Resistance supports a cause that represents the collectively expressed rationales for opposition to authority, as well as the individual motivations for participating in such a group. The rationale for resistance consists of either a desire for sweeping changes in authority, transformation in society, or specified changes for individual groups and communities. Cosgrove and Hahn, *Conceptual Typology of Resistance*, 17–24.

82. Mir-Hossein Mousavi, "Outline of a New Program for the Green Movement," June 15, 2010, https://irandataportal.syr.edu/outline-of-a-new-program-for-the-green-movement-15-june-2010, accessed April 9, 2024.

83. For 2017–2020 protests see, Harvard Dataverse, Mass Mobilization Protest Data.

84. Shirazi and Wilkins, "Mapping the Political Discourse," 39.

85. Emily Blout, "Society in Flux: Exploring Iran's Social Movements and Women's Rights," Paper presented at The Iran Enigma, University of South Florida, Global and National Security Institute, April 10, 2010, https://www.usf.edu/gnsi/events/policy-dialogues-gnsi/. Information presented by Blout derives from her book, *Media and Power in Modern Iran*, (I.B. Tauris, 2023).

86. Parsa, *Democracy in Iran*, 206-243. Also see Majd's "Think Again: Iran's Green Movement." Also see Blout, "Society in Flux." Personnel numbers of Iranian organizations taken from Wikipedia, April 10, 2024.

87. Parsa, *Democracy in Iran*, 320.

88. Paul J. Tompkins Jr., and Robert R. Leonhard, *Undergrounds in Insurgent, Revolutionary, and Resistance Warfare*, (U.S. Army Special Operations Command, 2012), 10–12.

89. Tompkins Jr. and Leonhard, *Undergrounds in Insurgent, Revolutionary, and Resistance Warfare*.

90. Army Technical Publication 3-05.1: *Unconventional Warfare at the Combined Joint Special Operations Task Force Level*, (U.S. Army Special Operations Command, Apr. 2021).

91. See "The Universal Declaration of Human Rights," General Assembly Resolution 217A, United Nations, December 10, 1948.

92. Harvard Dataverse, Mass Mobilization Protest Data.

93. Parsa, *Democracy in Iran*, 247.

94. Jones, "Containing Iran," 58-60.

95. William Walldorf, "The U.S. Public Has Never Been More Anti-War. Biden Isn't Taking Note," *Time* 203, February 26, 2024, 5–6.

96. Rubio, "Why Protests Matter."

97. Parsa, *Democracy in Iran*, 255–275.

98. Colin Agur and Nicholas Frisch. "Digital Disobedience and the Limits of Persuasion: Social Media Activism in Hong Kong's 2014 Umbrella Movement," *Social Media + Society* 5 (February 1, 2019).

99. Paul J. Tompkins Jr., and Robert R. Leonhard, *Undergrounds in Insurgent, Revolutionary, and Resistance Warfare*, 10-12.

100. Gene Sharp's doctrine and his 198 methods of nonviolent action also provide a sound basis for understanding possible tactics. See, Robert L. Helvey, *On Strategic Nonviolent Conflict: Thinking About the Fundamentals* (The Albert Einstein Institution, 2004).

101. Parsa, *Democracy* in Iran, 259–263.

102. Notably, Israel, the United States, and Canada remain the only major Western nations without diplomatic relations with the Islamic Republic.

103. The international array of sanctions to raise the level of economic pain has made little progress to change Iranian behavior. Jones, "Containing Iran," 59.

104. Jones, "Containing Iran," 59.

APPENDIX B: Utilizing *Resilience and Resistance* as a Textbook

By Robert S. Burrell

INTRODUCTION

This book was designed to fulfill U.S Congressional intent on supplying data-centric irregular warfare education and designed for international security professionals, particularly in the U.S. Government. Accordingly, *Resilience and Resistance* has great potential for use as a textbook. It can be divided into roughly 20 lessons that can be delivered in a seminar-like setting over 160 hours (equating to four continuous weeks). Or *Resilience and Resistance* can facilitate a semester-long seminar by assigning and discussing the chapters as readings over a typical 16-week semester. Additionally, utilizing the methodologies presented within, a student could fulfill a master's degree writing requirement (compiling a 10,000-15,000-word thesis) by following the data-centric approach presented in *Resilience and Resistance.*

Completing the two following practicums and merging the analysis provides the foundation of a comprehensive thesis. This type of thesis approach facilitates understanding of a country or region uniquely laid out in terms of resilience and resistance, the results of which could potentially inform policy and strategy. These two practicums include: (a) one carried out mid-semester after completing Part I in the book, and (b) the other comprising the final product at the end of the semester and completed after finishing Part II of the book. A description of these two recommended practicums follows.

PART I: FOUNDATIONS IN RESILIENCY AND RESISTANCE

Purpose. Following the completion of Part I in Resilience and Resistance, begin a strategic-operational response practicum on a case study of student's choice.

Overview. Students should develop a strategic-operational assessment of a specific intrastate or trans-regional issue utilizing various design applications. The recommended method is to measure resistance and resilience factors of a particular state using the mnemonic introduced in Part I. The key categories in this methodology include (a) measuring a state's **R**esiliency, (b) analyzing the potential for **S**tate-sponsored resistance, (c) evaluating potential to **R**einforce resiliency, (d) measuring **R**esistance potential to current governance or occupation, and (e) evaluating potential for external **S**upport to resistance. The mnemonic is **RSRRS**.
Learning Outcomes. Upon completion of Part I in *Resilience and Resistance*, students should begin a comprehensive strategic-operational response practicum on a case study of student's choice, incorporating the following sub-components:

1. Forecast emergent cases of contentious politics by applying social movement theory to historical cases.
2. Evaluate SOF's support to resilience and resistance as it relates to future employment of SOF.
3. Analyze factors/root causes of state fragility and vulnerability, adversary frames and narratives, and strategies.
4. Conduct a partial (or full in the case of a master's thesis) strategic estimate for a particular problem set or area of responsibility.

PART II: APPLICATION OF RESILIENCY AND RESISTANCE

Purpose. Following the completion of Part II in *Resilience and Resistance*, complete a strategic-operational response practicum on a country case study of student's choice.

Overview. Building upon the practicum from Part I, students should complete their strategic-operational response to a specific intrastate or trans-regional issue utilizing various design applications, which includes examination of non-state actors, particularly resistance movements. In developing a response, students might consider various types of resistance organizations within the specified area; how those resistance movements operate in terms of leadership, cause, environment, organization, and actions; and what potential options exist for an external sponsor to inform national policy or military strategy in that region. As outlined in Table B.1, Phase One of this assessment incorporates the work completed in the first practicum. Building upon this, students should utilize Phase Two, Phase Three, and Phase Four outlined in Table B.1. and detailed in *Resilience and Resistance*.

TABLE B.1. THE 12-STEP RESILIENCE AND RESISTANCE METHODOLOGY

PHASE	STEPS
One	1. Measuring the state's resiliency.
	2. Identifying the potential for a state-sponsored resistance strategy (if applicable).
	3. Measuring the potential for external support to resiliency.
	4. Measuring the potential for resistance to current authority.
	5. Measuring the potential for external support to resistance.
Two	6. Identifying the prevalent resistance groups within the state and placing them on the Resistance Continuum.
Three	7. Assessing one or more resistance groups in terms of leadership,
	8. cause,
	9. environment,
	10. organization,
	11. and actions.
Four	12. Making a recommendation concerning potential external support to resiliency or resistance, which normally proposes one of three options: (a) support the resiliency of the current governing authority, (b) support resistance to the same, or (c) prepare the environment for a future policy supporting resiliency or resistance.

Source: Author

After completing the first three phases, the final step in Phase Four is making an evaluation of the data and recommending a course of action. The typical suggestion in Phase Four proposes one of three options: (a) support the resiliency of current governance, (b) support resistance to the same, or (c) prepare the environment for a future policy supporting resiliency or resistance. However, a full proposal might also consider the underlying societal factors that have given space for resistance to operate; how to counter the frames and narratives of the adversary (either an external actor, state authority, or the resistance); consideration of timings for various aspects of the response; measures of effectiveness; and risk assessment and mitigation.

Learning Outcomes. Upon completion of Part II in ***Resilience and Resistance***, students should complete a comprehensive strategic-operational response practicum on a current case study of student's choice, incorporating the following sub-components:

1. **Analyze** historical and emergent approaches to irregular warfare as practiced by malign actors (state and non-state).
2. **Evaluate** allied and interagency capabilities in irregular approaches to strategy and planning.
3. **Evaluate** SOF-unique operations, actions, and investments in support of integrated deterrence.
4. **Apply** design-informed campaigning utilizing integrative statecraft solutions.
5. **Conduct** a partial (or full, in the case of a master's thesis) strategic estimate for a particular problem set or area of responsibility.

APPENDIX C: ACRONYMS

A2/AD – anti-access/area denial

ACEOA – actors, causes, environment, organization, and actions (attributes of a resistance)

ACLED – Armed Conflict Location & Event Data Project

ADC – asymmetric defense component

AED – Aviation Enterprise Development

AI – artificial intelligence

ALF – Animal Liberation Front

APT – advanced persistent threat

AQAP – Al-Qaeda

ARIS – Assessing Revolutionary and Insurgent Strategies

ARSOF – U.S. Army Special Operations Forces

ARVN – Army of the Republic of Vietnam

ASCOPE – areas, structures, capabilities, organizations, people, and events.

ASEAN – Association of Southeast Asian Nations

AUC – *Autodefensas Unida Columbia* (United Self-Defense Groups of Columbia)

BCG – *Batallones contraguerrillas* (Columbian counterguerrilla battalions)

BPACS – Building Partner Air Capacity Seminar

BRIM – *brigades moviles* (Columbian mobile brigades)

C2 – command and control

CBRNE – chemical, bacteriological, radiological, nuclear, and explosive

CCP – Chinese Communist Party

ChON – *Chasti osobogo naznachenya* (Russian special purpose brigades)

CISA – U.S. Cybersecurity and Infrastructure Agency

CNP – Columbian National Police (*Policia nacional*)

COIN – counterinsurgency

COLAR – Columbian Army (*Ejerctio Nacional*)

CSIS – Center for Strategic and International Studies

CT – counterterrorism

D4M – deny, degrade, disrupt, destroy, and manipulate

DA – direct action

DARPA – Defense Advanced Research Project Agency

DIB – defense industrial base

DIGEST – drugs, infrastructure, governance, economics, structures, and tribes

DIME – diplomatic, informational, military, and economic

DOD – Department of Defense

DPP – Democratic Progressive Party (Taiwan)

DPR – Donetsk People's Republic (Russia/Ukraine)

DPRK – Democratic People's Republic of Korea

ELF – Earth Liberation Front

ELN – (Columbian) National Liberation Army (*Ejercito de Liberacion Nacional*)

EU – European Union

FARC – Revolutionary Armed Forces of Columbia (*Fuerzas Armadas Revolucionarias de Columbia*)

FATA – Iranian cyber police

FID – foreign internal defense

FLN – Front de Liberation Nationale

FMLN – Farabundo Marti National Liberation Front (El Salvador)

FSA – Free Syrian Army

FSB – Federal Security Service (Russia)

FTO – foreign terrorist organization

GCC – Gulf Cooperation Council

GDP – gross domestic product

GEO – geosynchronous orbit

GRU – main intelligence directorate (Russia)

GSOF – Georgian Special Operations Forces

GTEP – Georgian Train and Equip Program

GWOT – Global War on Terror

HN – host nation

HUMINT – human intelligence

IDF – Israeli Defense Forces

IOT – internet of things

IR – International relations

IRGC - Islamic Revolutionary Guard Corps

IRGC – Islamic Revolutionary Guard Corps

IRI – Islamic Republic of Iran

ISAF – International Security Force Assistance (NATO)

ISIS – Islamic State of Iraq and Syria

ISIY – Islamic State in Yemen

ISR – intelligence, surveillance, and reconnaissance

ITN – Iranian threat network

IW – irregular warfare

JOC – Joint operating construct

JPC – Joint Phasing Construct

JPR – joint personnel recovery

JSOU – Joint Special Operations University

KDPI – Democratic Party of Iranian Kurdistan

KEII – Korean Escapee (Defector) Information Institute

KGB – Committee on State Security (Russia)

KSSO – Special Operations Forces Command (Russia)

LEO – low earth orbit

LH – Lebanese Hezbollah

LPR – Luhansk People's Republic (Russia/Ukraine)

LSCO –large scale combat operations

MA – military assistance

MC – Military Committee

MDN – Ministry of Defense (Columbia; *Ministerio de Defense Nacional*)

MEK – Mujahideen e Khaleq

MENA – Middle Eastern and

MEO – middle earth orbit

METT-TC – mission, enemy, terrain, terrain and weather, troops, time available, and civil considerations

MIDLIFE – military, information, diplomatic, law enforcement, intelligence, financial, and economic

MISO – military information support operations

MVD – Ministry of Internal Affairs (Russia)

NAR – nonconventional assisted recovery

NAVCO – Nonviolent and Violent Campaigns and Outcomes

NCC – NATO SOF Coordination Centre

NCO – non-comissioned officer

NCSM(IM) – national Socialist Council of Nagalim-Isak-Muivah (India)

NDS – National Defense Strategy

NRO – National Reconnaissance Office

NSHQ – NATO Special Operations Headquarters

NSOS – NATO Special Operations School

NSS – National Security Strategy

NVA – North Vietnamese Army

NVDA – National Volunteer Defense Army

NVVKU – Novosibirsk Higher Military Command School (Russia)

OCO – offensive cyberspace operations

OECD – Organisation for Economic Co-Operation and Development

OPCW – Organization for the Prohibition of Chemical Weapons

OPEC – Organization of Petroleum Exporting Companies

OSO – Office for Special Operations

PARU – Police Aerial Reinforcement Unit

PEEV – *Plan Energetico y Vial* (Columbia, Energy and Road Plan)

PIRA – Provisional Irish Republican Army

PJAK – Kurdistan Free Life Party

PKK – Kurdistan Worker's Party

PLA – People's Liberation Army (China)

PLAN – People's Liberation Army Navy (China)

PM – Prime Minister

PMC – private military company (Russia)

PMESII-PT – political, military, economic, social, information, infrastructure, physical environment, and time

PMF – Popular Mobilization Forces

PNT – precision, navigation, and timing

PRC – People's Republic of China

PUK – Patriotic Union of Kurdistan

QF – Quds Forces

RAP – rapid assessment program

RFA – Radio Free Asia

ROC – Resistance Operating Concept

ROK – Republic of Korea

RSRRS – resiliency, state-sponsored resilience, reinforcing resiliency, resistance potential, support to resistance.

RVVDKU – Ryazan Higher Airborne Command School (Russia)

SACEUR – Supreme Allied Commander Europe (NATO)

SCS – South China Sea

SFA – security force assistance

SGU – Special Guerilla Unit

SHAPE – Supreme Headquarters Allied Powers Europe (NATO)

SNA – social network analysis

SNGP – Significant NATO-Georgia Package

SOBR – Special Rapid Response Unit (Russia)

SOCEUR – U.S. Special Operations Command Europe

SOE – state-owned enterprise

SOFAD – SOF advisor (NATO)

SOIC – Study of Internal Conflict

SOP – standard operating procedure

SORO – Special Operations Research Office

SR – special reconnaissance

SRR – Support to Resilience and Resistance

SSF – Strategic Support Force

SSO – *sily spetsial'nykh operatsii (Russian SOF units)*

STOL – short takeoff and landing

SVR – Russian intelligence service

SWEATMSG – sewer, water, electricity, agriculture, trash, medical, schools, and governance

SWOT – strengths, weaknesses, opportunities, and threats

TTP – tactics, techniques, and procedures

UAE – United Arab Emirates

UAR – unconventional assisted recovery

UCDP – Uppsala Conflict Data Program

UN – United Nations

UROK – Unified Republic of Korea

USAID – United States Agency for International Development

Made in United States
Cleveland, OH
06 November 2026

47771497R00310